T0134849

Lecture Notes in Statistics

Volume 225

Lecture Notes in Statistics (LNS) includes research work on topics that are more specialized than volumes in Springer Series in Statistics (SSS). The series editors are currently Peter Bühlmann, Peter Diggle, Ursula Gather, and Scott Zeger. Peter Bickel, Ingram Olkin, and Stephen Fienberg were editors of the series for many years.

Gregory C. Reinsel • Raja P. Velu • Kun Chen

Multivariate Reduced-Rank Regression

Theory, Methods and Applications

Second Editon

Gregory C. Reinsel
Statistics
University of Wisconsin–Madison
Madison, WI, USA

Raja P. Velu
Managerial Statistics and Finance
Syracuse University
Syracuse, NY, USA

Kun Chen
Statistics
University of Connecticut
Storrs, CT, USA

ISSN 0930-0325 ISSN 2197-7186 (electronic)
Lecture Notes in Statistics
ISBN 978-1-0716-2791-4 ISBN 978-1-0716-2793-8 (eBook)
https://doi.org/10.1007/978-1-0716-2793-8

This Springer imprint is published by the registered company Springer Science+Business Media, LLC, part of Springer Nature.
The registered company address is: 1 New York Plaza, New York, NY 10004, U.S.A.

In memory of our teachers
Ted Anderson and Greg Reinsel

Preface to the Second Edition

The first version of the book was published in 1998. It covered the developments in the field as of that period. The pioneering work of Ted Anderson has provided a framework for rich extensions of the reduced-rank model not only in methodology but also in applications. Since the publication of the book, the interest in big data has grown exponentially and thus there is a need for models that are relevant for the analysis of large dimensional data. The basic reduced-rank models described in the book provide a natural way to handle large dimensional data without using any *a priori* theory. These models have been mainly extended via various forms of regularization techniques that found their way in applications varying from natural science areas such as genetics to social science fields including economics and finance. These techniques lead to further dimension-reduction via variable selection beyond what is implied by the reduced-rank models. They also add a great value in terms of elegant, simplified interpretations of the model coefficients and in the context of time series data, provide more accurate forecasts. Many of the developments have come not only from the field of statistics but also from machine learning, economics, etc., and thus, the interest in these extended topics comes from various areas of orientation. With emphasis on computational aspects, many of the research papers are also usually supplemented with coded algorithms for ready implementation.

Because of the specialized nature of the topics, the first edition of the book was in the form of a research monograph. But the book has been received very well; the latest Google search indicates that there are more than 560 citations of the book with most of the citations in articles published in top journals in statistics, econometrics, and machine learning. Several doctoral theses have followed as well, extending the models discussed in the book. Given the significance of the big data methodology and its applications and given the continued interest in the book, the coverage in the revision is intended to be extensive and timely.

What Is New in This Revision?

The first edition of the book had nine chapters; it started with the multivariate regression model, established the basics, and then introduced the reduced-rank aspect of the coefficient matrix, outlining its practical significance both in terms of dimension-reduction and in terms of model interpretation. Two notable features of this book are the theoretical developments in the context of models for chronological data as well as panel data. The book covered applications in finance and in macroeconomics where well-founded theories lead naturally to a reduced-rank model that is core to the book. The book concluded with a chapter on alternate models that were available at that time. Select methods were illustrated with numerical examples. In this revision, while most of the book (the first eight chapters) has remained intact, some sections are expanded as at least two chapters had resulted in journal publications and hence required some changes from the knowledge gained in the process of their publications. Many of the chapters are also updated with recent applications and extensions in various fields.

We begin the revision with Chapter 9 from the first edition that provided brief descriptions of alternate methods to dimension reduction; the content of this chapter is now moved to Section 2.9 and is expanded to include some new related methods such as envelope models. In the new Chapter 9, we discuss partially reduced-rank regression with grouped responses. In multivariate regression setup, there have been many strategies to perform predictor variable selection with predictor group structure, while the response group structure is largely overlooked. To close this gap, we introduce the partially reduced-rank regression model, where only a subset of response variables in its relationship with predictor set has reduced-rank structure. In many applications, this method can handle potential block structures both in the response and in the predictor variables and dimension reduction on partial subsets more naturally.

In the era of data explosion, reduced-rank estimation, as a general dimension reduction strategy, has been cast into the celebrated regularized learning paradigm and flourished as a key ingredient in modern multivariate learning with high-dimensional, large-scale data. We add four new chapters on these topics: (a) high-dimensional reduced-rank regression, (b) unbiased risk estimation in reduced-rank regression, (c) generalized reduced-rank regression, and (d) sparse reduced-rank regression. The materials in these chapters feature a unified high-dimensional multivariate learning framework, rigorous non-asymptotic analysis, thorough exposition on computation and optimization, and a versatile set of real-world applications.

In Chapter 10, our focus is on high-dimensional reduced-rank regression. We start with an overview of the regularized learning paradigm, illustrated by a few fundamental sparse regression methods. This part acts as a bridge between the classical regime of large T (sample size), small n (number of predictors or responses), and the high-dimensional regime of large n, small T. With this paradigm shift, we formulate the reduced-rank regression models under the framework of singular value penalization. A prototype approach, adaptive nuclear norm penalized regression, is discussed in detail, with a non-asymptotic analysis on rank selection and prediction. We then show that sparse models and reduced-rank models can be understood in a unified framework of integrative reduced-rank regression. This method aims to recover group-specific reduced-rank structures from different predictor groups and thus can be regarded as a multi-view extension of the reduced-rank regression with two sets of regressors covered in Chapter 3.

In Chapter 11, we study unbiased risk estimation of reduced-rank regression for model assessment and selection. Under Stein's unbiased risk estimation framework, a foremost task in estimating the true prediction error (risk) of a reduced-rank model is to obtain an accurate estimate of model complexity. This concept is more broadly known as the model degrees of freedom, and a naive estimator is simply the number of free parameters. However, for non-linear estimation procedures, including reduced-rank regression, the exact correspondence between the model complexity and the number of free parameters may not hold. In this chapter, we provide a finite-sample unbiased estimator of the degrees of freedom for a general class of reduced-rank estimators. The finding that the unbiased estimator admits an explicit form as a function of the estimated singular values is quite elegant. The results cover a significant gap in the literature and advocate the use of the unbiased degrees of freedom in constructing information criterion for a more accurate selection of reduced-rank models.

Chapter 12 introduces several generalized reduced-rank regression models to handle complex data issues. Specifically, we discuss robust estimation and outlier detection in reduced-rank models with possibly contaminated data, reduced-rank matrix completion with incomplete data, and mixed-response with a mix of continuous and discrete data reduced-rank regression. The effect of outliers on the estimation and inference is illustrated with practical examples. In the end, we introduce a simple yet effective framework that simultaneously accounts for mixed data, potential outliers, and missing values. These methods go beyond the conventional setup and greatly enhance the applicability of reduced-rank methodology in real-world applications.

In Chapter 13, we consider the integration of the reduced-rank structure with the sparsity structure in high-dimensional multivariate regression. It has been realized that these two types of low-dimensional structures could very much complement each other: while the reduced-rank structure exploits the dependence among variables to achieve parsimony, the sparseness helps in variable selection. In this chapter, we first consider the situation where both, reduced-rank structure and column sparsity in the coefficient matrix, are present. This leads to the *sparse reduced-rank regression* for dimension reduction and predictor selection. We then consider a more challenging situation where the sparsity is imposed on the (generalized) singular value decomposition of the reduced-rank coefficient matrix, which leads to the *co-sparse factor regression* for achieving factor-specific, two-way variable selection.

The focus of this revised edition is not only the expanded list of topics covered with in-depth theory, but also the intuitions and insights that will illuminate further developments and novel applications. We expand on the examples bringing in more applications in genetics, environmental science etc. The illustrative examples from economics and finance are augmented with new, novel applications. We also attempt to provide the computational tools and resources necessary to apply the methods. We have carefully selected examples to show the unique value of the reduced-rank regression methods whether to uncovering latent relationships which are not readily observable with the usual methods or to large dimensional forecasting for chronological data where the traditional methods break down. An R package called `rrpack` with the new tools developed in the book is made publicly available.

Acknowledgments We want to thank Caleb McWhorter who has helped us immensely in getting the book to its current form by his skilled knowledge in LaTeX. We also want to thank our doctoral students, Zhaoque Zhou for his help on Chapter 8 and Jianmin Chen who helped in carrying out several numerical examples. Jianmin Chen, Xin Yang, and Bruce Jin all helped with proofreading Chapters 10–13. The reviews of the proposal for the revision of the book were quite constructive, and we want to thank the reviewers. Professor Guofu Zhou at Washington University at St. Louis, read through Chapter 8 and offered valuable suggestions; we want to express our appreciation. Raja Velu would like to thank Whitman School of Management for the support toward the production of this book.

Special Recognition A big project like this is not possible without the support of the family. We owe our gratitude to our spouses, Yasodha and Tingting, and our children for their everlasting love and support.

Syracuse, NY, USA — Raja Velu
Storrs, CT, USA — Kun Chen

Preface to the First Edition

In the area of multivariate analysis, there are two broad themes that have emerged over time. The analysis typically involves exploring the variations in a set of interrelated variables or investigating the simultaneous relationships between two or more sets of variables. In either case, the themes involve explicit modeling of the relationships or dimension-reduction of the sets of variables. The multivariate regression methodology and its variants are the preferred tools for the parametric modeling and descriptive tools such as principal components or canonical correlations are the tools used for addressing the dimension-reduction issues. Both act as complementary to each other and data analysts typically want to make use of these tools for a thorough analysis of multivariate data. A technique that combines the two broad themes in a natural fashion is the method of reduced-rank regression. This method starts with the classical multivariate regression model framework but recognizes the possibility for the reduction in the number of parameters through a restriction on the rank of the regression coefficient matrix. This feature is attractive because regression methods, whether they are in the context of a single response variable or in the context of several response variables, are popular statistical tools. The technique of reduced-rank regression and its encompassing features are the primary focus of this book.

The book develops the method of reduced-rank regression starting from the classical multivariate linear regression model. The major developments in the area are presented along with the historical connections to concepts and topics such as latent variables, errors-in-variables, and discriminant analysis that arise in various science areas and a variety of statistical areas. Applications of reduced-rank regression methods in fields such as economics, psychology, and biometrics are reviewed. Illustrative examples are presented with details sufficient for readers to construct their own applications. The computational algorithms are described so that readers can potentially use them for their analysis, estimation, and inference.

The book is broadly organized into three areas: regression modeling for general multi-response data, for multivariate time series data, and for longitudinal data. Chapter 1 gives basic results pertaining to analysis for the classical multivariate linear regression model. Chapters 2 and 3 contain the important developments and results for analysis and applications for the multivariate reduced-rank regression model. In addition to giving the fundamental methods and technical results, we also provide relationships between these methods and other familiar statistical topics such as linear structural models, linear discriminant analysis, canonical correlation analysis, and principal components. Practical numerical examples are also presented to illustrate the methods. Together, Chapters 1 through 3 cover the basic theory and applications related to modeling of multi-response data, such as multivariate cross-sectional data that may occur commonly in business and economics.

Chapters 4 and 5 focus on modeling of multivariate time series data. The fourth chapter deals with multivariate regression models where the errors are autocorrelated, a situation that is commonly considered in econometrics and other areas involving time series data. Chapter 5 is devoted to multiple autoregressive time series modeling. In addition to natural extensions of the reduced-rank models introduced in Chaps. 1 through 3 to the time series setting, we also elaborate on nested reduced-rank models for multiple autoregressive time series. A particular feature that may occur in time series is unit-root nonstationarity, and we show how issues about nonstationarity can be investigated through reduced-rank regression methods and associated canonical correlation methods. We also devote a section to the related topic of cointegration among multiple nonstationary time series, the associated modeling, estimation, and testing. Numerical examples using multivariate time series data are provided to illustrate several of the main topics in Chaps. 4 and 5.

Chapter 6 deals with the analysis of balanced longitudinal data, especially through use of the popular growth curve or generalized multivariate analysis of variance (GMANOVA) model. Results for the basic growth curve model are first developed and presented. Then we illustrate how the model can be modified and extended to accommodate reduced-rank features in the multivariate regression structure. As a result, we obtain an expanded collection of flexible models that can be useful for more parsimonious modeling of growth curve data. The important results on estimation and hypothesis testing are developed for these different forms of growth curve models with reduced-rank features, and a numerical example is presented to illustrate the methods. The use of reduced-rank growth curve methods in linear discriminant analysis is also briefly discussed.

In Chapter 7, we consider seemingly unrelated regression (SUR) equations models, which are widely used in econometrics and other fields. These models are more general than the traditional multivariate regression models because they allow for the values of the predictor variables associated with different response variables to be different. After a brief review of estimation and inference aspects of SUR models, we investigate the possibility of reduced rank for the coefficient matrix pooled from the SUR equations. Although the predictor variables take on different values for the different response variables, in practical applications the predictors generally represent similar variables in nature. Therefore we may expect certain commonality among the vectors of regression coefficients that arise from the different equations. The reduced-rank regression approach also provides a type of canonical analysis in the context of the SUR equations models. Estimation and inference results are derived for the SUR model with reduced-rank structure, including details on the computational algorithms, and an application in marketing is presented for numerical illustration.

In Chapter 8, the use of reduced-rank methods in the area of financial economics is discussed. It is seen that, in this setting, several applications of the reduced-rank regression model arise in a fairly natural way from economic theories. The context in which these models arise, particularly in the area of asset pricing, the particular form of the models and some empirical results are briefly surveyed. This topic provides an illustration of the use of reduced-rank models not mainly as an empirical statistical modeling tool for dimension-reduction but more as arising somewhat naturally in relation to certain economic theories.

The book concludes with a final chapter that highlights some other techniques associated with recent developments in multivariate regression modeling, many of which represent areas and topics for further research. This includes brief surveys and discussion of other related multivariate regression modeling methodologies that have similar parameter reduction objectives as reduced-rank regression, such as multivariate ridge regression, partial least squares, joint continuum regression, and other shrinkage and regularization techniques.

Because of the specialized nature of the topic of multivariate reduced-rank regression, a certain level of background knowledge in mathematical and statistical concepts is assumed. The reader will need to be familiar with basic matrix theory and have some limited exposure to multivariate methods. As noted above, the classical multivariate regression model is reviewed in Chapter 1 and many basic estimation and hypothesis testing results pertaining to that model are derived and presented. However, it is still anticipated that the reader will have some previous experience with the fundamentals of univariate linear regression modeling.

We want to acknowledge the useful comments by reviewers of an initial draft of this book, which were especially helpful in revising the earlier chapters. In particular, they provided perspectives for the material in Chapter 2 on historical developments of the topic. We are grateful to T. W. Anderson for his constructive comments on many general aspects of the subject matter contained in the book. We also want to thank Guofu Zhou for his helpful comments, in particular on material in Chapter 8 devoted to applications in financial economics. The scanner data used for illustration in Chapter 7 was provided by A. C. Nielsen and we want to thank David Bohl for his assistance. Raja Velu would also like to thank the University of Wisconsin-Whitewater for granting him a sabbatical during the 1996–97 academic year and the University of Wisconsin-Madison for hosting him; the sabbatical was partly devoted to writing this book.

Finally, we want to express our indebtedness to our wives, Sandy and Yasodha, and our children, for their support and love which were essential to complete this book.

Madison, WI, USA
Syracuse, NY, USA

Gregory C. Reinsel
Raja Velu

Contents

About the Authors

Gregory C. Reinsel (now deceased) was Professor of Statistics at the University of Wisconsin, Madison. He was a fellow of the American Statistical Association. He is also author of the book *Elements of Multivariate Time Series Analysis, Second Edition*, and coauthor, with G.E.P. Box and G.M. Jenkins, of the book *Time Series Analysis: Forecasting and Control, Third Edition.*
Greg will remain the first author, in our gratitude.

Raja Velu taught business analytics and finance at Syracuse University. The first version of the book was mainly based on his thesis written under the supervision of Professor Reinsel and Professor Dean Wichern. He works in the big data models area with interest in high-dimensional time series and forecasting applications. His book, *Algorithmic Trading and Quantitative Strategies*, co-authored with practitioners from CITI and JP Morgan Chase, is published by Taylor and Francis. He has also served in the tech industry with companies such as Yahoo!, IBM-Almaded, and Microsoft Research. He was recently (2021–2022) a visiting researcher at Google working with the Resource Efficiency Data Science team.

Kun Chen is an associate professor in the Department of Statistics at the University of Connecticut. He is a Fellow of the American Statistical Association and an Elected Member of the International Statistical Institute. The first version of the book has had profound influence on his research since his PhD study at the University of Iowa under the supervision of Professor Kung-Sik Chan. In recent years, he has been pursuing a comprehensive investigation of theory, methodologies, and computational approaches of reduced-rank modeling in large-scale statistical learning. His related work has resulted in many publications in statistics, machine learning, and scientific journals and the developed methods have been applied to tackle consequential problems in various fields including public health, ecology, and biological sciences.

Chapter 1
Multivariate Linear Regression

1.1 Introduction

Regression methods are perhaps the most widely used statistical tools in data analysis. When several response variables are studied simultaneously, we are in the sphere of multivariate regression. The usual description of the multivariate regression model, which relates the set of m multiple responses to a set of n predictor variables, assumes implicitly that the $m \times n$ regression coefficient matrix is of full rank. It can then be demonstrated that the simultaneous estimation of the elements of the coefficient matrix, by least squares or maximum likelihood estimation methods, yields the same results as a set of m multiple regressions, where each of the m individual response variables is regressed separately on the predictor variables. Hence, the fact that the multiple responses are likely to be related is not involved in the estimation of the regression coefficients as no information about the correlations among the response variables is taken into account. Any new knowledge gained by recognizing the multivariate nature of the problem and the fact that the responses are related is not incorporated when estimating the regression parameters jointly. There are two practical concerns regarding this general multivariate regression model. First, the accurate estimation of all the regression coefficients may require a relatively large number of observations, which might involve some practical limitations. Second, even if the data are available, interpreting simultaneously a large number of the regression coefficients can become unwieldy. Achieving parsimony in the number of unknown parameters may be desirable both from the estimation and interpretation points of view of multivariate regression analysis. We will address these concerns by recognizing a big data feature that enters when modeling via multivariate regression.

G. C. Reinsel et al., *Multivariate Reduced-Rank Regression*, Lecture Notes in Statistics 225, https://doi.org/10.1007/978-1-0716-2793-8_1

The special feature that can enter the multivariate linear regression model is that we admit the possibility that the rank of the regression coefficient matrix can be deficient. This implies that there are linear restrictions on the coefficient matrix, and these restrictions themselves are often not known a priori. Alternately, the coefficient matrix can be written as a product of two component matrices of lower dimension. Such a model is called a *reduced-rank* regression model, and the model structure and estimation method for this model will be described in detail later. It follows from this estimation method that the assumption of lower rank for the regression coefficient matrix leads to estimation results which take into account the interrelations among the multiple responses and which use the entire set of predictor variables in a systematic fashion, compared to some ad hoc procedures which may use only some portion of the predictor variables in an attempt to achieve parsimony. The abovementioned features both have practical implications. When we model with a large number of response and predictor variables, the implication in terms of restrictions serves a useful purpose. Certain linear combinations of response variables can, eventually, be ignored for the regression modeling purposes, since these combinations will be found to be unrelated to the predictor variables. The alternate implication indicates that only certain linear combinations of the predictor variables need to be used in the regression model, since any remaining linear combinations will be found to have no influence on the response variables given the first set of linear combinations. Thus, the reduced-rank regression procedure takes care of the dimension reduction aspect of multivariate regression model building through the assumption of lower rank for the regression coefficient matrix. This feature of reduced-rank regression and its variations are particularly useful in modeling big data.

In the case of multivariate linear regression, where the dependence of response variables among each other is utilized through the assumption of lower rank, the connection between the regression coefficients and the descriptive tools of principal components and canonical variates can be demonstrated. There is a correspondence between the rank of the regression coefficient matrix and a specified number of principal components or canonical correlations. Therefore, the use of these descriptive tools in multivariate model building is explicitly demonstrated through the reduced-rank assumption. Before we proceed with the consideration of various aspects of reduced-rank models in detail in subsequent chapters, we present some basic results concerning the classical full-rank multivariate linear regression model in the remainder of this chapter. More detailed and in-depth presentations of results for the full-rank multivariate linear regression model may be found in the books by Anderson (1984b, Chap. 8), Izenman (2008, Chap. 6), Muirhead (1982, Chap. 10), and Srivastava and Khatri (1979, Chaps. 5 and 6), among others.

1.2 Multivariate Linear Regression Model and Least Squares Estimator

We consider the general multivariate linear model,

$$Y_k = CX_k + \epsilon_k, \qquad k = 1, \ldots, T, \tag{1.1}$$

where $Y_k = (y_{1k}, \ldots, y_{mk})'$ is an $m \times 1$ vector of response variables, $X_k = (x_{1k}, \ldots, x_{nk})'$ is an $n \times 1$ vector of predictor variables, C is an $m \times n$ regression coefficient matrix, and $\epsilon_k = (\epsilon_{1k}, \ldots, \epsilon_{mk})'$ is the $m \times 1$ vector of random errors, with mean vector $E(\epsilon_k) = 0$ and covariance matrix $\text{Cov}(\epsilon_k) = \Sigma_{\epsilon\epsilon}$, an $m \times m$ positive-definite matrix. The ϵ_k are assumed to be independent for different k. We shall assume that T vector observations are available and define the $m \times T$ and $n \times T$ data matrices, respectively, as

$$\mathbf{Y} = [Y_1, \ldots, Y_T] \qquad \text{and} \qquad \mathbf{X} = [X_1, \ldots, X_T]. \tag{1.2}$$

To begin with, we assume that $m + n \leq T$ and that $\mathbf{X}$ is of full rank $n = \text{rank}(\mathbf{X}) < T$. This last condition is required to obtain a unique (least squares) solution to the first order equations. These conditions will be relaxed in later chapters. We arrange the error vectors $\epsilon_k, k = 1, \ldots, T$, into an $m \times T$ matrix $\boldsymbol{\epsilon} = [\epsilon_1, \epsilon_2, \ldots, \epsilon_T]$, similar to the data matrices given by (1.2). We will also consider the errors arranged into an $mT \times 1$ vector $e = (\boldsymbol{\epsilon}'_{(1)}, \ldots, \boldsymbol{\epsilon}'_{(m)})'$, where $\boldsymbol{\epsilon}_{(j)} = (\epsilon_{j1}, \ldots, \epsilon_{jT})'$ is the vector of errors corresponding to the jth response variable in the regression equation (1.1). Then we have that

$$E(e) = 0, \qquad \text{Cov}(e) = \Sigma_{\epsilon\epsilon} \otimes I_T, \tag{1.3}$$

where $A \otimes B$ denotes the Kronecker product of two matrices A and B. That is, if A is an $m \times n$ matrix and B is a $p \times q$ matrix, then the Kronecker product $A \otimes B$ is the $mp \times nq$ matrix with block elements $A \otimes B = (a_{ij}B)$, where a_{ij} is the (i, j)th element of A.

The unknown parameters in model (1.1) are the elements of the regression coefficient matrix C and the error covariance matrix $\Sigma_{\epsilon\epsilon}$. These can be estimated by the method of least squares, or equivalently, the method of maximum likelihood under the assumption of normality of the ϵ_k. Before we present the details on the estimation method, we rewrite the model (1.1) in terms of the complete data matrices $\mathbf{Y}$ and $\mathbf{X}$ as

$$\mathbf{Y} = C\mathbf{X} + \boldsymbol{\epsilon}. \tag{1.4}$$

The least squares criterion is given as

$$e'e = \text{tr}(\boldsymbol{\epsilon}\boldsymbol{\epsilon}') = \text{tr}[(\mathbf{Y} - C\mathbf{X})(\mathbf{Y} - C\mathbf{X})'] = \text{tr}\left[\sum_{k=1}^{T} \epsilon_k \epsilon_k'\right], \tag{1.5}$$

where $\mathrm{tr}(A)$ denotes the trace of a square matrix A, that is, the sum of the diagonal elements of A. The least squares (LS) estimate of C is the value $\widetilde{C}$ that minimizes the criterion in (1.5). Using results on matrix derivatives (e.g., Rao 1973, p. 72), in particular $\partial\mathrm{tr}(CZ)/\partial C = Z'$, we have the first order equations to obtain a minimum as

$$\frac{\partial\mathrm{tr}(\boldsymbol{\epsilon}\boldsymbol{\epsilon}')}{\partial C} = 2[-\mathbf{Y}\mathbf{X}' + C(\mathbf{X}\mathbf{X}')] = 0, \tag{1.6}$$

which yields a unique solution for C as

$$\widetilde{C} = \mathbf{Y}\mathbf{X}'(\mathbf{X}\mathbf{X}')^{-1} = \left(\frac{1}{T}\mathbf{Y}\mathbf{X}'\right)\left(\frac{1}{T}\mathbf{X}\mathbf{X}'\right)^{-1}. \tag{1.7}$$

If we let $C'_{(j)}$ denote the jth row of the matrix C, it is directly seen from (1.7) that the corresponding jth row of the LS estimate $\widetilde{C}$ is

$$\widetilde{C}'_{(j)} = \mathbf{Y}'_{(j)}\mathbf{X}'(\mathbf{X}\mathbf{X}')^{-1} = \left(\frac{1}{T}\mathbf{Y}'_{(j)}\mathbf{X}'\right)\left(\frac{1}{T}\mathbf{X}\mathbf{X}'\right)^{-1}, \tag{1.8}$$

for $j = 1, \ldots, m$, where $\mathbf{Y}'_{(j)} = (y_{j1}, \ldots, y_{jT})$ denotes the jth row of the matrix $\mathbf{Y}$, that is, $\mathbf{Y}_{(j)}$ is the $T \times 1$ vector of observations on the jth response variable. Thus, it is noted here that $\mathbf{Y}'_{(j)}\mathbf{X}' = \sum_{k=1}^{T} y_{jk}X'_k$ refers to the cross products between the jth response variable and the n predictor variables, summed over all T observations. Hence, the rows of the matrix C are estimated in (1.7) by the least squares regression of each response variable on the predictor variables and therefore the covariances among the response variables do not enter into the estimation.

From the presentation of model (1.1), it follows that our focus here is on the regression coefficient matrix C of the response variables $\mathbf{Y}$ on the predictor variables $\mathbf{X}$; the intercept term typically will not be explicitly represented in the model. With a constant term explicitly included, the model would take the form $Y_k = d + CX_k + \epsilon_k$, where d is an $m \times 1$ vector of unknown parameters. For convenience, in such a model, it will be assumed that the predictor variables have been centered by subtraction of appropriate sample means, so that the resulting data matrix $\mathbf{X}$ has zero means, i.e., $\bar{X} = \frac{1}{T}\sum_{k=1}^{T} X_k = 0$. These mean-corrected observations would typically be the basis of all the calculations that follow. Under this setup, the basic results concerning the LS estimate of C would continue to hold, while the LS estimate of the intercept term is simply $\widetilde{d} = \bar{Y} = \frac{1}{T}\sum_{k=1}^{T} Y_k$. For practical applications, an intercept is almost always needed in modeling, and in terms of estimation of C and $\Sigma_{\epsilon\epsilon}$ the inclusion of a constant term is equivalent to performing calculations with response variables Y_k that have (also) been corrected for the sample mean $\bar{Y}$.

The maximum likelihood method of estimation, under the assumption that values of the predictor variables X_k are known constant vectors and the assumption of normality of the error terms ϵ_k in addition to the properties given in (1.3), yields the same results as least squares. The likelihood is equal to

$$f(\boldsymbol{\epsilon}) = (2\pi)^{-mT/2} |\Sigma_{\epsilon\epsilon}|^{-T/2} \exp\left[-(1/2)\mathrm{tr}\left(\Sigma_{\epsilon\epsilon}^{-1}\boldsymbol{\epsilon}\boldsymbol{\epsilon}'\right)\right], \tag{1.9}$$

where $\boldsymbol{\epsilon} = \mathbf{Y} - C\mathbf{X}$. For maximum likelihood (ML) estimation, the criterion that must be minimized to estimate C is seen from (1.9) to be

$$\mathrm{tr}\left(\Sigma_{\epsilon\epsilon^{-1}}\boldsymbol{\epsilon}\boldsymbol{\epsilon}'\right) = \mathrm{tr}\left[\Sigma_{\epsilon\epsilon}^{-1/2}(\mathbf{Y} - C\mathbf{X})(\mathbf{Y} - C\mathbf{X})'\,\Sigma_{\epsilon\epsilon}^{-1/2}\right], \tag{1.10}$$

which can also be expressed as $e'(\Sigma_{\epsilon\epsilon}^{-1} \otimes I_T)e$. From (1.10), the first order equations to obtain a minimum are $\partial\mathrm{tr}(\Sigma_{\epsilon\epsilon}^{-1}\boldsymbol{\epsilon}\boldsymbol{\epsilon}')/\partial C = 2\Sigma_{\epsilon\epsilon}^{-1}[-\mathbf{Y}\mathbf{X}' + C(\mathbf{X}\mathbf{X}')] = 0$, which yields the same solution for C as given in (1.7). In the later chapters, we shall consider a more general criterion,

$$\mathrm{tr}(\Gamma\boldsymbol{\epsilon}\boldsymbol{\epsilon}') = \mathrm{tr}[\Gamma^{1/2}(\mathbf{Y} - C\mathbf{X})(\mathbf{Y} - C\mathbf{X})'\Gamma^{1/2}], \tag{1.11}$$

where Γ is a positive-definite matrix. The criteria (1.5) and (1.10) follow as special cases of (1.11) by appropriate choices of the matrix Γ. The ML estimator of the error covariance matrix $\Sigma_{\epsilon\epsilon}$ is obtained from the least squares residuals $\widehat{\boldsymbol{\epsilon}} = \mathbf{Y} - \widetilde{C}\mathbf{X}$ as $\widetilde{\Sigma}_{\epsilon\epsilon} = \frac{1}{T}(\mathbf{Y} - \widetilde{C}\mathbf{X})(\mathbf{Y} - \widetilde{C}\mathbf{X})' \equiv \frac{1}{T}S$, where $S = (\mathbf{Y} - \widetilde{C}\mathbf{X})(\mathbf{Y} - \widetilde{C}\mathbf{X})' = \sum_{k=1}^{T} \widehat{\epsilon}_k \widehat{\epsilon}_k'$ is the error sum of squares matrix, and $\widehat{\epsilon}_k = Y_k - \widetilde{C}X_k$, $k = 1, \ldots, T$, are the least squares residual vectors. An unbiased estimator of $\Sigma_{\epsilon\epsilon}$ is obtained as

$$\overline{\Sigma}_{\epsilon\epsilon} = \frac{1}{T - n}\,\widehat{\boldsymbol{\epsilon}}\widehat{\boldsymbol{\epsilon}}' = \frac{1}{T - n}\sum_{k=1}^{T} \widehat{\epsilon}_k \widehat{\epsilon}_k'. \tag{1.12}$$

1.3 Further Inference Properties in the Multivariate Regression Model

Under the assumption that the matrix of predictor variables $\mathbf{X}$ is fixed and nonstochastic, properties of the least squares estimator $\widetilde{C}$ in (1.7) can readily be derived. In practice, the X_k in (1.1) may consist of observations on a stochastic vector, in which case the (finite sample) distributional results to be presented for the LS estimator $\widetilde{C}$ can be viewed as conditional on fixed observed values of the X_k. First, $\widetilde{C}$ is an unbiased estimator of C since from (1.4)

$$E(\widetilde{C}) = E[\mathbf{Y}\mathbf{X}'(\mathbf{X}\mathbf{X}')^{-1}] = E[\mathbf{Y}]\mathbf{X}'(\mathbf{X}\mathbf{X}')^{-1} = C\mathbf{X}\mathbf{X}'(\mathbf{X}\mathbf{X}')^{-1} = C.$$

From (1.8), we find that

$$\mathrm{Cov}(\widetilde{C}_{(i)}, \widetilde{C}_{(j)}) = (\mathbf{X}\mathbf{X}')^{-1}\mathbf{X}\,\mathrm{Cov}(\mathbf{Y}_{(i)}, \mathbf{Y}_{(j)})\,\mathbf{X}'(\mathbf{X}\mathbf{X}')^{-1} = \sigma_{ij}(\mathbf{X}\mathbf{X}')^{-1}$$

for $i, j = 1, \ldots, m$, where σ_{ij} denotes the (i, j)th element of the covariance matrix $\Sigma_{\epsilon\epsilon}$, since $\text{Cov}(\mathbf{Y}_{(i)}, \mathbf{Y}_{(j)}) = \sigma_{ij} I_T$. The distributional properties of $\widetilde{C}$ follow easily from multivariate normality of the error terms ϵ_k. Specifically, we consider the distribution of $\text{vec}(\widetilde{C}')$, where the "vec" operator transforms an $n \times m$ matrix into an nm-dimensional column vector by stacking the columns of the matrix below each other. Thus, for an $m \times n$ matrix $\widetilde{C}$, $\text{vec}(\widetilde{C}') = (\widetilde{C}'_{(1)}, \ldots, \widetilde{C}'_{(m)})'$, where $\widetilde{C}'_{(j)}$ refers to the jth row of the matrix $\widetilde{C}$. A useful property that relates the vec operation with the Kronecker product of matrices is that if A, B, and C are matrices of appropriate dimensions so that the product ABC is defined, then $\text{vec}(ABC) = (C' \otimes A)\text{vec}(B)$ [see Neudecker (1969)]. Another useful property is $\text{tr}(ABCB') = \{\text{vec}(B)\}'(C' \otimes A)\,\text{vec}(B)$.

Now notice that the model (1.4) may be expressed in the vector form as

$$y = \text{vec}(\mathbf{Y}') = \text{vec}(\mathbf{X}'C') + \text{vec}(\boldsymbol{\epsilon}') = (I_m \otimes \mathbf{X}')\text{vec}(C') + e, \tag{1.13}$$

where $y = \text{vec}(\mathbf{Y}') = (\mathbf{Y}'_{(1)}, \ldots, \mathbf{Y}'_{(m)})'$, and $e = \text{vec}(\boldsymbol{\epsilon}') = (\boldsymbol{\epsilon}'_{(1)}, \ldots, \boldsymbol{\epsilon}'_{(m)})'$ is similar in form to y with $\text{Cov}(e) = \Sigma_{\epsilon\epsilon} \otimes I_T$. In addition, the least squares estimator $\widetilde{C}$ in (1.7) can be written in vector form as

$$\text{vec}(\widetilde{C}') = \text{vec}\big((\mathbf{X}\mathbf{X}')^{-1}\mathbf{X}\mathbf{Y}'\big) = (I_m \otimes (\mathbf{X}\mathbf{X}')^{-1}\mathbf{X})\,\text{vec}(\mathbf{Y}'), \tag{1.14}$$

where $\text{vec}(\mathbf{Y}')$ is distributed as multivariate normal with mean vector $(I_m \otimes \mathbf{X}')\text{vec}(C')$ and covariance matrix $\Sigma_{\epsilon\epsilon} \otimes I_T$. It then follows directly that $\text{vec}(\widetilde{C}')$ is also distributed as multivariate normal, with

$$E[\text{vec}(\widetilde{C}')] = (I_m \otimes (\mathbf{X}\mathbf{X}')^{-1}\mathbf{X})\,E[\text{vec}(\mathbf{Y}')] = \text{vec}(C'),$$

and

$$\begin{aligned}\text{Cov}[\text{vec}(\widetilde{C}')] &= (I_m \otimes (\mathbf{X}\mathbf{X}')^{-1}\mathbf{X})\,\text{Cov}[\text{vec}(\mathbf{Y}')]\,(I_m \otimes \mathbf{X}'(\mathbf{X}\mathbf{X}')^{-1})\\ &= \Sigma_{\epsilon\epsilon} \otimes (\mathbf{X}\mathbf{X}')^{-1}.\end{aligned}$$

We summarize the main results on the LS estimate $\widetilde{C}$ in the following.

Result 1.1 For the model (1.1) under the normality assumption on the ϵ_k, the least squares estimator $\widetilde{C} = \mathbf{Y}\mathbf{X}'(\mathbf{X}\mathbf{X}')^{-1}$ is the same as the maximum likelihood estimator of C, and the distribution of the least squares estimator $\widetilde{C}$ is such that

$$\text{vec}(\widetilde{C}') \sim N\big(\text{vec}(C'), \Sigma_{\epsilon\epsilon} \otimes (\mathbf{X}\mathbf{X}')^{-1}\big). \tag{1.15}$$

Note, in particular, that this result implies that the jth row of $\widetilde{C}$, $\widetilde{C}'_{(j)}$, which is the vector of least squares estimates of regression coefficients for the jth response variable, has the distribution $\widetilde{C}_{(j)} \sim N(C_{(j)}, \sigma_{jj}(\mathbf{X}\mathbf{X}')^{-1})$.

The inference on the elements of the matrix C can be made using the result in (1.15). In practice, because $\Sigma_{\epsilon\epsilon}$ is unknown, a reasonable estimator such as the unbiased estimator in (1.12) is substituted for the covariance matrix in (1.15). All the necessary calculations for the above linear least squares inference procedures can be carried out using standard computer software, such as Statistical Analysis System (SAS) or using programs based on the "R" language such as "`rrpack`."

We shall also indicate the likelihood ratio procedure for testing certain simple linear hypotheses regarding the regression coefficient matrix. Consider $\mathbf{X}$ partitioned as $\mathbf{X} = [\mathbf{X}_1', \mathbf{X}_2']'$ and corresponding $C = [C_1, C_2]$, so that the model (1.4) is written as $\mathbf{Y} = C_1\mathbf{X}_1 + C_2\mathbf{X}_2 + \boldsymbol{\epsilon}$, where C_1 is $m \times n_1$ and C_2 is $m \times n_2$ with $n_1 + n_2 = n$. It may be of interest to test the null hypothesis $H_0\colon C_2 = 0$ against the alternative $C_2 \neq 0$. The null hypothesis implies that the predictor variables $\mathbf{X}_2$ do not have any (additional) influence on the response variables $\mathbf{Y}$, given the impact of the variables $\mathbf{X}_1$. Using the likelihood ratio (LR) testing approach, it is easy to see from (1.9) that the LR test statistic is $\lambda = U^{T/2}$, where $U = |S|/|S_1|$, $S = (\mathbf{Y} - \widetilde{C}\mathbf{X})(\mathbf{Y} - \widetilde{C}\mathbf{X})'$ and $S_1 = (\mathbf{Y} - \widehat{C}_1\mathbf{X}_1)(\mathbf{Y} - \widehat{C}_1\mathbf{X}_1)'$. The matrix S is the residual sum of squares matrix from maximum likelihood (i.e., least squares) fitting of the full model, while S_1 is the residual sum of squares matrix obtained from fitting the reduced model with $C_2 = 0$ and $\widehat{C}_1 = \mathbf{Y}\mathbf{X}_1'(\mathbf{X}_1\mathbf{X}_1')^{-1}$. This LR statistic result follows directly by noting that the value of the multivariate normal likelihood function evaluated at the maximum likelihood estimates $\widetilde{C}$ and $\widetilde{\Sigma}_{\epsilon\epsilon} = (1/T)S$ is equal to a constant times $|S|^{-T/2}$, because $\mathrm{tr}(\widetilde{\Sigma}_{\epsilon\epsilon}^{-1}\widetilde{\boldsymbol{\epsilon}}\widetilde{\boldsymbol{\epsilon}}') = \mathrm{tr}(T\,\widetilde{\Sigma}_{\epsilon\epsilon}^{-1}\widetilde{\Sigma}_{\epsilon\epsilon}) = Tm$ is a constant when evaluated at the ML estimate. Thus, the likelihood ratio for testing $H_0\colon C_2 = 0$ is $\lambda = |S_1|^{-T/2}/|S|^{-T/2} = U^{T/2}$. It has been shown (see Anderson (1984b, Chapter 8)) that for moderate and large sample size T, the test statistic

$$\mathcal{M} = -[T - n + (n_2 - m - 1)/2]\log(U) \tag{1.16}$$

is approximately distributed as chi-squared with $n_2 m$ degrees of freedom ($\chi^2_{n_2 m}$), and the hypothesis is rejected when $\mathcal{M}$ is greater than a constant determined by the $\chi^2_{n_2 m}$ distribution.

It is useful to express the LR statistic $\lambda = U^{T/2}$ in an alternate equivalent form. Write $\mathbf{X}\mathbf{X}' = A = [(A_{ij})]$, where $A_{ij} = \mathbf{X}_i\mathbf{X}_j'$, $i, j = 1, 2$, and let $A_{22.1} = A_{22} - A_{21}A_{11}^{-1}A_{12}$. Then the LS estimate of C_1 under the reduced model with $C_2 = 0$ can be expressed as $\widehat{C}_1 = \mathbf{Y}\mathbf{X}_1'(\mathbf{X}_1\mathbf{X}_1')^{-1} = \widetilde{C}_1 + \widetilde{C}_2 A_{21}A_{11}^{-1}$ because of the full model LS normal equations (1.6), $\mathbf{Y}\mathbf{X}_1' = \widetilde{C}_1\mathbf{X}_1\mathbf{X}_1' + \widetilde{C}_2\mathbf{X}_2\mathbf{X}_1'$. Hence,

$$\mathbf{Y} - \widehat{C}_1\mathbf{X}_1 = (\mathbf{Y} - \widetilde{C}\mathbf{X}) + \widetilde{C}_2\,(\mathbf{X}_2 - A_{21}A_{11}^{-1}\,\mathbf{X}_1),$$

where the two terms on the right are orthogonal since $(\mathbf{Y} - \widetilde{C}\mathbf{X})\mathbf{X}' = 0$ because $\widetilde{C}$ is a solution to the first order equations (1.6). It thus follows that

$$\begin{aligned} S_1 &= (\mathbf{Y} - \widehat{C}_1\mathbf{X}_1)(\mathbf{Y} - \widehat{C}_1\mathbf{X}_1)' \\ &= (\mathbf{Y} - \widetilde{C}\mathbf{X})(\mathbf{Y} - \widetilde{C}\mathbf{X})' + \widetilde{C}_2(A_{22} - A_{21}A_{11}^{-1}A_{12})\widetilde{C}_2' \\ &= S + \widetilde{C}_2\,A_{22.1}\,\widetilde{C}_2', \end{aligned} \tag{1.17}$$

and also note that $A_{22.1}^{-1}$ is the $n_2 \times n_2$ lower diagonal block of the matrix $(\mathbf{XX}')^{-1}$ (e.g., see Rao (1973, p. 33)). Hence, we see that

$$U = |S|/|S_1| = |S|/|S + H| = 1/|I + S^{-1}H|,$$

where $H = \widetilde{C}_2 A_{22.1} \widetilde{C}_2'$ is referred to as the hypothesis sum of squares matrix (associated with the null hypothesis $C_2 = 0$). Therefore, we have $-\log(U) = \sum_{i=1}^{m} \log(1 + \widehat{\lambda}_i^2)$, where $\widehat{\lambda}_i^2$ are the eigenvalues of $S^{-1}H$, and the LR statistic $\mathcal{M}$ in (1.16) is directly expressible in terms of these eigenvalues. In addition, it can be shown (e.g., see Reinsel (1997, Section A4.3)) that the $\widehat{\lambda}_i^2$ are related to the sample partial canonical correlations $\widehat{\rho}_i$ between $\mathbf{Y}$ and $\mathbf{X}_2$, eliminating $\mathbf{X}_1$, as $\widehat{\lambda}_i^2 = \widehat{\rho}_i^2/(1 - \widehat{\rho}_i^2)$, $i = 1, \ldots, m$. We summarize the above in the following result.

Result 1.2 For the model (1.1) under the normality assumption on the ϵ_k, the likelihood ratio for testing H_0: $C_2 = 0$ is $\lambda = U^{T/2}$, where $U = |S|/|S_1| = 1/|I + S^{-1}H|$ with $H = \widetilde{C}_2 A_{22.1} \widetilde{C}_2'$. Therefore, $-\log(U) = \sum_{i=1}^{m} \log(1 + \widehat{\lambda}_i^2)$, where $\widehat{\lambda}_i^2$ are the eigenvalues of $S^{-1}H$. For large sample size T, the asymptotic distribution of the test statistic $\mathcal{M} = -[T - n + (n_2 - m - 1)/2] \log(U)$ is chi-squared with $n_2 m$ degrees of freedom.

Known Restrictions: Finally, we briefly consider testing the more general null hypothesis of the form H_0: $F_1 C G_2 = 0$, where F_1 and G_2 are known full-rank matrices of appropriate dimensions $m_1 \times m$ and $n \times n_2$, respectively. In the hypothesis $F_1 C G_2 = 0$, the matrix G_2 allows for restrictions among the coefficients in C across (subsets of) the predictor variables, whereas F_1 provides for restrictions in the coefficients of C across the different response variables. For instance, the hypothesis $C_2 = 0$ considered earlier is represented in this form with $G_2 = [0', I_{n_2}']'$ and $F_1 = I_m$, and the hypothesis of the form $F_1 C G_2 = 0$ with the same matrix G_2 and $F_1 = [I_{m_1}, 0]$ would postulate that the predictor variables $\mathbf{X}_2$ do not enter into the linear regression relationship *only* for the first $m_1 < m$ components of the response variables $\mathbf{Y}$. Using the LR approach, the likelihood ratio is given by $\lambda = U^{T/2}$, where $U = |S|/|S_1|$ and $S_1 = (\mathbf{Y} - \widehat{C}\mathbf{X})(\mathbf{Y} - \widehat{C}\mathbf{X})'$ with $\widehat{C}$ denoting the ML estimator of C obtained subject to the constraint that $F_1 C G_2 = 0$. In fact, using Lagrange multiplier methods, it can be shown that the constrained ML estimator under $F_1 C G_2 = 0$ can be expressed as

$$\widehat{C} = \widetilde{C} - S F_1' (F_1 S F_1')^{-1} F_1 \widetilde{C} G_2 [G_2' (\mathbf{XX}')^{-1} G_2]^{-1} G_2' (\mathbf{XX}')^{-1}. \tag{1.18}$$

An alternate way to develop this last result is to consider matrices $F = [F_1', F_2']'$ and $G = [G_1, G_2]$ such that $F_1 F_2' = 0$ and $G_1' G_2 = 0$ (for convenience) and apply the transformation to the model (1.4) as $\mathbf{Y}^* = F\mathbf{Y} = FCGG^{-1}\mathbf{X} + F\boldsymbol{\epsilon} \equiv C^*\mathbf{X}^* + \boldsymbol{\epsilon}^*$, where $C^* = FCG$, $\mathbf{Y}^* = F\mathbf{Y}$, and $\mathbf{X}^* = G^{-1}\mathbf{X}$. With C^* partitioned into submatrices C_{ij}^*, $i, j = 1, 2$, the hypothesis $F_1 C G_2 = 0$ is equivalent to $C_{12}^* = 0$. Maximum likelihood estimation of the parameters C^* in this transformed "canonical form" model, subject to the restriction $C_{12}^* = 0$, and transformation of the resulting

ML estimate $\widehat{C}^*$ back to the parameters C of the original model as $\widehat{C} = F^{-1}\widehat{C}^* G^{-1}$ then lead to the result in (1.18). Using either method, from (1.18), we therefore have

$$\begin{aligned}\mathbf{Y} - \widehat{C}\mathbf{X} &= (\mathbf{Y} - \widetilde{C}\mathbf{X}) \\ &\quad + SF_1'(F_1SF_1')^{-1}F_1\widetilde{C}G_2[G_2'(\mathbf{XX}')^{-1}G_2]^{-1}G_2'(\mathbf{XX}')^{-1}\mathbf{X},\end{aligned}$$

where the two terms on the right are orthogonal. Following the above developments as in (1.17), we thus readily find that

$$\begin{aligned}S_1 &= S + SF_1'(F_1SF_1')^{-1}(F_1\widetilde{C}G_2)[G_2'(\mathbf{XX}')^{-1}G_2]^{-1} \\ &\quad \times (F_1\widetilde{C}G_2)'(F_1SF_1')^{-1}F_1S \equiv S + H, \end{aligned} \tag{1.19}$$

and $U = 1/|I + S^{-1}H|$ so that the LR test can be based on the eigenvalues of $S^{-1}H$.

The nonzero eigenvalues of the $m \times m$ matrix $S^{-1}H$ are the same as those of the $m_1 \times m_1$ matrix $S_*^{-1}H_*$, where $S_* = F_1SF_1'$ and $H_* = F_1HF_1' = F_1\widetilde{C}G_2[G_2'(XX')^{-1}G_2]^{-1}(F_1\widetilde{C}G_2)'$, so that $U^{-1} = |I + S_*^{-1}H_*| = \prod_{i=1}^{m_1}(1+\widehat{\lambda}_i^2)$, where the $\widehat{\lambda}_i^2$ are the eigenvalues of $S_*^{-1}H_*$. Analogous to (1.16), the test statistic used is $\mathcal{M} = -[T - n + \frac{1}{2}(n_2 - m_1 - 1)]\log(U)$, which has an approximate $\chi^2_{n_2m_1}$ distribution under H_0. Somewhat more accurate approximation of the null distribution of $\log(U)$ is available based on the F-distribution, and in some cases (notably when $n_2 = 1$ or 2 or $m_1 = 1$ or 2), exact F-distribution results for U are available (see Anderson (1984b, Chap. 8)). Values relating to the (exact) distribution of the statistic U for various values of m_1, n_2, and $T - n$ have been tabulated by Schatzoff (1966), Pillai and Gupta (1969), and Lee (1972). However, for most of our applications, we will simply use the large sample $\chi^2_{n_2m_1}$ approximation for the distribution of the LR statistic $\mathcal{M}$.

Concerning the exact distribution theory associated with the test procedure, under normality of the ϵ_k, it is established (e.g., Srivastava and Khatri (1979, Chap. 6)) that S_* has the $W(F_1\Sigma_{\epsilon\epsilon}F_1', T - n)$ distribution, the Wishart distribution with matrix $F_1\Sigma_{\epsilon\epsilon}F_1'$ and $T - n$ degrees of freedom, and H_* is distributed as Wishart under H_0 with matrix $F_1\Sigma_{\epsilon\epsilon}F_1'$ and n_2 degrees of freedom. The sum of squares matrices S_* and H_* is also independently distributed, which follows from the well-known fact that the residual sum of squares matrix S and the least squares estimator $\widetilde{C}$ are independent, and H_* is a function of $\widetilde{C}$ alone. Thus, in particular, in the special case $m_1 = 1$, we see that $S_*^{-1}H_* = H_*/S_*$ is distributed as $n_2/(T - n)$ times the F-distribution with n_2 and $T - n$ degrees of freedom.

Tests other than the LR procedure may also be considered for the multivariate linear hypothesis $F_1CG_2 = 0$. The most notable of these is the test developed by Roy (1953) based on the union-intersection principle of test construction. Roy's test rejects for large values of the test statistic which is equal to the largest eigenvalue of the matrix $S_*^{-1}H_*$. Charts and tables of the upper critical values for the largest root distribution have been computed by Heck (1960) and Pillai (1967).

Notice that the above hypotheses concerning C can be viewed as an assumption that the coefficient matrix C has a *reduced* rank, but of a particular *known* or specified form (i.e., the matrices F_1 and G_2 in the constraint $F_1CG_2 = 0$ are known). In the next chapter, we consider further investigating the possibility of a reduced-rank structure for the regression coefficient matrix C, but where the form of the reduced rank is not known. In particular, in such cases of reduced-rank structure for C, linear constraints of the form $F_1C = 0$ will hold but with the $m_1 \times m$ matrix F_1 unknown ($m_1 < m$). Notice from (1.18) that the appropriate ML estimator of C under (known) constraints on C of such a form is given by

$$\widehat{C} = (I_m - SF_1'(F_1SF_1')^{-1}F_1)\widetilde{C} \equiv P\widetilde{C}, \tag{1.20}$$

where $P = I_m - SF_1'(F_1SF_1')^{-1}F_1$ is an idempotent matrix of rank $r = m - m_1$. In Sections 2.3 and 2.6, we will find that the ML estimator $\widehat{C}$ of C subject to the constraint of having reduced rank r (but with the linear constraints $F_1C = 0$ unknown) can be expressed in a similar form as in (1.20), but in terms of an estimate $\widehat{F_1}$ whose rows are determined from examination of the eigenvectors of the matrix $S^{-1/2}\widetilde{C}(\mathbf{XX}')\widetilde{C}'S^{-1/2}$ associated with its m_1 smallest eigenvalues.

Similarly, the constraints across the predictor variables, $CG_2 = 0$, where the matrix $G_2(n \times n_2)$ is unknown, have applications in Finance (see Chapter 8). For a known G_2, the ML estimator of C takes the form,

$$\widehat{C} = \widetilde{C}[I_n - G_2(G_2'(XX')^{-1}G_2)^{-1}G_2'(XX')^{-1}] \equiv \widetilde{C}Q, \tag{1.21}$$

where $Q = I_n - G_2(G_2'(XX')^{-1}G_2)^{-1}G_2'(XX')^{-1}$ is an idempotent matrix of rank $n - n_2$. The ML estimator $\widehat{C}$ with reduced rank r can also be expressed in a form similar to (1.21), but the columns of G_2 are determined by the eigenvectors of the matrix $(XX')^{1/2}\widetilde{C}'S^{-1}\widetilde{C}(XX')^{1/2}$ associated with its n_2 smallest eigenvalues.

1.4 Prediction in the Multivariate Linear Regression Model

Another problem of interest in regard to the multivariate linear regression model (1.1) is confidence regions for the mean responses corresponding to a fixed value X_0 of the predictor variables. The mean vector of responses at X_0 is CX_0 and its estimate is $\widetilde{C}X_0$, where $\widetilde{C} = \mathbf{YX}'(\mathbf{XX}')^{-1}$ is the LS estimator of C. Under the assumption of normality of the ϵ_k, noting that $\widetilde{C}X_0 = \text{vec}(X_0'\widetilde{C}') = (I_m \otimes X_0')\text{vec}(\widetilde{C}')$, we see that $\widetilde{C}X_0$ is distributed as multivariate normal $N(CX_0, X_0'(\mathbf{XX}')^{-1}X_0\Sigma_{\epsilon\epsilon})$ from Result 1.1. Since $\widetilde{C}$ is distributed independently of $\overline{\Sigma}_{\epsilon\epsilon} = \frac{1}{T-n}S$ with $(T-n)\overline{\Sigma}_{\epsilon\epsilon}$ distributed as Wishart with matrix $\Sigma_{\epsilon\epsilon}$ and $T - n$ degrees of freedom, it follows that

$$\mathcal{T}^2 = (\widetilde{C}X_0 - CX_0)'\,\overline{\Sigma}_{\epsilon\epsilon}^{-1}\,(\widetilde{C}X_0 - CX_0)/\{X_0'(\mathbf{XX}')^{-1}X_0\} \tag{1.22}$$

has Hotelling's T^2-distribution with $T - n$ degrees of freedom (Anderson 1984b, Chap. 5). Thus, $\{(T - n - m + 1)/m(T - n)\}\mathcal{T}^2$ has the F-distribution with m and $T - n - m + 1$ degrees of freedom, so that a $100(1 - \alpha)\%$ confidence ellipsoid for the true mean response CX_0 at X_0 is given by

$$\begin{aligned}&(\widetilde{C}X_0 - CX_0)'\,\overline{\Sigma}_{\epsilon\epsilon}^{-1}\,(\widetilde{C}X_0 - CX_0)\\ &\quad \le \{X_0'(\mathbf{XX}')^{-1}X_0\}\left[\frac{m(T-n)}{T-n-m+1}\,F_{m,T-n-m+1}(\alpha)\right],\end{aligned} \tag{1.23}$$

where $F_{m,T-n-m+1}(\alpha)$ is the upper 100αth percentile of the F-distribution with m and $T - n - m + 1$ degrees of freedom. Because

$$(a'\widetilde{C}X_0 - a'CX_0)^2/(a'\overline{\Sigma}_{\epsilon\epsilon}a) \le (\widetilde{C}X_0 - CX_0)'\overline{\Sigma}_{\epsilon\epsilon}^{-1}(\widetilde{C}X_0 - CX_0)$$

for every nonzero $m \times 1$ vector a, by the Cauchy–Schwarz inequality, $100(1 - \alpha)\%$ *simultaneous* confidence intervals for all linear combinations $a'CX_0$ of the individual mean responses are given by

$$a'\widetilde{C}X_0 \pm \{(a'\overline{\Sigma}_{\epsilon\epsilon}a)X_0'(\mathbf{XX}')^{-1}X_0\}^{1/2}\left[\frac{m(T-n)}{T-n-m+1}\,F_{m,T-n-m+1}(\alpha)\right]^{1/2}.$$

Notice that the mean response vector CX_0 of interest represents a particular example of the general parametric function F_1CG_2 discussed earlier in the hypothesis testing context of Section 1.3, corresponding to $F_1 = I_m$, $G_2 = X_0$, and hence $n_2 = 1$.

A related problem of interest is the prediction of values of a new response vector $Y_0 = CX_0 + \epsilon_0$ corresponding to the predictor variables X_0, where it is assumed that Y_0 is independent of the values $Y_1, \ldots, Y_T$. The predictor of Y_0 is given by $\widehat{Y}_0 = \widetilde{C}X_0$, and it follows similarly to the above result (1.23) that a $100(1 - \alpha)\%$ prediction ellipsoid for Y_0 is

$$\begin{aligned}&(Y_0 - \widetilde{C}X_0)'\overline{\Sigma}_{\epsilon\epsilon}^{-1}(Y_0 - \widetilde{C}X_0)\\ &\quad \le \{1 + X_0'(\mathbf{XX}')^{-1}X_0\}\left[\frac{m(T-n)}{T-n-m+1}F_{m,T-n-m+1}(\alpha)\right].\end{aligned} \tag{1.24}$$

Another prediction problem that may be of interest on occasion is that of prediction of a subset of the components of Y_0 given that the remaining components have been observed. This problem would also involve the covariance structure $\Sigma_{\epsilon\epsilon}$ of Y_0, but it will not be discussed explicitly here. Besides the various dimension reduction interpretations, one additional purpose of investigating the possibility of a reduced-rank structure for C, as will be undertaken in the next chapter, is to obtain more precise predictions of future response vectors as a result of more accurate estimation of the regression coefficient matrix C in model (1.1). The advantage of reduced-rank models in prediction will become apparent when dealing with a large number of variables. This will be demonstrated in the next chapter.

1.5 Numerical Examples

In this section we present two examples to illustrate some of the methods and results described in the earlier sections of this chapter. We consider the basic multivariate regression analysis methods of the previous sections applied to these examples.

1.5.1 Biochemical Data

We consider biochemical data taken from a study by Smith et al. (1962). The data involve biochemical measurements on several characteristics of urine specimens of 17 men classified into two weight groups, overweight and underweight. Each subject, except for one, contributed two morning samples of urine, and the data considered in this example consist of the 33 individual samples (treated as independent with no provision for the possible correlations within the two determinations of each subject) for five response variables ($m = 5$), pigment creatinine (y_1), and concentrations of phosphate (y_2), phosphorous (y_3), creatinine (y_4), and choline (y_5), and the concomitant or predictor variables, volume (x_2), and specific gravity (x_3), in addition to the predictor variable for weight of the subject (x_1). The multivariate data set for this example is presented in Table A.1 of the Appendix.

We consider a multivariate linear regression model for the kth vector of responses of the form

$$Y_k = c_0 + c_1 x_{1k} + c_2 x_{2k} + c_3 x_{3k} + \epsilon_k \equiv C X_k + \epsilon_k, \quad k = 1, \ldots, T, \qquad (1.25)$$

with $T = 33$, where $Y_k = (y_{1k}, y_{2k}, y_{3k}, y_{4k}, y_{5k})'$ is a 5×1 vector, and $X_k = (1, x_{1k}, x_{2k}, x_{3k})'$ is a 4×1 vector (hence $n = 4$). With $\mathbf{Y}$ and $\mathbf{X}$ denoting 5×33 and 4×33 data matrices, respectively, the least squares estimate of the regression coefficient matrix is obtained as

$$\widetilde{C} = \mathbf{Y}\mathbf{X}'(\mathbf{X}\mathbf{X}')^{-1} = \begin{bmatrix} 15.2809 & -2.9090 & 1.9631 & 0.2043 \\ (4.3020) & (1.0712) & (0.6902) & (1.1690) \\ \\ 1.4159 & 0.6044 & -0.4816 & 0.2667 \\ (0.8051) & (0.2005) & (0.1292) & (0.2188) \\ \\ 2.0187 & 0.5768 & -0.4245 & -0.0401 \\ (0.6665) & (0.1659) & (0.1069) & (0.1811) \\ \\ 1.8717 & 0.6160 & -0.5781 & 0.3518 \\ (0.5792) & (0.1442) & (0.0929) & (0.1574) \\ \\ -0.8902 & 1.3798 & -0.6289 & 2.8908 \\ (7.0268) & (1.7496) & (1.1273) & (1.9095) \end{bmatrix}.$$

The unbiased estimate of the 5×5 covariance matrix of the errors ϵ_k (with correlations shown above the diagonal) is given by

$$\overline{\Sigma}_{\epsilon\epsilon} = \frac{1}{33-4}\widehat{\epsilon\epsilon'} = \begin{bmatrix} 11.2154 & -0.4225 & -0.3076 & -0.3261 & -0.0780 \\ -0.8867 & 0.3928 & 0.8951 & 0.6769 & 0.3418 \\ -0.5344 & 0.2911 & 0.2692 & 0.6605 & 0.2891 \\ -0.4924 & 0.1913 & 0.1545 & 0.2033 & 0.0343 \\ -1.4285 & 1.1718 & 0.8203 & 0.0845 & 29.9219 \end{bmatrix},$$

where $\widehat{\epsilon} = \mathbf{Y} - \widetilde{C}\mathbf{X}$. The estimated standard deviations of the individual elements of $\widetilde{C}$ are obtained as the square roots of the diagonal elements of $\overline{\Sigma}_{\epsilon\epsilon} \otimes (\mathbf{X}\mathbf{X}')^{-1}$, where

$$(\mathbf{X}\mathbf{X}')^{-1} = \begin{bmatrix} 1.6501 & -0.0807 & -0.1934 & -0.3798 \\ -0.0807 & 0.1023 & -0.0146 & -0.0243 \\ -0.1934 & -0.0146 & 0.0423 & 0.0424 \\ -0.3798 & -0.0243 & 0.0424 & 0.1219 \end{bmatrix}.$$

These estimated standard deviations are evaluated and displayed in parentheses below the corresponding estimates in $\widetilde{C}$. We find that many of the individual estimates in $\widetilde{C}$ are greater than twice their estimated standard deviations.

We illustrate the LR test procedure by testing the hypothesis that the two concomitant predictor variables x_2 and x_3 have no influence on the response variables, that is, $H_0\colon C_2 \equiv [c_2, c_3] = 0$. The matrix due to this null hypothesis is obtained as

$$H = \widetilde{C}_2 A_{22.1} \widetilde{C}_2' = \begin{bmatrix} 129.365 & -38.712 & -28.064 & -47.170 & -106.563 \\ -38.712 & 12.469 & 8.410 & 15.266 & 40.140 \\ -28.064 & 8.410 & 6.088 & 10.248 & 23.225 \\ -47.170 & 15.266 & 10.248 & 18.695 & 49.587 \\ -106.563 & 40.140 & 23.225 & 49.587 & 164.779 \end{bmatrix},$$

where

$$A_{22.1} = \begin{bmatrix} 0.042472 & 0.042358 \\ 0.042358 & 0.121855 \end{bmatrix}^{-1} = \begin{bmatrix} 36.0395 & -12.5278 \\ -12.5278 & 12.5613 \end{bmatrix}$$

and the error matrix is $S = \widehat{\boldsymbol{\epsilon}}\widehat{\boldsymbol{\epsilon}}' = (33-4)\widetilde{\Sigma}_{\epsilon\epsilon}$. The nonzero eigenvalues of the matrix $S^{-1}H = S^{-1}\widetilde{C}_2 A_{22.1}\widetilde{C}_2'$ are the same as those of

$$\widetilde{C}_2' S^{-1} \widetilde{C}_2 A_{22.1} = \begin{bmatrix} 2.5587 & -1.1269 \\ -2.1683 & 1.4912 \end{bmatrix}$$

and are found to equal $\widehat{\lambda}_1^2 = 3.67671$ and $\widehat{\lambda}_2^2 = 0.37319$. Hence, with $T = 33$, $n = 4$, $n_2 = 2$, and $m = 5$, the value of the test statistic in (1.16) is obtained as

$$\mathcal{M} = [33 - 4 + (2 - 5 - 1)/2][\log(1 + 3.6767) + \log(1 + 0.3732)] = 50.2126$$

with $n_2 m = 2 \cdot 5 = 10$ degrees of freedom. Since the critical value of the χ_{10}^2 distribution at level $\alpha = 0.05$ is 18.31, H_0 is clearly rejected and the implication is that the concomitant variables (x_2, x_3) have influence on (at least some of) the response variables y_1, y_2, y_3, y_4, and y_5. From informal examination of the estimates in $\widetilde{C}$ and their estimated standard deviations, it is expected that the significance of this hypothesis testing result is due mainly to the influence of the variable volume (x_2), with the variable x_3 having much less contribution. (In fact, the value of the test statistic in (1.16) for $H_0\colon c_3 = 0$ yields the value $\mathcal{M} = 14.6389$ with 5 degrees of freedom and indicates only mild influence for x_3 on the response variables.)

Similarly, we can test the hypothesis that the weight variable (x_1) has no influence on the response variables, $H_0\colon c_1 = 0$. For the LR test of this hypothesis, since $n_2 = 1$, the single nonzero eigenvalue of $S^{-1}H$ can be obtained as the scalar $\widetilde{c}_1' S^{-1}\widetilde{c}_1/0.10231 = 0.08184/0.10231 = 0.79995$, noting that 0.10231 represents the (2, 2)-element of the matrix $(\mathbf{XX}')^{-1}$. So the value of the test statistic in (1.16) is $\mathcal{M} = [33 - 4 + (1 - 5 - 1)/2]\log(1 + 0.79995) = 15.5756$ with 5 degrees of freedom. The null hypothesis H_0 is rejected at the 0.05 level. Because $n_2 = 1$ for this H_0, an "exact" test is available based on the F-distribution, similar to the form presented in relation to (1.22) with $X_0 = (0, 1, 0, 0)'$ so that $CX_0 = c_1$, using the statistic $\frac{33-4-5+1}{5(33-4)}\widetilde{c}_1'\widetilde{\Sigma}_{\epsilon\epsilon}^{-1}\widetilde{c}_1/0.10231 = 3.9998$ with $m = 5$ and $T - n - m + 1 = 33 - 4 - 5 + 1 = 25$ degrees of freedom. A similar conclusion is obtained as with the approximate chi-squared LR test statistic.

Concluding, we also point out that the last three columns of the LS estimate $\widetilde{C}$ (excluding the column of the constant terms) give some indications of a reduced-rank structure for the 5×3 matrix $[c_1, c_2, c_3]$ of regression coefficients in model (1.25). For example, notice the similarity of coefficient estimates among the second through fourth rows of the last three columns of $\widetilde{C}$. This type of reduced-rank structure and other forms will be examined formally in the subsequent chapters.

1.5.2 Sales Performance Data

We consider the data on sales staff given in Johnson and Wichern (2008). This data was also analyzed by Cook and Su (2013) in the context of envelope model fitting. The performance quality of a firm's sales staff is measured by growth, profitability, and new-account sales. These measures are then related to series of tests taken by the staff for creativity, mechanical reasoning, abstract reasoning, and mathematical ability. The number of observations is $T = 50$, the number of response variables, $m = 4$, and the number of predictors, $n = 3$. The data is listed in Table A.2 in the Appendix.

The multivariate regression model (1.4) is fit for this data. Observe that the matrix "Y" is of dimension 4×50 and "X" is of dimension 3×50; the coefficient matrix, C, is of dimension 4×3. The predictor variables are, x_1: sales growth, x_2: sales profitability, and x_3: new account sales. The response variables are, y_1: score on creativity test, y_2: score on mechanical reasoning test, y_3: score on abstract reasoning test, and y_4: score on mathematics test.

The scatterplot of the variables is given in Fig. 1.1. It is interesting to note that all three predictor variables are highly correlated among themselves and the response variable, y_4, is highly correlated with all predictors. The correlation matrix given below in Table 1.1 confirms that.

We subtract the mean for each series and the demeaned series form the data matrices Y and X given in the text. The regression coefficient matrix, $\widetilde{C}$, in (1.7) is

$$\widehat{C} = \begin{bmatrix} -0.086 & -0.027 & 0.753^* \\ (0.171) & (0.107) & (0.186) \\ & & \\ 0.062 & 0.213^* & -0.012 \\ (0.138) & (0.087) & (0.150) \\ & & \\ 0.428^* & -0.248^* & 0.152 \\ (0.077) & (0.048) & (0.084) \\ & & \\ 0.443^* & 0.610^* & 0.193 \\ (0.191) & (0.120) & (0.208) \end{bmatrix}$$

*Significant at 0.05 significant level.

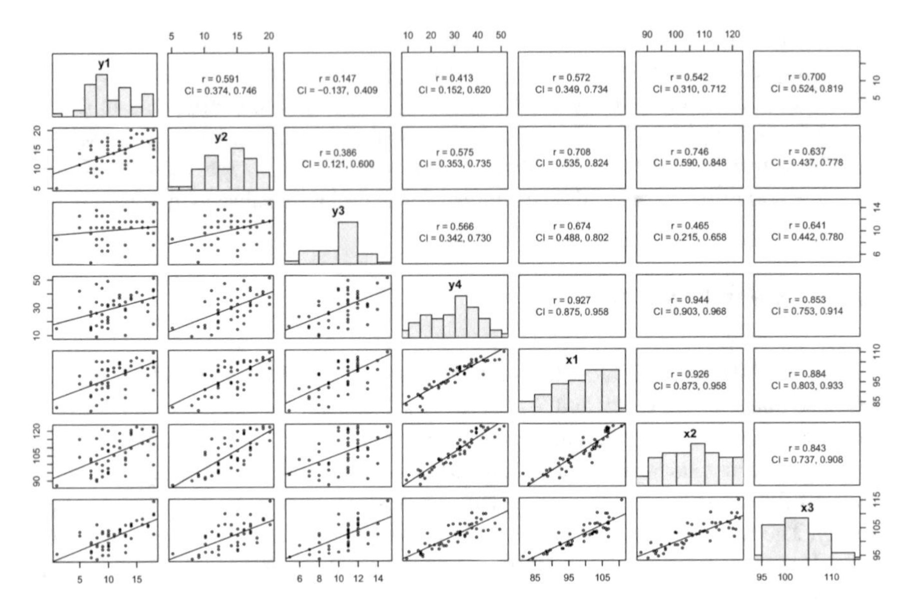

Fig. 1.1 Scatterplot of sales performance

Table 1.1 Correlations

	y_1	y_2	y_3	y_4	x_1	x_2	x_3
y_2	0.591						
y_3	0.147	0.386					
y_4	0.413	0.575	0.566				
x_1	0.572	0.708	0.674	0.927			
x_2	0.542	0.746	0.465	0.944	0.926		
x_3	0.700	0.637	0.641	0.853	0.884	0.843	1
Average	11.220	14.180	10.560	29.760	98.836	106.622	102.810
Std. Dev.	3.950	3.385	2.140	10.538	7.337	10.124	4.712

Although most of the elements of $\widetilde{C}$ appear to be significant, the matrix $\widetilde{C}$ could be of lower rank. The unbiased estimate of the error covariance matrix is given as

$$\overline{\Sigma}_{\epsilon\epsilon} = \begin{pmatrix} 8.113 & 0.387 & -0.701 & -0.679 \\ 2.530 & 5.273 & 0.027 & -0.693 \\ -2.556 & 0.079 & 1.639 & 0.176 \\ -6.140 & -5.054 & 0.715 & 10.085 \end{pmatrix}.$$

Here the entries above the main diagonal are correlations and the entries below are covariances. Observe that the residuals in Fig. 1.2 exhibit some high correlations indicating that some information from there can be useful for modeling the relationships. This and other aspects will be explored in Chapter 2.

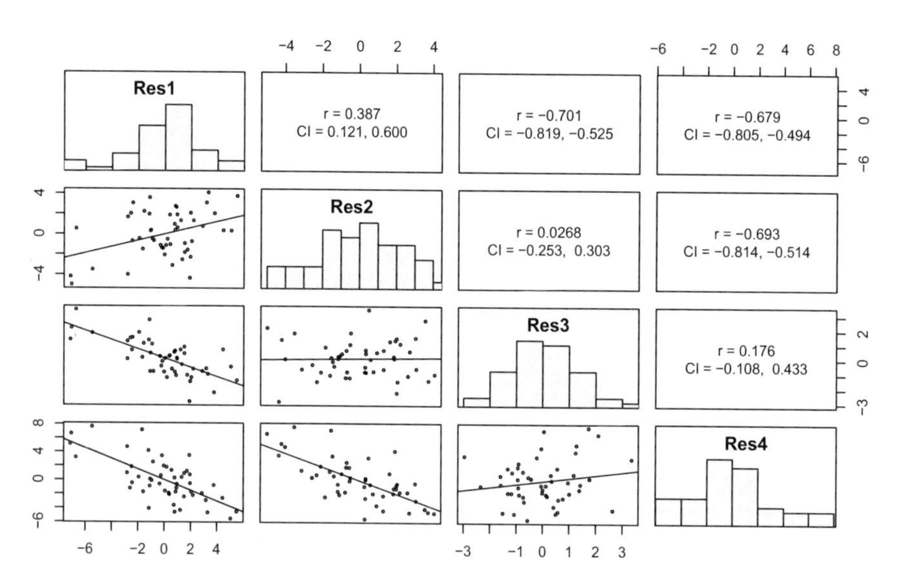

Fig. 1.2 Residual plots

Chapter 2
Reduced-Rank Regression Model

2.1 The Basic Reduced-Rank Model and Background

The classical multivariate regression model presented in Chapter 1, as noted before, does not make direct use of the fact that the response variables are likely to be correlated. A more serious practical concern is that even for a moderate number of variables whose interrelationships are to be investigated, the number of parameters in the regression matrix can be large. For example, in a multivariate analysis of economic variables (see Example 2.2), Gudmundsson (1977) uses $m = 7$ response variables and $n = 6$ predictor variables, thus totaling 42 regression coefficient parameters (excluding intercepts) to be estimated, in the classical regression setup. But the number of vector data points available for estimation is only $T = 36$; these are quarterly observations from 1948 to 1956 for the United Kingdom. Thus, in many practical situations, especially in Big Data settings where the number of variables can exceed the number of observations, there is a need to reduce the number of parameters in model (1.1). We approach this problem through the assumption of lower rank of the matrix C in model (1.1). More formally, in the model $Y_k = CX_k + \epsilon_k$, we assume that

$$\text{rank}(C) = r \leq \min(m, n). \tag{2.1}$$

The methodology and results that we will present for the multivariate regression model under this assumption of reduced rank apply equally to both the cases of dimension $m \leq n$ and $m > n$. However, for some of the initial discussion below, we shall assume $m \leq n < T$ for convenience of notation.

The rank condition (2.1) has two related practical implications. First, with $r < m$ it implies that there are $(m - r)$ linear restrictions on the regression coefficient matrix C of the form

G. C. Reinsel et al., *Multivariate Reduced-Rank Regression*, Lecture Notes in Statistics 225, https://doi.org/10.1007/978-1-0716-2793-8_2

$$\ell_i' C = 0, \quad i = 1, 2, \ldots, (m - r), \tag{2.2}$$

and these restrictions themselves are often not known a priori in the sense that $\ell_1, \ldots, \ell_{m-r}$ are unknown. Premultiplying (1.1) by ℓ_i', we have

$$\ell_i' Y_k = \ell_i' \epsilon_k. \tag{2.3}$$

The linear combinations, $\ell_i' Y_k$, $i = 1, 2, \ldots, (m - r)$, could be modeled without any reference to the predictor variables X_k and depend only on the distribution of the error term ϵ_k. Otherwise, these linear combinations can be isolated and can be investigated separately.

The second implication that is somewhat complementary to (2.2) is that with assumption (2.1), C can be written as a product of two lower dimensional matrices that are of full ranks. Specifically, C can be expressed as

$$C = AB, \tag{2.4}$$

where A is of dimension $m \times r$ and B is of dimension $r \times n$, but both have rank r. Note that the r columns of the left matrix factor A in (2.4) can be viewed as a basis for the column space of C, while the r rows of B form a basis for the row space. The model (1.1) can then be written as

$$Y_k = A(BX_k) + \epsilon_k, \qquad k = 1, \ldots, T, \tag{2.5}$$

where BX_k is of reduced dimension with only r components. A practical use of (2.5) is that the r linear combinations of the predictor variables X_k are sufficient to model the variation in the response variables Y_k and there may not be a need for all n linear combinations or otherwise for all n variables, as would be the case when no single predictor variable can be discarded from the full-rank analysis. Hence, there is a gain in simplicity and in interpretation through the reduced-rank regression modeling, although from a practitioner point of view one would still need to have measurements on all n predictor variables to perform the reduced-rank analysis leading to the construction of the smaller number of relevant linear combinations. In the later chapters (10–13), we discuss sparse methods that will address selecting or discarding predictor variables in the reduced-rank regression setup.

We shall describe now how the reduced-rank model arises in various contexts from different disciplines and this will also provide a historical background. Anderson (1951) was the first to consider in detail the reduced-rank regression problem, for the case where X_k, the set of predictor variables, is fixed. In an application to economics, he called Y_k the economic variables, but X_k is taken to be the vector of noneconomic variables that can be manipulated. The linear combinations $\ell' Y_k$ are called structural relations. These are stable relationships among the macroeconomic variables that could not be manipulated. Estimating the restrictions "ℓ" is thus important to understand the basic structure of the economy. Izenman (1975) introduced the term "reduced-rank regression" and examined this model in detail. The reduced-rank regression model and its statistical properties were examined further by Robinson (1973, 1974), Tso (1981), Davies and Tso (1982), Zhou (1994), and Geweke (1996), among others.

Subsequent to but separate from the initial work of Anderson (1951), reduced-rank regression concepts were considered in various contexts and under different terminologies by several authors, as noted by van der Leeden (1990, Chapter 2–3). Rao (1964) studied principal components and presented results that can be related to reduced-rank regression, referring to the most useful linear combinations of the predictor variables as *principal components of instrumental variables*. Fortier (1966) considered reduced-rank modeling, which he referred to as *simultaneous linear prediction* modeling, and van den Wollenberg (1977) discussed basically the same procedure as an alternative to canonical correlation analysis, which is known as *redundancy analysis*. The basic results in these works are essentially the same and are related to a particular version of reduced-rank estimation, which we note explicitly in Section 2.3. Keller and Wansbeek (1983), starting from an errors-in-variable model form, demonstrated the formulation of a reduced-rank model both in terms of linear restrictions on the regression coefficient matrix and in terms of lower-dimensional component matrices, referred to as the *principal relations* and the *principal factors* specifications, respectively. In the machine learning community, the reduced-rank model falls in the category of multi-task learning.

An appealing physical interpretation of the reduced-rank model was offered by Brillinger (1969). This is to regard a situation where the n component vector X_k represents information which is to be used to send a message Y_k having m components ($m \leq n$), but such a message can only be transmitted through r channels ($r \leq m$). Thus, BX_k acts as a code and on receipt of the code, form the vector ABX_k by premultiplying the code, which it is hoped, would be as close as possible to the desired Y_k.

Aside from this attractiveness as a statistical technique for parameter reduction, it is also known that a reduced-rank regression structure has an interpretation in terms of underlying hypothetical constructs that drive the two sets of variables. The concept of unobservable 'latent' variables has a long history in economics and the social sciences [see Zellner (1970) and Goldberger (1972)]. These latent variables are hypothetical constructs and the observable variables may appear as both effects and causes of these latent variables. The latent models are extensively used in sociology and in economics. To illustrate, let $Y_k = AY_k^* + U_k$, where Y_k^* is a scalar latent variable, A is a column vector, and Y_k is the set of indicators of the underlying hypothetical construct. Suppose Y_k^* is determined by a set of causal variables X_k through $Y_k^* = BX_k + V_k$, where B is a row vector. By combining the two models, we have the reduced-form equation connecting the multiple causes to the multiple indicators (denoted as MIMIC model),

$$Y_k = ABX_k + (AV_k + U_k) = CX_k + \epsilon_k. \tag{2.6}$$

The above multivariate regression model falls in the reduced-rank framework and it follows that the coefficient matrix is principally constrained to have rank one.

Jöreskog and Goldberger (1975) describe the above model with an application in sociology. Clearly, this situation can be extended to allow for $r > 1$ latent variables, with the resulting model (2.6) having coefficient matrix of reduced rank r.

In the context of modeling macroeconomic time series $\{Y_t\}$, Sargent and Sims (1977) discuss the notion of "index" models, which fit into the above framework. The indexes, $X_t^* = BX_t$, which are smaller in number are constructed from a large set of time series predictor variables X_t and are found to drive the vector of observed response variables Y_t. They are further interpreted through a theory of macroeconomics. More recent developments of reduced-rank models applied to vector autoregressive time series analysis, for example, where the predictor variables X_t in (2.5) represent lagged values Y_{t-1} of the time series Y_t, have been pursued by Reinsel (1983), Velu, Reinsel and Wichern (1986), Ahn and Reinsel (1988, 1990), and Johansen (1988, 1991), among others. In the area of financial economics, the reduced-rank regression models naturally arise from a priori economic theory (see Chapter 8; Velu and Zhou (1999)).

Models closely related to the reduced-rank regression structure in (2.5) have been studied and applied in various other areas. For example, in the work preceding that of Anderson's, Tintner (1945, 1950) and Geary (1948) studied a similar problem within an economic setting in which the (unknown) "systematic parts" W_k of the Y_k's are not necessarily specified to be linear functions of a known set of predictor variables X_k but were still assumed to satisfy a set of $m - r$ linear relationships $\ell_i' W_k = 0$. That is, the model considered was

$$Y_k - \mu = W_k + \epsilon_k, \tag{2.7}$$

where W_k is unobservable and is taken to be the systematic part and ϵ_k is the random error. It is assumed that the systematic part $\{W_k\}$ varies in a lower-dimensional linear space of dimension r $(< m)$, so that the W_k satisfy $LW_k = 0$ for all k (assuming the W_k are "centered" at zero) with L denoting an unknown $(m-r)\times m$ full-rank matrix having rows ℓ_i' as in (2.3). Typically, analysis and estimation of the parameters L in such a model require some special assumptions on the covariance structure $\Sigma_{\epsilon\epsilon}$ of the error part ϵ_k, such as $\Sigma_{\epsilon\epsilon}$ known or $\Sigma_{\epsilon\epsilon} = \sigma^2 \boldsymbol{\Psi}_0$, where σ^2 is an unknown scalar parameter, but $\boldsymbol{\Psi}_0$ is a known $m \times m$ matrix. This problem falls within the framework of models for functional relationships or structural relationships in the systematic part $\{W_k\}$, that is, the relationships are represented by $LW_k = 0$. The terminology "functional" is commonly used when the W_k are taken to be fixed and "structural" when the W_k are treated as random.

The above model can also be considered from the multivariate errors-in-variables regression model viewpoint. To give a more explicit representation in this form, write the $(m - r) \times m$ full-rank matrix L as $L = [L_1, L_2]$, where L_1 is $(m - r) \times (m - r)$ and assumed to be of full rank, and partition Y_k, W_k, and ϵ_k in a compatible fashion as $Y_k = (Y_{1k}', Y_{2k}')'$, $W_k = (W_{1k}', W_{2k}')'$, and $\epsilon_k = (\epsilon_{1k}', \epsilon_{2k}')'$. Then we can multiply the linear relation $LW_k = L_1 W_{1k} + L_2 W_{2k} = 0$ on the left by L_1^{-1} to obtain

$$W_{1k} = -L_1^{-1} L_2 W_{2k} \equiv K_2 W_{2k},$$

where $K_2 = -L_1^{-1} L_2$. So we have the multivariate regression relation

$$Y_{1k} - \mu_1 = W_{1k} + \epsilon_{1k} \equiv K_2 W_{2k} + \epsilon_{1k}, \tag{2.8}$$

where the "independent" variables W_{2k} in the regression model are not directly observed but are observed with error, in that $Y_{2k} - \mu_2 = W_{2k} + \epsilon_{2k}$ is observed. Analysis and estimation of the above models (2.7) and (2.8) have been studied and described by many authors including Sprent (1966), Moran (1971), Gleser and Watson (1973), Gleser (1981), Theobald (1975), Anderson (1976, 1984a), Bhargava (1979), and Amemiya and Fuller (1984). A few specific estimation results for these models will be presented briefly in relation to the topic of principal components analysis which is discussed in Section 2.4.

The model (2.7) with the linear restrictions $LW_k = 0$ can also be given a "parametric" or factor analysis representation as $Y_k - \mu = AF_k + \epsilon_k$ [for example, see Anderson (1984a)], where A is an $m \times r$ matrix of factor loadings of full rank $r < m$ (such that $LA = 0$) and F_k is an $r \times 1$ vector of unobservable common factors. (Model (2.5) can be viewed as a special case of this form with $F_k = BX_k$ being specified as r unknown linear combinations of the observable set of n predictor variables X_k, and (2.5) has been advocated for use in the factor analysis context as a special case of fixed effects factor analysis by van der Leeden (1990, Chapter 5).) Typically, for the factor analysis model, it is assumed that the F_k are random vectors with mean zero and covariance matrix I_r, and $\Sigma_{\epsilon\epsilon} = \text{diag}(\sigma_1^2, \ldots, \sigma_m^2)$ is diagonal. Interest in factor analysis modeling stems originally from its applications in psychology and education, and early developments were by Thurstone (1947), Reiersøl (1950), Lawley (1940, 1943, 1953), Rao (1955), Anderson and Rubin (1956), and Jöreskog (1967, 1969).

When the error covariance matrix in model (2.7) is assumed to be unknown and arbitrary, replicated vector observations are needed for estimation. For example, we may have $N \geq 2$ observations for each k with the model $Y_{kj} - \mu = W_k + \epsilon_{kj}$, $j = 1, \ldots, N$, $k = 1, \ldots, T$. This is a multivariate one-way ANOVA model, and the analysis of such a model under the reduced-rank or linear restriction assumptions for the (fixed effects) W_k has been considered by Anderson (1951, 1984a), Villegas (1961, 1982), Theobald (1975), Healy (1980), and Campbell (1984), among others, following earlier work by Fisher (1938) who considered a related problem in regard to discriminant analysis. This one-way ANOVA model situation will be discussed in more detail later in Section 3.2.

The above brief summary indicates that research and interest in applications of models that directly involve reduced-rank regression concepts or that are closely related have a long history and that the scope is quite broad. However, because the focus of this book is directly on the (reduced-rank) multivariate regression modeling, primarily, it will not be possible to explore all of the above discussed related areas in detail. Articles by Anderson (1984a, 1991) provide an excellent summary of the developments and results in some of these areas, such as estimating linear functional and structural relationships, factor analysis models, and the relation to estimation of simultaneous equations models in econometrics. Interested readers may refer to these articles for a more detailed and complete survey of developments in these areas.

2.2 Some Examples of Application of the Reduced-Rank Model

In this section we introduce and discuss some practical examples where reduced-rank methods have been applied.

Example: 2.1 Davies and Tso (1982) have applied reduced-rank regression methods to study the experimental properties of hydrocarbon fuel mixtures in relating response to composition. The responses are the engine test ratings and they are related to the composition of hydrocarbon fuels, which is characterized by the technique of gas–liquid chromatography (g.l.c.). The responses that are correlated among themselves are known to be dependent on the same portion of g.l.c. chromatograms. The precision of the g.l.c. determination of composition is quite high, so that the resultant data vector of explanatory variables may be regarded as deterministic, but the responses exhibit a wide variation. From the description of the problem, it is possible that a multivariate regression model with rank constraints might be appropriate. A small number of certain features of the chromatogram could be used to predict fuel properties. Because the experimentation in this area is quite expensive, Davies and Tso (1982) point out the usefulness of the model in reducing the number of components.

A total of 30 vector observations, corresponding to different gasoline blends, are divided into two data sets of 14 and 16 blends, respectively, for validation purposes. On each blend, five gasoline distillation measurements $y_1, \ldots, y_5$ were considered, each representing percentage weight evaporated by a specified temperature, and ten x variables were considered to represent the corresponding blend composition by condensing the blend's low-resolution chromatogram into a 10-component vector by grouping of peak areas. Thus, the classical multivariate regression would call for estimating fifty regression coefficient parameters (excluding intercepts) with thirty vector observations. Using the mean square error criterion for model fit, $\frac{1}{Tm}\|\mathbf{Y} - \widehat{\mathbf{Y}}_{(r)}\|^2$, where $\widehat{\mathbf{Y}}_{(r)}$ is the matrix of predicted values of the responses when the rank of the regression coefficient matrix is taken to be r and is estimated by minimizing the overall error sum of squares criterion, Davies and Tso demonstrate that $r = 2$ is fairly sufficient. In fact, in both data sets, over 95% of the variation in Y_k is explained by considering a model with $r = 2$ linear combinations of the X_k. Thus, one may notice the substantial reduction in the number of parameters. Moreover, the similarity in the two linear combination or two factor linear subspace of the X_k

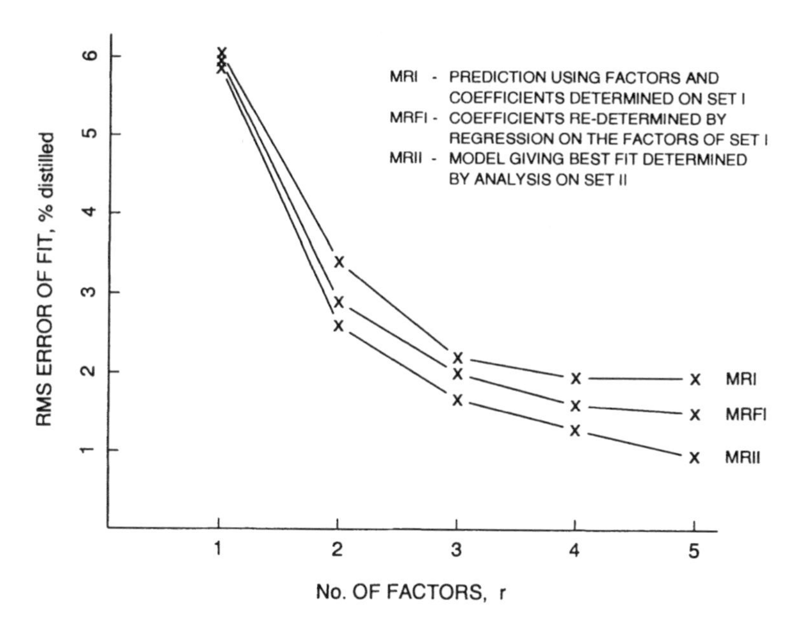

Fig. 2.1 Prediction and fitting to distillation data for Set II gasolines

between the two data sets was fairly high, suggesting that the factors identified have some physical significance.

Figure 2.1 shows the results, in terms of residual mean square (RMS) error, of predicting the response measurements for gasolines in Set II from a knowledge of their composition estimates from both Set I and Set II. The upper curve (MRI) corresponds to predictions on Set II using coefficients of the linear combinations (factors) and corresponding regression coefficients estimated from Set I. The second curve (MRFI) corresponds to a fit for Set II obtained using coefficients of the linear combinations (factors) of the X_k estimated from Set I but re-estimating the regression coefficients of the Y_k on these factors using Set II. The lower curve (MRII) corresponds to the model estimated entirely using data from Set II.

Example: 2.2 The second example we discuss here is based on the work of Gudmundsson (1977), mentioned earlier in Section 2.1. Klein et al. (1961) constructed a detailed econometric model of the United Kingdom. The data consist of quarterly observations from 1948 to 1956 of 37 time series of response variables and 32 time series of predictor variables, some of which may be lagged values of response variables. This could be conceptually handled through the reduced-rank regression model, but the computational burden would be enormous. Therefore, some grouping of variables based on economic considerations was applied. But the focus of Gudmundsson's paper was to construct some linear combinations of variables, representing various aspects of the general economic situation. Because most series are likely to be related to the state of the economy in general, it is anticipated that some useful linear combinations are possible.

The following subset of time series variables was used in the reduced-rank regression calculations. The dependent variables are

y_{1t} Index of total industrial production
y_{2t} Consumption of food, drinks, and tobacco at constant prices
y_{3t} Price index of total consumption
y_{4t} Total civilian labor force
y_{5t} Index of weekly wage rates
y_{6t} Index of volume of total imports
y_{7t} Index of volume of total exports

and the predictor variables are $x_{1t} = y_{1t-1}$, $x_{2t} = y_{2t-1}$, $x_{3t} = y_{6t-1}$, $x_{4t} = y_{7t-1}$, $x_{5t} =$ price index of total imports lagged by two time periods and $x_{6t} = t$, the time index.

The predictor variables are normalized before fitting the reduced-rank regression model. Gudmundsson chose to estimate the linear combinations of predictor variables, $x_{it}^* = \beta_i' X_t$, $i = 1, \ldots, r$, as the coefficients β_i which maximize the sum of the squared sample correlation coefficients of x_i^* and the endogenous variables $y_1, \ldots, y_m$, subject to normalization and orthogonality conditions for the x_i^*. The first three linear combinations were judged by Gudmundsson to be sufficient for use in explaining all the endogenous variables, but no formal justification was provided for this choice. This implicitly is assuming that the rank of the regression coefficient matrix is three. The method used for estimating the β_i was mildly criticized by Theobald (1978). From the estimation results to be presented in Section 2.3, it follows that these estimates can be viewed as ML estimates under normality and the (possibly unrealistic) assumption that the covariance matrix of the error vector ϵ_t takes the form $\Sigma_{\epsilon\epsilon} = \sigma^2 I_m$.

Table 2.1 presents the information on the resulting (estimates of the) linear combinations ($X_t^* = BX_t$) and the sample correlations of the linear combinations with all the response variables. It can be seen from the first part of the table that the linear combinations are dominated by x_6, the time trend variable, and by the variable x_1, the lagged index of industrial production. But from the correlations, it would appear that the first linear combination, which accounts for most of the variation, is fairly highly correlated with all the response variables. Such a combination can also be taken as an index that represents the general economic situation. Gudmundsson also compared the performance of the three linear combinations in predicting all 37 variables used in Klein et al. (1961). The comparisons were made using the root mean square error criterion. The reduced-rank model performed better than the econometric model for production variables, roughly comparable for consumption, imports and exports, but not as good as the econometric model for prices and employment series. This has to be expected because the three linear combinations were constructed from a subset of the independent variables that did not have representation by price or employment. But considering the magnitude of dimension reduction, this is somewhat a small price to pay.

Table 2.1 Coefficients in the linear combinations $X_t^* = BX_t$ and sample correlations with response variables Y_t for econometric data example

Coefficients of independent variables in $X_t^* = BX_t$

	x_{1t}	x_{2t}	x_{3t}	x_{4t}	x_{5t}	x_{6t}
x_{1t}^*	0.148	0.095	0.086	0.013	−0.058	0.720
x_{2t}^*	−0.797	−0.193	−0.013	−0.226	0.302	0.936
x_{3t}^*	−0.406	−0.271	−0.238	0.319	−0.271	0.767

Correlation coefficients of X_t^* with endogenous variables (Y_t)

	y_{1t}	y_{2t}	y_{3t}	y_{4t}	y_{5t}	y_{6t}	y_{7t}	Sum of squared Corr. coefs.
x_{1t}^*	0.974	0.941	0.966	0.984	0.979	0.897	0.868	6.252
x_{2t}^*	−0.083	−0.008	0.225	−0.042	0.138	−0.110	−0.142	0.110
x_{3t}^*	0.034	0.156	−0.065	0.015	0.016	−0.192	0.028	0.068

Example: 2.3 Glasbey (1992) used the reduced-rank model in a recent application to demonstrate the relationship between measurements on solar radiation taken over various sites in Scotland and the physical characteristics of the sites. The aim of the analysis was to summarize the differences in solar radiation among sites and if possible to relate them to other physical characteristics of the sites. The sites, radiometric stations ten in number, were located on the range of Pentland Hills to the south of Edinburgh (Fig. 2.2) and monthly averages of solar radiation measurements on a log-scale for a period of 19 months form the 19×10 data matrix $\mathbf{Y}$. The log-scale was used to make variability over the seasons to be approximately the same. Because of the proximity of these stations, one might expect these response variables to be interrelated over sites, but any possible spatial correlations over the sites will be ignored. The $\mathbf{X}$ matrix is formed with five site characteristics, latitude (x_1), longitude (x_2), altitude (x_3), and percentage sun loss in summer (x_4) and in winter (x_5). Thus, the data matrix $\mathbf{X}$ is of dimension 5×10. This is a novel application of the reduced-rank model and does not conform to the dimensional details presented earlier (i.e., $m = 19$, $n = 5$, and $T = 10$ in this example, so that $m > T$). At the first stage after adjusting for monthly means, simple models for the response variables at each site that account for most of the variability in $\mathbf{Y}$ are constructed via the method of principal components and singular value decomposition (SVD). Two terms or components were considered to be sufficient in the singular value representation for $\mathbf{Y}$ on the basis that they account for a high percent (85%) of the overall variation of the site values about the monthly mean over all sites. Then these two summary vectors (1×10) of the responses were regressed on $\mathbf{X}$, the matrix of physical characteristics. The result can be shown to be in a form of reduced-rank regression, where the rank of the matrix is assumed to be equal to two. Over 80% of the variation in month-corrected $\mathbf{Y}$ is supposedly accounted by the two linear combinations of $\mathbf{X}$:

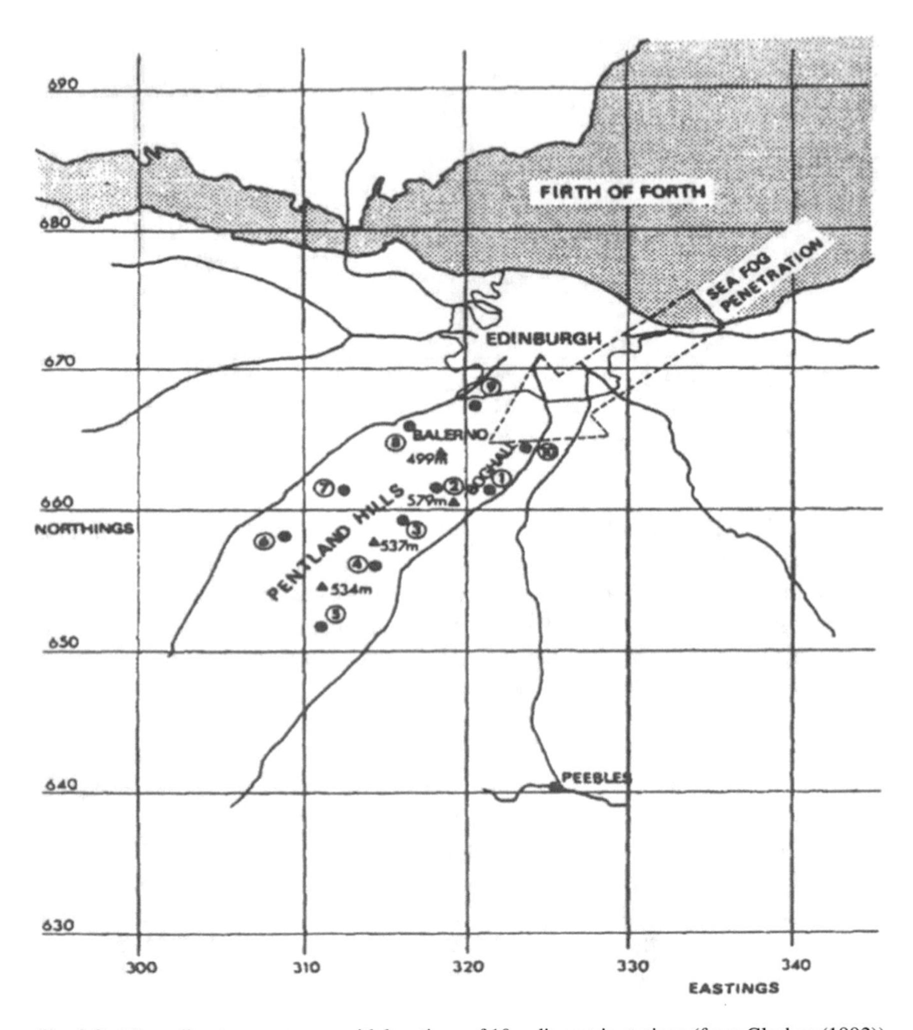

Fig. 2.2 Microclimate survey area with locations of 10 radiometric stations (from Glasbey (1992))

$$\begin{aligned}
x_1^* &= -319.3 + \underset{(1.2)}{5.7x_1} + \underset{(0.6)}{0.2x_2} - \underset{(0.0013)}{0.0022x_3} + \underset{(0.012)}{0.004x_4} - \underset{(0.0019)}{0.0076x_5} \\
x_2^* &= 160.2 + \underset{(2.1)}{2.6x_1} + \underset{(1.1)}{4.8x_2} - \underset{(0.0024)}{0.0037x_3} + \underset{(0.022)}{0.017x_4} + \underset{(0.0034)}{0.0019x_5},
\end{aligned}$$

where the numbers in parentheses are the estimated standard errors.

The first combination x_1^* is primarily a latitude effect, with some negative adjustments for altitude and percentage sun loss in winter. The second combination x_2^* is a longitude effect but adjusted for altitude. The more northerly sites (see

Fig. 2.2) have positive values and the southern sites have negative values for the regression coefficient on x_1^* and western sites have the positive values on x_2^*. In general, Glasbey (1992) concludes that the sites at higher altitudes received less radiation; winter sun loss reduced radiation levels but the summer loss is not known to be effective.

Based on the above examples alone, it would appear that the reduced-rank methods have strong potential for both theoretical study and applications. A few other recent applications of reduced-rank regression include the following. In a study by Ryan et al. (1992), reduced-rank methods were applied to the analysis of response data (DNA, RNA, and protein) from a joint toxicity bioassay experiment to determine the joint effects of toxic compounds that include copper and zinc on the growth of larval fathead minnows. In the field of financial economics, several asset pricing theories have been proposed for testing the efficiency of portfolios, as described by Zhou (1991, 1995). These theories generally imply certain rank restrictions on the regression coefficient matrix that relates asset returns to market factors. Empirical verification using asset returns data on industry portfolios has been made through tests for reduced-rank regression. Other recent applications will be covered throughout the book.

2.3 Estimation of Parameters in the Reduced-Rank Model

The reduced-rank regression model described in Section 2.1 has the following parameters to be estimated from a sample of observations: the matrix A which is of dimension $m \times r$, the matrix B which is of dimension $r \times n$, and the $m \times m$ covariance matrix $\Sigma_{\epsilon\epsilon}$ of the error term. Estimation of the constraint vectors $\ell_1, \ldots, \ell_{m-r}$ will be taken up following these. Before we proceed with the details, we note that in the reduced-rank model with $C = AB$ the component matrices A and B are determined only up to nonsingular linear transformations. To obtain a particular unique set of parameter values for A and B from the elements of C, certain normalization conditions are typically imposed. These conditions, as we have chosen here, will follow from the estimation criterion. To clearly exhibit the algebraic features of the problem, we will initially formulate the criterion for determination of A and B in terms of "population" quantities, and later we will demonstrate how the maximum likelihood estimates from sample data are obtained through use of the initial population results.

As we shall see shortly, reduced-rank estimation will be obtained as a certain reduced-rank approximation of the full-rank least squares estimate of the regression coefficient matrix. Therefore, for presentation of reduced-rank estimation, we first need an important matrix result called the Householder–Young or Eckart–Young theorem (Eckart and Young 1936), which presents the main tool for approximating a full-rank matrix by a matrix of lower rank. The solution is related to the singular value decomposition of the full-rank matrix. As a preliminary to this main theorem, we state a needed result given in Rao (1973, p. 63).

Lemma 2.1 *Let A be an $m \times m$ symmetric matrix with eigenvalues $\lambda_1 \geq \cdots \geq \lambda_m$, and denote the corresponding normalized eigenvectors as $P_1, \ldots, P_m$. The supremum of $\sum_{i=1}^{r} X_i' A X_i = \text{tr}(X'AX)$, with $X = [X_1, \ldots, X_r]$, over all sets of $r \leq m$ mutually orthonormal vectors $X_1, \ldots, X_r$ is equal to $\sum_{i=1}^{r} \lambda_i$ and is attained when the $X_i = P_i$, $i = 1, \ldots, r$.*

Proof Let $P = [P_1, \ldots, P_m]$, so that P is an orthogonal matrix with $P'AP = \Lambda = \text{diag}(\lambda_1, \ldots, \lambda_m)$. Any set of r mutually orthonormal vectors $X_1, \ldots, X_r$ can be expressed as $X_i = \sum_{j=1}^{m} c_{ij} P_j \equiv P c_i$, $i = 1, \ldots, r$, where $c_i = (c_{i1}, \ldots, c_{im})'$, with $X_i' X_i = c_i' P' P c_i = c_i' c_i = 1$ and $X_i' X_l = c_i' P' P c_l = c_i' c_l = 0$, for $i \neq l$. Then

$$\sum_{i=1}^{r} X_i' A X_i = \sum_{i=1}^{r} c_i' P' A P c_i = \sum_{i=1}^{r} c_i' \Lambda c_i = \sum_{i=1}^{r} \sum_{j=1}^{m} \lambda_j c_{ij}^2 = \sum_{j=1}^{m} \lambda_j \left(\sum_{i=1}^{r} c_{ij}^2 \right),$$

where the coefficient of each λ_j, $\sum_{i=1}^{r} c_{ij}^2$, is ≤ 1 and the sum of these coefficients is r, since $\sum_{j=1}^{m} \sum_{i=1}^{r} c_{ij}^2 = \sum_{i=1}^{r} c_i' c_i = r$. To maximize the sum above, the optimum choice is thus to make the coefficients of the λ_j equal to 1 for $j = 1, \ldots, r$, and hence 0 for $j > r$. This is attained only by the choice $X_i = P_i$, $i = 1, \ldots, r$, that is, $c_{ij} = 1$ for $i = j$ and $c_{ij} = 0$ for $i \neq j$, where P_i is a normalized eigenvector of A corresponding to the eigenvalue λ_i. The eigenvectors P_i are (essentially) unique, of course, if the eigenvalues λ_i are distinct, but the choice of eigenvectors will not be unique if some among the first r eigenvalues are not distinct.

Theorem 2.1 *Let S be a matrix of order $m \times n$ and of rank m. The Euclidean norm,* $\text{tr}[(S - P)(S - P)']$, *is minimum among matrices P of the same order as S but of rank $r (\leq m)$, when $P = MM'S$, where M is $m \times r$ and the columns of M are the first r (normalized) eigenvectors of SS', that is, the normalized eigenvectors corresponding to the r largest eigenvalues of SS'.*

Proof Let $P = MN$, where M is an $m \times r$ matrix and N is an $r \times n$ matrix. Without loss of generality, assume that M is orthonormal with $M'M = I_r$. Minimizing the criterion given in the theorem over N, for a given M, yields $N = (M'M)^{-1} M'S = M'S$, by similar argument that led to the least squares solution in (1.7). Substituting this in the criterion yields, after some simplification,

$$\text{tr}[(S - P)(S - P)'] = \text{tr}[SS'(I_m - MM')] = \text{tr}(SS') - \text{tr}(M'SS'M). \tag{2.9}$$

Minimizing the above quantity with respect to M is equivalent to maximizing

$$\text{tr}(M'SS'M) \quad \text{subject to} \quad M'M = I_r. \tag{2.10}$$

Based on the result in Lemma 2.1, it follows that the above goal is achieved by choosing the columns of M to be the (orthonormal) eigenvectors of SS' that correspond to the first (largest) r eigenvalues.

Remark—Singular Value Decomposition The positive square roots of the eigenvalues of SS' are referred to as the singular values of the matrix S. In general, an $m \times n$ matrix S, of rank s, can be expressed in the *singular value decomposition* as $S = V \Lambda U'$, where $\Lambda = \text{diag}(\lambda_1, \ldots, \lambda_s)$ with $\lambda_1^2 \geq \cdots \geq \lambda_s^2 > 0$ being the nonzero eigenvalues of SS', $V = [V_1, \ldots, V_s]$ is an $m \times s$ matrix such that $V'V = I_s$, and $U = [U_1, \ldots, U_s]$ is $n \times s$ such that $U'U = I_s$. The columns V_i are the normalized eigenvectors of SS' corresponding to the eigenvalue λ_i^2, and $U_i = \frac{1}{\lambda_i} S'V_i$, $i = 1, \ldots, s$. For the situation in Theorem 2.1 above, from the singular value decomposition of S, we see that the criterion to be minimized over rank r approximations $P = MN$ is $\text{tr}[(V \Lambda U' - MN)(V \Lambda U' - MN)'] = \text{tr}[(\Lambda - V'MNU)(\Lambda - V'MNU)']$. The result in Theorem 2.1 states that the minimizing matrix factors M and N are given by $M = [V_1, \ldots, V_r] \equiv V_{(r)}$ and $N = M'S = V_{(r)}'S = V_{(r)}'V \Lambda U' = [\lambda_1 U_1, \ldots, \lambda_r U_r]' \equiv \Lambda_{(r)} U_{(r)}'$, with $U_{(r)} = [U_1, \ldots, U_r]$ and $\Lambda_{(r)} = \text{diag}(\lambda_1, \ldots, \lambda_r)$. So $P = MN = V_{(r)} \Lambda_{(r)} U_{(r)}'$ yields the "optimum" rank r approximation to $S = V \Lambda U'$, and the resulting minimum value of the trace criterion is $\sum_{i=r+1}^{m} \lambda_i^2$.

We restate the multivariate regression model before we proceed with the estimation of the relevant parameters. From (2.5), the reduced-rank regression model is

$$Y_k = ABX_k + \epsilon_k, \qquad k = 1, \ldots, T, \tag{2.11}$$

where A is of dimension $m \times r$, B of dimension $r \times n$, and the vectors ϵ_k are assumed to be independent with mean zero and covariance matrix $\Sigma_{\epsilon\epsilon}$, where $\Sigma_{\epsilon\epsilon}$ is positive definite. Estimation of A and B in (2.11) is based on the following result (Brillinger 1981, Section 10.2), which essentially uses Theorem 2.1. For this "population" result, the vector of predictor variables X_k in (2.11) will be treated as a stochastic vector for ease of exposition.

Theorem 2.2 *Suppose the* $(m+n)$*-dimensional random vector* $(Y_k', X_k')'$ *has mean vector* 0 *and covariance matrix with* $\Sigma_{yx} = \Sigma_{xy}' = \text{Cov}(Y_k, X_k)$ *and* $\Sigma_{xx} = \text{Cov}(X_k)$ *nonsingular. Then, for any positive-definite matrix* Γ, *an* $m \times r$ *matrix* A *and* $r \times n$ *matrix* B, *for* $r \leq \min(m, n)$, *which minimize*

$$\text{tr}\{E[\Gamma^{1/2}(Y_k - ABX_k)(Y_k - ABX_k)'\Gamma^{1/2}]\}, \tag{2.12}$$

are given by

$$A^{(r)} = \Gamma^{-1/2}[V_1, \ldots, V_r] = \Gamma^{-1/2}V, \qquad B^{(r)} = V'\Gamma^{1/2}\Sigma_{yx}\Sigma_{xx}^{-1}, \tag{2.13}$$

where $V = [V_1, \ldots, V_r]$ and V_j is the (normalized) eigenvector that corresponds to the jth largest eigenvalue λ_j^2 of the matrix $\Gamma^{1/2}\Sigma_{yx}\Sigma_{xx}^{-1}\Sigma_{xy}\Gamma^{1/2}$ $(j = 1, 2, \ldots, r)$.

Proof Setting $S = \Gamma^{1/2}\Sigma_{yx}\Sigma_{xx}^{-1/2}$ and $P = \Gamma^{1/2}AB\Sigma_{xx}^{1/2}$, the result follows easily from Theorem 2.1 because the criterion (2.12) can be expressed as

$$\begin{aligned}
&\mathrm{tr}\{\Gamma^{1/2}(\Sigma_{yy} - AB\Sigma_{xy} - \Sigma_{yx}B'A' + AB\Sigma_{xx}B'A')\Gamma^{1/2}\} \\
&\quad = \mathrm{tr}\{\Gamma^{1/2}(\Sigma_{yy} - \Sigma_{yx}\Sigma_{xx}^{-1}\Sigma_{xy})\Gamma^{1/2}\} \\
&\qquad + \mathrm{tr}\{\Gamma^{1/2}(\Sigma_{yx}\Sigma_{xx}^{-1/2} - AB\Sigma_{xx}^{1/2})(\Sigma_{yx}\Sigma_{xx}^{-1/2} - AB\Sigma_{xx}^{1/2})'\Gamma^{1/2}\}, \qquad (2.12')
\end{aligned}$$

where $\Sigma_{yy} = \mathrm{Cov}(Y_k)$. Hence we see that minimization of this criterion, with respect to A and B, is equivalent to minimization of the criterion considered in Theorem 2.1. In terms of quantities used in Theorem 2.1, $M = \Gamma^{1/2}A^{(r)}$ and $N = B^{(r)}\Sigma_{xx}^{1/2}$. At the minimum the criterion (2.12) has the value $\mathrm{tr}(\Sigma_{yy}\Gamma) - \sum_{j=1}^{r}\lambda_j^2$.

Several remarks follow regarding Theorem 2.2 [see Izenman (1975, 1980)]. As mentioned before, nothing appears to be gained by estimating the equations jointly in the multivariate regression model. But a true multivariate feature enters the model when we assume that the coefficient matrix may not have full rank. Hence, the display of superscripts for A and B in Theorem 2.2 is meant to denote the prescribed rank r. We also observed earlier that the decomposition $C = AB$ is not unique because for any nonsingular $r \times r$ matrix P, $C = AP^{-1}PB$, and hence to determine A and B uniquely, we must impose some normalization conditions. For the expressions in (2.13), the eigenvectors V_j are normalized to satisfy $V_j'V_j = 1$, and this is equivalent to normalization for A and B as follows:

$$B\,\Sigma_{xx}B' = \Lambda^2, \quad A'\Gamma A = I_r, \qquad (2.14)$$

where $\Lambda^2 = \mathrm{diag}(\lambda_1^2, \ldots, \lambda_r^2)$ and I_r is an $r \times r$ identity matrix. Thus, the number of independent regression parameters in the reduced-rank model (2.11) is $r(m+n-r)$ compared to mn parameters in the full-rank model. Hence, a substantial reduction in the number of parameters is possible.

The elements of the reduced-rank approximation of the matrix C are given as

$$C^{(r)} = A^{(r)}B^{(r)} = \Gamma^{-1/2}\left(\sum_{j=1}^{r} V_jV_j'\right)\Gamma^{1/2}\,\Sigma_{yx}\,\Sigma_{xx}^{-1} = P_\Gamma\,\Sigma_{yx}\,\Sigma_{xx}^{-1}, \qquad (2.15)$$

where P_Γ is an idempotent matrix for any Γ, but need not be symmetric. Observe that, analogous to (1.7), $\Sigma_{yx}\Sigma_{xx}^{-1}$ is the usual full-rank (population) regression coefficient matrix. When $r = m$, $\sum_{j=1}^{m} V_jV_j' = I_m$, and therefore, $C^{(r)}$ reduces to the full-rank coefficient matrix.

Rao (1979) has shown a stronger result that the solution for A and B in Theorem 2.2 also simultaneously minimizes all the eigenvalues of the matrix in (2.12) provided that $\Gamma = \Sigma_{yy}^{-1}$. This was proved using the Poincairé separation theorem. Robinson (1974) has proved a similar result, which indicates that the solutions of A and B in Theorem 2.2, when the corresponding sample quantities are substituted and with $\Gamma = \widetilde{\Sigma}_{\epsilon\epsilon}^{-1}$, are Gaussian estimates under the reduced-rank model (2.11), that is, maximum likelihood estimates under the assumption of normality of the ϵ_k. In the remainder of this book, we may often refer to such estimates derived from the multivariate normal likelihood as maximum likelihood estimates, even when the distributional assumption of normality is not being made explicitly. We shall sketch the details below to establish the maximum likelihood estimates of the parameters in the reduced-rank regression model (2.11). Before we proceed with this, we first state a needed result from Rao (1979, Theorem 2.3) which provides, in some sense, a stronger form of the result in Theorem 2.1.

Lemma 2.2 *Let S be an $m \times n$ matrix of rank m and P be an $m \times n$ matrix of rank $\leq r(\leq m)$. Then*

$$\lambda_i(S - P) \geq \lambda_{r+i}(S) \quad \textit{for any } i,$$

where $\lambda_i(S)$ denotes the ith largest singular value of the matrix S (as defined in the remark following the proof of Theorem 2.1), and $\lambda_{r+i}(S)$ is defined to be zero for $r + i > m$. The equality is attained for all i if and only if $P = V_{(r)}\Lambda_{(r)}U'_{(r)}$, where $S = V\Lambda U'$ represents the singular value decomposition of S as described in the remark.

For ML estimation of model (2.11), let $\mathbf{Y} = [Y_1, \ldots, Y_T]$ and $\mathbf{X} = [X_1, \ldots, X_T]$, where T denotes the number of vector observations. Assuming the ϵ_k are independent and identically distributed (i.i.d.), following a multivariate normal distribution with mean 0 and covariance matrix $\Sigma_{\epsilon\epsilon}$, the log-likelihood, apart from irrelevant constants, is

$$L(C, \Sigma_{\epsilon\epsilon}) = \left(\frac{T}{2}\right)\left[\log|\Sigma_{\epsilon\epsilon}^{-1}| - \operatorname{tr}(\Sigma_{\epsilon\epsilon}^{-1}W)\right], \tag{2.16}$$

where $|\Sigma_{\epsilon\epsilon}^{-1}|$ is the determinant of the matrix $\Sigma_{\epsilon\epsilon}^{-1}$ and $W = (1/T)(\mathbf{Y} - C\mathbf{X})(\mathbf{Y} - C\mathbf{X})'$. When $\Sigma_{\epsilon\epsilon}$ is unknown, the solution obtained by maximizing the above log-likelihood is $\widehat{\Sigma}_{\epsilon\epsilon} = W$. Hence, the concentrated log-likelihood is $L(C, \widehat{\Sigma}_{\epsilon\epsilon}) = -(T/2)(\log|W| + m)$. Maximizing this expression (with the structure $C = AB$, of course) is equivalent to minimizing $|W|$ and hence minimizing $|\widetilde{\Sigma}_{\epsilon\epsilon}^{-1}W|$ since $|\widetilde{\Sigma}_{\epsilon\epsilon}^{-1}|$ is a positive constant, where

$$\widetilde{\Sigma}_{\epsilon\epsilon} = (1/T)(\mathbf{YY}' - \mathbf{YX}'(\mathbf{XX}')^{-1}\mathbf{XY}') \equiv (1/T)(\mathbf{Y} - \widetilde{C}\mathbf{X})(\mathbf{Y} - \widetilde{C}\mathbf{X})'$$

and $\widetilde{C} = \mathbf{YX}'(\mathbf{XX}')^{-1}$ is the (full-rank) LS estimate of C.

Robinson (1974) has shown that the solution in Theorem 2.2 when sample quantities are substituted, namely

$$\widehat{A}^{(r)} = \Gamma^{-1/2}[\widehat{V}_1, \ldots, \widehat{V}_r], \quad \widehat{B}^{(r)} = [\widehat{V}_1, \ldots, \widehat{V}_r]' \, \Gamma^{1/2} \, \widehat{\Sigma}_{yx} \, \widehat{\Sigma}_{xx}^{-1}, \tag{2.17}$$

where $\widehat{\Sigma}_{yx} = (1/T)\mathbf{YX}'$, $\widehat{\Sigma}_{xx} = (1/T)\mathbf{XX}'$, and $\widehat{V}_j$ is the eigenvector that corresponds to the jth largest eigenvalue $\widehat{\lambda}_j^2$ of $\Gamma^{1/2}\widehat{\Sigma}_{yx}\widehat{\Sigma}_{xx}^{-1}\widehat{\Sigma}_{xy}\Gamma^{1/2}$, with the choice $\Gamma = \widetilde{\Sigma}_{\epsilon\epsilon}^{-1}$, minimizes simultaneously all the eigenvalues of $\widetilde{\Sigma}_{\epsilon\epsilon}^{-1} W$ and hence minimizes $|\widetilde{\Sigma}_{\epsilon\epsilon}^{-1} W|$. To establish this, write

$$\begin{aligned} W &= \frac{1}{T}(\mathbf{Y} - \widetilde{C}\mathbf{X} + (\widetilde{C} - AB)\mathbf{X})(\mathbf{Y} - \widetilde{C}\mathbf{X} + (\widetilde{C} - AB)\mathbf{X})' \\ &= \frac{1}{T}(\mathbf{Y} - \widetilde{C}\mathbf{X})(\mathbf{Y} - \widetilde{C}\mathbf{X})' + \frac{1}{T}(\widetilde{C} - AB)\mathbf{XX}'(\widetilde{C} - AB)' \\ &\equiv \widetilde{\Sigma}_{\epsilon\epsilon} + (\widetilde{C} - AB)\, \widehat{\Sigma}_{xx} (\widetilde{C} - AB)'. \end{aligned}$$

Thus $|\widetilde{\Sigma}_{\epsilon\epsilon}^{-1} W| = |I_m + \widetilde{\Sigma}_{\epsilon\epsilon}^{-1}(\widetilde{C} - AB)\widehat{\Sigma}_{xx}(\widetilde{C} - AB)'| = \prod_{i=1}^{m}(1 + \delta_i^2)$, where δ_i^2 are the eigenvalues of the matrix $\widetilde{\Sigma}_{\epsilon\epsilon}^{-1}(\widetilde{C} - AB)\widehat{\Sigma}_{xx}(\widetilde{C} - AB)'$. So minimizing $|\widetilde{\Sigma}_{\epsilon\epsilon}^{-1} W|$ is equivalent to simultaneously minimizing all the eigenvalues of $\widetilde{\Sigma}_{\epsilon\epsilon}^{-1/2}(\widetilde{C} - AB)\widehat{\Sigma}_{xx}(\widetilde{C} - AB)'\widetilde{\Sigma}_{\epsilon\epsilon}^{-1/2} \equiv (S - P)(S - P)'$, with $S = \widetilde{\Sigma}_{\epsilon\epsilon}^{-1/2}\widetilde{C}\widehat{\Sigma}_{xx}^{1/2}$ and $P = \widetilde{\Sigma}_{\epsilon\epsilon}^{-1/2} AB\widehat{\Sigma}_{xx}^{1/2}$. From the result of Lemma 2.2, we see that the simultaneous minimization is achieved with P chosen as the rank r approximation of S obtained through the singular value decomposition of S, that is, $P = \widehat{V}_{(r)}\widehat{V}_{(r)}' S = \widehat{V}_{(r)}\widehat{V}_{(r)}' \widetilde{\Sigma}_{\epsilon\epsilon}^{-1/2}\widetilde{C}\widehat{\Sigma}_{xx}^{1/2} \equiv \widetilde{\Sigma}_{\epsilon\epsilon}^{-1/2}\widehat{A}^{(r)}\widehat{B}^{(r)}\widehat{\Sigma}_{xx}^{1/2}$. So $\widehat{C}^{(r)} \equiv \widehat{A}^{(r)}\widehat{B}^{(r)} = \widetilde{\Sigma}_{\epsilon\epsilon}^{1/2}\widehat{V}_{(r)}\widehat{V}_{(r)}'\widetilde{\Sigma}_{\epsilon\epsilon}^{-1/2}\widetilde{C}$, with $\widehat{A}^{(r)}$ and $\widehat{B}^{(r)}$ as given in (2.17), is the ML estimate.

Now the correspondence between the maximum likelihood estimators and the quantities $A^{(r)}$ and $B^{(r)}$ given in Theorem 2.2 is evident. It must be noted here that $\widetilde{\Sigma}_{\epsilon\epsilon}$ is the ML estimate of $\Sigma_{\epsilon\epsilon}$ obtained in the full-rank regression model, that is, $\widetilde{\Sigma}_{\epsilon\epsilon} = (1/T)(\mathbf{Y} - \widetilde{C}\mathbf{X})(\mathbf{Y} - \widetilde{C}\mathbf{X})'$ with $\widetilde{C} = \widehat{\Sigma}_{yx}\widehat{\Sigma}_{xx}^{-1}$ equal to the full-rank estimate of C as given in Eq. (1.7). The ML estimate of $\Sigma_{\epsilon\epsilon}$ under the reduced-rank structure is given by $\widehat{\Sigma}_{\epsilon\epsilon} = (1/T)(\mathbf{Y} - \widehat{C}^{(r)}\mathbf{X})(\mathbf{Y} - \widehat{C}^{(r)}\mathbf{X})'$, with $\widehat{C}^{(r)} = \widehat{A}^{(r)}\widehat{B}^{(r)}$ obtained from (2.17). From the developments above, we see that $\widehat{\Sigma}_{\epsilon\epsilon}$ can be represented as

$$\begin{aligned} \widehat{\Sigma}_{\epsilon\epsilon} &= \widetilde{\Sigma}_{\epsilon\epsilon} + (\widetilde{C} - \widehat{C}^{(r)})\, \widehat{\Sigma}_{xx}\, (\widetilde{C} - \widehat{C}^{(r)})' \\ &= \widetilde{\Sigma}_{\epsilon\epsilon} + (I_m - P)\, \widetilde{C}\, \widehat{\Sigma}_{xx}\, \widetilde{C}'\, (I_m - P)' \\ &= \widetilde{\Sigma}_{\epsilon\epsilon} + \widetilde{\Sigma}_{\epsilon\epsilon}^{1/2}\, (I_m - \widehat{V}_{(r)}\, \widehat{V}_{(r)}')\, \widehat{R}\, (I_m - \widehat{V}_{(r)}\, \widehat{V}_{(r)}')\, \widetilde{\Sigma}_{\epsilon\epsilon}^{1/2}, \end{aligned} \tag{2.18}$$

where $P = \widetilde{\Sigma}_{\epsilon\epsilon}^{1/2}\, \widehat{V}_{(r)}\, \widehat{V}_{(r)}'\widetilde{\Sigma}_{\epsilon\epsilon}^{-1/2}$, $\widehat{R} = \widetilde{\Sigma}_{\epsilon\epsilon}^{-1/2}\, \widehat{\Sigma}_{yx}\, \widehat{\Sigma}_{xx}^{-1}\, \widehat{\Sigma}_{xy}\, \widetilde{\Sigma}_{\epsilon\epsilon}^{-1/2}$, and $\widehat{V}_{(r)} = [\widehat{V}_1, \ldots, \widehat{V}_r]$. Note, also, that the sample version of the criterion in (2.12) and of the term on the right-hand side of the corresponding identity in (2.12′) (with $\Gamma = \widetilde{\Sigma}_{\epsilon\epsilon}^{-1}$) is

$$\operatorname{tr}\left\{\widetilde{\Sigma}_{\epsilon\epsilon}^{-1}\frac{1}{T}(\mathbf{Y}-AB\mathbf{X})(\mathbf{Y}-AB\mathbf{X})'\right\} = m + \operatorname{tr}\left\{\widetilde{\Sigma}_{\epsilon\epsilon}^{-1}(\widetilde{C}-AB)\widehat{\Sigma}_{xx}(\widetilde{C}-AB)'\right\}.$$

So the ML estimates $\widehat{A}^{(r)}$ and $\widehat{B}^{(r)}$ can be viewed as providing the "optimum" reduced-rank approximation of the full-rank least squares estimator $\widetilde{C} = \widehat{\Sigma}_{yx}\widehat{\Sigma}_{xx}^{-1}$ corresponding to this criterion. The choice $\Gamma = \widetilde{\Sigma}_{\epsilon\epsilon}^{-1}$ in the criterion, which leads to the ML estimates $\widehat{A}^{(r)}$ and $\widehat{B}^{(r)}$, provides an asymptotically efficient estimator under the assumptions that the errors ϵ_k are independent and normally distributed as $N(0, \Sigma_{\epsilon\epsilon})$.

Remark—Equivalence of Two Estimators Let the solution (estimates) for the component matrices A and B given by Theorem 2.2 when sample quantities are substituted, with the choice $\Gamma = \widehat{\Sigma}_{yy}^{-1}$, where $\widehat{\Sigma}_{yy} = (1/T)\mathbf{YY}'$, be denoted as $\widehat{A}_*^{(r)}$ and $\widehat{B}_*^{(r)}$, respectively. Then we mention that while this results in different component estimates from the estimates $\widehat{A}^{(r)}$ and $\widehat{B}^{(r)}$ provided in (2.17) for the choice $\Gamma = \widetilde{\Sigma}_{\epsilon\epsilon}^{-1}$, it yields the *same* ML estimate for the overall coefficient matrix $C = AB$, that is, $\widehat{C}^{(r)} = \widehat{A}^{(r)}\widehat{B}^{(r)} \equiv \widehat{A}_*^{(r)}\widehat{B}_*^{(r)}$. This connection in solutions between the choices $\Gamma = \widehat{\Sigma}_{yy}^{-1}$ and $\Gamma = \widetilde{\Sigma}_{\epsilon\epsilon}^{-1}$ will be discussed in more detail in Section 2.4.

It must be further observed that the $(m-r)$ linear restrictions on $C^{(r)} = A^{(r)}B^{(r)}$ are

$$\ell_j' = V_j'\Gamma^{1/2}, \quad j = r+1, \ldots, m, \tag{2.19}$$

because

$$V_j'\Gamma^{1/2}C^{(r)} = V_j'\Gamma^{1/2}\Gamma^{-1/2}[V_1, \ldots, V_r]B^{(r)} = 0', \quad j = r+1, \ldots, m.$$

Thus, the last $(m-r)$ eigenvectors of the matrix given in Theorem 2.2 also provide the solution to the complementary part of the problem. Recall that the linear combination $\ell_j'Y_k = \ell_j'\epsilon_k$ in (2.3) does not depend on the predictor variables X_k. The necessary calculations for both aspects of the (population) problem can be achieved through the computation of all eigenvalues and eigenvectors of the matrix $\Gamma^{1/2}\Sigma_{yx}\Sigma_{xx}^{-1}\Sigma_{xy}\Gamma^{1/2}$, given in Theorem 2.2.

A comment worth noting here refers to the solution to the reduced-rank problem. In the construction of Theorem 2.1, observe that the optimal matrix N was derived in terms of M and then the optimal matrix M was determined. This is equivalent to deriving the matrix B in terms of A and then determining the optimal matrix A. Conversely, one could fix B and solve for A in terms of B first. This is the regression step in the calculation procedure. Because

$$Y_k = A(BX_k) + \epsilon_k = AX_k^* + \epsilon_k,$$

given $X_k^* = BX_k$, $A = \Sigma_{yx} B'(B\Sigma_{xx}B')^{-1}$, the usual regression coefficient matrix obtained by regressing Y_k on X_k^*. If we substitute A in terms of B in the criterion (2.12) given in Theorem 2.2, it reduces to minimizing $\text{tr}\{\Sigma_{yy}\,\Gamma - (B\Sigma_{xx}\,B')^{-1}\,B\,\Sigma_{xy}\,\Gamma\,\Sigma_{yx}B'\}$. From this, it follows that the columns of $\Sigma_{xx}^{1/2}\,B'$ may be chosen as the eigenvectors corresponding to the r largest eigenvalues of $\Sigma_{xx}^{-1/2}\,\Sigma_{xy}\,\Gamma\,\Sigma_{yx}\,\Sigma_{xx}^{-1/2}$. To see all this in a convenient way, notice that, in terms of the notation of Theorem 2.1, the criterion in (2.12′) to be minimized, $\text{tr}[(S - P)(S - P)']$, can also be expressed as $\text{tr}[(S - P)'(S - P)] \equiv \text{tr}[(S' - P')(S' - P')']$ with $P' = N'M'$. Thus, we may treat this as simply interchanging S with S' in the reduced-rank approximation problem. Hence, following the proof of Theorem 2.1, the minimizing value of M', for given N', is obtained as $M' = (NN')^{-1}NS'$. With the correspondences $S = \Gamma^{1/2}\Sigma_{yx}\Sigma_{xx}^{-1/2}$, $M = \Gamma^{1/2}A$, and $N = B\Sigma_{xx}^{1/2}$ used in the proof of Theorem 2.2, this leads to the optimal value $M' \equiv A'\Gamma^{1/2} = (B\Sigma_{xx}B')^{-1}B\Sigma_{xy}\Gamma^{1/2}$, and hence $A = \Sigma_{yx}\,B'\,(B\Sigma_{xx}\,B')^{-1}$ as indicated above. Then as in Theorem 2.1, the optimal value of $N' \equiv \Sigma_{xx}^{1/2}B'$ is determined as the matrix whose columns are the eigenvalues of $S'S = \Sigma_{xx}^{-1/2}\,\Sigma_{xy}\,\Gamma\Sigma_{yx}\,\Sigma_{xx}^{-1/2}$ corresponding to its r largest eigenvalues. Hence we see that this approach calls for eigenvectors of $S'S$ instead of SS' as in the result of Theorem 2.1. The approach presented here, of fixing B first, is appealing as various hypothesized values can be set for B and also it is easier to calculate quantities needed for making inferences, such as standard errors of the estimator of A, and so on, as in the full-rank setup.

The result presented in Theorem 2.2 and the above remark describe how an explicit solution for the component matrices A and B in (2.12) can be obtained through computation of eigenvalues and eigenvectors of either SS' or $S'S$. For the extended reduced-rank models to be considered in later chapters such an explicit solution is not possible, and iterative procedures need to be used. Although the reduced-rank model (2.11) is nonlinear in the parameter matrices A and B, the structure can be regarded as bilinear, and thus certain simplifications in the computations are possible. For the population problem, consider the first order equations resulting from minimization of the criterion (2.12), that is, consider the first partial derivatives of the criterion (2.12) with respect to A and B, set equal to zero. These first order equations are readily found to be

$$\Sigma_{yx}B' - AB\Sigma_{xx}B' = 0 \quad \text{and} \quad A'\Gamma\Sigma_{yx} - (A'\Gamma A)B\Sigma_{xx} = 0.$$

As observed previously, from the first equation above, we see that for given B, the solution for A can be calculated as $A = \Sigma_{yx}\,B'\,(B\,\Sigma_{xx}\,B')^{-1}$. In addition, similarly, from the second equations above, we find that for given A, the solution for B can be obtained as $B = (A'\,\Gamma\,A)^{-1}\,A'\,\Gamma\Sigma_{yx}\,\Sigma_{xx}^{-1}$, or $B = A'\,\Gamma\,\Sigma_{yx}\Sigma_{xx}^{-1}$ assuming that the normalization conditions in (2.14) for A are imposed. The iterative procedure, similar to calculations involved in a method known as *partial least squares*, calls for iterating between the two solutions (i.e., the solution for A in terms of B and the solution for B in terms of A) and at each step of the iteration imposing the normalization conditions (2.14). A similar iterative procedure was suggested by Lyttkens (1972) for the computation of canonical variates and canonical correlations.

It might also be noted at this point that alternate normalizations could be considered. One alternate normalization, with B partitioned as $B = [B_1, B_2]$, where B_1 is $r \times r$, is obtained by assuming that, since B is of rank r, its leading $r \times r$ submatrix B_1 is nonsingular (the components of X_k can always be arranged so that this holds). Then we can write $C = AB = (AB_1)[I_r, B_1^{-1}B_2] \equiv A_0[I_r, B_0]$ and the component parameter matrices A_0 and B_0 are uniquely identified. (A similar alternate normalization could be imposed by specifying the upper $r \times r$ submatrix of A equal to I_r). The appeal of using partial least squares increases under either of these alternate normalizations, since either normalization can be directly imposed on the corresponding first order equations above. These notions also carry over to the sample situation of constructing estimates $\widehat{A}^{(r)}$ and $\widehat{B}^{(r)}$, iteratively, by using sample versions of the above partial least squares estimating equations. An additional appeal in the sample setting is that it is very convenient to accommodate extra zero constraints on the parameter elements of A and B, if desired, in the partial least squares estimation whereas this is not possible in the eigenvalue–eigenvector solutions as in (2.17).

Remark—Case of Special Covariance Structure The preceding results on ML estimation apply to the typical case where the error covariance matrix $\Sigma_{\epsilon\epsilon}$ is unknown and is a completely general positive-definite matrix. We may also briefly consider the special case where the error covariance matrix in model (2.11) is assumed to have the specified form $\Sigma_{\epsilon\epsilon} = \sigma^2 \boldsymbol{\Psi}_0$, where σ^2 is an unknown scalar parameter, but $\boldsymbol{\Psi}_0$ is a known (specified) $m \times m$ matrix. Then from (2.16), the ML estimates of A and B are obtained by minimizing

$$\begin{aligned}&\frac{1}{\sigma^2}\,\mathrm{tr}[\boldsymbol{\Psi}_0^{-1}(\mathbf{Y} - AB\mathbf{X})(\mathbf{Y} - AB\mathbf{X})']\\ &\quad = \frac{1}{\sigma^2}\,\mathrm{tr}[\boldsymbol{\Psi}_0^{-1}(\mathbf{Y} - \widetilde{C}\mathbf{X})(\mathbf{Y} - \widetilde{C}\mathbf{X})'] + \frac{T}{\sigma^2}\,\mathrm{tr}\left[\boldsymbol{\Psi}_0^{-1}(\widetilde{C} - AB)\widehat{\Sigma}_{xx}(\widetilde{C} - AB)'\right].\end{aligned}$$

So from Theorem 2.1, it follows immediately that the ML solution is $\widehat{A} = \boldsymbol{\Psi}_0^{1/2}\,\widehat{V}$, $\widehat{B} = \widehat{V}'\,\boldsymbol{\Psi}_0^{-1/2}\,\widetilde{C}$, and so $\widehat{C} = \widehat{A}\,\widehat{B} = \boldsymbol{\Psi}_0^{1/2}\,\widehat{V}\,\widehat{V}'\,\boldsymbol{\Psi}_0^{-1/2}\,\widetilde{C}$, where $\widehat{V} = [\widehat{V}_1, \ldots, \widehat{V}_r]$ and the $\widehat{V}_j$ are the normalized eigenvectors of the matrix

$$\boldsymbol{\Psi}_0^{-1/2}\widetilde{C}\,\widehat{\Sigma}_{xx}\,\widetilde{C}'\boldsymbol{\Psi}_0^{-1/2} \equiv \boldsymbol{\Psi}_0^{-1/2}\widehat{\Sigma}_{yx}\,\widehat{\Sigma}_{xx}^{-1}\,\widehat{\Sigma}_{xy}\,\boldsymbol{\Psi}_0^{-1/2}$$

corresponding to the r largest eigenvalues $\widehat{\lambda}_1^2 > \cdots > \widehat{\lambda}_r^2$. The corresponding ML estimate of σ^2 is given by

$$\widehat{\sigma}^2 = \frac{1}{Tm} \operatorname{tr}[(\mathbf{Y} - \widehat{A}\widehat{B}\mathbf{X})' \boldsymbol{\Psi}_0^{-1} (\mathbf{Y} - \widehat{A}\widehat{B}\mathbf{X})]$$

$$\equiv \frac{1}{m} \left\{ \operatorname{tr}[\boldsymbol{\Psi}_0^{-1} \widetilde{\Sigma}_{\epsilon\epsilon}] + \sum_{i=r+1}^{m} \widehat{\lambda}_i^2 \right\}. \tag{2.20}$$

In particular, assuming $\Sigma_{\epsilon\epsilon} = \sigma^2 I_m$, so $\boldsymbol{\Psi}_0 = I_m$, corresponds to minimizing the sum of squares criterion $\operatorname{tr}[(\mathbf{Y} - AB\mathbf{X})(\mathbf{Y} - AB\mathbf{X})']$. This gives the reduced-rank ML or least squares estimators as $\widehat{A} = \widehat{V}$, $\widehat{B} = \widehat{V}'\widetilde{C}$, and $\widehat{C} = \widehat{V}\widehat{V}'\widetilde{C}$, with $\widehat{\sigma}^2 = \frac{1}{Tm} \operatorname{tr}[(\mathbf{Y} - \widehat{A}\widehat{B}\mathbf{X})'(\mathbf{Y} - \widehat{A}\widehat{B}\mathbf{X})] \equiv \frac{1}{m}\{\operatorname{tr}[\widetilde{\Sigma}_{\epsilon\epsilon}] + \sum_{i=r+1}^{m} \widehat{\lambda}_i^2\}$, where the $\widehat{V}_j$ are normalized eigenvectors of the matrix $\widetilde{C}\widehat{\Sigma}_{xx}\widetilde{C}' \equiv \widehat{\Sigma}_{yx}\widehat{\Sigma}_{xx}^{-1}\widehat{\Sigma}_{xy}$. This reduced-rank LS estimator was used by Davies and Tso (1982) and Gudmundsson (1977), in particular. The corresponding population reduced-rank LS result for random vectors Y and X amounts to setting $\Gamma = I_m$ in the criterion (2.12) of Theorem 2.2, with the problem reduced to finding A and B which minimize $\operatorname{tr}\{E[(Y - ABX)(Y - ABX)']\} \equiv E[(Y - ABX)'(Y - ABX)]$. The optimal (LS) solution is $A = V_{(r)}$, $B = V_{(r)}' \Sigma_{yx}\Sigma_{xx}^{-1}$, where the V_j are normalized eigenvectors of $\Sigma_{yx}\Sigma_{xx}^{-1}\Sigma_{xy}$. In Section 2.1, we referenced the related works by Rao (1964), Fortier (1966), and van den Wollenberg (1977) that used different terminologies, such as redundancy analysis, regarding the reduced-rank problem, but the procedures discussed in these papers can be recognized basically as being equivalent to the use of this reduced-rank LS method. It was also noted by Fortier (1966) that the linear predictive factors, that is, the transformed predictor variables $BX = V'\Sigma_{yx}\Sigma_{xx}^{-1}X$, resulting from the LS approach can be viewed simply as the principal components (to be discussed in the next section) of the unrestricted LS prediction vector $\widehat{Y} = \Sigma_{yx}\Sigma_{xx}^{-1}X$, noting that $\operatorname{Cov}(\widehat{Y}) = \Sigma_{yx}\Sigma_{xx}^{-1}\Sigma_{xy}$ and the V_j in $V'\Sigma_{yx}\Sigma_{xx}^{-1}X \equiv V'\widehat{Y}$ are the eigenvectors of this covariance matrix.

It should be mentioned, perhaps, that estimation results under the LS criterion are sensitive to the choice of scaling of the response variables Y_k, and in applications of the LS method it is often suggested that the component variables y_{ik} be standardized to have unit variances before applying the LS procedure. Of course, if Y_k is "standardized" in the more general way as $\Sigma_{yy}^{-1/2}Y_k$ so that the standardized vector has the identity covariance matrix, then the LS procedure applied to $\Sigma_{yy}^{-1/2}Y_k$ is equivalent to the canonical correlation analysis results for the original variables Y_k. Even when the form $\Sigma_{\epsilon\epsilon} = \sigma^2 I_m$ does not hold, the reduced-rank LS estimator may be useful relative to the reduced-rank ML estimator in situations where the number of vector observations T is not large relative to the dimensions m and n of the response and predictor vectors Y_k and X_k, since then the estimator $\widetilde{\Sigma}_{\epsilon\epsilon}$ may not be very accurate or may not even be nonsingular (e.g., as in the case of Example 2.3 where $T = 10 < m = 19$).

2.4 Relation to Principal Components and Canonical Correlation Analysis

2.4.1 Principal Components Analysis

Principal component analysis, originally developed by Hotelling (1933), is usually concerned with explaining the covariance structure of a large number of interdependent variables through a small number of linear combinations of the original variables. The two objectives of the analysis are data reduction and interpretation which coincide with the objectives of the reduced-rank regression. A major difference between the two approaches is that the principal components are descriptive tools that do not usually rest on any modeling or distributional assumptions.

Let the $m \times 1$ random vector Y have the covariance matrix Σ_{yy} with eigenvalues $\lambda_1^2 \geq \lambda_2^2 \geq \cdots \geq \lambda_m^2 \geq 0$. Consider the linear combinations $y_1^* = \ell_1' Y, \ldots, y_m^* = \ell_m' Y$, normalized so that $\ell_i'\ell_i = 1$, $i = 1, \ldots, m$. These linear combinations have

$$\begin{aligned} \text{Var}(y_i^*) &= \ell_i' \, \Sigma_{yy} \, \ell_i, \quad i = 1, 2, \ldots, m, \\ \text{Cov}(y_i^*, y_j^*) &= \ell_i' \, \Sigma_{yy} \, \ell_j, \quad i, j = 1, 2, \ldots, m. \end{aligned} \tag{2.21}$$

In a principal components analysis, the linear combinations y_i^* are chosen so that they are not correlated but their variances are as large as possible. That is, $y_1^* = \ell_1' Y$ is chosen to have the largest variance among linear combinations with $\ell_1'\ell_1 = 1$, $y_2^* = \ell_2' Y$ is chosen to have the largest variance subject to $\ell_2'\ell_2 = 1$ and the condition that $\text{Cov}(y_2^*, y_1^*) = \ell_2'\Sigma_{yy}\ell_1 = 0$, and so on. It can be easily shown (e.g., from Lemma 2.1) that the ℓ_i's are chosen to be the (normalized) eigenvectors that correspond to the eigenvalues λ_i^2 of the matrix Σ_{yy}, so that the normalized eigenvectors satisfy $\ell_i'\Sigma_{yy}\ell_i = \lambda_i^2 \equiv \text{Var}(y_i^*)$. Often, the first $r (r < m)$ components y_i^*, corresponding to the r largest eigenvalues, will contain nearly all the variation that arises from all m variables, in that $\sum_{i=1}^r \text{Var}(y_i^*) \equiv \sum_{i=1}^r \ell_i'\Sigma_{yy}\ell_i$ represents a very large proportion of the "total variance" $\sum_{i=1}^m \text{Var}(y_i^*) \equiv \text{tr}(\Sigma_{yy})$. For such cases, in studying variations of Y, attention can be directed on these r linear combinations resulting in a reduction in dimension of the variables involved.

The principal component problem can be represented as a reduced-rank problem, in the sense that a reduced-rank model can be written for this situation as

$$Y_k = ABY_k + \epsilon_k \tag{2.22}$$

(with $X_k \equiv Y_k$). From the solutions of Theorem 2.2, by setting $\Gamma = I_m$, we have $A^{(r)} = [V_1, \ldots, V_r]$ and $B^{(r)} = V'$ because $\Sigma_{yx} = \Sigma_{yy}$. The V_js refer to the eigenvectors of Σ_{yy}. Thus, a correspondence is easily established. The ℓ_i in (2.21) are the same as the V_i that result from the solution for the matrices A and B in (2.22). If both approaches provide the same solutions, one might ask, then, what is the advantage of formulating the principal components analysis in terms of reduced-rank regression model. The inclusion of the error term ϵ_k appears to be superfluous, but the model (2.22) could provide a useful framework to perform the residual analysis to detect outliers, influential observations, and so on, for the principal components method.

2.4.2 Application to Functional and Structural Relationships Models

Consider further the linear functional and structural relationships model discussed briefly in Section 2.1,

$$Y_k - \mu = W_k + \epsilon_k, \qquad k = 1, \ldots, T, \tag{2.23}$$

where the $\{W_k\}$ are assumed to satisfy the linear relations $LW_k = 0$ for all k, with L an unknown $(m - r) \times m$ matrix of full rank and the normalization $LL' = I_{m-r}$ is assumed. In the functional case where the W_k are treated as fixed, with $E(Y_k - \mu) = W_k$, it is assumed that $\sum_{k=1}^{T} W_k = 0$, whereas in the structural case the W_k are assumed to be random vectors with $E(W_k) = 0$ and $\text{Cov}(W_k) = \Theta = \Lambda\Lambda'$, where Λ is an $m \times r$ full-rank matrix $(r < m)$. In both cases, it is also assumed that the error covariance matrix has the structure $\Sigma_{\epsilon\epsilon} = \sigma^2 I_m$, where σ^2 is unknown. Thus, in the structural case, the Y_k have covariance matrix of the form $\text{Cov}(Y_k) = \Theta + \Sigma_{\epsilon\epsilon} = \Lambda\Lambda' + \sigma^2 I_m$.

Let $\bar{Y} = (1/T)\sum_{k=1}^{T} Y_k$ and $\widetilde{\Sigma}_{yy} = (1/T)\sum_{k=1}^{T}(Y_k - \bar{Y})(Y_k - \bar{Y})'$ denote the sample mean vector and sample covariance matrix of the Y_k. Also let $\widehat{\lambda}_1^2 > \widehat{\lambda}_2^2 > \cdots > \widehat{\lambda}_m^2 > 0$ denote the eigenvalues of $\widetilde{\Sigma}_{yy}$ with corresponding normalized eigenvectors $\widehat{\ell}_i$, $i = 1, \ldots, m$. Then $\widehat{y}_{ik}^* = \widehat{\ell}_i' Y_k$ are the sample principal components of the Y_k. In the functional case, it has been shown originally by Tintner (1945), also see Anderson (1984a), that an ML estimate of the unknown constraints matrix L is given by the matrix $\widehat{L} = \widehat{L}_2' \equiv [\widehat{\ell}_{r+1}, \ldots, \widehat{\ell}_m]'$ whose rows correspond to the last (smallest) $m - r$ principal components in the sample. We also set $\widehat{L}_1 = [\widehat{\ell}_1, \ldots, \widehat{\ell}_r]$, corresponding to the first (largest) r sample principal components. Then, the ML estimates of the unknown but fixed values W_k are given by

$$\widehat{W}_k = \widehat{L}_1\widehat{L}_1'(Y_k - \bar{Y}), \qquad k = 1, \ldots, T, \tag{2.24}$$

with $\widehat{\mu} = \bar{Y}$, and $\widehat{\sigma}^2 = \frac{1}{m}\sum_{i=r+1}^{m} \widehat{\lambda}_i^2$. Thus, in effect, under ML estimation the variability described by the first r sample principal components (as measured by $\widehat{\lambda}_1^2, \ldots, \widehat{\lambda}_r^2$) is assigned to the variation in the mean values W_k and the remaining variability, $\sum_{i=r+1}^{m} \widehat{\lambda}_i^2$, is assigned to the variance σ^2 of the errors ϵ_k. The above estimator $\widehat{\sigma}^2$ of σ^2 is biased downward, however.

We verify these results by using the framework of the ML estimation results established in Section 2.3 for the reduced-rank regression model. As discussed in Section 2.1, the functional model $Y_k - \mu = W_k + \epsilon_k$, $LW_k = 0$, can be written equivalently as

$$Y_k - \mu = AF_k + \epsilon_k, \qquad k = 1, \ldots, T; \tag{2.25}$$

that is, $W_k = AF_k$, where A is $m \times r$ such that $LA = 0$ and the F_k are $r \times 1$ vectors of unknown constants. Hence, in matrix form we have

$$\mathbf{Y} = \mu\mathbf{1}' + \mathbf{W} + \boldsymbol{\epsilon} = \mu\mathbf{1}' + AB\mathbf{X} + \boldsymbol{\epsilon} \equiv \mu\mathbf{1}' + C\mathbf{X} + \boldsymbol{\epsilon}, \tag{2.26}$$

where $\mathbf{X} = I_T$, $\mathbf{W} = [W_1, \ldots, W_T] \equiv C = AB$ with $B = [F_1, \ldots, F_T]$, and $\mathbf{1} = (1, \ldots, 1)'$ denotes a $T \times 1$ vector of ones. For estimation, the condition $\mathbf{W1} = \sum_{k=1}^{T} W_k = 0$, equivalently, $\sum_{k=1}^{T} F_k = 0$, is imposed. The unrestricted LS estimators of the model are $\widetilde{\mu} = \bar{Y}$, $\widetilde{W}_k = Y_k - \bar{Y}$, $k = 1, \ldots, T$, or $\widetilde{C} = \mathbf{Y} - \bar{Y}\mathbf{1}'$ (with $\widetilde{\Sigma}_{\epsilon\epsilon} \equiv 0$ since $Y_k - \widetilde{\mu} - \widetilde{W}_k \equiv 0$). Then notice that $\widetilde{C}\widehat{\Sigma}_{xx}\widetilde{C}' = \frac{1}{T}(\mathbf{Y} - \bar{Y}\mathbf{1}')(\mathbf{Y} - \bar{Y}\mathbf{1}')' = \widetilde{\Sigma}_{yy}$, with eigenvalues $\widehat{\lambda}_1^2 > \widehat{\lambda}_2^2 > \cdots > \widehat{\lambda}_m^2 > 0$ and corresponding normalized eigenvectors $\widehat{\ell}_i$, $i = 1, \ldots, m$. Then with $\widehat{L}_1 = [\widehat{\ell}_1, \ldots, \widehat{\ell}_r]$, according to the ML estimation results for the special case $\Sigma_{\epsilon\epsilon} = \sigma^2 I_m$ presented at the end of Section 2.3, we have that the ML estimates are $\widehat{A} = \widehat{L}_1$, $\widehat{B} = \widehat{L}_1'\widetilde{C} \equiv \widehat{L}_1'(\mathbf{Y} - \bar{Y}\mathbf{1}')$, so that $\widehat{\mathbf{W}} = \widehat{C} = \widehat{A}\,\widehat{B} = \widehat{L}_1\widehat{L}_1'(\mathbf{Y} - \bar{Y}\mathbf{1}')$. That is,

$$\widehat{W}_k = \widehat{L}_1\widehat{L}_1'(Y_k - \bar{Y}), \qquad k = 1, \ldots, T,$$

as in (2.24), with $\widehat{\mu} = \widetilde{\mu} = \bar{Y}$. In addition, the ML estimate of σ^2, from (2.20), is given by $\widehat{\sigma}^2 = \frac{1}{m}\sum_{i=r+1}^{m} \widehat{\lambda}_i^2$ as stated above. (Actually, the above obtained results follow from a slight extension of results for the basic model considered in this chapter, which explicitly allows for a constant term. The general case of this extension will be considered in Section 3.1.)

In the structural case of the model (2.23), $\widehat{L} = \widehat{L}_2'$ is also an ML estimate of L, whereas the ML estimate of σ^2 is $\widehat{\sigma}^2 = \frac{1}{m-r}\sum_{i=r+1}^{m} \widehat{\lambda}_i^2$. The ML estimate of $\mathrm{Cov}(W_k) = \Theta = \Lambda\Lambda'$ is given by

$$\widehat{\Theta} = \widehat{L}_1\widehat{D}_1\widehat{L}_1' - \widehat{\sigma}^2\widehat{L}_1\widehat{L}_1' = \widehat{L}_1(\widehat{D}_1 - \widehat{\sigma}^2 I_m)\widehat{L}_1',$$

where $\widehat{D}_1 = \mathrm{diag}(\widehat{\lambda}_1^2, \ldots, \widehat{\lambda}_r^2)$. So the ML estimate of $\mathrm{Cov}(Y_k) = \Sigma_{yy}$ is

$$\widehat{\Sigma}_{yy} = \widehat{\Theta} + \widehat{\sigma}^2 I_m \equiv \widehat{L}_1\widehat{D}_1\widehat{L}_1' + \widehat{\sigma}^2\,\widehat{L}_2\widehat{L}_2',$$

noting that $\widehat{L}_1\widehat{L}_1' + \widehat{L}_2\widehat{L}_2' = I_m$. These estimators were obtained by Lawley (1953) and by Theobald (1975), also see Anderson (1984a).

Thus, in the above linear relationship models, we see that the sample principal components arise fundamentally in the ML solution. The first r principal components are the basis for estimation of the systematic part W_k of Y_k, e.g., $\widehat{W}_k = \widehat{L}_1\widehat{L}_1'(Y_k - \bar{Y})$ in the functional case, and $(Y_k - \bar{Y}) - \widehat{L}_1\widehat{L}_1'(Y_k - \bar{Y}) = \widehat{L}_2\widehat{L}_2'(Y_k - \bar{Y})$ can be considered as the estimate of the error. However, a distinction in this application of principal components analysis is that the error in approximation of Y_k through its first r principal components $\widehat{L}_1' Y_k$ is not necessarily small but should have the attributes of the error term ϵ_k, namely mean 0 and covariance matrix with equal variances σ^2. So in this analysis the remaining last $(m - r)$ values $\widehat{\lambda}_{r+1}^2, \ldots, \widehat{\lambda}_m^2$ are not necessarily "negligible" relative to $\widehat{\lambda}_1^2, \ldots, \widehat{\lambda}_r^2$, as would be hoped in some typical uses of principal components analysis. In the linear functional or structural analysis, the representation of Y_k by such a model form might be considered to be good provided only that $\widehat{\lambda}_{r+1}^2, \ldots, \widehat{\lambda}_m^2$ are roughly equal (but not necessarily small).

2.4.3 Canonical Correlation Analysis

Canonical correlation analysis was introduced by Hotelling (1935, 1936) as a method of summarizing relationships between two sets of variables. The objective is to find that linear combination of one set of variables which is most highly correlated with any linear combination of a second set of variables. The technique has received attention from theoretical statisticians as early as Bartlett (1938), but ter Braak (1990) identifies that because of difficulty in interpretation of the canonical variates, its use is not widespread in some disciplines. By exploring the connection between the reduced-rank regression and canonical correlations, we provide an additional method of interpreting the canonical variates. Because the role of any descriptive tool such as canonical correlations is in better model building, the connection we display below provides an avenue for interpreting and further using the canonical correlations.

More formally, in canonical correlation analysis, for random vectors X_k and Y_k of dimensions n and m, respectively, the purpose is to obtain an $r \times n$ matrix G and an $r \times m$ matrix H, so that the r-vector variates $\xi_k = GX_k$ and $\omega_k = HY_k$ will capture as much of the covariances between X_k and Y_k as possible. The following results from Izenman (1975) summarize how this is done.

Theorem 2.3 *We assume the same conditions as given in Theorem 2.2 and further we assume Σ_{yy} is nonsingular. Then the $r \times n$ matrix G and the $r \times m$ matrix H, with $H\Sigma_{yy}H' = I_r$, that minimize simultaneously all the eigenvalues of*

$$E[(HY - GX)(HY - GX)'] \tag{2.27}$$

are given by

$$G = V_*' \Sigma_{yy}^{-1/2} \Sigma_{yx} \Sigma_{xx}^{-1}, \qquad H = V_*' \Sigma_{yy}^{-1/2},$$

where $V_* = [V_1^*, \ldots, V_r^*]$ *and* V_j^* *is the (normalized) eigenvector corresponding to the* j*th largest eigenvalue* ρ_j^2 *of the matrix*

$$R_* = \Sigma_{yy}^{-1/2} \Sigma_{yx} \Sigma_{xx}^{-1} \Sigma_{xy} \Sigma_{yy}^{-1/2}. \tag{2.28}$$

Denoting by g_j' and h_j' the jth rows of the matrices G and H, respectively, the jth pair of canonical variates is defined to be $\xi_j = g_j' X_k$ and $\omega_j = h_j' Y_k$, $j = 1, 2, \ldots, m$ $(m \leq n)$. The correlation between ξ_j and ω_j is the jth canonical correlation coefficient, and hence, since $\text{Cov}(\xi_j, \omega_j) = V_j^{*\prime} R_* V_j^* = \rho_j^2$, $\text{Var}(\xi_j) = V_j^{*\prime} R_* V_j^* = \rho_j^2$, and $\text{Var}(\omega_j) = V_j^{*\prime} V_j^* = 1$, we have $\text{Corr}(\xi_j, \omega_j) = \rho_j$ and the ρ_js are the canonical correlations between X_k and Y_k. The canonical variates possess the property that ξ_1 and ω_1 have the largest correlation among all possible linear combinations of X_k and Y_k, ξ_2 and ω_2 have the largest possible correlation among all linear combinations of X_k and Y_k that are uncorrelated with ξ_1 and ω_1, and so on.

We shall compare the results of Theorem 2.3 with those of Theorem 2.2. When $\Gamma = \Sigma_{yy}^{-1}$, the matrix $B_*^{(r)} = V_*' \Gamma^{1/2} \Sigma_{yx} \Sigma_{xx}^{-1}$ from (2.13) is identical to G and $H = V_*' \Sigma_{yy}^{-1/2}$ is equal to $A_*^{(r)-}$, a reflexive generalized inverse of $A_*^{(r)} = \Sigma_{yy}^{1/2} V_*$ Rao (1973, p. 26). The transformed variates $\xi_k^{(r)} = B_*^{(r)} X_k$ and $\omega_k^{(r)} = A_*^{(r)-} Y_k$ have $V_*' R_* V_* = \Lambda_*^2 \equiv \text{diag}(\rho_1^2, \ldots, \rho_r^2)$ as their cross-covariance matrix. The matrix R_* is a multivariate version of the simple squared correlation coefficient between two variables $(m = n = 1)$ and also of the squared multiple correlation coefficient between a single dependent variable $(m = 1)$ and a number of predictor variables.

The model, with the choice of $\Gamma = \Sigma_{yy}^{-1}$, is referred to as the reduced-rank model corresponding to the canonical variates situation by Izenman (1975). Although a number of choices for Γ are possible, we want to comment on the choice of $\Gamma = \Sigma_{\epsilon\epsilon}^{-1}$, which leads to maximum likelihood estimates (see Robinson (1974) and Section 2.3). Because $\Sigma_{\epsilon\epsilon} = \Sigma_{yy} - \Sigma_{yx} \Sigma_{xx}^{-1} \Sigma_{xy}$, the correspondence with results for the choice $\Gamma = \Sigma_{yy}^{-1}$ is fairly simple. In the first choice of $\Gamma = \Sigma_{yy}^{-1}$, the squared canonical correlations are indeed the eigenvalues of $\Sigma_{yx} \Sigma_{xx}^{-1} \Sigma_{xy}$ with respect to Σ_{yy}. For the choice of $\Gamma = \Sigma_{\epsilon\epsilon}^{-1}$, we compute eigenvalues, say λ_j^2, of the same matrix $\Sigma_{yx} \Sigma_{xx}^{-1} \Sigma_{xy}$ but with respect to $\Sigma_{yy} - \Sigma_{yx} \Sigma_{xx}^{-1} \Sigma_{xy}$, that is, roots of the determinantal equation $|\lambda_j^2 (\Sigma_{yy} - \Sigma_{yx} \Sigma_{xx}^{-1} \Sigma_{xy}) - \Sigma_{yx} \Sigma_{xx}^{-1} \Sigma_{xy}| = 0$. Since the ρ_j^2 for the first choice $\Gamma = \Sigma_{yy}^{-1}$ satisfy $|\rho_j^2 \Sigma_{yy} - \Sigma_{yx} \Sigma_{xx}^{-1} \Sigma_{xy}| = 0$, algebraically it follows that

$$\rho_j^2 = \lambda_j^2 / (1 + \lambda_j^2). \tag{2.29}$$

The canonical variates can be recovered from the calculation of eigenvalues and eigenvectors for the choice $\Gamma = \Sigma_{\epsilon\epsilon}^{-1}$ through

$$\xi^{(r)} = B^{(r)} X_k = V' \Sigma_{\epsilon\epsilon}^{-1/2} \Sigma_{yx} \Sigma_{xx}^{-1} X_k \quad \text{and} \quad \omega^{(r)} = A^{(r)-} Y_k = V' \Sigma_{\epsilon\epsilon}^{-1/2} Y_k,$$

where $V = [V_1, \ldots, V_r]$, λ_j^2 and V_j are the eigenvalues and (normalized) eigenvectors, respectively, of the matrix $\Sigma_{\epsilon\epsilon}^{-1/2} \Sigma_{yx} \Sigma_{xx}^{-1} \Sigma_{xy} \Sigma_{\epsilon\epsilon}^{-1/2}$, and $\lambda_j = \rho_j/(1-\rho_j^2)^{1/2}$, where ρ_j is the jth largest canonical correlation. It must be noted that the corresponding solutions from (2.13) for the choice $\Gamma = \Sigma_{\epsilon\epsilon}^{-1}$ satisfy $A^{(r)} = \Sigma_{\epsilon\epsilon}^{1/2} V = \Sigma_{\epsilon\epsilon}^{1/2} \Sigma_{\epsilon\epsilon}^{1/2} \Sigma_{yy}^{-1/2} V_*(I_r - \Lambda_*^2)^{-1/2} = \Sigma_{\epsilon\epsilon} \Sigma_{yy}^{-1/2} V_*(I_r - \Lambda_*^2)^{-1/2}$ and $B^{(r)} = V' \Sigma_{\epsilon\epsilon}^{-1/2} \Sigma_{yx} \Sigma_{xx}^{-1} = (I_r - \Lambda_*^2)^{-1/2} V_*' \Sigma_{yy}^{-1/2} \Sigma_{yx} \Sigma_{xx}^{-1}$ because $V_* = \Sigma_{yy}^{1/2} \Sigma_{\epsilon\epsilon}^{-1/2} V (I_r - \Lambda_*^2)^{1/2}$ and so $V = \Sigma_{\epsilon\epsilon}^{1/2} \Sigma_{yy}^{-1/2} V_* (I_r - \Lambda_*^2)^{-1/2}$, where $V_* = [V_1^*, \ldots, V_r^*]$ and the V_j^* are normalized eigenvectors of the matrix R_* in (2.28). Also notice then that

$$\begin{aligned}
A^{(r)} &= \Sigma_{\epsilon\epsilon} \Sigma_{yy}^{-1/2} V_*(I_r - \Lambda_*^2)^{-1/2} \\
&= (\Sigma_{yy} - \Sigma_{yx} \Sigma_{xx}^{-1} \Sigma_{xy}) \Sigma_{yy}^{-1/2} V_* (I_r - \Lambda_*^2)^{-1/2} \\
&= \Sigma_{yy}^{1/2} (I_m - \Sigma_{yy}^{-1/2} \Sigma_{yx} \Sigma_{xx}^{-1} \Sigma_{xy} \Sigma_{yy}^{-1/2}) V_* (I_r - \Lambda_*^2)^{-1/2} \\
&= \Sigma_{yy}^{1/2} V_* (I_r - \Lambda_*^2)^{1/2}
\end{aligned}$$

and $B^{(r)} = (I_r - \Lambda_*^2)^{-1/2} V_*' \Sigma_{yy}^{-1/2} \Sigma_{yx} \Sigma_{xx}^{-1}$, whereas the solutions of A and B for the choice $\Gamma = \Sigma_{yy}^{-1}$ are given by $A_*^{(r)} = \Sigma_{yy}^{1/2} V_*$ and $B_*^{(r)} = V_*' \Sigma_{yy}^{-1/2} \Sigma_{yx} \Sigma_{xx}^{-1}$. Thus, compared to these latter solutions, only the columns and rows of the solution matrices A and B, respectively, are scaled differently for the choice $\Gamma = \Sigma_{\epsilon\epsilon}^{-1}$, but the solution for the overall coefficient matrix remains the same with $A^{(r)} B^{(r)} = A_*^{(r)} B_*^{(r)}$.

The connection described above between reduced-rank regression quantities in Theorem 2.2 and canonical correlation analysis in the population also carries over to the sample situation of ML estimation of the reduced-rank model. In particular, let $\widehat{\rho}_j^2$ denote the sample squared canonical correlations between the X_k and Y_k, that is, the eigenvalues of $\widehat{R}_* = \widehat{\Sigma}_{yy}^{-1/2} \widehat{\Sigma}_{yx} \widehat{\Sigma}_{xx}^{-1} \widehat{\Sigma}_{xy} \widehat{\Sigma}_{yy}^{-1/2}$, where $\widehat{\Sigma}_{yy} = \frac{1}{T} \mathbf{Y}\mathbf{Y}'$, $\widehat{\Sigma}_{yx} = \frac{1}{T} \mathbf{Y}\mathbf{X}'$, and $\widehat{\Sigma}_{xx} = \frac{1}{T} \mathbf{X}\mathbf{X}'$, and let $\widehat{V}_j^*$ denote the corresponding (normalized) eigenvectors, $j = 1, \ldots, m$. (Recall from the remark in Section 1.2 that, in practice, in the regression analysis the predictor and response variables $\mathbf{X}_k$ and $\mathbf{Y}_k$ will typically be centered by subtraction of appropriate sample mean vectors before the sample canonical correlation calculations above are performed.) Then the ML estimators of A and B in the reduced-rank model can be expressed in an equivalent form to (2.17) as

$$\begin{aligned}
\widehat{A}^{(r)} &= \widetilde{\Sigma}_{\epsilon\epsilon} \widehat{\Sigma}_{yy}^{-1/2} \widehat{V}_*(I_r - \widehat{\Lambda}_*^2)^{-1/2} = \widehat{\Sigma}_{yy}^{1/2} \widehat{V}_*(I_r - \widehat{\Lambda}_*^2)^{1/2}, \\
\widehat{B}^{(r)} &= (I_r - \widehat{\Lambda}_*^2)^{-1/2} \widehat{V}_*' \widehat{\Sigma}_{yy}^{-1/2} \widehat{\Sigma}_{yx} \widehat{\Sigma}_{xx}^{-1},
\end{aligned}$$

with $\widehat{V}_* = [\widehat{V}_1^*, \ldots, \widehat{V}_r^*]$, $\widehat{\Lambda}_*^2 = \text{diag}(\widehat{\rho}_1^2, \ldots, \widehat{\rho}_r^2)$, and $\widetilde{\Sigma}_{\epsilon\epsilon} = \widehat{\Sigma}_{yy} - \widehat{\Sigma}_{yx}\widehat{\Sigma}_{xx}^{-1}\widehat{\Sigma}_{xy}$.

Through the linkage between the two procedures, we have demonstrated how the descriptive tool of canonical correlation can also be used for modeling and prediction purposes. Such descriptive tools are not easily available for extended models considered in the book.

2.5 Asymptotic Distribution of Estimators in Reduced-Rank Model

The small sample distribution of the reduced-rank estimators associated with the canonical correlations, although available, is somewhat difficult to use. Therefore, we focus on the large sample behavior of the estimators and we derive in this section the asymptotic theory for the estimators of the parameters A and B in model (2.11). The asymptotic results follow from Robinson (1973). From Theorem 2.2, for any positive-definite Γ, the population quantities can be expressed as

$$A = \Gamma^{-1/2}[V_1, \ldots, V_r] = \Gamma^{-1/2}V = [\alpha_1, \ldots, \alpha_r], \tag{2.30a}$$

$$B = V'\,\Gamma^{1/2}\,\Sigma_{yx}\,\Sigma_{xx}^{-1} = [\beta_1, \ldots, \beta_r]'. \tag{2.30b}$$

The estimators of A and B, from (2.17), corresponding to these populations quantities are

$$\widehat{A} = \Gamma^{-1/2}[\widehat{V}_1, \ldots, \widehat{V}_r] = \Gamma^{-1/2}\widehat{V} = [\widehat{\alpha}_1, \ldots, \widehat{\alpha}_r], \tag{2.31a}$$

$$\widehat{B} = \widehat{V}'\,\Gamma^{1/2}\,\widehat{\Sigma}_{yx}\,\widehat{\Sigma}_{xx}^{-1} = [\widehat{\beta}_1, \ldots, \widehat{\beta}_r]', \tag{2.31b}$$

where $\widehat{V}_j$ is the normalized eigenvector corresponding to the jth largest eigenvalue $\widehat{\lambda}_j^2$ of $\Gamma^{1/2}\,\widehat{\Sigma}_{yx}\widehat{\Sigma}_{xx}^{-1}\,\widehat{\Sigma}_{xy}\,\Gamma^{1/2}$, with $\widehat{\Sigma}_{yx} = (1/T)\mathbf{YX}'$ and $\widehat{\Sigma}_{xx} = (1/T)\mathbf{XX}'$. We assume that $\Sigma_{xx} = \lim_{T\to\infty} \dfrac{1}{T}\mathbf{XX}'$ exists and is positive definite, and $\Sigma_{yx} = C\Sigma_{xx}$. In the following result, we assume $\Gamma = \Sigma_{\epsilon\epsilon}^{-1}$ is known, for convenience. The argument and asymptotic results for estimators $\widehat{A}$ and $\widehat{B}$ would be somewhat affected by the use of $\Gamma = \widetilde{\Sigma}_{\epsilon\epsilon}^{-1}$, where $\widetilde{\Sigma}_{\epsilon\epsilon} = (1/T)(\mathbf{YY}' - \mathbf{YX}'(\mathbf{XX}')^{-1}\mathbf{XY}')$, mainly because the normalization restrictions on $\widehat{A}$ then depend on $\widetilde{\Sigma}_{\epsilon\epsilon}^{-1}$. These effects will be discussed subsequently. Recall that for the choice of $\Gamma = \widetilde{\Sigma}_{\epsilon\epsilon}^{-1}$ the estimators $\widehat{A}$ and $\widehat{B}$ in (2.31) are the maximum likelihood estimators of A and B under the assumption of normality of the error terms ϵ_k (Robinson (1974, Theorem 1), and Section 2.3).

Our main result on the asymptotic distribution of $\widehat{A}$ and $\widehat{B}$ when $\Gamma = \Sigma_{\epsilon\epsilon}^{-1}$ is assumed to be known and the predictor variables are non-stochastic is contained in the next theorem.

Theorem 2.4 *For the model* (2.11) *where the* ϵ_k *satisfy the usual assumptions, let* $(\text{vec}(A)', \text{vec}(B)')' \in \Theta$, *a compact set defined by the normalization conditions* (2.14). *Then, with* $\widehat{A}$, $\widehat{B}$ *and* A, B *given by* (2.31) *and* (2.30), *respectively,* $\text{vec}(\widehat{A})$ *and* $\text{vec}(\widehat{B})$ *converge in probability to* $\text{vec}(A)$ *and* $\text{vec}(B)$, *respectively, as* $T \to \infty$. *Also, the vector variates* $T^{1/2}(\widehat{\alpha}_j - \alpha_j)$ *and* $T^{1/2}(\widehat{\beta}_j - \beta_j)$ $(j = 1, \ldots, r)$ *have a joint limiting distribution as* $T \to \infty$, *which is singular multivariate normal with null mean vectors and the following asymptotic covariance matrices:*

$$E[T(\widehat{\alpha}_j - \alpha_j)(\widehat{\alpha}_\ell - \alpha_\ell)'] \to \begin{cases} \displaystyle\sum_{i \neq j = 1}^{m} \frac{\lambda_j^2 + \lambda_i^2}{(\lambda_j^2 - \lambda_i^2)^2} \alpha_i \alpha_i' & j = \ell \\ -\displaystyle\frac{\lambda_j^2 + \lambda_\ell^2}{(\lambda_j^2 - \lambda_\ell^2)^2} \alpha_\ell \alpha_j' & j \neq \ell, \end{cases}$$

$$E[T(\widehat{\beta}_j - \beta_j)(\widehat{\beta}_\ell - \beta_\ell)'] \to \begin{cases} \displaystyle\sum_{i \neq j = 1}^{r} \frac{3\lambda_j^2 - \lambda_i^2}{(\lambda_j^2 - \lambda_i^2)^2} \beta_i \beta_i' + \Sigma_{xx}^{-1} & j = \ell, \\ -\displaystyle\frac{\lambda_j^2 + \lambda_\ell^2}{(\lambda_j^2 - \lambda_\ell^2)^2} \beta_\ell \beta_j' & j \neq \ell, \end{cases}$$

$$E[T(\widehat{\alpha}_j - \alpha_j)(\widehat{\beta}_\ell - \beta_\ell)'] \to \begin{cases} 2\lambda_j^2 \displaystyle\sum_{i \neq j = 1}^{r} \frac{1}{(\lambda_j^2 - \lambda_i^2)^2} \alpha_i \beta_i' & j = \ell, \\ -\displaystyle\frac{2\lambda_\ell^2}{(\lambda_j^2 - \lambda_\ell^2)^2} \alpha_\ell \beta_j' & j \neq \ell, \end{cases}$$

where the λ_j^2, $j = 1, \ldots, r$, *are the nonzero eigenvalues, assumed to be distinct, of the matrix* $\Gamma^{1/2}\Sigma_{yx}\Sigma_{xx}^{-1}\Sigma_{xy}\Gamma^{1/2}$, *with* $\Gamma = \Sigma_{\epsilon\epsilon}^{-1}$.

Proof With the use of perturbation expansion of matrices (Izenman 1975), the eigenvectors $\widehat{V}_j$ of the matrix $M = \Sigma_{\epsilon\epsilon}^{-1/2}\,\widehat{\Sigma}_{yx}\,\widehat{\Sigma}_{xx}^{-1}\,\widehat{\Sigma}_{xy}\,\Sigma_{\epsilon\epsilon}^{-1/2}$ can be expanded around the eigenvectors V_j of $N = \Sigma_{\epsilon\epsilon}^{-1/2}\,\Sigma_{yx}\,\Sigma_{xx}^{-1}\,\Sigma_{xy}\,\Sigma_{\epsilon\epsilon}^{-1/2}$ to give

$$\widehat{V}_j \sim V_j + \sum_{i \neq j}^{m} \frac{1}{(\lambda_j^2 - \lambda_i^2)} V_i[V_j'(M - N)V_i] \qquad (j = 1, \ldots, r). \tag{2.32}$$

Then since $\widehat{\Sigma}_{xx}$ and $\widehat{\Sigma}_{yx}$ converge in probability to Σ_{xx} and Σ_{yx} (Anderson 1971, p. 195), M converges to N, and hence $\widehat{V}_j$ converges in probability to V_j as $T \to \infty$. Consistency of $\widehat{A}$ and $\widehat{B}$ then follows directly from the relations (2.30) and (2.31).

To establish asymptotic normality, using $\widehat{\Sigma}_{xy} = \widehat{\Sigma}_{xx}C' + T^{-1}\sum_{k=1}^{T} X_k\epsilon_k'$, we can write

$$\widehat{\Sigma}_{yx}\widehat{\Sigma}_{xx}^{-1}\widehat{\Sigma}_{xy} - \Sigma_{yx}\Sigma_{xx}^{-1}\Sigma_{xy} = CU_T + U_T'C' + U_T'\widehat{\Sigma}_{xx}^{-1}U_T + C(\widehat{\Sigma}_{xx} - \Sigma_{xx})C',$$

where $U_T = T^{-1}\sum_{k=1}^{T} X_k\epsilon_k'$. Hence, following Robinson (1973), from the perturbation expansion in (2.32), we have (based on the first two terms, as predictors are assumed to be non-stochastic)

$$\begin{aligned} T^{1/2}(\widehat{V}_j - V_j) &= T^{1/2}\sum_{i\neq j}^{m} \frac{1}{(\lambda_j^2-\lambda_i^2)} V_i[(V_i'\Sigma_{\epsilon\epsilon}^{-1/2}\otimes V_j'VB) \\ &\quad + (V_j'\Sigma_{\epsilon\epsilon}^{-1/2}\otimes V_i'VB)]\operatorname{vec}(U_T) + O_p(T^{-1/2}) \\ &\equiv T^{1/2}D_j'\operatorname{vec}(U_T) + O_p(T^{-1/2}) \quad (j=1,\ldots,r), \end{aligned} \tag{2.33}$$

where we have used the relation $\operatorname{vec}(ABC) = (C'\otimes A)\operatorname{vec}(B)$ and the fact that $\Sigma_{\epsilon\epsilon}^{-1/2}C = \Sigma_{\epsilon\epsilon}^{-1/2}AB = VB$.

Define $D = [D_1,\ldots,D_r]$, so that $T^{1/2}\operatorname{vec}(\widehat{V}-V) \sim T^{1/2}D'\operatorname{vec}(U_T)$. Then since $T^{1/2}\operatorname{vec}(U_T) = \operatorname{vec}(T^{-1/2}\sum_{k=1}^{T} X_k\epsilon_k')$ converges in distribution to $N(0, \Sigma_{\epsilon\epsilon}\otimes\Sigma_{xx})$ as $T\to\infty$ (Anderson 1971, p. 200), the asymptotic distribution of $T^{1/2}\operatorname{vec}(\widehat{V}-V)$ is $N(0,G)$, where $G = D'(\Sigma_{\epsilon\epsilon}\otimes\Sigma_{xx})D$ has blocks $G_{j\ell} = D_j'(\Sigma_{\epsilon\epsilon}\otimes\Sigma_{xx})D_\ell$, with

$$G_{jj} = \sum_{i\neq j=1}^{m} \frac{\lambda_j^2+\lambda_i^2}{(\lambda_j^2-\lambda_i^2)^2} V_iV_i' \qquad (j=1,\ldots,r),$$

$$G_{j\ell} = -\frac{\lambda_j^2+\lambda_\ell^2}{(\lambda_j^2-\lambda_\ell^2)^2} V_\ell V_j' \qquad (j\neq \ell),$$

if we note that $V_i'V_j = 0$ for $i\neq j$, $V_i'V_i = 1$, and $B\Sigma_{xx}B' = \Lambda^2$. The asymptotic distribution of $T^{1/2}\operatorname{vec}(\widehat{A}-A)$ follows easily from the above and relation (2.31) with $T^{1/2}\operatorname{vec}(\widehat{A}-A) \simeq T^{1/2}(I_r\otimes\Sigma_{\epsilon\epsilon}^{1/2})\operatorname{vec}(\widehat{V}-V)$ converging in distribution to $N\{0, (I_r\otimes\Sigma_{\epsilon\epsilon}^{1/2})G(I_r\otimes\Sigma_{\epsilon\epsilon}^{1/2})\}$. Hence, if we note the relation $\Sigma_{\epsilon\epsilon}^{1/2}V_i = \alpha_i$, the individual covariance terms that correspond to the above distribution are as given in the theorem.

To evaluate the asymptotic distribution of $T^{1/2}\operatorname{vec}(\widehat{B}'-B')$, observe that

$$\begin{aligned} T^{1/2}\operatorname{vec}(\widehat{B}'-B') &= T^{1/2}\operatorname{vec}(\widehat{\Sigma}_{xx}^{-1}\widehat{\Sigma}_{xy}\Sigma_{\epsilon\epsilon}^{-1/2}\widehat{V} - \Sigma_{xx}^{-1}\Sigma_{xy}\Sigma_{\epsilon\epsilon}^{-1/2}V) \\ &\simeq T^{1/2}\operatorname{vec}\{\Sigma_{xx}^{-1}\Sigma_{xy}\Sigma_{\epsilon\epsilon}^{-1/2}(\widehat{V}-V)\} \\ &\quad + T^{1/2}\operatorname{vec}\{(\widehat{\Sigma}_{xx}^{-1}\widehat{\Sigma}_{xy}\Sigma_{\epsilon\epsilon}^{-1/2} - \Sigma_{xx}^{-1}\Sigma_{xy}\Sigma_{\epsilon\epsilon}^{-1/2})V\} \\ &= T^{1/2}(I_r\otimes\Sigma_{xx}^{-1}\Sigma_{xy}\Sigma_{\epsilon\epsilon}^{-1/2})\operatorname{vec}(\widehat{V}-V) \\ &\quad + T^{1/2}(V'\Sigma_{\epsilon\epsilon}^{-1/2}\otimes\widehat{\Sigma}_{xx}^{-1})\operatorname{vec}\left(T^{-1}\sum_{k=1}^{T} X_k\epsilon_k'\right) \end{aligned}$$

$$\simeq \{(I_r \otimes B'V')D' + (V'\Sigma_{\epsilon\epsilon}^{-1/2} \otimes \Sigma_{xx}^{-1})\}T^{1/2} \operatorname{vec}(U_T), \tag{2.34}$$

where we again have used the fact that $\Sigma_{\epsilon\epsilon}^{-1/2}\Sigma_{yx}\Sigma_{xx}^{-1} = \Sigma_{\epsilon\epsilon}^{-1/2}C = VB$. Hence $T^{1/2}\operatorname{vec}(\widehat{B}' - B')$ converges in distribution to $N(0, H)$, where

$$\begin{aligned} H &= (I_r \otimes B'V')\, D'\, (\Sigma_{\epsilon\epsilon} \otimes \Sigma_{xx})\, D\, (I_r \otimes VB) + (I_r \otimes \Sigma_{xx}^{-1}) \\ &\quad + (I_r \otimes B'V')\, D'\, (\Sigma_{\epsilon\epsilon} \otimes \Sigma_{xx})(\Sigma_{\epsilon\epsilon}^{-1/2}V \otimes \Sigma_{xx}^{-1}) \\ &\quad + (V'\Sigma_{\epsilon\epsilon}^{-1/2} \otimes \Sigma_{xx}^{-1})(\Sigma_{\epsilon\epsilon} \otimes \Sigma_{xx})\, D\, (I_r \otimes VB), \end{aligned} \tag{2.35}$$

and we used the fact that $V'V = I_r$.

The elements of the matrix H can be evaluated more explicitly as follows: the (j, ℓ)th block of $(I_r \otimes B'V')D'(\Sigma_{\epsilon\epsilon} \otimes \Sigma_{xx})D(I_r \otimes VB)$ is

$$B'V'G_{j\ell}VB = \begin{cases} \sum(\lambda_j^2 + \lambda_i^2)(\lambda_j^2 - \lambda_i^2)^{-2}\beta_i\,\beta_i', & j = \ell \\ -(\lambda_j^2 + \lambda_\ell^2)(\lambda_j^2 - \lambda_\ell^2)^{-2}\beta_\ell\,\beta_j', & j \neq \ell, \end{cases}$$

while the (j, ℓ)th block of $(I_r \otimes B'V')D'(\Sigma_{\epsilon\epsilon} \otimes \Sigma_{xx})(\Sigma_{\epsilon\epsilon}^{-1/2}V \otimes \Sigma_{xx}^{-1})$ is

$$B'V'D_j'(\Sigma_{\epsilon\epsilon}^{1/2}V_\ell \otimes I_n) = \begin{cases} \sum(\lambda_j^2 - \lambda_i^2)^{-1}\beta_i\,\beta_i', & j = \ell \\ (\lambda_j^2 - \lambda_\ell^2)^{-1}\beta_\ell\,\beta_j', & j \neq \ell, \end{cases}$$

where the sums are over $i = 1, \ldots, r, i \neq j$. Combining terms and simplifying, we obtain the results stated in the theorem.

Finally, from (2.33) and (2.34), asymptotic covariances between the elements of $T^{1/2}\operatorname{vec}(\widehat{A} - A)$ and $T^{1/2}\operatorname{vec}(\widehat{B}' - B')$ are given by the elements of the matrix

$$(I_r \otimes \Sigma_{\epsilon\epsilon}^{1/2})D'(\Sigma_{\epsilon\epsilon} \otimes \Sigma_{xx})[D(I_r \otimes VB) + (\Sigma_{\epsilon\epsilon}^{-1/2}V \otimes \Sigma_{xx}^{-1})].$$

The (j, ℓ)th block of this matrix can be determined explicitly in a manner similar to the previous derivation for the blocks of the matrix H and can be shown to have elements as given in the statement of the theorem.

Robinson (1974) develops the asymptotic theory for estimates of the coefficients in a general multivariate regression model which includes as a special case the reduced-rank regression model (2.5). The results presented here do not agree with the distribution results of Robinson. The difference occurs in the distribution of $T^{1/2}\operatorname{vec}(\widehat{B}' - B')$. Robinson claims that $T^{1/2}\operatorname{vec}(\widehat{B}' - B')$ has asymptotic covariance matrix $(I_r \otimes \Sigma_{xx}^{-1})$. However, $(I_r \otimes \Sigma_{xx}^{-1})$ is the asymptotic covariance matrix for the estimator $\widetilde{B} = V'\Sigma_{\epsilon\epsilon}^{-1/2}\widehat{\Sigma}_{yx}\widehat{\Sigma}_{xx}^{-1}$ of B, which would be appropriate if V, or equivalently A, were known. As can be seen from (2.34) and (2.35), the actual asymptotic covariance matrix of the estimator $\widehat{B}$ includes additional contributions besides $(I_r \otimes \Sigma_{xx}^{-1})$, which correspond to the presence of the first term on the right-hand side of (2.34).

When $\operatorname{rank}(C) = 1$, that is, $C = \alpha\beta'$, the asymptotic distributions of the estimators $\widehat{\alpha}$ and $\widehat{\beta}$ implied by Theorem 2.4 agree with those of Robinson (1974). The results agree in the rank one situation because of the asymptotic independence of $\widehat{\alpha}$, $\widehat{\beta}$ for this case; the first term on the right-hand side of (2.34) vanishes in this case since $(I_r \otimes B'V')D' = \beta V_1' D_1' = 0$. This asymptotic independence does not hold in general.

The asymptotic theory for the estimators $\widehat{A}$ and $\widehat{B}$ is useful for inferences concerning whether certain components of X_k contribute in the formation of explanatory indexes BX_k and also whether certain components of Y_k are influenced by certain of these indices. Because of the connection between the component matrices and the quantities involved in the canonical correlation analysis, inferences regarding the latter quantities can also be made via Theorem 2.4. Anderson (1999a,b) presents the asymptotic distribution of the canonical variates and also of the reduced-rank regression estimators when predictors are stochastic and non-stochastic. For the stochastic case, additional terms are needed for the asymptotic distribution of $\widehat{\alpha}$ and $\widehat{\beta}$ in Theorem 2.4. As the details are quite involved, we do not pursue them here.

An additional quantity of interest is the overall regression coefficient matrix $C = AB$ and the asymptotic distribution of the estimator $\widehat{C} = \widehat{A}\widehat{B}$ follows directly from the results of Theorem 2.4 [see Stoica and Viberg (1996) and Lütkepohl (1993, p. 199)]. It is easily observed, under conditions of Theorem 2.4, that $\widehat{C}$ is a consistent estimator of C. The asymptotic distribution of $\widehat{C}$ can be derived in the following way. Based on the relation

$$T^{1/2}(\widehat{C} - C) = T^{1/2}(\widehat{A}\widehat{B} - AB) = T^{1/2}(\widehat{A} - A)B + T^{1/2}A(\widehat{B} - B) + o_p(1),$$

it follows that

$$\begin{aligned} T^{1/2}\operatorname{vec}(\widehat{C}' - C') &= T^{1/2}[(I_m \otimes B'), (A \otimes I_n)] \begin{bmatrix} \operatorname{vec}(\widehat{A}' - A') \\ \operatorname{vec}(\widehat{B}' - B') \end{bmatrix} + o_p(1) \\ &= T^{1/2} M \begin{bmatrix} \operatorname{vec}(\widehat{A}' - A') \\ \operatorname{vec}(\widehat{B}' - B') \end{bmatrix} + o_p(1), \end{aligned}$$

where $M = [(I_m \otimes B'), (A \otimes I_n)]$. Collecting the terms in Theorem 2.4 for the variances and covariances in the limiting distribution of the components in the vectors $T^{1/2}(\widehat{\alpha}_j - \alpha_j)$ and $T^{1/2}(\widehat{\beta}_j - \beta_j)$, $j = 1, \ldots, r$, we denote this covariance matrix as

$$W = \lim_{T\to\infty} \operatorname{Cov}\begin{bmatrix} T^{1/2}\operatorname{vec}(\widehat{A}' - A') \\ T^{1/2}\operatorname{vec}(\widehat{B}' - B') \end{bmatrix} = \begin{bmatrix} W_{11} & W_{12} \\ W_{21} & W_{22} \end{bmatrix},$$

where W_{11} denotes the asymptotic covariance matrix of $T^{1/2}\,\mathrm{vec}(\widehat{A}' - A')$, W_{12} is the asymptotic covariance matrix between $T^{1/2}\,\mathrm{vec}(\widehat{A}' - A')$ and $T^{1/2}\,\mathrm{vec}(\widehat{B}' - B')$, and W_{22} is the asymptotic covariance matrix of $T^{1/2}\,\mathrm{vec}(\widehat{B}' - B')$; for example, $W_{22} = H$, the expression in (2.35) from the proof of Theorem 2.4. It must be noted that care is needed in identifying the elements of W_{11} and W_{12} because the results in the proof of Theorem 2.4 refer to $\mathrm{vec}(\widehat{A} - A)$ not $\mathrm{vec}(\widehat{A}' - A')$. The asymptotic distribution of $\widehat{C}$ now follows directly as

$$T^{1/2}\,\mathrm{vec}(\widehat{C}' - C') \xrightarrow{d} N(0, MWM'). \tag{2.36}$$

When the rank of the matrix C is assumed to be full, the distribution in (2.36) is equivalent to that in (1.15).

We want to demonstrate the equivalence between the asymptotic distribution of $\widehat{C}$ given in Anderson (1999a) and the distribution in (2.36). Here it should be observed that the covariance matrix is $W = B_\theta^-$, which is a (reflexive) generalized inverse of the asymptotic information matrix B_θ. Observer that

$$B_\theta = M'(\Sigma_{\epsilon\epsilon}^{-1} \otimes \Sigma_{\epsilon\epsilon})M \equiv \overline{M}'\overline{M} = \begin{bmatrix} \Sigma_{\epsilon\epsilon}^{-1} \otimes B\Sigma_{\epsilon\epsilon}B' & \Sigma_{\epsilon\epsilon}^{-1}A \otimes B\Sigma_{\epsilon\epsilon} \\ A'\Sigma_{\epsilon\epsilon}^{-1} \otimes \Sigma_{\epsilon\epsilon}B' & A'\Sigma_{\epsilon\epsilon}^{-1}A \otimes \Sigma_{\epsilon\epsilon} \end{bmatrix},$$

where $\overline{M} = (\Sigma_{\epsilon\epsilon}^{-1/2} \otimes \Sigma_{\epsilon\epsilon}^{1/2})M$. Now, from a result in Searle (1971, p. 20), the matrix quantity $\overline{M}(\overline{M}'\overline{M})^-\overline{M}' = \overline{M}B_\theta^-\overline{M}'$ is invariant to the choice of generalized inverse B_θ^- of B_θ. Hence, it follows that MB_θ^-M' is also the same for all choices of generalized inverse B_θ^-. There are many convenient choices for B_θ^-, but we consider one given by Searle (1971, p. 27) for partitioned matrices. If we denote the blocks of B_θ above as F_{ij}, $i, j = 1, 2$, then this choice of B_θ^- is given by

$$B_\theta^- = \begin{bmatrix} F_{11}^{-1} + F_{11}^{-1}F_{12}Q^-F_{12}'F_{11}^{-1} & -F_{11}^{-1}F_{12}Q^- \\ -Q^-F_{12}'F_{11}^{-1} & Q^- \end{bmatrix},$$

where $Q = F_{22} - F_{12}'F_{11}^{-1}F_{12} = \{A'\Sigma_{\epsilon\epsilon}^{-1}A\} \otimes [\Sigma_{\epsilon\epsilon} - \Sigma_{xx}B'\{B\Sigma_{xx}B'\}^{-1}B\Sigma_{xx}]$, $F_{11}^{-1} = \Sigma_{\epsilon\epsilon} \otimes \{B\Sigma_{xx}B'\}^{-1}$, and we can take a generalized inverse of Q as $Q^- = \{A'\Sigma_{\epsilon\epsilon}^{-1}A\}^{-1} \otimes \Sigma_{xx}^{-1}$. Therefore, the chosen generalized inverse of B_θ takes the form

$$B_\theta^- = \begin{bmatrix} b_{11} & b_{12} \\ b_{21} & b_{22} \end{bmatrix},$$

where

$$b_{11} = (\Sigma_{\epsilon\epsilon} \otimes \{B\Sigma_{xx}B'\}^{-1}) + A\{A'\Sigma_{\epsilon\epsilon}^{-1}A\}^{-1}A' \otimes \{B\Sigma_{xx}B'\}^{-1}),$$

$$b_{12} = -A\{A'\Sigma_{\epsilon\epsilon}^{-1}A\}^{-1} \otimes \{B\Sigma_{xx}B'\}^{-1}B,$$

$$b_{21} = -\{A'\Sigma_{\epsilon\epsilon}^{-1}A\}^{-1}A' \otimes B'\{B\Sigma_{xx}B'\}^{-1},$$

$$b_{22} = \{A'\Sigma_{\epsilon\epsilon}^{-1}A\}^{-1} \otimes \Sigma_{xx}^{-1}.$$

Thus, we obtain the asymptotic covariance matrix for the reduced-rank ML estimator as

$$
\begin{aligned}
MB_{\theta}^{-}M' &= [I_m \otimes B', A \otimes I_n]B_{\theta}^{-}[I_m \otimes B', A \otimes I_n]' \\
&= [\Sigma_{\epsilon\epsilon} - A\{A'\Sigma_{\epsilon\epsilon}^{-1}A\}^{-1}A'] \otimes B'\{B\Sigma_{xx}B'\}^{-1}B \\
&\quad + A\{A'\Sigma_{\epsilon\epsilon}^{-1}A\}^{-1}A' \otimes \Sigma_{xx}^{-1}.
\end{aligned}
$$

Notice that this is the *same* result as given in (3.20) of the paper by Anderson (1999a), apart from a reverse in the order of the Kronecker products which is due to the convention used here for arranging the vector version of C by rows instead of columns. Also this is an equivalent form of the asymptotic covariance matrix result as given by Stoica and Viberg (1996).

The result in (2.36) for the reduced-rank estimator $\widehat{C}$ can be compared with the asymptotic covariance matrix of $T^{1/2}\,\mathrm{vec}(\widetilde{C}' - C')$ for the full-rank estimator $\widetilde{C}$ (which ignores the reduced-rank structure). This covariance matrix is given by $\Sigma_{\epsilon\epsilon} \otimes \Sigma_{xx}^{-1}$ from Result 1.1 of Section 1.3. For the general rank r case, the covariance matrix in (2.36) does not take a particularly convenient form for comparison with that of the full-rank estimator. However, we consider the special case of $r = 1$ for which explicit results are easily obtained. For this case, we have already noted that $(I_r \otimes B'V')D' = 0$ and so we see from (2.35) that W_{22} reduces to $W_{22} = H = \Sigma_{xx}^{-1}$. It also follows that $W_{12} = 0$ and we can find that $W_{11} = \Sigma_{\epsilon\epsilon}^{1/2}(I_m - V_1V_1')\,\Sigma_{\epsilon\epsilon}^{1/2}(\beta'\Sigma_{xx}\beta)^{-1}$, with $C = \alpha\beta'$. Then using $\alpha = \Sigma_{\epsilon\epsilon}^{1/2}V_1$, it follows that the asymptotic covariance matrix in (2.36) for the rank one case simplifies to

$$
\begin{aligned}
MWM' &= \{(\Sigma_{\epsilon\epsilon} - \Sigma_{\epsilon\epsilon}^{1/2}V_1V_1'\Sigma_{\epsilon\epsilon}^{1/2}) \otimes \beta(\beta'\Sigma_{xx}\beta)^{-1}\beta'\} \\
&\quad + (\Sigma_{\epsilon\epsilon}^{1/2}V_1V_1'\Sigma_{\epsilon\epsilon}^{1/2} \otimes \Sigma_{xx}^{-1}) \\
&= (\Sigma_{\epsilon\epsilon} \otimes \Sigma_{xx}^{-1}) \\
&\quad - \{(\Sigma_{\epsilon\epsilon} - \Sigma_{\epsilon\epsilon}^{1/2}V_1V_1'\Sigma_{\epsilon\epsilon}^{1/2}) \otimes (\Sigma_{xx}^{-1} - \beta(\beta'\Sigma_{xx}\beta)^{-1}\beta')\}.
\end{aligned}
$$

This expression can be directly compared with the corresponding covariance matrix expression $\Sigma_{\epsilon\epsilon} \otimes \Sigma_{xx}^{-1}$ of the full-rank estimator, and the reduction in covariance matrix of the reduced-rank estimator is readily revealed for this special case.

One can also look at the effect of reduced-rank estimation on prediction. The prediction error of the predictor $\widehat{Y}_0 = \widetilde{C}X_0$ at X_0 for full-rank LS estimation is $Y_0 - \widetilde{C}X_0 = \epsilon_0 - (\widetilde{C} - C)X_0$, so that the prediction error covariance matrix is $(1 + X_0'(\mathbf{XX}')^{-1}X_0)\Sigma_{\epsilon\epsilon}$. In the rank one case, using (2.36), the covariance matrix of the prediction error $Y_0 - \widehat{C}X_0 = \epsilon_0 - (\widehat{C} - C)X_0 \equiv \epsilon_0 - (I_m \otimes X_0')\,\mathrm{vec}(\widehat{C}' - C')$ using the reduced-rank estimator $\widehat{C} = \widehat{\alpha}\widehat{\beta}'$ is (approximately)

$$
\begin{aligned}
&\Sigma_{\epsilon\epsilon} + \frac{1}{T}(I_m \otimes X_0')MWM'(I_m \otimes X_0) \\
&\quad = (1 + X_0'(\mathbf{XX}')^{-1}X_0)\,\Sigma_{\epsilon\epsilon} \\
&\quad - \left\{X_0'(\mathbf{XX}')^{-1}X_0 - \frac{(\beta' X_0)^2}{(\beta'(\mathbf{XX}')\beta)}\right\}\left(\Sigma_{\epsilon\epsilon} - \Sigma_{\epsilon\epsilon}^{1/2}\,V_1\,V_1'\,\Sigma_{\epsilon\epsilon}^{1/2}\right).
\end{aligned}
$$

So the prediction error covariance matrix is decreased (by the positive amount equal to the second term on the right side above) by use of the reduced-rank estimator relative to the full-rank LS estimator.

For the rank "r" case, the results, although are not explicit as in the rank one case, can be shown to hold. The efficiency gained from the assumptions of low rank if it is true can be demonstrated as follows: note that the asymptotic covariance matrix of $\widetilde{C}$ can be written as

$$
\begin{aligned}
\mathrm{MWM}' &= [\Sigma_{\epsilon\epsilon} \otimes \Sigma_{xx}^{-1}] - [(\Sigma_{\epsilon\epsilon} - A(A'\Sigma_{\epsilon\epsilon}^{-1}A)^{-1}A') \otimes (\Sigma_{xx}^{-1} - B'(B\Sigma_{xx}B')^{-1}B)] \\
&= [\Sigma_{\epsilon\epsilon} \otimes \Sigma_{xx}^{-1}] - \left[(\Sigma_{\epsilon\epsilon} - \Sigma_{\epsilon\epsilon}^{1/2}VV'\Sigma_{\epsilon\epsilon}^{1/2}) \otimes \left(\Sigma_{xx}^{-1} - \sum_{i=1}^{r}\frac{\beta_i\beta_i'}{\beta_i'\Sigma_{xx}\beta_i}\right)\right].
\end{aligned}
$$

It is observed that the two matrices involved in the Kronecker product in the second term are both positive definite. The covariance matrix of the prediction error at $X = X_0$ can be written as (approximately)

$$
\begin{aligned}
&\Sigma_{\epsilon\epsilon} + \frac{1}{T}(I_m \otimes X_0')\mathrm{MWM}'(I_m \otimes x_0) = (1 + X_0'(XX')^{-1}X_0)\Sigma_{\epsilon\epsilon} \\
&\quad - \left\{X_0'(XX')X_0 - \sum_{i=1}^{r}\frac{(\beta_i'X_0)^2}{\beta_i'(XX')\beta_i}\right\}\left\{\Sigma_{\epsilon\epsilon} - \Sigma_{\epsilon\epsilon}^{1/2}VV'\Sigma_{\epsilon\epsilon}^{1/2}\right\}.
\end{aligned}
$$

It can be easily seen that the last two terms are indeed positive, and therefore there is certainly a gain in using reduced-rank models.

Remark—Impact of Estimating Error Covariance Matrix We now discuss the effect on asymptotic distribution results from the use of the estimator $\Gamma = \widetilde{\Sigma}_{\epsilon\epsilon}^{-1}$ in construction of the estimators $\widehat{A}$ and $\widehat{B}$ in (2.31). Restricting consideration to the rank one case, the perturbation methods applied in the proof of Theorem 2.4 can also be used to obtain the asymptotic covariance matrix of estimators $\widehat{\alpha}$ and $\widehat{\beta}$ based on $\widetilde{\Sigma}_{\epsilon\epsilon}^{-1}$, subject to $\widehat{\alpha}'\widetilde{\Sigma}_{\epsilon\epsilon}^{-1}\widehat{\alpha} = 1$. First, define $\widetilde{M} = \widetilde{\Sigma}_{\epsilon\epsilon}^{-1/2}\widehat{\Sigma}_{yx}\widehat{\Sigma}_{xx}^{-1}\widehat{\Sigma}_{xy}\widetilde{\Sigma}_{\epsilon\epsilon}^{-1/2}$ and let $\widetilde{V}_1$ be the normalized eigenvector of $\widetilde{M}$ corresponding to its largest eigenvalue $\widetilde{\lambda}_1^2$. Similar to (2.32), we have the expansion of $\widetilde{V}_1$ around V_1 as

$$\widetilde{V}_1 \sim V_1 + \sum_{i=2}^{m} \frac{1}{\lambda_1^2} V_i [V_1'(\widetilde{M} - N) V_i] \equiv V_1 + \frac{1}{\lambda_1^2}(I_m - V_1 V_1')(\widetilde{M} - N) V_1.$$

Then write $\widetilde{M} - N = (\widetilde{M} - M) + (M - N)$, where $M - N$ is treated as in the proof of Theorem 2.4, and

$$\begin{aligned}
\widetilde{M} - M &= \widetilde{\Sigma}_{\epsilon\epsilon}^{-1/2} \widehat{\Sigma}_{yx} \widehat{\Sigma}_{xx}^{-1} \widehat{\Sigma}_{xy} \widetilde{\Sigma}_{\epsilon\epsilon}^{-1/2} - \Sigma_{\epsilon\epsilon}^{-1/2} \widehat{\Sigma}_{yx} \widehat{\Sigma}_{xx}^{-1} \widehat{\Sigma}_{xy} \Sigma_{\epsilon\epsilon}^{-1/2} \\
&= (\widetilde{\Sigma}_{\epsilon\epsilon}^{-1/2} - \Sigma_{\epsilon\epsilon}^{-1/2}) \Sigma_{yx} \Sigma_{xx}^{-1} \Sigma_{xy} \Sigma_{\epsilon\epsilon}^{-1/2} \\
&\quad + \Sigma_{\epsilon\epsilon}^{-1/2} \Sigma_{yx} \Sigma_{xx}^{-1} \Sigma_{xy} (\widetilde{\Sigma}_{\epsilon\epsilon}^{-1/2} - \Sigma_{\epsilon\epsilon}^{-1/2}) + O_p(T^{-1}).
\end{aligned}$$

Also notice that $\Sigma_{yx} \Sigma_{xx}^{-1} \Sigma_{xy} = \alpha\beta' \Sigma_{xx} \beta\alpha' = \lambda_1^2 \alpha\alpha'$ in the rank one case.

Because $\alpha = \Sigma_{\epsilon\epsilon}^{1/2} V_1$ and $C = \alpha\beta'$, we have $(I_m - V_1 V_1') \Sigma_{\epsilon\epsilon}^{-1/2} C = 0$. Using this and the approximation $\widetilde{\Sigma}_{\epsilon\epsilon}^{-1/2} - \Sigma_{\epsilon\epsilon}^{-1/2} = -\Sigma_{\epsilon\epsilon}^{-1/2} (\widetilde{\Sigma}_{\epsilon\epsilon}^{1/2} - \Sigma_{\epsilon\epsilon}^{1/2}) \Sigma_{\epsilon\epsilon}^{-1/2} + O_p(T^{-1})$, we obtain a representation analogous to (2.33) as

$$\begin{aligned}
T^{1/2}(\widetilde{V}_1 - V_1) &= \frac{1}{\lambda_1^2}(I_m - V_1 V_1') \left\{ \Sigma_{\epsilon\epsilon}^{-1/2} T^{1/2} U_T' \beta + \lambda_1^2 T^{1/2} \left(\widetilde{\Sigma}_{\epsilon\epsilon}^{-1/2} - \Sigma_{\epsilon\epsilon}^{-1/2} \right) \alpha \right\} \\
&= \frac{1}{\lambda_1^2} \left[(I_m - V_1 V_1') \Sigma_{\epsilon\epsilon}^{-1/2} \otimes \beta' \right] T^{1/2} \operatorname{vec}(U_T) \\
&\quad - (I_m - V_1 V_1') \Sigma_{\epsilon\epsilon}^{-1/2} T^{1/2} \left(\widetilde{\Sigma}_{\epsilon\epsilon}^{1/2} - \Sigma_{\epsilon\epsilon}^{1/2} \right) \Sigma_{\epsilon\epsilon}^{-1/2} \alpha + O_p(T^{-1/2}).
\end{aligned}$$

The estimator of $\alpha = \Sigma_{\epsilon\epsilon}^{1/2} V_1$ is $\widehat{\alpha} = \widetilde{\Sigma}_{\epsilon\epsilon}^{1/2} \widetilde{V}_1$, so that $\widehat{\alpha} - \alpha = \Sigma_{\epsilon\epsilon}^{1/2} (\widetilde{V}_1 - V_1) + (\widetilde{\Sigma}_{\epsilon\epsilon}^{1/2} - \Sigma_{\epsilon\epsilon}^{1/2}) V_1 + O_p(T^{-1})$. Therefore, from above, we have

$$\begin{aligned}
T^{1/2}(\widehat{\alpha} - \alpha) &= \frac{1}{\lambda_1^2} \left[\Sigma_{\epsilon\epsilon}^{1/2} (I_m - V_1 V_1') \Sigma_{\epsilon\epsilon}^{-1/2} \otimes \beta' \right] T^{1/2} \operatorname{vec}(U_T) \\
&\quad + \alpha\alpha' \Sigma_{\epsilon\epsilon}^{-1} T^{1/2} \left(\widetilde{\Sigma}_{\epsilon\epsilon}^{1/2} - \Sigma_{\epsilon\epsilon}^{1/2} \right) \Sigma_{\epsilon\epsilon}^{-1/2} \alpha + O_p(T^{-1/2}) \\
&= \frac{1}{\lambda_1^2} \left[\Sigma_{\epsilon\epsilon}^{1/2} (I_m - V_1 V_1') \Sigma_{\epsilon\epsilon}^{-1/2} \otimes \beta' \right] T^{1/2} \operatorname{vec}(U_T) \\
&\quad + \frac{1}{2} \left(\alpha' \Sigma_{\epsilon\epsilon}^{-1} \otimes \alpha\alpha' \Sigma_{\epsilon\epsilon}^{-1} \right) T^{1/2} \operatorname{vec}(\widetilde{\Sigma}_{\epsilon\epsilon} - \Sigma_{\epsilon\epsilon}), \qquad (2.37)
\end{aligned}$$

where we have used the relation $(\widetilde{\Sigma}_{\epsilon\epsilon}^{1/2} - \Sigma_{\epsilon\epsilon}^{1/2}) \Sigma_{\epsilon\epsilon}^{1/2} + \Sigma_{\epsilon\epsilon}^{1/2} (\widetilde{\Sigma}_{\epsilon\epsilon}^{1/2} - \Sigma_{\epsilon\epsilon}^{1/2}) = \widetilde{\Sigma}_{\epsilon\epsilon} - \Sigma_{\epsilon\epsilon} + O_p(T^{-1})$.

From Muirhead (1982, Chapter 3) and Schott (1997, Chapter 9), $T^{1/2} \operatorname{vec}(\widetilde{\Sigma}_{\epsilon\epsilon} - \Sigma_{\epsilon\epsilon})$ has asymptotic covariance matrix equal to $2D_m^* (\Sigma_{\epsilon\epsilon} \otimes \Sigma_{\epsilon\epsilon}) D_m^*$, where $D_m^* = D_m (D_m D_m')^{-1} D_m'$ and D_m is the $m^2 \times \frac{1}{2} m(m+1)$ duplication matrix such that $D_m \operatorname{vech}(\Sigma_{\epsilon\epsilon}) = \operatorname{vec}(\Sigma_{\epsilon\epsilon})$. Here, $\operatorname{vech}(\Sigma_{\epsilon\epsilon})$ represents the $\frac{1}{2} m(m+1) \times 1$ vector consisting of the distinct elements of $\Sigma_{\epsilon\epsilon}$ (i.e., those on and below the diagonal). In addition, we know that $T^{1/2}\,\mathrm{vec}(U_T)$ converges in distribution to $N(0, \Sigma_{\epsilon\epsilon} \otimes \Sigma_{xx})$, and U_T and $\widetilde{\Sigma}_{\epsilon\epsilon}$ are asymptotically independent. These results can be used in (2.37) to obtain the asymptotic covariance matrix of $\widehat{\alpha}$ as

$$\begin{aligned} T\,\mathrm{Cov}(\widehat{\alpha} - \alpha) &= \left(\frac{1}{\lambda_1^2}\right)^2 \left[\Sigma_{\epsilon\epsilon}^{1/2}\left(I_m - V_1 V_1'\right)\Sigma_{\epsilon\epsilon}^{-1/2} \otimes \beta'\right](\Sigma_{\epsilon\epsilon} \otimes \Sigma_{xx}) \\ &\quad \times \left[\Sigma_{\epsilon\epsilon}^{-1/2}\left(I_m - V_1 V_1'\right)\Sigma_{\epsilon\epsilon}^{1/2} \otimes \beta\right] + \frac{1}{4}\left(\alpha'\,\Sigma_{\epsilon\epsilon}^{-1} \otimes \alpha\alpha'\,\Sigma_{\epsilon\epsilon}^{-1}\right) \\ &\quad \times 2D_m^*(\Sigma_{\epsilon\epsilon} \otimes \Sigma_{\epsilon\epsilon})\,D_m^*\,(\Sigma_{\epsilon\epsilon}^{-1}\alpha \otimes \Sigma_{\epsilon\epsilon}^{-1}\alpha\alpha') \\ &= \frac{1}{\lambda_1^2}\,\Sigma_{\epsilon\epsilon}^{1/2}(I_m - V_1 V_1')\,\Sigma_{\epsilon\epsilon}^{1/2} + \frac{1}{2}\,\alpha\,\alpha', \end{aligned}$$

since $D_m^*(\Sigma_{\epsilon\epsilon}^{-1}\alpha \otimes \Sigma_{\epsilon\epsilon}^{-1}\alpha) = \Sigma_{\epsilon\epsilon}^{-1}\alpha \otimes \Sigma_{\epsilon\epsilon}^{-1}\alpha$.

Next we consider the estimator $\widehat{\beta}' = \widetilde{V}_1'\,\widetilde{\Sigma}_{\epsilon\epsilon}^{-1/2}\,\widehat{\Sigma}_{yx}\,\widehat{\Sigma}_{xx}^{-1} = \widehat{\alpha}'\,\widetilde{\Sigma}_{\epsilon\epsilon}^{-1}\,\widehat{\Sigma}_{yx}\,\widehat{\Sigma}_{xx}^{-1}$. From the approximation $\widetilde{\Sigma}_{\epsilon\epsilon}^{-1}\widehat{\alpha} - \Sigma_{\epsilon\epsilon}^{-1}\alpha = \Sigma_{\epsilon\epsilon}^{-1}(\widehat{\alpha} - \alpha) + (\widetilde{\Sigma}_{\epsilon\epsilon}^{-1} - \Sigma_{\epsilon\epsilon}^{-1})\alpha + O_p(T^{-1})$, we obtain the representation

$$\begin{aligned} T^{1/2}(\widehat{\beta} - \beta) &= \Sigma_{xx}^{-1}\,\Sigma_{xy}\,\{\Sigma_{\epsilon\epsilon}^{-1}T^{1/2}(\widehat{\alpha} - \alpha) + T^{1/2}(\widetilde{\Sigma}_{\epsilon\epsilon}^{-1} - \Sigma_{\epsilon\epsilon}^{-1})\,\alpha\} \\ &\quad + T^{1/2}(\widehat{\Sigma}_{xx}^{-1}\,\widehat{\Sigma}_{xy} - \Sigma_{xx}^{-1}\,\Sigma_{xy})\,\Sigma_{\epsilon\epsilon}^{-1}\alpha + O_p(T^{-1/2}) \\ &= \Sigma_{xx}^{-1}\,\Sigma_{xy}\,\{\Sigma_{\epsilon\epsilon}^{-1}T^{1/2}(\widehat{\alpha} - \alpha) - \Sigma_{\epsilon\epsilon}^{-1}\,T^{1/2}(\widetilde{\Sigma}_{\epsilon\epsilon} - \Sigma_{\epsilon\epsilon})\Sigma_{\epsilon\epsilon}^{-1}\alpha\} \\ &\quad + T^{1/2}(\widehat{\Sigma}_{xx}^{-1}\,\widehat{\Sigma}_{xy} - \Sigma_{xx}^{-1}\,\Sigma_{xy})\,\Sigma_{\epsilon\epsilon}^{-1}\alpha, \end{aligned}$$

using the approximation $\widetilde{\Sigma}_{\epsilon\epsilon}^{-1} - \Sigma_{\epsilon\epsilon}^{-1} = -\Sigma_{\epsilon\epsilon}^{-1}(\widetilde{\Sigma}_{\epsilon\epsilon} - \Sigma_{\epsilon\epsilon})\Sigma_{\epsilon\epsilon}^{-1} + O_p(T^{-1})$. Thus, from the approximation for $T^{1/2}(\widehat{\alpha} - \alpha)$ given in (2.37), we have

$$\begin{aligned} T^{1/2}(\widehat{\beta} - \beta) &= -\frac{1}{2}\,\beta\,\alpha'\,\Sigma_{\epsilon\epsilon}^{-1}\,T^{1/2}(\widetilde{\Sigma}_{\epsilon\epsilon} - \Sigma_{\epsilon\epsilon})\,\Sigma_{\epsilon\epsilon}^{-1}\,\alpha \\ &\quad + T^{1/2}(\widehat{\Sigma}_{xx}^{-1}\,\widehat{\Sigma}_{xy} - \Sigma_{xx}^{-1}\,\Sigma_{xy})\,\Sigma_{\epsilon\epsilon}^{-1}\,\alpha + O_p(T^{-1/2}) \\ &= -\frac{1}{2}\,(\alpha'\,\Sigma_{\epsilon\epsilon}^{-1} \otimes \beta\,\alpha'\,\Sigma_{\epsilon\epsilon}^{-1})T^{1/2}\,\mathrm{vec}(\widetilde{\Sigma}_{\epsilon\epsilon} - \Sigma_{\epsilon\epsilon}) \\ &\quad + (\alpha'\,\Sigma_{\epsilon\epsilon}^{-1} \otimes \Sigma_{xx}^{-1})\,T^{1/2}\,\mathrm{vec}(U_T). \end{aligned} \tag{2.38}$$

This representation yields the asymptotic covariance matrix for $\widehat{\beta}$ as

$$T\,\mathrm{Cov}(\widehat{\beta}-\beta) = (\alpha'\,\Sigma_{\epsilon\epsilon}^{-1} \otimes \Sigma_{xx}^{-1})(\Sigma_{\epsilon\epsilon} \otimes \Sigma_{xx})(\Sigma_{\epsilon\epsilon}^{-1}\,\alpha \otimes \Sigma_{xx}^{-1})$$
$$+ \frac{1}{4}\,(\alpha'\,\Sigma_{\epsilon\epsilon}^{-1} \otimes \beta\,\alpha'\,\Sigma_{\epsilon\epsilon}^{-1})\,2D_m^{*}\,(\Sigma_{\epsilon\epsilon} \otimes \Sigma_{\epsilon\epsilon})\,D_m^{*}\,(\Sigma_{\epsilon\epsilon}^{-1}\,\alpha \otimes \Sigma_{\epsilon\epsilon}^{-1}\,\alpha\,\beta')$$
$$= \Sigma_{xx}^{-1} + \frac{1}{2}\,\beta\,\beta'.$$

Furthermore, the asymptotic cross-covariance matrix between $\widehat{\alpha}$ and $\widehat{\beta}$ is also obtained from the representations (2.37) and (2.38) as

$$T\,\mathrm{Cov}(\widehat{\alpha}-\alpha, \widehat{\beta}-\beta) = \frac{1}{\lambda_1^2}\,[\Sigma_{\epsilon\epsilon}^{1/2}(I_m - V_1 V_1')\,\Sigma_{\epsilon\epsilon}^{-1/2} \otimes \beta']\,(\Sigma_{\epsilon\epsilon} \otimes \Sigma_{xx})$$
$$\times\,(\Sigma_{\epsilon\epsilon}^{-1}\alpha \otimes \Sigma_{xx}^{-1}) - \frac{1}{4}\,(\alpha'\,\Sigma_{\epsilon\epsilon}^{-1} \otimes \alpha\,\alpha'\,\Sigma_{\epsilon\epsilon}^{-1})$$
$$\times\,2D_m^{*}(\Sigma_{\epsilon\epsilon} \otimes \Sigma_{\epsilon\epsilon})\,D_m^{*}\,(\Sigma_{\epsilon\epsilon}^{-1}\,\alpha \otimes \Sigma_{\epsilon\epsilon}^{-1}\,\alpha\,\beta')$$
$$= \frac{1}{\lambda_1^2}\,[\Sigma_{\epsilon\epsilon}^{1/2}(I_m - V_1 V_1')\,\Sigma_{\epsilon\epsilon}^{-1/2}\,\alpha \otimes \beta'] - \frac{1}{2}\,\alpha\,\beta' = -\frac{1}{2}\,\alpha\,\beta'.$$

Therefore, we find that $T^{1/2}\{(\widehat{\alpha}-\alpha)', (\widehat{\beta}-\beta)'\}'$ has asymptotic covariance matrix given by

$$W = \begin{bmatrix} W_{11} & W_{12} \\ W_{21} & W_{22} \end{bmatrix} = \begin{bmatrix} (\Sigma_{\epsilon\epsilon} - \alpha\alpha')(\beta'\Sigma_{xx}\beta)^{-1} + \frac{1}{2}\alpha\alpha' & -\frac{1}{2}a\beta' \\ -\frac{1}{2}\beta\alpha' & \Sigma_{xx}^{-1} + \frac{1}{2}\beta\beta' \end{bmatrix}.$$

Compared to the previous case where $\Gamma = \Sigma_{\epsilon\epsilon}^{-1}$ is taken as known, the above includes the additional term $\frac{1}{2}\delta\delta'$ with $\delta = (\alpha', -\beta')'$. However, it is also readily seen that the asymptotic covariance matrix, MWM', where $M = [I_m \otimes \beta, \alpha \otimes I_n]$, of the overall regression coefficient matrix estimator $\widehat{C} = \widehat{\alpha}\widehat{\beta}'$ is not affected by use of $\widetilde{\Sigma}_{\epsilon\epsilon}^{-1}$ instead of $\Sigma_{\epsilon\epsilon}^{-1}$, because $M\delta = (\alpha \otimes \beta) - (\alpha \otimes \beta) = 0$. Therefore, whereas the asymptotic distribution of the individual component estimators $\widehat{\alpha}$ and $\widehat{\beta}$ is affected by use of $\widetilde{\Sigma}_{\epsilon\epsilon}^{-1}$ in place of $\Sigma_{\epsilon\epsilon}^{-1}$, that of the overall estimator $\widehat{C} = \widehat{\alpha}\widehat{\beta}'$ is not changed.

2.6 Identification of Rank of the Regression Coefficient Matrix

It is obviously important to be able to determine the rank of the coefficient matrix C for it is a key element in the structure of the reduced-rank regression problem. The asymptotic results are derived assuming that the rank condition is true. However, the relationship between canonical correlation analysis and reduced-rank regression allows one to check on the rank by testing if certain correlations are zero.

Bartlett (1947) suggested $T \sum_{j=r+1}^{m} \log(1 + \widehat{\lambda}_j^2)$ as the appropriate statistic for testing the significance of the last $(m - r)$ canonical correlations, where $\widehat{\lambda}_j = \widehat{\rho}_j/(1 - \widehat{\rho}_j^2)^{1/2}$ and $\widehat{\rho}_j$ is the jth largest sample canonical correlation between the Y_k and X_k. From the discussion presented in Section 2.4, it follows that if the hypothesis that the last $(m - r)$ population canonical correlations are zero is true, then it is equivalent to assuming that $\text{rank}(C) = r$ (strictly, that $\text{rank}(C) \leq r$). Bartlett's statistic follows from the likelihood ratio method of test construction [see Anderson (1951)], as we now indicate.

Under the likelihood ratio (LR) testing approach, it is easy to see, similar to the results presented at the end of Section 1.3, that the LR test statistic for testing $\text{rank}(C) = r$ is $\lambda = U^{T/2}$, where $U = |S|/|S_1|$,

$$S = (\mathbf{Y} - \widetilde{C}\mathbf{X})(\mathbf{Y} - \widetilde{C}\mathbf{X})', \qquad S_1 = (\mathbf{Y} - \widehat{C}^{(r)}\mathbf{X})(\mathbf{Y} - \widehat{C}^{(r)}\mathbf{X})',$$

S is the residual sum of squares matrix from fitting the full-rank regression model, while S_1 is the residual sum of squares matrix from fitting the model under the hypothesis of the rank condition on C. It must be noted that $\widetilde{C}$ and $\widehat{C}^{(r)}$ are related as $\widehat{C}^{(r)} = P\widetilde{C}$, where P is an idempotent matrix of rank r. Specifically, recall from Sections 2.3 and 2.4 that the ML estimates under the null hypothesis of $\text{rank}(C) = r$ are $\widehat{A}^{(r)} = \widetilde{\Sigma}_{\epsilon\epsilon}^{1/2}\widehat{V}$, $\widehat{B}^{(r)} = \widehat{V}'\,\widetilde{\Sigma}_{\epsilon\epsilon}^{-1/2}\,\widehat{\Sigma}_{yx}\,\widehat{\Sigma}_{xx}^{-1}$, with $\widehat{C}^{(r)} = \widehat{A}^{(r)}\widehat{B}^{(r)}$, where $\widehat{V} = [\widehat{V}_1, \ldots, \widehat{V}_r]$ and the $\widehat{V}_j$ are (normalized) eigenvectors of the matrix $\widehat{R} = \widetilde{\Sigma}_{\epsilon\epsilon}^{-1/2}\,\widehat{\Sigma}_{yx}\,\widehat{\Sigma}_{xx}^{-1}\,\widehat{\Sigma}_{xy}\,\widetilde{\Sigma}_{\epsilon\epsilon}^{-1/2}$ associated with the r largest eigenvalues $\widehat{\lambda}_j^2$, and $\widehat{\Sigma}_{\epsilon\epsilon} = (1/T)S_1 = (1/T)(\mathbf{Y} - \widehat{C}^{(r)}\mathbf{X})(\mathbf{Y} - \widehat{C}^{(r)}\mathbf{X})'$. The unrestricted ML estimates are, of course, $\widetilde{C} = \widehat{\Sigma}_{yx}\,\widehat{\Sigma}_{xx}^{-1}$ and $\widetilde{\Sigma}_{\epsilon\epsilon} = (1/T)(\mathbf{Y} - \widetilde{C}\mathbf{X})(\mathbf{Y} - \widetilde{C}\mathbf{X})'$ as derived in Section 1.2. Notice that, since $\sum_{j=1}^{m} \widehat{V}_j\widehat{V}_j' = I_m$, we have

$$\begin{aligned}
\widetilde{C} - \widehat{C}^{(r)} &= (I_m - \widetilde{\Sigma}_{\epsilon\epsilon}^{1/2}\,\widehat{V}\,\widehat{V}'\,\widetilde{\Sigma}_{\epsilon\epsilon}^{-1/2})\,\widehat{\Sigma}_{yx}\,\widehat{\Sigma}_{xx}^{-1} \\
&= \widetilde{\Sigma}_{\epsilon\epsilon}^{1/2}\left[\sum_{j=r+1}^{m} \widehat{V}_j\widehat{V}_j'\right]\widetilde{\Sigma}_{\epsilon\epsilon}^{-1/2}\,\widehat{\Sigma}_{yx}\,\widehat{\Sigma}_{xx}^{-1}.
\end{aligned}$$

Therefore, we find that

$$\begin{aligned}
\frac{1}{T}S_1 &= \frac{1}{T}(\mathbf{Y} - \widetilde{C}\mathbf{X} + (\widetilde{C} - \widehat{C}^{(r)})\mathbf{X})(\mathbf{Y} - \widetilde{C}\mathbf{X} + (\widetilde{C} - \widehat{C}^{(r)})\mathbf{X})' \\
&= \widetilde{\Sigma}_{\epsilon\epsilon} + \widetilde{\Sigma}_{\epsilon\epsilon}^{1/2}\left[\sum_{j=r+1}^{m} \widehat{V}_j\widehat{V}_j'\right]\widehat{R}\left[\sum_{j=r+1}^{m} \widehat{V}_j\widehat{V}_j'\right]\widetilde{\Sigma}_{\epsilon\epsilon}^{1/2} \\
&= \widetilde{\Sigma}_{\epsilon\epsilon} + \widetilde{\Sigma}_{\epsilon\epsilon}^{1/2}\left[\sum_{j=r+1}^{m} \widehat{\lambda}_j^2\widehat{V}_j\widehat{V}_j'\right]\widetilde{\Sigma}_{\epsilon\epsilon}^{1/2}, \qquad (2.39)
\end{aligned}$$

where the $\widehat{\lambda}_j^2$, $j = r+1, \ldots, m$, are the $(m-r)$ smallest eigenvalues of $\widehat{R}$. Hence, using (2.39), the LR statistic is

$$\lambda = \left(\frac{|S_1|}{|S|}\right)^{-T/2} = \left(\frac{|(1/T)S_1|}{|\widetilde{\Sigma}_{\epsilon\epsilon}|}\right)^{-T/2}$$
$$= \left(\frac{|\widetilde{\Sigma}_{\epsilon\epsilon}||I_m + \sum_{j=r+1}^{m} \widehat{\lambda}_j^2 \widehat{V}_j \widehat{V}_j'|}{|\widetilde{\Sigma}_{\epsilon\epsilon}|}\right)^{-T/2} = \left[\prod_{j=r+1}^{m} (1+\widehat{\lambda}_j^2)\right]^{-T/2}.$$

Therefore, the criterion $\lambda = U^{T/2}$ is such that

$$-2\log(\lambda) = T \sum_{j=r+1}^{m} \log(1+\widehat{\lambda}_j^2) = -T \sum_{j=r+1}^{m} \log(1-\widehat{\rho}_j^2), \tag{2.40}$$

since $1+\widehat{\lambda}_j^2 = 1/(1-\widehat{\rho}_j^2)$. This statistic, $-2\log(\lambda)$, follows asymptotically the $\chi^2_{(m-r)(n-r)}$ distribution under the null hypothesis [see Anderson (1951, Theorem 3)]. The test statistic is asymptotically equivalent to $T\sum_{j=r+1}^{m} \widehat{\lambda}_j^2 \equiv T\,\mathrm{tr}\{\widetilde{\Sigma}_{\epsilon\epsilon}^{-1}(\widetilde{C} - \widehat{C}^{(r)})\,\widehat{\Sigma}_{xx}(\widetilde{C} - \widehat{C}^{(r)})'\}$, which follows the above χ^2 distribution and this result was obtained independently by Hsu (1941). A simple correction factor for the LR statistic in (2.40), to improve the approximation to the $\chi^2_{(m-r)(n-r)}$ distribution, is given by $-2\{[T-n+(n-m-1)/2]/T\}\log(\lambda) = -[T-n+(n-m-1)/2]\sum_{j=r+1}^{m} \log(1-\widehat{\rho}_j^2)$. This is similar to the corrected form used in (1.16) for the LR test in the classical full-rank model.

Thus, to specify the appropriate rank of the matrix C, we use the test statistic $\mathcal{M} = -[T-(m+n+1)/2]\sum_{j=r+1}^{m} \log(1-\widehat{\rho}_j^2)$, for $r = 0, 1, \ldots, m-1$, and reject $H_0\colon \mathrm{rank}(C) = r$ when $\mathcal{M}$ is greater than an upper critical value determined by the $\chi^2_{(m-r)(n-r)}$ distribution. The smallest value of r for which H_0 is not rejected provides a reasonable choice for the rank. Additional guidance in selecting the appropriate rank could be obtained by considering the LR test of $H_0\colon \mathrm{rank}(C) = r$ versus the more refined alternative hypothesis $H_1\colon \mathrm{rank}(C) = r+1$. This test statistic is given by $-T\log(1-\widehat{\rho}_{r+1}^2)$.

Alternative procedures for testing the rank of a matrix in more general settings have been proposed by Cragg and Donald (1996) based on the LDU decomposition of the unrestricted estimate of the coefficient matrix and by Kleibergen and Paap (2006) using the singular value decomposition of the coefficient matrix. Recall that the rank test proposed in (2.40) is based on the singular value decomposition of $\widetilde{\Sigma}_{\epsilon\epsilon}^{-1/2}\widetilde{C}\,\Sigma_{\epsilon\epsilon}^{1/2}$, a weighted coefficient matrix. The weights come from the covariance matrix of $\mathrm{vec}(\widetilde{C}')$, which has a Kronecker product structure. In many applications of reduced-rank models, such a structure is not possible; therefore, these tests can be applied to more general models. Other useful tools for the specification of the rank include the use of information-theoretic model selection criteria, such as the AIC criterion of Akaike (1974) and the BIC criterion of Schwarz (1978), or cross-validation methods (Stone 1974) based on measures of the predictive performance of models of various ranks.

The importance of estimating the rank correctly and its associated bias-variance tradeoff are explored in detail later in Section 10.4.3 in the book. Anderson (2002) considers the properties of the reduced-rank estimator when the rank is not specified correctly. If the specified rank is less than the true rank, the estimator is likely to be biased, and when it is higher, the estimator tends to have larger variance. Later in the book, we demonstrate how a few outliers can distort the true rank.

Remark: Relation of LR Tests Between Known and Unknown Constraints We briefly compare the results relating to the LR test for specified constraints on the regression coefficient matrix C (see Section 1.3) with the above LR test results for the rank of C, that is, for a specified number of constraints, but where these constraints are not known a priori. From Section 1.3, the LR test of $H_0\colon F_1 C = 0$, where F_1 is a specified (known) $(m-r) \times m$ full-rank matrix, is obtained as $\mathcal{M} = [(T-n)+\frac{1}{2}(n-m_1-1)]\sum_{i=1}^{m_1} \log(1+\widehat{\lambda}_i^2)$, with $m_1 = m-r$, where $\widehat{\lambda}_i^2$ are the $m-r$ eigenvalues of $S_*^{-1}H_* = (F_1 S F_1')^{-1} F_1 \widetilde{C}\,\mathbf{X}\mathbf{X}'\widetilde{C}' F_1' \equiv (F_1 \widetilde{\Sigma}_{\epsilon\epsilon} F_1')^{-1} F_1 \widetilde{C} \widehat{\Sigma}_{xx} \widetilde{C}' F_1'$. The asymptotic distribution of $\mathcal{M}$ under H_0 is $\chi^2_{(m-r)n}$. In the case where F_1 is unknown, $F_1 C = 0$ (for some F_1) is equivalent to the hypothesis that the rank of C is r. As has been shown in this section, the LR test for this case is given by $\mathcal{M} = [(T-n)+\frac{1}{2}(n-m-1)]\sum_{j=r+1}^{m} \log(1+\widehat{\lambda}_j^2)$, where $\widehat{\lambda}_j^2$ are the $m-r$ smallest eigenvalues of $\widehat{R} = \widetilde{\Sigma}_{\epsilon\epsilon}^{-1/2}\,\widetilde{C}\,\widehat{\Sigma}_{xx}\,\widetilde{C}'\,\widetilde{\Sigma}_{\epsilon\epsilon}^{-1/2}$, and the asymptotic distribution of $\mathcal{M}$ is $\chi^2_{(m-r)(n-r)}$. In addition, the corresponding ML estimate of the unknown constraints matrix $F_1 = [\ell_{r+1}, \ldots, \ell_m]'$ in $F_1 C = 0$ is given by $\widehat{F}_1 = [\widehat{V}_{r+1}, \ldots, \widehat{V}_m]'\,\widetilde{\Sigma}_{\epsilon\epsilon}^{-1/2}$, where $\widehat{V}_j$ are the normalized eigenvectors of $\widehat{R}$ associated with the $\widehat{\lambda}_j^2$. Hence we see a strong similarity in the form of the LR test statistic results, where in the case of unknown constraints the LR statistic can be viewed as taking the same form as in the known constraints case, but with the constraints estimated by ML to have a particular form. (Note that if the ML estimate $\widehat{F}_1$ is substituted for F_1 in the expression $S_*^{-1}H_*$ for the case of known F_1, this expression reduces to $[\widehat{V}_{r+1}, \ldots, \widehat{V}_m]'\,\widetilde{\Sigma}_{\epsilon\epsilon}^{-1/2}\,\widetilde{C}\,\widehat{\Sigma}_{xx}\,\widetilde{C}'\,\widetilde{\Sigma}_{\epsilon\epsilon}^{-1/2}\,[\widehat{V}_{r+1}, \ldots, \widehat{V}_m]$.) Because the constraints are estimated, the degrees of freedom of $(m-r)n$ in the known constraints case are adjusted to $(m-r)(n-r)$ in the case of unknown constraints, a decrease of $(m-r)r$. This reflects the increase in the number of unknown (unconstrained) parameters from rn in the known constraints case to $r(m+n-r)$ in the unknown constraints case.

The correspondence between correlation and regression is known to be fully exploited in the modeling of data in the case of multiple regression. The results above demonstrate the usefulness of the canonical correlations as descriptive statistics in aiding the multivariate regression modeling of large data sets. However, we shall demonstrate in the future chapters that such a neat connection does not always appear to hold in such an explicit way for more complex models, but nevertheless canonical correlation methods can still be used as descriptive tools and in model specification for these more complicated situations.

2.7 Reduced-Rank Inverse Regression for Estimating Structural Dimension

Bura and Cook (2001) consider a more general nonnormal distributional setting in which study of the conditional distribution of $\mathbf{Y}_k$ given $\mathbf{X}_k$ is of interest, and the conditional distribution is assumed to depend on $\mathbf{X}_k$ only through the r linear combinations $\mathbf{X}_k^* = \mathbf{B}\mathbf{X}_k$. The linear subspace generated by the rows of $\mathbf{B}$ is then referred to as a *dimension reduction subspace* for the conditional distribution of $\mathbf{Y}_k$ given $\mathbf{X}_K$, and r is called the *structural dimension*. Our previous discussions related to model (2.5) can be viewed as the more specialized situation that focuses on the conditional mean or regression function $E(\mathbf{Y}_k \mid \mathbf{X}_k)$, which is assumed to be a *linear* function of $\mathbf{X}_k^* = \mathbf{B}\mathbf{X}_k$. In the more general setting, if the dimension reduction subspace can be identified, then simplifications occur in the study of the conditional distribution of $\mathbf{Y}_k$ given $\mathbf{X}_k$, such as use of summary plots of $\mathbf{Y}_k$ versus $\mathbf{B}\mathbf{X}_k$ as sufficient graphical displays of all the necessary information.

Consider the standardized version of $\mathbf{X}_k$, $\mathbf{Z}_k = \Sigma_{xx}^{-1/2}[\mathbf{X}_k - E(\mathbf{X}_k)]$. Then Bura and Cook (2001) show that examination of the inverse regression function $E(\mathbf{Z}_k\mathbf{Y}_k)$ can be used to help determine the dimension and the basis vectors of the dimension reduction subspace. They use parametric multivariate linear regression to represent the inverse regression function as

$$E(\mathbf{Z}_k \mid \mathbf{Y}_k) = \mathbf{G}\mathbf{F}_k, \tag{2.41}$$

where $\mathbf{F}_k = [f_1(\mathbf{Y}_k), f_2(\mathbf{Y}_k), \ldots, f_q(\mathbf{Y}_k)]'$ is q-dimensional, $\mathbf{G}$ is $n \times q$, and the f_i are q arbitrary real-valued linearly independent known functions of $\mathbf{Y}_k$, e.g., such as powers of the components of $\mathbf{Y}_k$. If q is chosen sufficiently large so that the linear model gives a valid representation of $E(\mathbf{Z}_k \mid \mathbf{Y}_k)$ and the distribution of $\mathbf{X}_k$ satisfies a certain linearity condition, then the theory from dimension reduction subspaces indicates that the coefficient matrix $\mathbf{G}$ in the inverse regression (2.41) will have reduced rank $r^* \leq r$ and that $E(\mathbf{Z}_k \mid \mathbf{Y}_k) = \mathbf{G}\mathbf{F}_k$ belongs to the linear subspace generated by the r columns of $\Sigma_{xx}^{1/2}\mathbf{B}'$. This yields that $r^* = \text{rank}(\mathbf{G})$ provides a lower bound on the dimension of the dimension reduction subspace and that $\mathbf{G}$ has the reduced-rank factorization $\mathbf{G} = \Sigma_{xx}^{1/2}\mathbf{B}'\mathbf{D} \equiv \mathbf{A}_*\mathbf{D}$ for an $r \times q$ matrix $\mathbf{D}$.

Therefore, it follows that the usual reduced-rank estimation and testing procedures can be applied to the inverse regression model (2.41) to determine the structural dimension r and the basis vectors (rows) of $\mathbf{B}$. Specifically, define the data matrices $\mathbf{F} = [\mathbf{F}_1, \ldots, \mathbf{F}_T]$ and $\mathbf{Z} = [\widetilde{\mathbf{Z}}_1, \ldots, \widetilde{\mathbf{Z}}_T]$, where $\widetilde{\mathbf{Z}}_k = \widehat{\Sigma}_{xx}^{-1/2}(\mathbf{X}_k - \overline{\mathbf{X}})$, $k = 1, \ldots, T$. Then let

$$\widetilde{\mathbf{G}} = \mathbf{Z}\mathbf{F}'(\mathbf{F}\mathbf{F}')^{-1} \equiv \widehat{\boldsymbol{\Sigma}}_{zf}\widetilde{\boldsymbol{\Sigma}}_{ff}^{-1} \quad \text{and} \quad \widetilde{\boldsymbol{\Sigma}}_{z|y} = \widetilde{\boldsymbol{\Sigma}}_{zz} - \widetilde{\boldsymbol{\Sigma}}_{zf}\widetilde{\boldsymbol{\Sigma}}_{ff}^{-1}\widetilde{\boldsymbol{\Sigma}}_{fz}$$

be the usual LS estimate of $\mathbf{G}$ and corresponding residual covariance matrix associated with the inverse linear regression model (2.41). Also, note that $\widehat{\Sigma}_{xx}^{1/2}\widehat{\Sigma}_{zf} = \widehat{\Sigma}_{xf}$

and that $\widehat{\Sigma}_{xx}^{1/2} \widehat{\Sigma}_{z|y} \widehat{\Sigma}_{xx}^{1/2} \equiv \widehat{\Sigma}_{x|y}$, the residual covariance matrix from the LS regression of $\mathbf{X}_k$ on $\mathbf{F}_k$. Then as in the usual procedure, we consider the eigenvalues $\widehat{\lambda}_j^2$ and (normalized) eigenvector $\widehat{\mathbf{V}}_j$ of the matrix

$$\widehat{\mathbf{R}} = \widehat{\Sigma}_{z|y}^{-1/2} \widehat{\Sigma}_{zf} \widehat{\Sigma}_{ff}^{-1} \widehat{\Sigma}_{fz} \widehat{\Sigma}_{z|y}^{-1/2} \equiv \widehat{\Sigma}_{x|y}^{-1/2} \widehat{\Sigma}_{xf} \widehat{\Sigma}_{ff}^{-1} \widehat{\Sigma}_{fx} \widehat{\Sigma}_{x|y}^{-1/2}. \quad (2.42)$$

Using a slightly different motivation, Bura and Cook (2001) suggested use of the "trace" statistic $T \sum_{j=r+1}^{\min(n,q)} \widehat{\lambda}_j^2$ to test the reduced-rank hypothesis that $\text{rank}(\mathbf{G}) = r$, which is asymptotically distributed as $\chi^2_{(n-r)(q-r)}$. From previous discussions, however, this is asymptotically equivalent to the Gaussian-based LR test statistic $T \sum_{j=r+1}^{\min(n,q)} \log(1 + \widehat{\lambda}_j^2) = -T \sum_{j=r+1}^{\min(n,q)} \log(1 - \widehat{\rho}_j^2)$. Furthermore, as analogous to results in Section 2.3, the reduced rank estimate of the factor $\mathbf{A}_* = \Sigma_{xx}^{1/2} \mathbf{B}'$ in $\mathbf{G} = \mathbf{A}_* \mathbf{D}$ is given by $\widehat{\mathbf{A}}_* = \widehat{\Sigma}_{z|y}^{1/2} \widehat{\mathbf{V}}_{(r)}$, and hence an estimate of $\mathbf{B}'$ is provided by $\widehat{\mathbf{B}}' = \widehat{\Sigma}_{xx}^{-1/2} \widehat{\Sigma}_{z|y}^{1/2} \widehat{\mathbf{V}}_{(r)} \equiv \widehat{\Sigma}_{xx}^{-1} \widehat{\mathbf{\Sigma}}_{x|y}^{1/2} \widehat{\mathbf{V}}_{(r)}$, where $\widehat{\mathbf{V}}_{(r)} = [\widehat{\mathbf{V}}_1, \ldots, \widehat{\mathbf{V}}_r]$. Thus, the required information is provided through use of the usual reduced-rank procedures applied to the inverse regression model.

2.8 Numerical Examples

We illustrate the reduced-rank regression methods developed in this chapter using the example on biochemical data and the sales performance data that were examined in Section 1.5.

Biochemical Data As indicated in the analysis presented in Section 1.5.1, these data suggest that there may be a reduced-rank structure for the regression coefficient matrix (excluding the constant term) of the predictor variables $X_k = (x_{1k}, x_{2k}, x_{3k})'$. To examine this feature, the rank of the coefficient matrix first needs to be determined. The LR test statistic $\mathcal{M} = -[(T-1) - (m+n+1)/2] \sum_{j=r+1}^{m} \log(1 - \widehat{\rho}_j^2)$ with $n = 3$, discussed in relation to (2.40), is used for $r = 0, 1, 2$, to test the rank, that is, to test H_0: $\text{rank}(C) \leq r$, and the results are given in Table 2.2. It would appear from these results that the possibility that the rank is either one or two could be entertained. Only estimation results for the rank 2 situation will be presented in detail here.

Table 2.2 Summary of results for LR tests on rank of coefficient matrix for biochemical data

Eigenvalues $\widehat{\lambda}_j^2$	Canonical Correlations $\widehat{\rho}_j$	Rank r	$\mathcal{M} = \text{LR}$ Statistic	d.f.	Critical Value (5%)
4.121	0.897		56.89	15	24.99
0.519	0.584	1	11.98	8	15.51
0.018	0.132	2	0.49	3	7.81

To obtain the maximum likelihood estimates, we find the normalized eigenvectors of the matrix $\widehat{R} = \widetilde{\Sigma}_{\epsilon\epsilon}^{-1/2} \widehat{\Sigma}_{yx} \widehat{\Sigma}_{xx}^{-1} \widehat{\Sigma}_{xy} \widetilde{\Sigma}_{\epsilon\epsilon}^{-1/2}$ associated with the two largest eigenvalues, $\widehat{\lambda}_1^2$ and $\widehat{\lambda}_2^2$, in Table 2.2. (Note that in constructing the sample covariance matrices $\widehat{\Sigma}_{xx}$ and $\widehat{\Sigma}_{xy}$, the variables Y_k and X_k are adjusted for sample means.) The normalized eigenvectors are

$$\widehat{V}_1' = (-0.290,\ 0.234,\ 0.025,\ 0.906,\ 0.199)$$

and

$$\widehat{V}_2' = (0.425,\ 0.269,\ -0.841,\ 0.047,\ 0.194),$$

respectively. We then compute the ML estimates from $\widehat{A}^{(2)} = \widetilde{\Sigma}_{\epsilon\epsilon}^{1/2} \widehat{V}_{(2)}$ and $\widehat{B}^{(2)} = \widehat{V}_{(2)}' \widetilde{\Sigma}_{\epsilon\epsilon}^{-1/2} \widehat{\Sigma}_{yx} \widehat{\Sigma}_{xx}^{-1}$, with $\widehat{V}_{(2)} = [\widehat{V}_1, \widehat{V}_2]$. Using the results of Theorem 2.4, the asymptotic standard errors of the elements of these matrix estimates are also obtained. These ML estimates and associated standard errors that follow from results in Section 2.5 are given below as

$$\widehat{A}' = \begin{pmatrix} -1.085 & 0.333 & 0.231 & 0.397 & 1.109 \\ (0.265) & (0.042) & (0.040) & (0.013) & (0.435) \\ & & & & \\ 1.335 & -0.126 & -0.266 & -0.092 & 0.875 \\ (0.655) & (0.120) & (0.085) & (0.049) & (1.246) \end{pmatrix},$$

$$\widehat{B} = \begin{pmatrix} 1.361 & -1.326 & 1.190 \\ (0.352) & (0.222) & (0.405) \\ & & \\ -0.908 & 0.494 & 1.240 \\ (0.302) & (0.179) & (0.337) \end{pmatrix}.$$

To appreciate the recovery of information by using only two linear combinations $\widehat{B} X_k$ of $X_k = (x_{1k}, x_{2k}, x_{3k})'$ to predict Y_k, the reduced-rank matrix $\widehat{C} = \widehat{A}\widehat{B}$ can be compared with the full-rank LS estimate $\widetilde{C}$ given in Section 1.5. Notice again that we are excluding the first column of intercepts from this comparison. The reduced-rank estimate of the regression coefficient matrix is

$$\widehat{C} = \begin{pmatrix} -2.6893 & 2.0981 & 0.3649 \\ 0.5679 & -0.5040 & 0.2400 \\ 0.5558 & -0.4374 & -0.0555 \\ 0.6248 & -0.5726 & 0.3583 \\ 0.7142 & -1.0379 & 2.4044 \end{pmatrix}.$$

Most of the coefficients of $\widetilde{C}$ that were found to be statistically significant are recovered in the reduced-rank estimate $\widehat{C}$ fairly well. The (approximate) unbiased estimate of the error covariance matrix from the reduced-rank analysis (with correlations shown above the diagonal) is given as

$$\widetilde{\Sigma}_{\epsilon\epsilon} = \frac{1}{33-3}\widehat{\epsilon}\,\widehat{\epsilon}' = \begin{pmatrix} 10.8813 & -0.4243 & -0.3090 & -0.3243 & -0.0841 \\ -0.8638 & 0.3808 & 0.8952 & 0.6743 & 0.3452 \\ -0.5204 & 0.2820 & 0.2606 & 0.6592 & 0.2912 \\ -0.4743 & 0.1845 & 0.1492 & 0.1966 & 0.0320 \\ -1.5012 & 1.1527 & 0.8045 & 0.0768 & 29.2890 \end{pmatrix},$$

where the residuals are obtained from the LS regression of Y_k on $\widehat{B}X_k$ with intercept included. The diagonal elements of the above matrix, when compared to the entries of the unbiased estimate of the error covariance matrix under the full-rank LS regression (given in Section 1.5), indicate that there is little or no loss in fit after discarding one linear combination of the predictor variables.

For comparison purposes, corresponding ML results when the rank of the coefficient matrix is assumed to be one can be obtained by using as regressor only the (single) linear combination of predictor variables formed from the first row of $\widehat{B}$. The diagonal elements of the resulting (approximate) unbiased estimate of the error covariance matrix from the rank one fit are given by 11.5149, 0.3773, 0.2913, 0.1949, and 28.7673. These values are very close to the residual variance estimates of both the rank 2 and full-rank models, suggesting that the rank one model could also provide an acceptable fit. Note that the single index or predictive factor in the estimated rank one model is roughly proportional to $x_1^* \approx (x_1 + x_3) - x_2$. Although a physical or scientific interpretation for this index is unclear, it does have statistical predictive importance. For illustration, Fig. 2.3 displays scatter plots of each of the response variables y_i versus the first predictive index $x_1^* = \widehat{\beta}_1' X$, with estimated regression line. It is seen that variables y_2, y_3, and y_4 have quite strong relationship with x_1^*, whereas the relationship with x_1^* is much weaker for y_5 and is of opposite sign for y_1.

Sales Performance Data The squared canonical correlations between the performance and the test scores of the sales staff are 0.9945, 0.8781, and 0.3836. The LR test to test for the rank$(C) \leq r$ is $M = -45\sum_{j=r+1}^{3}\log(1-\widehat{\rho}_j^2)$, which results in values, 276.4, 73.5, and 7.2, respectively, for $r = 0, 1, 2$. The chi-square critical values at 5% are 21.0, 12.6, and 5.99 for 12, 6, and 2 d.f. This clearly suggests that the rank of C matrix can be taken as two. The normalized eigenvectors are the columns of the matrix,

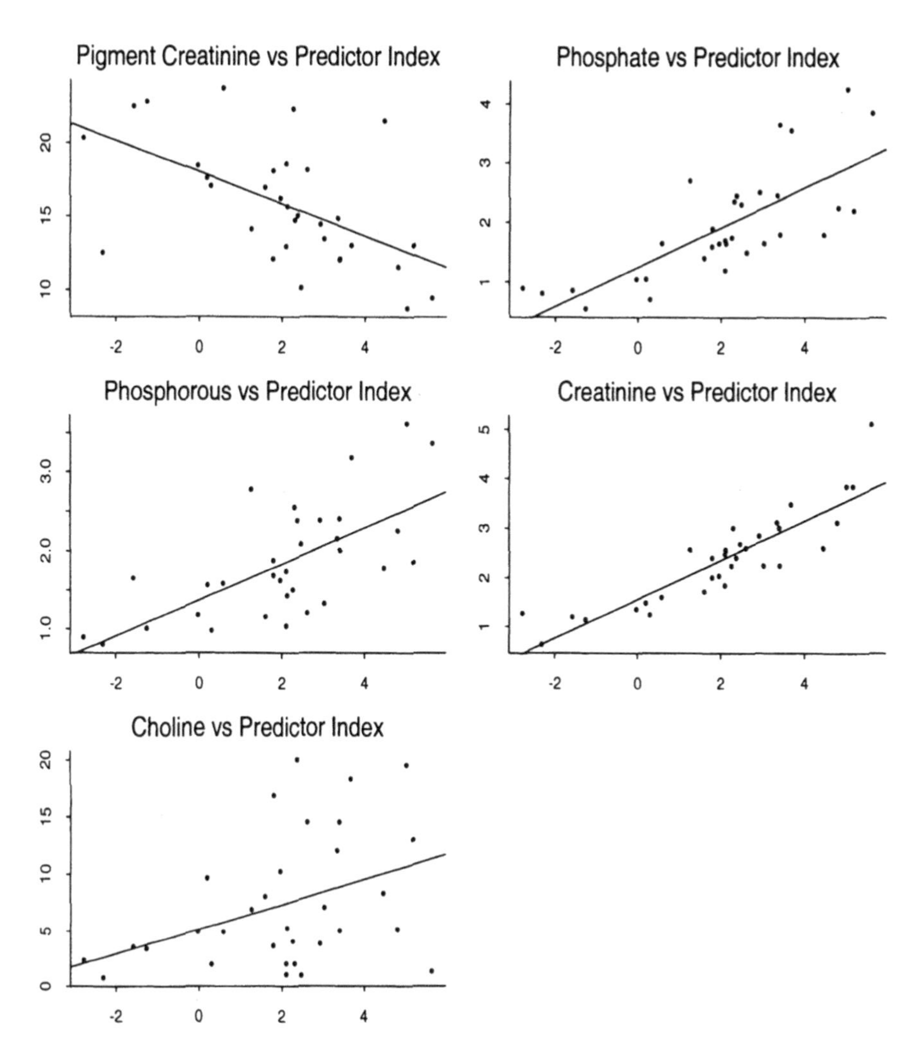

Fig. 2.3 Scatter plots of response variables $y_1, \ldots, y_5$ for biochemical data versus first predictive variable index x_1^*

$$\widehat{V} = \begin{pmatrix} 0.483 & 0.260 & -0.807 & -0.220 \\ 0.340 & -0.469 & 0.263 & -0.771 \\ 0.437 & 0.731 & 0.518 & -0.075 \\ 0.678 & -0.421 & 0.109 & 0.593 \end{pmatrix}.$$

The $\widehat{A}$ and $\widehat{B}$ component matrices when $\text{rank}(C) = 2$ follow:

$$\widehat{A}^{(2)} = \widetilde{\Sigma}_{\epsilon\epsilon}^{1/2}[\widehat{V}_1, \widehat{V}_2] = \begin{pmatrix} 0.262 & 0.404 \\ (0.041) & (0.222) \\ 0.253 & -0.381 \\ (0.033) & (0.168) \\ 0.144 & 0.508 \\ (0.019) & (0.094) \\ 1.033 & -0.986 \\ (0.044) & (0.213) \end{pmatrix},$$

$$\widehat{B}^{(2)} = \begin{pmatrix} \widehat{V}_1' \\ \widehat{V}_2' \end{pmatrix} \widetilde{\Sigma}_{\epsilon\epsilon}^{-1/2} \widetilde{C} = \begin{pmatrix} 0.597 & 0.200 & 0.749 \\ (0.061) & (0.040) & (0.067) \\ 0.323 & -0.449 & 0.442 \\ (0.059) & (0.038) & (0.065) \end{pmatrix}.$$

The product $\widehat{C}^{(2)} = \widehat{A}^{(2)} B^{(2)}$, the reduced rank estimate of C, is given as

$$\widehat{C}^{(2)} = \begin{pmatrix} 0.287 & -0.129 & 0.375 \\ 0.029 & 0.222 & 0.022 \\ 0.250 & -0.199 & 0.332 \\ 0.299 & 0.650 & 0.339 \end{pmatrix}.$$

It must be noted that the significant elements of the $\widetilde{C}$ matrix are recovered well with the rank two approximation. Note that the complementary version of the reduced-rank problem is estimating the linear constraint "l" so that $l'C = 0$; such a constraint implies $l'Y_k \sim l'e_k$, which is uncorrelated to the predictors. Note that $\widehat{l}_2' = \widehat{V}_3' \widetilde{\Sigma}_{\epsilon\epsilon}^{-1/2} = (-0.2697, 0.1552, 0.3065, -0.0124)$ and $\widehat{l}_1' = \widehat{V}_4' \widetilde{\Sigma}_{\epsilon\epsilon}^{-1/2} = (-0.0189, -0.3368, -0.0534, 0.0949)$. In general, the reasoning-related variables seem to contrast with other tests. The scatterplots of $Y_k^* = \widehat{l}' Y_k$ versus the predictor variables X_k are given in Fig. 2.4. The plots clearly indicate how these combinations are not related to the predictors.

2.9 Alternate Procedures for Analysis of Multivariate Regression Models

Although we mainly focus on the reduced-rank multivariate regression models in this book, it is helpful to briefly survey and discuss other related multivariate regression modeling methodologies that have similar parameter reduction objectives as reduced-rank regression, such as multivariate ridge regression, partial least squares, joint continuum regression, envelope models, and other shrinkage and regularization techniques. Some of these procedures are designed particularly for

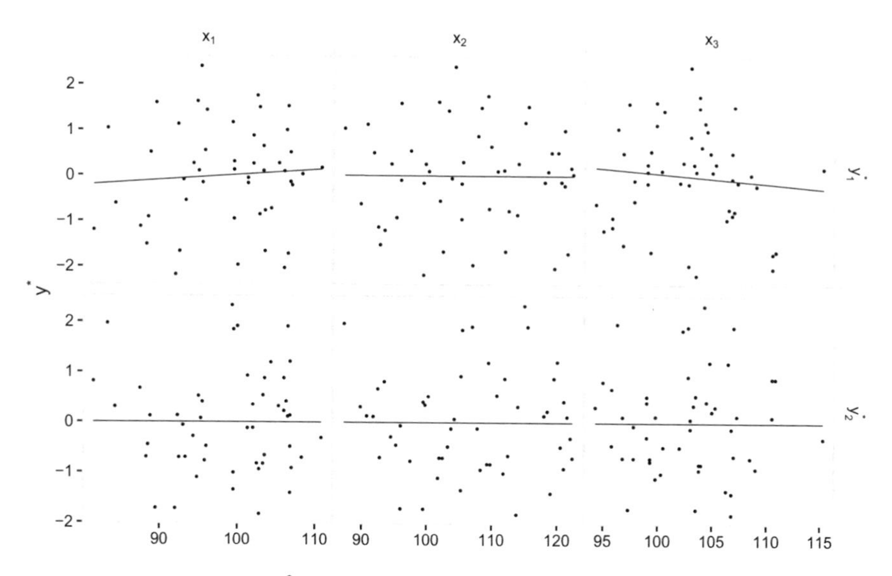

Fig. 2.4 Scatterplot of $Y^* = \widehat{l}'Y$ versus the predictor variables X

situations where there is a very large number of responses or predictors relative to the sample size. This situation where $m, n > T$ is addressed in the later chapters in the context of reduced-rank regression and the big data, as well.

Multivariate Ridge Regression and Other Shrinkage Estimates

Consider the standard multivariate linear regression model as in (1.4), $\mathbf{Y} = C\mathbf{X} + \boldsymbol{\epsilon}$, where $\mathbf{Y} = [Y_1, \ldots, Y_T]$ is $m \times T$, $\mathbf{X} = [X_1, \ldots, X_T]$ is $n \times T$, $C = [C_{(1)}, \ldots, C_{(m)}]'$ is $m \times n$, $\boldsymbol{\epsilon} = [\epsilon_1, \ldots, \epsilon_T]$, and the ϵ_k are mutually independent with $E(\epsilon_k) = 0$ and $\text{Cov}(\epsilon_k) = \Sigma_{\epsilon\epsilon}$. The ordinary LS estimator of C is $\widetilde{C} = \mathbf{Y}\mathbf{X}'(\mathbf{X}\mathbf{X}')^{-1}$, which can also be expressed in vector form as in (1.14),

$$\widetilde{c} = \text{vec}(\widetilde{C}') = (I_m \otimes (\mathbf{X}\mathbf{X}')^{-1}\mathbf{X})\,\text{vec}(\mathbf{Y}'). \tag{2.43}$$

As discussed in the introduction, Sclove (1971), among others, has argued that a deficiency of the LS estimator $\widetilde{C}$ is that it takes no account of $\Sigma_{\epsilon\epsilon}$, the error covariance matrix. To remedy this shortcoming and also to take advantage of possible improvements in the estimation of the individual $C_{(i)}$ even for the case where $\Sigma_{\epsilon\epsilon} = \sigma^2 I_m$ by using (univariate) ridge regression for that case, Brown and Zidek (1980, 1982), Haitovsky (1987), and others suggested multivariate ridge regression estimators of $c = \text{vec}(C')$.

The ridge regression estimator of c proposed by Brown and Zidek (1980) is of the form

$$\widehat{c}^*(K) = (I_m \otimes \mathbf{XX}' + K \otimes I_n)^{-1}(I_m \otimes \mathbf{X})\,\text{vec}(\mathbf{Y}')$$
$$= (I_m \otimes \mathbf{XX}' + K \otimes I_n)^{-1}(I_m \otimes \mathbf{XX}')\,\widetilde{c}, \tag{2.44}$$

where K is the $m \times m$ ridge matrix. Brown and Zidek (1980) motivated (2.44) as a Bayes estimator, with prior distribution for c such that $c \sim N(0, V \otimes I_n)$, where V is an $m \times m$ covariance matrix, with $K = \Sigma_{\epsilon\epsilon} V^{-1}$ in (2.44) (and $\Sigma_{\epsilon\epsilon}$ is treated as known). By letting the choice of the ridge matrix K depend on the data $\mathbf{Y}$ in certain ways, adaptive multivariate ridge estimators are obtained. Mean square error properties of these adaptive and empirical Bayes estimators are compared, and their dominance over the ordinary LS estimator (i.e., $K = 0$) is considered by Brown and Zidek (1980).

Haitovsky (1987) considered a slightly more general class of multivariate ridge estimators, developed by using a more general prior distribution for the regression coefficients c of the form $c \sim N(\mathbf{1}_m \otimes \xi, V \otimes \Omega)$, where $V > 0$ is the $m \times m$ covariance matrix, except for scale factor, between any two columns of C, and $\Omega > 0$ is the $n \times n$ covariance matrix, except for scale factor, between any two rows (i.e., $C'_{(i)}$ and $C'_{(j)}$) in C. This leads to multivariate ridge estimators of the form

$$\widehat{c}^* = (I_m \otimes \mathbf{XX}' + K_w \otimes \Omega^{-1})^{-1}(I_m \otimes \mathbf{X})\,\text{vec}(\mathbf{Y}')$$
$$= (I_m \otimes \mathbf{XX}' + K_w \otimes \Omega^{-1})^{-1}(I_m \otimes \mathbf{XX}')\,\widetilde{c}, \tag{2.45}$$

where $K_w = \Sigma_{\epsilon\epsilon} V^{-1} W$ and $W = I_m - (\mathbf{1}'_m V^{-1} \mathbf{1}_m)^{-1} \mathbf{1}_m \mathbf{1}'_m V^{-1}$, and the error covariance matrix $\Sigma_{\epsilon\epsilon}$ is assumed known. Estimation for the more realistic situation where $\Sigma_{\epsilon\epsilon}$ (and the prior covariance matrices V and Ω) are unknown was also discussed by Haitovsky (1987), using multivariate mixed-effects and covariance components model ML estimation methods. (The situation of unknown covariance matrix was also considered by Brown and Zidek (1982), for their class of ridge estimators.) Brown and Zidek (1980) and Haitovsky (1987) both noted particular multivariate shrinkage estimators, such as those of Efron and Morris (1972), result as special cases of their classes of multivariate ridge estimators. These estimators do not seem to have been used extensively in practical applications, although Brown and Payne (1975) used a specialized form of the estimators effectively in "election night forecasting."

Various other forms of estimators of the regression coefficient matrix C that "shrink" the usual LS estimator $\widetilde{C}$, in the spirit of Stein and Efron–Morris type shrinkage, were considered by Bilodeau and Kariya (1989), Honda (1991), and Konno (1991), among others. These authors examined the risk properties of the proposed classes of shrinkage estimators, and for certain risk functions they presented sufficient conditions for these alternate estimators to have improved risk relative to the LS (minimax) estimator.

Canonical Analysis Multivariate Shrinkage Methods
Different shrinkage methods for the multivariate linear regression model based on classical canonical correlation analysis between Y_k and X_k were proposed by van der Merwe and Zidek (1980) and Breiman and Friedman (1997). In some respects, these methods may be viewed as generalizations of reduced-rank regression modeling. Recall, as shown in Sections 2.3 and 2.4, that the ML reduced-rank estimator of the regression coefficient matrix C, of rank $r < \min(m, n)$, can be expressed in terms of the canonical correlation analysis between Y_k and X_k. Specifically, the ML reduced-rank estimator can be written as

$$\widehat{C}^{(r)} = \widehat{A}_*^{(r)}\widehat{B}_*^{(r)} = \widehat{\Sigma}_{yy}^{1/2}\widehat{V}_*^{(r)}\widehat{V}_*^{(r)\prime}\widehat{\Sigma}_{yy}^{-1/2}\widetilde{C}, \quad (2.46)$$

where $\widetilde{C} = \mathbf{YX}'(\mathbf{XX}')^{-1} \equiv \widehat{\Sigma}_{yx}\widehat{\Sigma}_{xx}^{-1}$ is the ordinary LS estimator of C, $\widehat{V}_*^{(r)} = [\widehat{V}_1^*, \ldots, \widehat{V}_r^*]$, and the $\widehat{V}_j^*$ are normalized eigenvectors of the matrix $\widehat{\Sigma}_{yy}^{-1/2}\widehat{\Sigma}_{yx}\widehat{\Sigma}_{xx}^{-1}\widehat{\Sigma}_{xy}\widehat{\Sigma}_{yy}^{-1/2}$ associated with its r largest eigenvalues $\widehat{\rho}_j^2$, the squared sample canonical correlations between the Y_k and X_k. Letting $\widehat{V}_* = [\widehat{V}_1^*, \ldots, \widehat{V}_m^*]$, an $m \times m$ orthogonal matrix, $\widehat{H} = \widehat{V}_*'\widehat{\Sigma}_{yy}^{-1/2}$, and $\Delta_* = \text{Diag}(1, \ldots, 1, 0, \ldots, 0)$, an $m \times m$ diagonal matrix with ones in the first r diagonals and zeros otherwise, the ML reduced-rank estimator in (2.46) can be expressed in the form

$$\widehat{C}^{(r)} = \widehat{H}^{-1}\Delta_*\widehat{H}\widetilde{C}. \quad (2.47)$$

As we have discussed previously in Sections 1.2 and 2.4, in practical application, the predictor variables X_k and response variables Y_k will typically already be centered and be expressed in terms of deviations from sample mean vectors before the canonical correlation and least squares calculations are performed.

In the canonical analysis-based multivariate shrinkage methods, the estimator of C is formed as in (2.47) but with the diagonal matrix Δ_* replaced by a diagonal matrix $\widehat{\Delta}$ of "shrinkage factors" $\{\widehat{d}_j\}$. For example, in the generalized cross-validation (GCV)-based approach of Breiman and Friedman (1997), the diagonal elements of $\widehat{\Delta}$ used are given by

$$\widehat{d}_j = \frac{(1-p)(\widehat{\rho}_j^2 - p)}{(1-p)^2\widehat{\rho}_j^2 + p^2(1-\widehat{\rho}_j^2)}, \qquad j = 1, \ldots, m, \quad (2.48)$$

where $p = n/T$ is the ratio of the number n of predictor variables to the sample size T. In practice, the positive-part shrinkage rule is used in that $\widehat{d}_j$ is replaced by 0 whenever $\widehat{d}_j < 0$ (i.e., $\widehat{\rho}_j^2 < n/T$) in (2.48). The form of the $\widehat{d}_j$ in (2.48) is derived by Breiman and Friedman (1997) so as to minimize the generalized cross-validation prediction mean square error. The diagonal elements $\widehat{d}_j$ may also be determined by a fully cross-validatory procedure. In the method proposed earlier by van der Merwe and Zidek (1980), the diagonal elements used for the matrix $\widehat{\Delta}$ in place of Δ_* in (2.47) are

$$f_j = \left(\widehat{\rho}_j^2 - \frac{n-m-1}{T}\right) / \widehat{\rho}_j^2 \left(1 - \frac{n-m-1}{T}\right),$$

with the positive-part rule also employed. Breiman and Friedman (1997) also presented and discussed generalizations of their procedure, which combine the canonical analysis shrinkage methods with principal components regression and (univariate) ridge regression, especially useful for situations involving underdetermined ($n > T$) or poorly determined ($n < T$ but large relative to T, e.g., $n \approx T$) samples. An alternate shrinkage procedure can also be motivated by the extended reduced-rank regression model in Chapter 3 and this procedure will be discussed explicitly therein.

Partial Least Squares

Partial least squares was introduced by Wold (1975) as an iterative computational algorithm for linear regression modeling, especially for situations where the number n of predictor variables is large. The multiple response ($m > 1$) version of partial least squares begins with a "canonical covariance" analysis, a sequential procedure to construct predictor variables as linear combinations of the original n-dimensional set of predictor variables X_k. At the first stage, the initial predictor variable $x_{1k}^* = b_1' X_k$ is determined as the first "canonical covariance" variate of the X_k, the linear combination of X_k that maximizes

$$\{(T-1) \times \text{ sample covariance of } b'X_k \text{ and } \ell' Y_k\} \equiv b' \mathbf{X} \mathbf{Y}' \ell$$

over unit vectors b (and over all associated unit vectors ℓ). For given vector b, we know that $\ell = \mathbf{Y}\mathbf{X}'b/(b'\mathbf{X}\mathbf{Y}'\mathbf{Y}\mathbf{X}'b)^{1/2}$ is the maximizing choice of unit vector for ℓ, so that we then need to maximize $b'\mathbf{X}\mathbf{Y}'\mathbf{Y}\mathbf{X}'b$ with respect to unit vectors b. Thus, the first linear combination vector b_1 is the (normalized) eigenvector of $\mathbf{X}\mathbf{Y}'\mathbf{Y}\mathbf{X}'$ corresponding to its largest eigenvalue. (In the univariate case, $m = 1$ and $\mathbf{Y}$ is $1 \times T$ and so this reduces to $b_1 = \mathbf{X}\mathbf{Y}'/(\mathbf{Y}\mathbf{X}'\mathbf{X}\mathbf{Y}')^{1/2}$ with associated "eigenvalue" equal to $\mathbf{Y}\mathbf{X}'\mathbf{X}\mathbf{Y}'$.)

At any subsequent stage j of the procedure, there are j predictor variables $x_{ik}^* = b_i' X_k$, $i = 1, \ldots, j$, determined and the next predictor variable $x_{j+1,k}^* = b_{j+1}' X_k$, that is, the next unit vector b_{j+1} needs to be found. It is to be chosen as the vector b to maximize the sample covariance of $b'X_k$ and $\ell' Y_k$ (with unit vector ℓ also chosen to maximize the quantity), that is, we maximize $b'\mathbf{X}\mathbf{Y}'\mathbf{Y}\mathbf{X}'b$, subject to b_{j+1} being a unit vector and $x_{j+1,k}^* = b_{j+1}' X_k$ being orthogonal to the previous set of j predictor variables x_{ik}^*, $i = 1, \ldots, j$ (i.e., $b_{j+1}'\mathbf{X}\mathbf{X}'b_i = 0$, for $i = 1, \ldots, j$). The number w of predictors chosen until the sequential process is stopped is a regularization parameter of the procedure; its value is determined through cross-validation methods in terms of prediction mean square error. The final estimated regression equations are constructed by ordinary LS calculation of the response vectors Y_k on the first w "canonical covariance" component predictor variables, $X_k^* = (x_{1k}^*, \ldots, x_{wk}^*)' \equiv \widehat{B} X_k$, where $\widehat{B} = [b_1, \ldots, b_w]'$, that is, the estimated regression equations are $\widehat{Y}_k = \widehat{A} X_k^* \equiv \widehat{A}\widehat{B} X_k$ with

$$\widehat{A} = \mathbf{Y}\mathbf{X}^{*\prime}(\mathbf{X}^*\mathbf{X}^{*\prime})^{-1} \equiv \mathbf{Y}\mathbf{X}'\widehat{B}'(\widehat{B}\mathbf{X}\mathbf{X}'\widehat{B}')^{-1}$$

and $\mathbf{X}^* = \widehat{B}\mathbf{X}$ is $w \times T$. In practice, the predictor variables X_k and response variables Y_k will typically already be expressed in terms of deviations from sample mean vectors, as we have discussed initially in Section 1.2.

Discussion of the sequential computational algorithms for the construction of the "canonical covariance" components $x_{jk}^* = b_j' X_k$ required in the partial least squares regression procedure and of various formulations and interpretations of the partial least squares method is given by Wold (1984), Helland (1988, 1990), Hoskuldsson (1988), Stone and Brooks (1990), Frank and Friedman (1993), and Garthwaite (1994). In particular, for univariate partial least squares, Garthwaite (1994) gives the interpretation of the linear combinations of predictor variables, $x_{jk}^* = b_j' X_k$, as weighted averages of predictors, where each individual predictor holds the residual information in an explanatory variable that is not contained in earlier linear combination components (x_{ik}^*, $i < j$), and the quantity to be predicted is the $1 \times T$ vector of residuals obtained from regressing y_k on the earlier components. Also, for the univariate case, Helland (1988, 1990) and Stone and Brooks (1990) showed that the linear combination vectors $\widehat{B} = [b_1, \ldots, b_w]'$ are obtainable from nonsingular $(w \times w)$ linear transformation of $[\mathbf{X}\mathbf{Y}', (\mathbf{X}\mathbf{X}')\mathbf{X}\mathbf{Y}', \ldots, (\mathbf{X}\mathbf{X}')^{w-1}\mathbf{X}\mathbf{Y}']'$.

Joint Continuum Regression

Continuum regression, proposed by Stone and Brooks (1990) for the univariate response case and extended by Brooks and Stone (1994) to the multivariate case, may be viewed as a generalization of the method of partial least squares regression modeling and also of principal components regression. Like partial least squares, it is a sequential construction method, involving cross-validatory choice of certain control parameters. At a given stage j, j predictor variables $x_{ik}^* = b_i' X_k$, $i = 1, \ldots, j$, have been determined and the next predictor $x_{j+1,k}^* = b_{j+1}' X_k$, that is, the next linear combination vector b_{j+1}, is chosen to maximize a certain criterion T_α (defined below) that expresses the potentiality of $x_{j+1,k}^*$ for predicting a similarly maximized linear combination of the response vector Y_k, subject to b_{j+1} being of unit length and $\{x_{j+1,k}^*\}$ being orthogonal to $\{x_{1k}^*, \ldots, x_{jk}^*\}$. The choice of the number w of predictor variables $x_{1k}^*, \ldots, x_{wk}^*$ and of the parameter α involved in the criterion T_α can be made on the basis of cross-validatory indices constructed from LS regression of Y_k on $X_k^* = (x_{1k}^*, \ldots, x_{wk}^*)'$. As in partial least squares, the predictor variables X_k and response variables Y_k may already be expressed in terms of deviations from sample mean vectors.

For $0 \leq \alpha < 1$, the form of the criterion T_α to determine the vector b at a given stage is

$$T_\alpha = \max_{\|\ell\|=1} \{(T-1) \times \text{ sample covariance of } b'X_k \text{ and } \ell'Y_k\}^2$$
$$\times\, (b'\mathbf{XX}'b)^{\{\alpha/(1-\alpha)\}-1}. \tag{2.49}$$

The first factor in (2.49) is an indicator of the predictive ability of the linear combination $b'X_k$ for the Y_k, and the second factor is a power of the sum of squares of the $b'X_k$ relevant to principal components analysis of the X_k. The maximization (with respect to ℓ for given vector b) built into the criterion (2.49) is straightforward, with the factor in brackets in (2.49) equal to $\{b'\mathbf{XY}'\ell\}^2$, leading to the maximizing unit vector $\ell = \mathbf{YX}'b/(b'\mathbf{XY}'\mathbf{YX}'b)^{1/2}$. Thus, with $\gamma = \alpha/(1-\alpha)$, this gives the criterion as

$$T_\alpha = (b'\mathbf{XY}'\mathbf{YX}'b)(b'\mathbf{XX}'b)^{\gamma-1}.$$

Numerical details of the method for determination of the vector b maximizing T_α (subject to the orthogonality constraints) are discussed by Stone and Brooks (1990) and Brooks and Stone (1994). As with partial least squares, the final estimated regression equations are obtained by ordinary LS calculation of the response vectors Y_k on the first w constructed predictor variables, $X_k^* = (x_{1k}^*, \ldots, x_{wk}^*)'$.

Brooks and Stone (1994) note that as $\alpha \to 1$ (so that $\gamma \to \infty$ and hence the criterion tends to be dominated by the second factor $(b'\mathbf{XX}'b)^{\gamma-1}$), the predictors constructed by this method become the principal components of $\mathbf{X}$. For the case $\alpha = 0$, the criterion becomes $T_0 = b'\mathbf{XY}'\mathbf{YX}'b/b'\mathbf{XX}'b$. For this case, no more than m predictor variables $x_{jk}^* = b_j'X_k$ are constructible; when $m \leq n < T$, it follows that the vectors b_j are directly related to the eigenvectors $\widehat{V}_j$ of the matrix $\mathbf{YX}'(\mathbf{XX}')^{-1}\mathbf{XY}'$ corresponding to the largest eigenvalues, with b_j' equal to a normalized version of $\widehat{V}_j'\mathbf{YX}'(\mathbf{XX}')^{-1}$. Hence, this case is equivalent to reduced-rank regression calculations with $\Sigma_{\epsilon\epsilon}$ taken to be of the form $\sigma^2 I_m$, the "least squares" method used by Davies and Tso (1982) as discussed in Section 2.3. In addition, the choice $\alpha = 1/2$ in the criterion gives simply $T_{1/2} = (b'\mathbf{XY}'\mathbf{YX}'b)$ and leads to a modification due to de Jong (1993) of a version of partial least squares regression determination. We note that for the univariate case, $m = 1$, Sundberg (1993) established a direct relation between first factor continuum regression (i.e., using only the first $w = 1$ predictor $x_{1k}^* = b_1'X_k$) and ridge regression and discussed its implications.

Envelope Regression

Recall the general multivariate linear model as presented in (1.1),

$$Y = CX + \epsilon,$$

where Y is an $m \times 1$ vector of response variables, X is an $n \times 1$ vector of predictor variables, C is an $m \times n$ regression coefficient matrix, and ϵ is the $m \times 1$ vector of random errors, with mean vector $E(\epsilon) = 0$ and covariance matrix $\text{Cov}(\epsilon) = \Sigma_{\epsilon\epsilon}$, an $m \times m$ positive-definite matrix. In the sequel, let $\mathcal{A}$ be the subspace spanned by the columns of the matrix A, i.e., $\mathcal{A} = \text{span}(A)$. We use either $\mathcal{P}_{\mathcal{A}}$ or $\mathcal{P}_A$ to denote the projection matrix onto $\text{span}(A)$ and also use either $\mathcal{Q}_{\mathcal{A}}$ or $\mathcal{Q}_A$ to denote the projection onto the orthogonal complement of $\mathcal{A}$, i.e., $\mathcal{Q}_{\mathcal{A}} = \mathcal{Q}_A = I - \mathcal{P}_A = I - \mathcal{P}_{\mathcal{A}}$.

Envelope method is a paradigm for dimension reduction introduced by Cook et al. (2010). In the context of multivariate linear regression, the main idea is to distinguish the "inmaterial" information in Y that is irrelevant to the association between Y and X. This leads to a reduction of the dimensionality of the problem and could potentially improve efficiency in model estimation and prediction. Consider a subspace $\mathcal{A} \subseteq \mathbb{R}^m$ that satisfies the following two conditions:

$$\text{(I) } \mathcal{Q}_{\mathcal{A}}Y \mid X \sim \mathcal{Q}_{\mathcal{A}}Y,$$

$$\text{(II) } \mathcal{P}_{\mathcal{A}}Y \perp \mathcal{Q}_{\mathcal{A}}Y \mid X.$$

Condition (I) indicates that the distribution of $\mathcal{Q}_{\mathcal{A}}Y$ does not depend on X and thus does not carry "direct" information on C, while condition (II) ensures that, given X, $\mathcal{Q}_{\mathcal{A}}Y$ cannot provide any "indirect" information on C through an association with $\mathcal{P}_{\mathcal{A}}Y$. Putting together these two conditions, we know that all the information on how X impacts Y (the information on C) is contained solely in $\mathcal{P}_{\mathcal{A}}Y$, while $\mathcal{Q}_{\mathcal{A}}Y$ contains only the so-called immaterial information.

We can now proceed to further characterize $\mathcal{Q}_{\mathcal{A}}Y$ under the multivariate linear regression model. Let $\mathcal{C} = \text{span}(C)$. Condition (I) holds if and only if $\mathcal{C} \subseteq \mathcal{A}$ and condition (II) holds if and only if $\mathcal{A}$ is a reducing subspace of $\Sigma_{\epsilon\epsilon}$; that is, $\mathcal{A}$ must be decomposed as

$$\Sigma_{\epsilon\epsilon} = \mathcal{P}_{\mathcal{A}}\Sigma_{\epsilon\epsilon}\mathcal{P}_{\mathcal{A}} + \mathcal{Q}_{\mathcal{A}}\Sigma_{\epsilon\epsilon}\mathcal{Q}_{\mathcal{A}}.$$

The intersection of all the subspaces with the above properties is, by construction, the smallest reducing subspace of $\Sigma_{\epsilon\epsilon}$ that contains $\mathcal{A}$, which is called the $\Sigma_{\epsilon\epsilon}$-envelope of $\mathcal{C}$, denoted as $\mathcal{E}_{\Sigma}(\mathcal{C})$.

Let $\Gamma \in \mathbb{R}^{m\times u}$ be an orthogonal basis of $\mathcal{E}_{\Sigma}(\mathcal{C})$, where $u \leq m$ is the dimension of the envelope. Since $C \subseteq \text{span}(\Gamma)$, we have that $C = \Gamma\xi$ for some $\xi \in \mathbb{R}^{u\times n}$. The multivariate linear regression model can then be expressed in its envelope form as follows:

$$\begin{aligned} Y &= \Gamma\xi X + \epsilon, \qquad \epsilon \sim N(0, \Sigma_{\epsilon\epsilon}), \\ \Sigma_{\epsilon\epsilon} &= \Gamma\Omega\Gamma' + \Gamma_0\Omega_0\Gamma_0', \end{aligned} \tag{2.50}$$

where $[\Gamma,\ \Gamma_0] \in \mathbb{R}^{m\times m}$ is orthogonal, and both $\Omega = \Gamma'\Sigma_{\epsilon\epsilon}\Gamma$ and $\Omega_0 = \Gamma_0'\Sigma_{\epsilon\epsilon}\Gamma_0$ are positive definite.

The idea of envelopes can be incorporated into reduced-rank regression, giving rise to a hybrid reduced-rank envelope model (Cook et al. 2015). The resulting estimator is at least as efficient as either the reduced-rank regression estimator or the ordinary envelope estimator. To formulate, recall that in reduced-rank regression, we allow that $\mathrm{rank}(C) = r \le \min(m, n)$, so the coefficient matrix C can be parameterized as

$$C = AB, \qquad A \in \mathbb{R}^{m\times r}, B \in \mathbb{R}^{r\times n},$$

where A and B are both of rank r. We recognize that $\mathcal{E}_\Sigma(C) = \mathcal{E}_\Sigma(A)$, and the dimension of the envelope u is always greater than or equal to the rank r because $\dim\{\mathcal{E}_\Sigma(C)\} \ge \dim\{\mathrm{span}(C)\} = \mathrm{rank}(C) = r$. The reduced-rank envelope model then specifies that

$$C = AB = \Gamma\xi = \Gamma\eta B, \qquad \Sigma_{\epsilon\epsilon} = \Gamma\Omega\Gamma' + \Gamma_0\Omega_0\Gamma_0',$$

where $\Omega \in \mathbb{R}^{u\times u}$ and $\Omega_0 \in \mathbb{R}^{(m-u)\times(m-u)}$ are positive-definite matrices, and $\eta \in \mathbb{R}^{u\times r}$ is of rank r. When the envelope dimension $u = m$, there is no immaterial information to be reduced, so the reduced-rank envelope model degenerates to the reduced-rank regression model. When the coefficient matrix C is full rank, the reduced-rank envelope model degenerates to the ordinary envelope model. The main difference between the two approaches is that the information of the component matrix, Γ, comes from both the conditional mean and the variance terms in the envelope modeling.

To see the potential gain of utilizing reduced-rank regression and envelopes, we can compare different models by the number of free parameters in the coefficient matrix C and the error covariance matrix $\Sigma_{\epsilon\epsilon}$; see Table 2.3. The total number of free parameters is reduced by $(m - r)(n - r)$ from the least squares method to reduced-rank regression and is further reduced by $(m - u)r$ from reduced-rank regression to reduced-rank envelope regression. As a numerical example, Cook et al. (2015) applied reduced-rank envelope regression on the salesman data ($m = 4$ and $n = 3$) presented in Section 2.8 and selected a model with $r = 2$ and $u = 4$. Compared to reduced-rank regression, the standard deviations of the elements in the reduced-rank envelope estimator were 1% to 24% smaller, where 24% smaller standard deviation corresponds to a doubling of the observations for reduced-

Table 2.3 Comparing number of free parameters in the coefficient matrix and the error covariance matrix

Method	Number of free parameters
Least squares	$mn + m(m+1)/2$
Reduced-rank regression	$(m+n-r)r + m(m+1)/2$
Envelope regression	$nu + m(m+1)/2$
Reduced-rank envelope regression	$(n+u-r)r + m(m+1)/2$

rank regression to achieve the same performance as its reduced-rank envelope counterpart.

Cook et al. (2015) showed that once we know the envelope, the coefficient matrix C can be obtained as the rank-r reduced-rank regression estimator from regressing $\Gamma' Y$ on X. In general, the estimation of the envelope models can be conducted via maximum likelihood under the normality assumption on ϵ. Various optimization algorithms have been developed, including the one-direction-at-a-time algorithm (Cook et al. 2010) and the faster envelope coordinate descent algorithm (Cook and Zhang 2018). We refer to Cook (2018) for details and more recent developments.

In conclusion, we hope that the reduced-rank and alternate dimension reduction tools for multivariate regression, as surveyed above, will develop further interest and find broader use in the future as large volumes of data are routinely being collected for analysis in many areas of application. A systematic comparison of these methods in terms of their performance in capturing various features of the data would be valuable.

Chapter 3
Reduced-Rank Regression Models with Two Sets of Regressors

3.1 Reduced-Rank Model of Anderson

In Chapter 2, we have demonstrated the utility of the reduced-rank model for analyzing data on a large number of variables. The interrelationship between the dependent and independent variables can be explained parsimoniously through the assumption of a lower rank for the regression coefficient matrix. The basic model that was described in Chapter 2 assumes that the predictor variables are all grouped into one set, and therefore, they are all subject to the same canonical calculations. In this chapter we broaden the scope of the reduced-rank model by entertaining the possibility that the predictor variables can be divided into two distinct sets with separate reduced-rank structures. Such an extension will be shown to have some interesting applications.

An extension of the basic model to include more than one set of regressors was already suggested in the seminal work of Anderson (1951). Anderson considered the following model where the set of dependent variables is influenced by two sets of regressor variables, one set having a reduced-rank coefficient matrix and the other having a full-rank coefficient matrix. Formally, let

$$Y_k = CX_k + DZ_k + \epsilon_k, \qquad k = 1, 2, \ldots, T, \tag{3.1}$$

where X_k is a vector of regressor variables, C is of reduced rank, and Z_k is a vector of additional variables that influence Y_k, but the matrix D is of full rank.

This model corresponds to a latent structure as mentioned by Robinson (1974). The set of regressor variables X_k supposedly influences the dependent variables Y_k through a few unobservable latent variables, while the set of regressor variables Z_k directly influences Y_k. That is, one might suppose the model $Y_k = AY_k^* + DZ_k + U_k$, where Y_k^* is an $r_1 \times 1$ vector of unobservable latent variables which influence Y_k, with Y_k^* determined in part through a set of causal variables X_k as $Y_k^* = BX_k +$

G. C. Reinsel et al., *Multivariate Reduced-Rank Regression*, Lecture Notes in Statistics 225, https://doi.org/10.1007/978-1-0716-2793-8_3

V_k. Then we obtain the reduced-form model equation as $Y_k = ABX_k + DZ_k + (AV_k + U_k) = CX_k + DZ_k + \epsilon_k$ with $C = AB$ of reduced rank. An alternative interpretation of the model is that Z_k may contain important variables, and hence the rank of its coefficient matrix is likely to be full, but X_k contains a large number of potential explanatory variables and possibly the dimension of X_k can be reduced. One elementary example of model (3.1) is its use to explicitly allow for a constant term in the reduced-rank model of Chapter 2 for the "true" predictor variables X_k, which can be done formally by taking Z_k to be the scalar 1 with D being the $m \times 1$ vector of constants in model (3.1).

The formulation of the model by Anderson was in terms of constraints as in (2.2) acting upon the reduced-rank coefficient matrix C. The focus was on estimation of these constraints and the inferential aspects of these estimated constraints. Because our focus here is on the component matrices A and B, we shall outline only their associated inferential results. In model (3.1), the vector of response variables Y_k is of dimension $m \times 1$, X_k is of dimension $n_1 \times 1$, the vector of additional predictor variables Z_k is of dimension $n_2 \times 1$, and the ϵ_k are $m \times 1$ independent random vectors of errors with mean 0 and positive-definite covariance matrix $\Sigma_{\epsilon\epsilon}$, for $k = 1, 2, \ldots, T$. Assume that

$$\text{rank}(C) = r_1 \leq \min(m, n_1), \tag{3.2}$$

and hence, as in (2.4), $C = AB$ where A is an $m \times r_1$ matrix and B is an $r_1 \times n_1$ matrix. Hence the model (3.1) can be written as

$$Y_k = ABX_k + DZ_k + \epsilon_k, \qquad k = 1, \ldots, T. \tag{3.3}$$

Define the data matrices as $\mathbf{Y} = [Y_1, \ldots, Y_T]$, $\mathbf{X} = [X_1, \ldots, X_T]$, and $\mathbf{Z} = [Z_1, \ldots, Z_T]$.

Estimation of A and B in (3.3) and estimation of D follow from the population result given below. The result is similar to Theorem 2.2 in Chapter 2.

Theorem 3.1 *Suppose the $(m + n_1 + n_2)$-dimensional random vector $(Y', X', Z')'$ has mean 0 and covariance matrix with $\Sigma_{yx} = \Sigma_{xy}' = \text{Cov}(Y, X)$, $\Sigma_{yz} = \Sigma_{zy}' = \text{Cov}(Y, Z)$, and $\begin{bmatrix} \Sigma_{xx} & \Sigma_{xz} \\ \Sigma_{zx} & \Sigma_{zz} \end{bmatrix} = \text{Cov}\begin{bmatrix} X \\ Z \end{bmatrix}$ is nonsingular. Then for any positive-definite matrix Γ, an $m \times r_1$ matrix A and an $r_1 \times n_1$ matrix B, for $r_1 \leq \min(m, n_1)$, which minimize*

$$\text{tr}\{E[\Gamma^{1/2}(Y - ABX - DZ)(Y - ABX - DZ)'\Gamma^{1/2}]\} \tag{3.4}$$

are given by

$$\begin{aligned} A^{(r_1)} &= \Gamma^{-1/2}[V_1, \ldots, V_{r_1}] = \Gamma^{-1/2}V^{(r_1)}, \\ B^{(r_1)} &= V^{(r_1)\prime}\Gamma^{1/2}\Sigma_{yx.z}\Sigma_{xx.z}^{-1} \end{aligned} \tag{3.5}$$

and $D = \Sigma_{yz}\Sigma_{zz}^{-1} - C^{(r_1)}\Sigma_{xz}\Sigma_{zz}^{-1}$ *with* $C^{(r_1)} = A^{(r_1)}B^{(r_1)}$*, where* $V^{(r_1)} = [V_1, \ldots, V_{r_1}]$ *and* V_j *is the (normalized) eigenvector that corresponds to the* j*th largest eigenvalue* λ_{1j}^2 *of the matrix* $\Gamma^{1/2}\Sigma_{yx.z}\Sigma_{xx.z}^{-1}\Sigma_{xy.z}\Gamma^{1/2}$ $(j = 1, 2, \ldots, r_1)$*. It must be observed that* $\Sigma_{yx.z} = \Sigma_{yx} - \Sigma_{yz}\Sigma_{zz}^{-1}\Sigma_{zx}$*,* $\Sigma_{xy.z} = \Sigma_{yx.z}'$*,* $\Sigma_{xx.z} = \Sigma_{xx} - \Sigma_{xz}\Sigma_{zz}^{-1}\Sigma_{zx}$ *are the partial covariance matrices obtained by eliminating the linear effects of* Z *from both* Y *and* X*.*

Proof With the following transformation,

$$X^* = X - \Sigma_{xz}\Sigma_{zz}^{-1}Z, \qquad D^* = D + C\Sigma_{xz}\Sigma_{zz}^{-1}, \tag{3.6}$$

we can write

$$Y - CX - DZ = Y - CX^* - D^*Z, \tag{3.7}$$

where X^* and Z are orthogonal, that is, $\mathrm{Cov}(X^*, Z) = 0$. The criterion in (3.4), which must then be observed, is the same as

$$\mathrm{tr}\{E[\Gamma^{1/2}(Y - ABX^* - D^*Z)(Y - ABX^* - D^*Z)'\Gamma^{1/2}]\}.$$

Because X^* and Z are orthogonal, the above criterion can be decomposed into two parts:

$$\mathrm{tr}\{E[\Gamma^{1/2}(Y - ABX^*)(Y - ABX^*)'\Gamma^{1/2}]\}$$
$$- (\mathrm{tr}\{E[\Gamma^{1/2}YY'\Gamma^{1/2}]\} - \mathrm{tr}\{E[\Gamma^{1/2}(Y - D^*Z)(Y - D^*Z)'\Gamma^{1/2}]\}).$$

The first part is similar to (2.12) in Theorem 2.2, and therefore the solutions for A and B follow directly. To see this clearly, observe that $\Sigma_{yx^*} = \Sigma_{yx.z}$ and $\Sigma_{x^*x^*} = \Sigma_{xx.z}$ are the partial covariance matrices. As far as the determination of D^* is concerned, observe that the part where D^* appears in the above criterion is simply the least squares criterion and therefore $D^* = \Sigma_{yz}\Sigma_{zz}^{-1}$. The optimal value of the solution for D is recovered from (3.6).

We shall now indicate several useful remarks in relation to the result of Theorem 3.1. As in the reduced-rank regression model (2.11) without the Z_k variables, there are some implied normalization conditions on the component matrices B and A in the solution of Theorem 3.1 as follows:

$$B\Sigma_{xx.z}B' = \Lambda_1^2, \qquad A'\Gamma A = I_{r_1}. \tag{3.8}$$

The solutions A and B in (3.5) are related to partial canonical vectors and λ_{1j}^2 are related to partial canonical correlations. For the most direct relation, the appropriate choice of Γ is the inverse of the partial covariance matrix of Y, adjusted for Z, $\Gamma = \Sigma_{yy.z}^{-1}$, where $\Sigma_{yy.z} = \Sigma_{yy} - \Sigma_{yz}\Sigma_{zz}^{-1}\Sigma_{zy}$. Then, $\lambda_{1j} = \rho_{1j}$ represent the partial canonical correlations between Y and X after eliminating Z. Another choice for Γ is $\Gamma = \Sigma_{yy.(x,z)}^{-1} = \Sigma_{\epsilon\epsilon}^{-1}$, and for this choice, it can be demonstrated, similar to the derivation in Section 2.4, that the associated eigenvalues λ_{1j}^2 in Theorem 3.1 are related to the partial canonical correlations ρ_{1j} as

$$\lambda_{1j}^2 = \rho_{1j}^2/(1 - \rho_{1j}^2). \tag{3.9}$$

The maximum likelihood estimates are obtained by substituting the sample quantities in the solutions of Theorem 3.1 and by the choice of $\Gamma = \widetilde{\Sigma}_{\epsilon\epsilon}^{-1}$, where

$$\widetilde{\Sigma}_{\epsilon\epsilon} \equiv \widehat{\Sigma}_{yy.(x,z)} = \widehat{\Sigma}_{yy} - [\widehat{\Sigma}_{yx}, \widehat{\Sigma}_{yz}] \begin{bmatrix} \widehat{\Sigma}_{xx} & \widehat{\Sigma}_{xz} \\ \widehat{\Sigma}_{zx} & \widehat{\Sigma}_{zz} \end{bmatrix}^{-1} \begin{bmatrix} \widehat{\Sigma}_{xy} \\ \widehat{\Sigma}_{zy} \end{bmatrix} \tag{3.10}$$

is the estimate of the error covariance matrix $\Sigma_{\epsilon\epsilon}$ based on full-rank regression. In (3.10), the estimates are defined to be $\widehat{\Sigma}_{yy} = \frac{1}{T}\mathbf{Y}\mathbf{Y}'$, $\widehat{\Sigma}_{yx} = \frac{1}{T}\mathbf{Y}\mathbf{X}'$, $\widehat{\Sigma}_{yz} = \frac{1}{T}\mathbf{Y}\mathbf{Z}'$, $\widehat{\Sigma}_{xx} = \frac{1}{T}\mathbf{X}\mathbf{X}'$, $\widehat{\Sigma}_{xz} = \frac{1}{T}\mathbf{X}\mathbf{Z}'$, and $\widehat{\Sigma}_{zz} = \frac{1}{T}\mathbf{Z}\mathbf{Z}'$, where $\mathbf{Y}$, $\mathbf{X}$, and $\mathbf{Z}$ are data matrices. This yields the ML estimates

$$\widehat{A}^{(r_1)} = \widetilde{\Sigma}_{\epsilon\epsilon}^{1/2}\widehat{V}, \qquad \widehat{B}^{(r_1)} = \widehat{V}'\widetilde{\Sigma}_{\epsilon\epsilon}^{-1/2}\widehat{\Sigma}_{yx.z}\widehat{\Sigma}_{xx.z}^{-1} \tag{3.11}$$

and $\widehat{D} = \widehat{\Sigma}_{yz}\widehat{\Sigma}_{zz}^{-1} - \widehat{C}^{(r_1)}\widehat{\Sigma}_{xz}\widehat{\Sigma}_{zz}^{-1}$ with $\widehat{C}^{(r_1)} = \widehat{A}^{(r_1)}\widehat{B}^{(r_1)}$, where $\widehat{V} = [\widehat{V}_1, \ldots, \widehat{V}_{r_1}]$ and the $\widehat{V}_j$ are the (normalized) eigenvectors of the matrix

$$\widehat{R} = \widetilde{\Sigma}_{\epsilon\epsilon}^{-1/2}\widehat{\Sigma}_{yx.z}\widehat{\Sigma}_{xx.z}^{-1}\widehat{\Sigma}_{xy.z}\widetilde{\Sigma}_{\epsilon\epsilon}^{-1/2} \equiv \widetilde{\Sigma}_{\epsilon\epsilon}^{-1/2}\widetilde{C}\widehat{\Sigma}_{xx.z}\widetilde{C}'\widetilde{\Sigma}_{\epsilon\epsilon}^{-1/2}$$

associated with the r_1 largest eigenvalues. In the above, $\widehat{\Sigma}_{yx.z} = \widehat{\Sigma}_{yx} - \widehat{\Sigma}_{yz}\widehat{\Sigma}_{zz}^{-1}\widehat{\Sigma}_{zx}$, $\widehat{\Sigma}_{xx.z} = \widehat{\Sigma}_{xx} - \widehat{\Sigma}_{xz}\widehat{\Sigma}_{zz}^{-1}\widehat{\Sigma}_{zx}$, and $\widetilde{C} = \widehat{\Sigma}_{yx.z}\widehat{\Sigma}_{xx.z}^{-1}$ is the full-rank LS estimate of C.

The above results on the form of ML estimates can be verified by similar arguments as in Section 2.3 for the single regressor set model (2.11). Similar to Section 2.3, the ML estimates of A, B, and D are obtained by minimizing $|W|$, or equivalently, $|\widetilde{\Sigma}_{\epsilon\epsilon}^{-1}W|$, where $W = (1/T)(\mathbf{Y} - AB\mathbf{X} - D\mathbf{Z})(\mathbf{Y} - AB\mathbf{X} - D\mathbf{Z})'$. Now with $C = AB$, we can write

$$\begin{aligned} \mathbf{Y} - C\mathbf{X} - D\mathbf{Z} &= (\mathbf{Y} - \widetilde{C}\mathbf{X} - \widetilde{D}\mathbf{Z}) + (\widetilde{C} - C)\mathbf{X} + (\widetilde{D} - D)\mathbf{Z} \\ &= (\mathbf{Y} - \widetilde{C}\mathbf{X} - \widetilde{D}\mathbf{Z}) + (\widetilde{C} - C)(\mathbf{X} - \widehat{\Sigma}_{xz}\widehat{\Sigma}_{zz}^{-1}\mathbf{Z}) \\ &\quad + \{(\widetilde{D} - D) + (\widetilde{C} - C)\widehat{\Sigma}_{xz}\widehat{\Sigma}_{zz}^{-1}\}\mathbf{Z}, \end{aligned}$$

where $\widetilde{C}$ and $\widetilde{D}$ are the unrestricted full-rank LS estimators, and the three terms on the right-hand side of the last expression are orthogonal. So we obtain the corresponding decomposition

$$W = \widetilde{\Sigma}_{\epsilon\epsilon} + (\widetilde{C} - AB)\widehat{\Sigma}_{xx.z}(\widetilde{C} - AB)'$$

$$+ (\widetilde{D}^* - D - C\widehat{\Sigma}_{xz}\widehat{\Sigma}_{zz}^{-1})\widehat{\Sigma}_{zz}(\widetilde{D}^* - D - C\widehat{\Sigma}_{xz}\widehat{\Sigma}_{zz}^{-1})',$$

where $\widetilde{D}^* = \widetilde{D} + \widetilde{C}\widehat{\Sigma}_{xz}\widehat{\Sigma}_{zz}^{-1} \equiv \widehat{\Sigma}_{yz}\widehat{\Sigma}_{zz}^{-1}$, with the last equality holding from the normal equations for the unrestricted LS estimators $\widetilde{C}$ and $\widetilde{D}$. For any C, the third component in the above decomposition can be uniformly minimized, as the zero matrix, by the choice of $\widehat{D} = \widetilde{D}^* - C\widehat{\Sigma}_{xz}\widehat{\Sigma}_{zz}^{-1} = \widehat{\Sigma}_{yz}\widehat{\Sigma}_{zz}^{-1} - C\widehat{\Sigma}_{xz}\widehat{\Sigma}_{zz}^{-1}$. Thus, the ML estimation of minimizing $|\widetilde{\Sigma}_{\epsilon\epsilon}^{-1}W|$ is reduced to minimizing

$$|I_m + \widetilde{\Sigma}_{\epsilon\epsilon}^{-1/2}(\widetilde{C} - AB)\widehat{\Sigma}_{xx.z}(\widetilde{C} - AB)'\widetilde{\Sigma}_{\epsilon\epsilon}^{-1/2}|$$

with respect to A and B. This is the same minimization problem as for ML estimation in Section 2.3, and it again follows similarly from Lemma 2.2 that all eigenvalues of the matrix $\widetilde{\Sigma}_{\epsilon\epsilon}^{-1/2}(\widetilde{C} - AB)\widehat{\Sigma}_{xx.z}(\widetilde{C} - AB)'\widetilde{\Sigma}_{\epsilon\epsilon}^{-1/2}$ will be simultaneously minimized by the choice of estimators given in (3.11), and hence the above determinant will also be minimized. Thus, these are ML estimators of A and B, and the corresponding ML estimator of D is as noted previously, $\widehat{D} = \widehat{\Sigma}_{yz}\widehat{\Sigma}_{zz}^{-1} - \widehat{C}^{(r_1)}\widehat{\Sigma}_{xz}\widehat{\Sigma}_{zz}^{-1}$ with $\widehat{C}^{(r_1)} = \widehat{A}^{(r_1)}\widehat{B}^{(r_1)}$. The ML estimate of $\Sigma_{\epsilon\epsilon}$ under the reduced-rank structure is given by $\widehat{\Sigma}_{\epsilon\epsilon} = (1/T)(\mathbf{Y} - \widehat{C}^{(r_1)}\mathbf{X} - \widehat{D}\mathbf{Z})(\mathbf{Y} - \widehat{C}^{(r_1)}\mathbf{X} - \widehat{D}\mathbf{Z})'$. Similar to the result in (2.18) of Section 2.3, the ML estimate $\widehat{\Sigma}_{\epsilon\epsilon}$ can be represented as

$$\begin{aligned}\widehat{\Sigma}_{\epsilon\epsilon} &= \widetilde{\Sigma}_{\epsilon\epsilon} + (\widetilde{C} - \widehat{C}^{(r_1)})\widehat{\Sigma}_{xx.z}(\widetilde{C} - \widehat{C}^{(r_1)})' \\ &= \widetilde{\Sigma}_{\epsilon\epsilon} + (I_m - P)\widetilde{C}\widehat{\Sigma}_{xx.z}\widetilde{C}'(I_m - P)',\end{aligned}$$

where $P = \widetilde{\Sigma}_{\epsilon\epsilon}^{1/2}\widehat{V}\widehat{V}'\widetilde{\Sigma}_{\epsilon\epsilon}^{-1/2}$, and $\widetilde{C} = \widehat{\Sigma}_{yx.z}\widehat{\Sigma}_{xx.z}^{-1}$ is the full-rank LS estimate of C.

Because of the connection displayed in (3.9), for example, the rank of the matrix C can be identified by the number of nonzero partial canonical correlations between Y_k and X_k, eliminating Z_k. More precisely, based on a sample of T observations, the likelihood ratio test procedure for testing $H_0\colon \text{rank}(C) \leq r_1$ yields the test statistic

$$-2\log(\lambda) = -T \sum_{j=r_1+1}^{m} \log(1 - \widehat{\rho}_{1j}^2), \tag{3.12}$$

where $\widehat{\rho}_{1j}$ are the sample partial canonical correlations between Y_k and X_k, eliminating Z_k, that is, $\widehat{\rho}_{1j}^2$ are the eigenvalues of the matrix

$$\widehat{R}_1 = \widehat{\Sigma}_{yy.z}^{-1/2}\widehat{\Sigma}_{yx.z}\widehat{\Sigma}_{xx.z}^{-1}\widehat{\Sigma}_{xy.z}\widehat{\Sigma}_{yy.z}^{-1/2}.$$

The derivation of the form of the LR statistic in (3.12) is similar to the derivation given in Section 2.6 for the reduced-rank model with one set of regressor variables. The restricted ML estimators are $\widehat{C}^{(r_1)}$ and $\widehat{D}$ as given above, and we let $S_1 = (\mathbf{Y} - \widehat{C}^{(r_1)}\mathbf{X} - \widehat{D}\mathbf{Z})(\mathbf{Y} - \widehat{C}^{(r_1)}\mathbf{X} - \widehat{D}\mathbf{Z})'$ denote the corresponding residual sum of squares matrix under H_0. Because $\widetilde{D} = \widehat{\Sigma}_{yz}\widehat{\Sigma}_{zz}^{-1} - \widetilde{C}\widehat{\Sigma}_{xz}\widehat{\Sigma}_{zz}^{-1}$, where $\widetilde{C} = \widehat{\Sigma}_{yx.z}\widehat{\Sigma}_{xx.z}^{-1}$ and $\widetilde{D}$ denote the full-rank LS estimators, it follows that

$$(\widetilde{C} - \widehat{C}^{(r_1)})\mathbf{X} + (\widetilde{D} - \widehat{D})\mathbf{Z} = (\widetilde{C} - \widehat{C}^{(r_1)})(\mathbf{X} - \widehat{\Sigma}_{xz}\widehat{\Sigma}_{zz}^{-1}\mathbf{Z}),$$

and we also have that

$$\begin{aligned}\widetilde{C} - \widehat{C}^{(r_1)} &= (I_m - \widetilde{\Sigma}_{\epsilon\epsilon}^{1/2}\widehat{V}\widehat{V}'\widetilde{\Sigma}_{\epsilon\epsilon}^{-1/2})\widetilde{C} \\ &= \widetilde{\Sigma}_{\epsilon\epsilon}^{1/2}\left[\sum_{j=r_1+1}^{m} \widehat{V}_j\widehat{V}_j'\right]\widetilde{\Sigma}_{\epsilon\epsilon}^{-1/2}\widehat{\Sigma}_{yx.z}\widehat{\Sigma}_{xx.z}^{-1}.\end{aligned}$$

Therefore, similar to the result in (2.39) of Section 2.6, we find that

$$\begin{aligned}\tfrac{1}{T}S_1 &= \tfrac{1}{T}S + \tfrac{1}{T}[(\widetilde{C} - \widehat{C}^{(r_1)})\mathbf{X} + (\widetilde{D} - \widehat{D})\mathbf{Z}][(\widetilde{C} - \widehat{C}^{(r_1)})\mathbf{X} + (\widetilde{D} - \widehat{D})\mathbf{Z}]' \\ &= \widetilde{\Sigma}_{\epsilon\epsilon} + \widetilde{\Sigma}_{\epsilon\epsilon}^{1/2}\left[\sum_{j=r_1+1}^{m} \widehat{V}_j\widehat{V}_j'\right]\widehat{R}_1\left[\sum_{j=r_1+1}^{m} \widehat{V}_j\widehat{V}_j'\right]\widetilde{\Sigma}_{\epsilon\epsilon}^{1/2} \\ &= \widetilde{\Sigma}_{\epsilon\epsilon} + \widetilde{\Sigma}_{\epsilon\epsilon}^{1/2}\left[\sum_{j=r_1+1}^{m} \widehat{\lambda}_{1j}^2\widehat{V}_j\widehat{V}_j'\right]\widetilde{\Sigma}_{\epsilon\epsilon}^{1/2},\end{aligned}$$

where the $\widehat{\lambda}_{1j}^2$, $j = r_1 + 1, \ldots, m$, are the $(m - r_1)$ smallest eigenvalues of the matrix $\widehat{R} = \widetilde{\Sigma}_{\epsilon\epsilon}^{-1/2}\widehat{\Sigma}_{yx.z}\widehat{\Sigma}_{xx.z}^{-1}\widehat{\Sigma}_{xy.z}\widetilde{\Sigma}_{\epsilon\epsilon}^{-1/2}$. From results similar to those presented in the discussion at the end of Section 2.4 [e.g., see (2.29)], we know that $\widehat{\lambda}_{1j}^2 = \widehat{\rho}_{1j}^2/(1 - \widehat{\rho}_{1j}^2)$ and so $1 + \widehat{\lambda}_{1j}^2 = 1/(1 - \widehat{\rho}_{1j}^2)$. Therefore,

$$U^{-1} = |S_1|/|S| = \left|\tfrac{1}{T}S_1\right|/|\widetilde{\Sigma}_{\epsilon\epsilon}| = \prod_{j=r_1+1}^{m}(1 + \widehat{\lambda}_{1j}^2) = \prod_{j=r_1+1}^{m}(1 - \widehat{\rho}_{1j}^2)^{-1}.$$

Thus, it follows that the LR criterion $-2\log(\lambda)$, where $\lambda = U^{T/2}$, is given as in (3.12). Anderson (1951) has shown that under the null hypothesis that $\text{rank}(C) \leq r_1$, the asymptotic distribution of the LR statistic is such that $-2\log(\lambda) \sim \chi^2_{(m-r_1)(n_1-r_1)}$. The LR statistic in (3.12) can be used to test the hypothesis H_0 : $\text{rank}(C) \leq r_1$ and can be considered for different values of r_1 to obtain an appropriate choice of the rank of C. As in Section 2.6, a correction factor is commonly used for the LR statistic in (3.12), yielding the test statistic $\mathcal{M} = -[T - n + (n_1 - m - 1)/2]\sum_{j=r_1+1}^{m}\log(1 - \widehat{\rho}_{1j}^2)$, whose null distribution is more closely approximated by the asymptotic $\chi^2_{(m-r_1)(n_1-r_1)}$ distribution than is that of $-2\log(\lambda)$ in (3.12).

The model (3.1) written as

$$Y_k = ABX_k + DZ_k + \epsilon_k, \qquad k = 1, \ldots, T, \tag{3.13}$$

provides an alternative way of interpreting the estimation procedure. Given B, therefore BX_k, estimates of A and D are obtained by the usual multivariate least squares regression of Y_k on BX_k and Z_k. When these estimates are substituted in the likelihood criterion as functions of B, the resulting quantity will directly lead to the estimation of B, via the canonical vectors associated with the sample partial canonical correlations between Y_k and X_k after eliminating Z_k. This interpretation will be shown to be useful for a more general model that we consider later in this chapter.

As in Section 2.3 for the single regressor reduced-rank model, the above discussion leads to an alternate form of the population solution to the reduced-rank problem for model (3.1) besides that given in Theorem 3.1. For given BX, we can substitute the least squares values of A and D (in particular, $A = \Sigma_{yx.z}B'(B\Sigma_{xx.z}B')^{-1}$) from the regression of Y on BX and Z into the criterion (3.4) or alternatively into its decomposition given in the proof of Theorem 3.1. This criterion then simplifies to $\text{tr}\{\Sigma_{yy.z}\Gamma - (B\Sigma_{xx.z}B')^{-1}B\Sigma_{xy.z}\Gamma\Sigma_{yx.z}B'\}$, and similar to the discussion at the end of Section 2.3, it follows that the solution for B is such that the columns of $\Sigma_{xx.z}^{1/2}B'$ can be chosen as the eigenvectors corresponding to the r_1 largest eigenvalues of $\Sigma_{xx.z}^{-1/2}\Sigma_{xy.z}\Gamma\Sigma_{yx.z}\Sigma_{xx.z}^{-1/2}$.

Although the results presented in Theorem 3.1 and above provide an explicit solution for the component matrices A, B, and for D, using the eigenvalues and eigenvectors of appropriate matrices, an iterative partial least squares approach similar to that discussed for the single regressor model in Section 2.3 can also be useful and is computationally convenient. For the population version of the problem, the first order equations that result from minimization of the criterion (3.4) are

$$\begin{aligned} \Sigma_{yx}B' - AB\Sigma_{xx}B' - D\Sigma_{zx}B' &= 0 \\ A'\Gamma\Sigma_{yx} - (A'\Gamma A)B\Sigma_{xx} - A'\Gamma D\Sigma_{zx} &= 0 \\ \Sigma_{yz} - AB\Sigma_{xz} - D\Sigma_{zz} &= 0. \end{aligned}$$

For given A and B, the first step in the iterative process is to obtain D from the third equation as $D = (\Sigma_{yz} - AB\Sigma_{xz})\Sigma_{zz}^{-1}$. For given B (and D), the solution for A can be calculated from the first equation as $A = (\Sigma_{yx} - D\Sigma_{zx})B'(B\Sigma_{xx}B')^{-1}$, while for given A (and D), from the second equation, the solution for B can be calculated as $B = (A'\Gamma A)^{-1}A'\Gamma(\Sigma_{yx} - D\Sigma_{zx})$. Alternatively, the optimal expression for D (from the third equation) can be substituted into the first equation, and the solution for A is then obtained for given B as $A = \Sigma_{yx.z}B'(B\Sigma_{xx.z}B')^{-1}$, and similarly, substituting the optimal expression for D into the second equation yields the solution of B for given A as $B = (A'\Gamma A)^{-1}A'\Gamma\Sigma_{yx.z}\Sigma_{xx.z}^{-1}$. Thus, in effect, D can be eliminated from the first two equations and the partial least squares iterations can be performed in terms of A and B only, completely analogous to the procedure in Section 2.3 but using partial covariance matrices. The solution for D can finally be recovered as given above. At each step of the iterations, one can consider using the normalization conditions (3.8) for A and B. This partial least squares procedure can also be carried over directly to the sample situation with the appropriate sample quantities being used to iteratively construct estimates $\widehat{A}^{(r_1)}$, $\widehat{B}^{(r_1)}$, and $\widehat{D}$, which are given more explicitly in (3.11).

Recall the original motivation behind Anderson's article. While A and B, the components of C, are formed out of the largest partial canonical correlations and the associated vectors, the constraints on C which were of Anderson's focus can be obtained via the $(m - r_1)$ smallest (possibly zero) partial canonical correlations and the associated vectors. It may be clear by now that the reduced-rank model in (3.13) of this chapter can be accommodated through essentially the same set of quantities as defined for the reduced-rank model in Chapter 2, with the difference being that the covariance matrices used in Chapter 2 are replaced by partial covariance matrices. For this reason, we do not repeat the asymptotic theory for the estimators that was presented in Section 2.5. The results of Theorem 2.4 in that section could be restated here for the case $\Gamma = \Sigma_{\epsilon\epsilon}^{-1}$ known by substituting the covariance matrices with corresponding partial covariance matrices. It is assumed that the predictors are non-stochastic as well. Results for the case of stochastic predictors follow from the asymptotic distribution presented later in Section 3.5.1 for an extended model.

3.2 Application to One-Way ANOVA and Linear Discriminant Analysis

We consider the reduced-rank model (3.1) applied to the multivariate one-way ANOVA situation and exhibit its relation to linear discriminant analysis. Suppose we have independent random samples from n multivariate normal distributions $N(\mu_i, \Sigma)$, denoted by $Y_{i1}, \ldots, Y_{iT_i}$, $i = 1, 2, \ldots, n$, and set $T = T_1 + \cdots + T_n$. It may be hypothesized that the n mean vectors μ_i lie in some (unknown) r-dimensional linear subspace, with $r \leq \min(n - 1, m)$. (We will assume $n < m$ for convenience of notation.) This is equivalent to the hypothesis that the $m \times (n - 1)$ matrix $(\mu_1 - \mu, \ldots, \mu_{n-1} - \mu)$, or $(\mu_1 - \mu_n, \ldots, \mu_{n-1} - \mu_n)$, is of reduced rank r, where $\mu = \sum_{i=1}^{n} T_i \mu_i / T$.

Because $Y_{ij} = (\mu_i - \mu_n) + \mu_n + \epsilon_{ij}$, $j = 1, 2, \ldots, T_i$, $i = 1, 2, \ldots, n$, the model can be easily written in the reduced-rank model form (3.1), $Y_{ij} = CX_{ij} + DZ_{ij} + \epsilon_{ij}$. The correspondences are $C = (\mu_1 - \mu_n, \ldots, \mu_{n-1} - \mu_n)$, $D = \mu_n$, $Z_{ij} \equiv 1$, and X_{ij} is an $(n - 1) \times 1$ vector with 1 in the ith position and zeros elsewhere for $i < n$ and $X_{nj} \equiv 0$. The reduced-rank restrictions imply $C = AB$ with $B = (\xi_1, \ldots, \xi_{n-1})$ of dimension $r \times (n - 1)$, or $\mu_i - \mu_n = A\xi_i$, $i = 1, 2, \ldots, n - 1$; equivalently, $L(\mu_i - \mu_n) = 0$, $i = 1, 2, \ldots, n-1$, where $L = (\ell_1', \ldots, \ell_{m-r}')'$ is $(m-r) \times m$ with row vectors ℓ_i' as in (2.1). In terms of data matrices, we have $\mathbf{Y} = C\mathbf{X} + D\mathbf{Z} + \boldsymbol{\epsilon}$, where $\mathbf{Y} = [Y_{11}, \ldots, Y_{1T_1}, \ldots, Y_{n1}, \ldots, Y_{nT_n}]$, $\mathbf{Z} = (1, \ldots, 1) \equiv \mathbf{1}_T'$, and $\mathbf{X} = [\mathbf{X}_{(1)}, \ldots, \mathbf{X}_{(n-1)}, 0]$ with $\mathbf{X}_{(i)} = [0, \ldots, 0, \mathbf{1}T_i, 0, \ldots, 0]'$ where $\mathbf{1}_n$ denotes an N-dimensional vector of ones, so that $\mathbf{X}_{(i)}$ is an $(n-1) \times T_i$ matrix with ones in the ith row and zeros elsewhere. From the LS estimation results of Sections 1.2, 1.3, and 3.1, the full-rank or unrestricted LS estimates of C and D are given as

$$\widetilde{C} = \widehat{\Sigma}_{yx.z}\widehat{\Sigma}_{xx.z}^{-1} = [\bar{Y}_1 - \bar{Y}_n, \ldots, \bar{Y}_{n-1} - \bar{Y}_n], \tag{3.14a}$$

$$\widetilde{D} = (\widehat{\Sigma}_{yz} - \widetilde{C}\widehat{\Sigma}_{xz})\widehat{\Sigma}_{zz}^{-1} = \bar{Y}_n, \tag{3.14b}$$

where $\bar{Y}_i = \sum_{j=1}^{T_i} Y_{ij}/T_i$ is the ith sample mean vector, $i = 1, 2, \ldots, n$. The corresponding ML estimate of the error covariance matrix $\Sigma_{\epsilon\epsilon} \equiv \Sigma$ is given by $\widetilde{\Sigma}_{\epsilon\epsilon} = \frac{1}{T}S$ with $S = \sum_{i=1}^{n}\sum_{j=1}^{T_i}(Y_{ij} - \bar{Y}_i)(Y_{ij} - \bar{Y}_i)'$ being the usual "within-group" sum of squares and cross-products matrix.

For the reduced-rank or restricted estimates of C and D, we need to consider the eigenvectors associated with the r largest eigenvalues of the matrix $\widehat{\Sigma}_{yx.z}\widehat{\Sigma}_{xx.z}^{-1}\widehat{\Sigma}_{xy.z} \equiv \widetilde{C}\widehat{\Sigma}_{xx.z}\widetilde{C}'$ with respect to the (unrestricted) residual covariance matrix estimate $\widetilde{\Sigma}_{\epsilon\epsilon}$. Observe that $\widehat{\Sigma}_{zz} = 1$ and $\widehat{\Sigma}_{xz} = \frac{1}{T}(T_1, \ldots, T_{n-1})'$, and hence

$$\widehat{\Sigma}_{xx.z} = \widehat{\Sigma}_{xx} - \widehat{\Sigma}_{xz}\widehat{\Sigma}_{zz}^{-1}\widehat{\Sigma}_{zx} = \frac{1}{T}[\mathrm{diag}(T_1, \ldots, T_{n-1}) - \frac{1}{T}NN'],$$

where $N = (T_1, \ldots, T_{n-1})'$. Therefore, it follows that

$$\begin{aligned}\widetilde{C}\widehat{\Sigma}_{xx.z}\widetilde{C}' &= \frac{1}{T}\left[\sum_{i=1}^{n-1} T_i(\bar{Y}_i - \bar{Y}_n)(\bar{Y}_i - \bar{Y}_n)' - T(\bar{Y} - \bar{Y}_n)(\bar{Y} - \bar{Y}_n)'\right] \\ &= \frac{1}{T}\left[\sum_{i=1}^{n} T_i(\bar{Y}_i - \bar{Y})(\bar{Y}_i - \bar{Y})'\right] \equiv \frac{1}{T}S_B,\end{aligned} \tag{3.15}$$

where S_B is the "between-group" sum of squares and cross-products matrix, with $\bar{Y} = \sum_{i=1}^{n} T_i\bar{Y}_i/T$. Let $\widehat{V}_k$ denote the normalized eigenvector of $\widetilde{\Sigma}_{\epsilon\epsilon}^{-1/2}\widetilde{C}\widehat{\Sigma}_{xx.z}\widetilde{C}'\widetilde{\Sigma}_{\epsilon\epsilon}^{-1/2} = \frac{1}{T}\widetilde{\Sigma}_{\epsilon\epsilon}^{-1/2}S_B\widetilde{\Sigma}_{\epsilon\epsilon}^{-1/2}$ corresponding to the kth largest eigenvalue $\widehat{\lambda}_k^2$. Then $\widehat{V}_k^* = \widetilde{\Sigma}_{\epsilon\epsilon}^{-1/2}\widehat{V}_k$ satisfies $(\widehat{\lambda}_k^2\widetilde{\Sigma}_{\epsilon\epsilon} - \frac{1}{T}S_B)\widehat{V}_k^* = 0$ or $(\widehat{\lambda}_k^2 S - S_B)\widehat{V}_k^* = 0$, with $\widehat{V}_k^*$ normalized so that $\widehat{V}_k^{*\prime}\widetilde{\Sigma}_{\epsilon\epsilon}\widehat{V}_k^* = 1$. Let $\widehat{V}_* = [\widehat{V}_1^*, \ldots, \widehat{V}_r^*] = \widetilde{\Sigma}_{\epsilon\epsilon}^{-1/2}\widehat{V}$ with $\widehat{V} = [\widehat{V}_1, \ldots, \widehat{V}_r]$. Then, from the results of Section 3.1, the ML reduced-rank estimate of C is

$$\widehat{C} = \widehat{A}^{(r)}\widehat{B}^{(r)} = \widetilde{\Sigma}_{\epsilon\epsilon}^{1/2}\widehat{V}\widehat{V}'\widetilde{\Sigma}_{\epsilon\epsilon}^{-1/2}\widetilde{C} = \widetilde{\Sigma}_{\epsilon\epsilon}\widehat{V}_*\widehat{V}_*'\widetilde{C}. \tag{3.16}$$

The corresponding restricted estimate of $D = \mu_n$ is then

$$\widehat{D} \equiv \widehat{\mu}_n = (\widehat{\Sigma}_{yz} - \widehat{C}\widehat{\Sigma}_{xz})\widehat{\Sigma}_{zz}^{-1} = \bar{Y} - \widehat{C}\frac{1}{T}N$$

$$= \bar{Y} - \widetilde{\Sigma}_{\epsilon\epsilon}\widehat{V}_*\widehat{V}_*'\frac{1}{T}\sum_{i=1}^{n-1} T_i(\bar{Y}_i - \bar{Y}_n) = \bar{Y} + \widetilde{\Sigma}_{\epsilon\epsilon}\widehat{V}_*\widehat{V}_*'(\bar{Y}_n - \bar{Y}). \tag{3.17}$$

From (3.16), it is now easy to see that the ML reduced-rank estimates $\widehat{\mu}_i$ of the individual population means μ_i are related as $\widehat{\mu}_i - \widehat{\mu}_n = \widetilde{\Sigma}_{\epsilon\epsilon}\widehat{V}_*\widehat{V}_*'(\bar{Y}_i - \bar{Y}_n)$, so that using (3.17), the ML reduced-rank estimates of the μ_i are

$$\widehat{\mu}_i = \bar{Y} + \widetilde{\Sigma}_{\epsilon\epsilon}\widehat{V}_*\widehat{V}_*'(\bar{Y}_i - \bar{Y}), \qquad i = 1, \ldots, n. \tag{3.18}$$

The reduction in dimension for estimation of the $\mu_i - \mu_n$ appears through the use of only the r linear combinations $\widehat{V}_*'(\bar{Y}_i - \bar{Y}_n)$ of the differences in sample mean vectors to estimate the $\mu_i - \mu_n$ under the reduced-rank restrictions. In addition, using the results in Sections 2.3 and 3.1, the ML estimate of the error covariance matrix $\Sigma_{\epsilon\epsilon}$ under the reduced-rank assumptions is given by

$$\widehat{\Sigma}_{\epsilon\epsilon} = \widetilde{\Sigma}_{\epsilon\epsilon} + (I_m - \widetilde{\Sigma}_{\epsilon\epsilon}\widehat{V}_*\widehat{V}_*')\left(\frac{1}{T}S_B\right)(I_m - \widehat{V}_*\widehat{V}_*'\widetilde{\Sigma}_{\epsilon\epsilon}) \tag{3.19}$$

and also note that $I_m - \widetilde{\Sigma}_{\epsilon\epsilon}\widehat{V}_*\widehat{V}_*' = \widetilde{\Sigma}_{\epsilon\epsilon}\widehat{V}_-\widehat{V}_-'$ with $\widehat{V}_- = [\widehat{V}_{r+1}^*, \ldots, \widehat{V}_m^*]$.

Now we indicate the relation with linear discriminant analysis. The setting of linear discriminant analysis is basically that of the one-way multivariate ANOVA model. The objectives of discriminant analysis are to obtain representations of the data that best separate the n groups or populations and based on this to construct a procedure to classify a new vector observation to one of the groups with reasonable accuracy. Discrimination is based on the separations among the m-dimensional mean vectors μ_i, $i = 1, \ldots, n$, of the n populations. If the mean vectors μ_i were to lie in some r-dimensional linear subspace, $r < \min(n-1, m)$, then only r linear functions of the μ_i (linear discriminant functions) would be needed to describe the separations among the n groups. The Fisher–Rao approach to discriminant analysis in the sample is to construct linear combinations (linear discriminants) $z_{ij} = a'Y_{ij}$ which successively maximize the ratio

$$\frac{a'S_Ba}{a'Sa} = \sum_{i=1}^{n} T_i(\bar{z}_i - \bar{z})^2 / \sum_{i=1}^{n}\sum_{j=1}^{T_i}(z_{ij} - \bar{z}_i)^2. \tag{3.20}$$

Note that this is the ratio of the between-group (i.e., between means) to the within-group variation in the sample. Thus, the sample *first linear discriminant function* is the linear combination $a_1'Y_{ij}$ that maximizes the ratio in (3.20), the sample *second linear discriminant function* is the linear combination $a_2'Y_{ij}$ that maximizes the ratio subject to $a_2'\widetilde{\Sigma}_{\epsilon\epsilon}a_1 = 0$ (i.e., the sample covariance between $a_1'Y_{ij}$ and $a_2'Y_{ij}$ is zero), and so on.

Now recall that the $\widehat{V}_k^*$ in the solution to the above one-way ANOVA model reduced-rank estimation problem are the eigenvectors of the matrix $S^{-1}S_B$ corresponding to its nonzero eigenvalues $\widehat{\lambda}_k^2$, for $k = 1, \ldots, n-1$, that is, they satisfy $(\widehat{\lambda}_k^2 S - S_B)\widehat{V}_k^* = 0$. From this, it is known that the first eigenvector $\widehat{V}_1^*$, for example, provides the linear combination $z_{ij} = a'Y_{ij}$ which maximizes the ratio (3.20), that is, $\widehat{V}_1^*$ yields the sample first linear discriminant function in the linear discriminant analysis of Fisher–Rao, and $\widehat{\lambda}_1^2$ represents the maximum value attained for this ratio. Then $\widehat{V}_2^*$ gives the second linear discriminant function, in the sense of providing the linear combination which gives the maximum of the above ratio subject to the orthogonality condition $\widehat{V}_1^{*\prime} S \widehat{V}_2^* = 0$. In this same fashion, the collection of vectors $\widehat{V}_k^*$ provides the set of $(n-1)$ sample linear discriminant functions. In linear discriminant analysis of n multivariate normal populations, these discriminant functions are generally applied to the sample mean vectors $\bar{Y}_i$, to consider $\widehat{V}_k^{*\prime}(\bar{Y}_i - \bar{Y})$, in an attempt to provide information that best separates or discriminates between the means of the n groups or populations. In practice, it may be desirable to choose a subspace of rank $r < n-1$ and use only the corresponding first r linear discriminant functions that maximize the separation of the group means. From the above discussion, these would be obtained as the first r eigenvectors $\widehat{V}_1^*, \ldots, \widehat{V}_r^*$, and they might contain nearly all the available information for discrimination. We thus see that restriction to consideration of only these r linear discriminant functions can be formalized by formulating the one-way ANOVA model with a *reduced-rank* model structure for the population mean vectors μ_i.

The above results on the ML estimates in the one-way ANOVA model under the reduced-rank structure were obtained by Anderson (1951) and later by Healy (1980), Campbell (1984), and Srivastava (1997) [also by Theobald (1975) with $\Sigma_{\epsilon\epsilon}$ assumed known], while the related problem in discriminant analysis was considered earlier by Fisher (1938). The connections between reduced-rank regression and discriminant analysis were also discussed by Campbell (1984), and a more direct formulation of the structure on the population means along the lines of (3.18), i.e., $\mu_i = \mu + \Sigma_{\epsilon\epsilon} V_* \xi_i \equiv \mu + \Sigma_{\epsilon\epsilon} V_* V_*'(\mu_i - \mu)$ was considered there. The reduced-rank linear discriminant analysis method forms the basis of recent extensions of discriminant analysis that are useful in analyzing more complex pattern recognition problems (Hastie and Tibshirani 1996) and in analyzing longitudinal data through growth curve models (Albert and Kshirsagar 1993).

3.3 Numerical Example Using Chemometrics Data

In the example of this section, we apply the multivariate reduced-rank regression methods of Chapter 2 and Section 3.1 to chemometrics data obtained from simulation of a low-density polyethylene tubular reactor. The data are taken from Skagerberg, MacGregor and Kiparissides (1992), who used partial least squares

(PLS) multivariate regression modeling discussed in Section 2.9 applied to these data both for predicting properties of the produced polymer and for multivariate process control. The data were also considered by Breiman and Friedman (1997) to illustrate the relative performance of different multivariate prediction methods.

The data set consists of $T = 56$ multivariate observations, with $m = 6$ response variables and $n = 22$ predictor (or "process") variables. The response variables are the following output properties of the polymer produced:

y_1, number-average molecular weight
y_2, weight-average molecular weight
y_3, frequency of long chain branching
y_4, frequency of short chain branching
y_5, content of vinyl groups in the polymer chain
y_6, content of vinylidene groups in the polymer chain

The process variable measurements employed consist of 20 different temperatures measured at equal distances along the reactor ($x_1 - x_{20}$), complemented with the wall temperature of the reactor (x_{21}) and the feed rate of the solvent (x_{22}) that also acts as a chain transfer agent. The temperature measurements in the temperature profile along the reactor are expected to be highly correlated, and therefore reduced-rank methods may be especially useful to reduce the dimensionality and complexity of the input or predictor set of data. For interpretational convenience, the response variables y_3, y_5, and y_6 were rescaled by the multiplicative factors 10^2, 10^3, and 10^2, respectively, before performing the analysis presented here, so that all six response variables would have variability of the same order of magnitude. The predictor variable x_{21} was also rescaled by the factor 10^{-1}.

As usual, both the response and predictor variables were first centered by subtraction of appropriate sample means prior to the multivariate analysis, and we let $\mathbf{Y}$ and $\mathbf{X}$ denote the resulting data matrices of dimensions 6×56 and 22×56, respectively. To obtain a preliminary "benchmark" model for comparison with subsequent analyses, a multivariate regression model was fit by LS to the vector of response variables using only the two predictor variables x_{21} and x_{22}, the wall temperature of the reactor and the solvent feed rate. The LS estimate of the regression coefficient matrix (with estimated standard errors given in parentheses) from this preliminary model is obtained as

$$\widetilde{C}_1' = (\mathbf{X}_1\mathbf{X}_1')^{-1}\mathbf{X}_1\mathbf{Y}'$$

$$= \begin{bmatrix} 0.1357 & 1.0956 & -1.5618 & 1.4234 & 0.5741 & 0.5339 \\ (0.0475) & (0.1230) & (0.1660) & (0.2045) & (0.0813) & (0.0783) \\ \\ 5.8376 & 28.3353 & -0.1305 & 0.1715 & 0.1065 & 0.0596 \\ (0.1704) & (0.4414) & (0.5953) & (0.7336) & (0.2917) & (0.2809) \end{bmatrix},$$

where $\mathbf{X}_1$ is 2×56. The unbiased estimate of the covariance matrix of the errors ϵ_k from the multivariate regression model (with correlations shown above the diagonal and the covariances below the diagonal) is given by

$$\overline{\Sigma}_{\epsilon\epsilon} = \frac{1}{56-3}\widehat{\epsilon\epsilon}'$$

$$= \begin{bmatrix} 0.15323 & 0.79248 & 0.26439 & 0.88877 & 0.89995 & 0.90231 \\ 0.31458 & 1.02835 & -0.10505 & 0.77472 & 0.75447 & 0.76567 \\ 0.14156 & -0.14571 & 1.87093 & 0.10493 & 0.14821 & 0.13324 \\ 0.58637 & 1.32412 & 0.24190 & 2.84066 & 0.95115 & 0.95700 \\ 0.23610 & 0.51276 & 0.13587 & 1.07438 & 0.44916 & 0.95458 \\ 0.22792 & 0.50104 & 0.11760 & 1.04082 & 0.41282 & 0.41640 \end{bmatrix},$$

where $\widehat{\epsilon} = \mathbf{Y} - \widetilde{C}_1\mathbf{X}_1$. We also have $\mathrm{tr}(\overline{\Sigma}_{\epsilon\epsilon}) = 6.75873$, and the determinant of the ML error covariance matrix estimate, $\widetilde{\Sigma}_{\epsilon\epsilon} = \frac{1}{56}\widehat{\epsilon\epsilon}'$, is $|\widetilde{\Sigma}_{\epsilon\epsilon}| = 0.25318 \times 10^{-4}$.

Clearly, from the results above, we see that the response variables y_1 and y_2 are very highly related to x_{22} and only more mildly related to x_{21}, whereas response variables $y_3 - y_6$ are more strongly related to x_{21} with little or no dependence on x_{22}. It is also found that the errors from the fitted regressions are very strongly positively correlated among the variables y_1 and $y_4 - y_6$, the errors for y_2 are rather highly positively correlated with each of y_1 and $y_4 - y_6$, and the errors for y_3 show relatively small correlations with the errors for all the other response variables. Thus, there is a structure in the relationship between the response variables and the two predictors.

We now incorporate the 20 measurements of the temperature profile along the reactor into the multivariate regression analysis. The LS estimation of the response variables regressed on the entire set of 22 predictor variables ($x_1 - x_{22}$) is performed and yields an ML estimate $\widetilde{\Sigma}_{\epsilon\epsilon} = \frac{1}{56}\widehat{\epsilon\epsilon}'$ of the error covariance matrix with $|\widetilde{\Sigma}_{\epsilon\epsilon}| = 0.97557 \times 10^{-8}$, where $\widehat{\epsilon} = \mathbf{Y} - \widetilde{C}\mathbf{X}$. A requisite LR test of the hypothesis that the regression coefficients for all temperature profile variables ($x_1 - x_{20}$) are equal to zero for all response variables gives a test statistic value, from (1.16), equal to $\mathcal{M} = -[56-23+(20-6-1)/2]\log(0.97557 \times 10^{-8}/0.25318 \times 10^{-4}) = 310.53$, with $6 * 20 = 120$ degrees of freedom. This leads to clear rejection of the null hypothesis and indicates that the set of 20 temperature variables is useful in explaining variations of the response variables. In addition, the "unbiased" estimate of the error covariance matrix, $\overline{\Sigma}_{\epsilon\epsilon} = \frac{1}{56-23}\widehat{\epsilon\epsilon}'$, has diagonal elements equal to 0.03297, 0.38851, 0.45825, 0.21962, 0.05836, 0.03717, with trace equal to 1.19487. Comparison to the corresponding values for the simple two predictor variable model above indicates that the inclusion of the temperature variables has its most dramatic effect in reduction of the error variances for response variables y_4, y_5, and y_6. A rather high proportion of the t-statistics for the individual regression coefficients of the temperature variables in the LS fitted model are not "significant," however, partly due to the high level of correlations among many of the 20 temperature profile variables. Because of this high degree of correlation among the temperature measurements, it is obviously desirable to search for alternatives to the ordinary least squares estimation results based on all 22 predictor variables. One common approach to resolve the problem of dealing with a large number of predictor variables might be to attempt to select a smaller subset of "key variables" as predictors for the regression, by using various variable selection methods, principal components methods, or other more ad hoc procedures. Unfortunately, this type of approach could lead to discarding important information that relates the response variables to the predictors, which might never be detected, and also to loss of precision and quality of the regression model. Therefore, to avoid this possibility, we consider the more systematic approach of using reduced-rank regression methods.

Because of the nature of the set of predictor variables, as a "natural" reduced-rank model for these data we consider the model (3.13) of Section 3.1, $Y_k = ABX_k + DZ_k + \epsilon_k$. Here, we take the "primary" or full-rank set of predictor variables as $Z = (1, x_{21}, x_{22})'$, the wall temperature and solvent feed rate variables, and the reduced-rank set as $X = (x_1, \dots, x_{20})'$, the 20 temperature measurements along the reactor. To determine an appropriate rank for the coefficient matrix $C = AB$ of the variables X_k, we perform a partial canonical correlation analysis of Y_k on X_k, after eliminating the effects of Z_k, and let $\widehat{\rho}_{1j}$, $j = 1, \dots, m$, denote the sample partial canonical correlations. The LR test statistic $\mathcal{M} = -[T - n + (n_1 - m - 1)/2] \sum_{j=r_1+1}^{m} \log(1 - \widehat{\rho}_{1j}^2)$, with $n = 23$ and $n_1 = 20$, discussed in Section 3.1 in relation to (3.12), is used for $r_1 = 0, 1, \dots, 5$, to test the rank of C, H_0: $\text{rank}(C) \leq r_1$. The results are presented in Table 3.1, and they strongly indicate that the rank of C can be taken as equal to two.

We then obtain the ML estimates for the model with rank $r_1 = 2$ by first finding the normalized eigenvectors of the matrix

$$\widehat{R} = \widetilde{\Sigma}_{\epsilon\epsilon}^{-1/2} \widehat{\Sigma}_{yx.z} \widehat{\Sigma}_{xx.z}^{-1} \widehat{\Sigma}_{xy.z} \widetilde{\Sigma}_{\epsilon\epsilon}^{-1/2}$$

associated with its two largest eigenvalues, $\widehat{\lambda}_{11}^2$ and $\widehat{\lambda}_{12}^2$, in Table 3.1. These normalized eigenvectors are found to be

$$\widehat{V}_1 = (0.15799, 0.26926, -0.37958, 0.57403, 0.46378, 0.46244)',$$

$$\widehat{V}_2 = (0.13435, 0.03954, 0.81760, 0.43827, -0.21363, 0.27241)'.$$

Maximum likelihood estimates of the matrix factors A and B are then computed as $\widehat{A}^{(2)} = \widetilde{\Sigma}_{\epsilon\epsilon}^{1/2} \widehat{V}_{(2)}$ and $\widehat{B}^{(2)} = \widehat{V}_{(2)}' \widetilde{\Sigma}_{\epsilon\epsilon}^{-1/2} \widehat{\Sigma}_{yx.z} \widehat{\Sigma}_{xx.z}^{-1}$, with $\widehat{V}_{(2)} = [\widehat{V}_1, \widehat{V}_2]$,

Table 3.1 Summary of results for LR tests on rank of coefficient matrix for chemometrics data

Eigenvalues $\widehat{\lambda}_{1j}^2$	Partial canonical correlations $\widehat{\rho}_{1j}$	Rank $\leq r_1$	$\mathcal{M}$ = LR statistic	d.f.	Critical value (5%)
57.198	0.991		310.53	120	146.57
6.427	0.930	1	150.00	95	118.75
1.119	0.727	2	0.80	72	92.81
0.667	0.633	3	41.14	51	68.67
0.396	0.533	4	20.95	32	46.19
0.217	0.422	5	7.76	15	25.00

and the ML estimate of the coefficient matrix D is obtained from $\widehat{D} = (\widehat{\Sigma}_{yz} - \widehat{A}^{(2)}\widehat{B}^{(2)}\widehat{\Sigma}_{xz})\widehat{\Sigma}_{zz}^{-1}$. The ML estimates for A and D are found to be

$$\widehat{A}^{(2)\prime} = \begin{bmatrix} 0.039611 & 0.108042 & -0.047137 & 0.196351 & 0.076314 & 0.074153 \\ 0.067271 & 0.037769 & 0.452448 & 0.229329 & 0.082096 & 0.091330 \end{bmatrix}$$

and

$$\widehat{D}' = \begin{bmatrix} 24.5251 & 18.9297 & 109.5563 & 97.6315 & 41.2813 & 40.7216 \\ -0.3685 & -0.2129 & -1.2063 & -1.0284 & -0.3756 & -0.3943 \\ 5.4768 & 27.6520 & -0.8007 & -1.4021 & -0.4906 & -0.5444 \end{bmatrix}.$$

Essentially, all of the coefficient estimates in $\widehat{A}$ and $\widehat{D}$ are highly significant. On examination of the two rows of the ML estimate $\widehat{B}^{(2)}$, it is found that the two linear combinations in the "transformed set of predictor variables" of reduced dimension two, $X_k^* = (x_{1k}^*, x_{2k}^*)' = \widehat{B}^{(2)} X_k$, are roughly approximated by

$$\begin{aligned} x_1^* &\approx -114x_1 - 121x_2 + 154x_3 + 58x_{14} - 85x_{15} \\ &\quad + 124x_{16} - 245x_{17} - 108x_{18} + 191x_{19} + 52x_{20}, \\ x_2^* &\approx -175x_2 + 93x_3 - 42x_{11} - 63x_{13} + 94x_{14} - 59x_{15} + 49x_{19} - 108x_{20}. \end{aligned}$$

Thus, it is found that certain of the temperature variables, particularly those in the range $x_5 - x_{10}$, do not play a prominent role in the modeling of the response variables, although a relatively large number among the 20 temperature variables do contribute substantially to the two linear combination predictor variables x_1^* and x_2^*. The total number of estimated coefficients in the regression model with the reduced-rank of two imposed on C is $6 * 3 + 6 * 2 + 2 * 20 - 4 = 66$ as compared to $6 * 23 = 138$ in the full-rank ordinary LS fitted model. The "average" number of parameters per regression equation in the reduced-rank model is thus 11, as opposed to 23 in the full-rank model. An "approximate unbiased" estimator of the error covariance matrix $\Sigma_{\epsilon\epsilon}$ under the reduced-rank model may be constructed as

$$\begin{aligned} \overline{\Sigma}_{\epsilon\epsilon} &= \frac{1}{56-11}\widetilde{\widehat{\epsilon}}\widetilde{\widehat{\epsilon}}' \\ &= \begin{bmatrix} 0.03260 & 0.04556 & 0.05621 & 0.01363 & 0.01874 & 0.01023 \\ 0.04556 & 0.36888 & 0.05422 & -0.01976 & -0.00776 & -0.00774 \\ 0.05621 & 0.05422 & 0.40822 & 0.11388 & 0.11901 & 0.05683 \\ 0.01363 & -0.01976 & 0.11388 & 0.18082 & 0.04824 & 0.02197 \\ 0.01874 & -0.00776 & 0.11901 & 0.04824 & 0.06057 & 0.02345 \\ 0.01023 & -0.00774 & 0.05683 & 0.02197 & 0.02345 & 0.03232 \end{bmatrix}, \end{aligned}$$

with $\text{tr}(\overline{\Sigma}_{\epsilon\epsilon}) = 1.08341$, where $\widehat{\epsilon} = \mathbf{Y} - \widehat{A}\widehat{B}\mathbf{X} - \widehat{D}\mathbf{Z}$. The diagonal elements of the estimate $\overline{\Sigma}_{\epsilon\epsilon}$ above, when compared to the corresponding entries of the unbiased estimate of $\Sigma_{\epsilon\epsilon}$ under full-rank LS as reported earlier, show substantial reductions in the estimated error variances for the response variables y_3, y_4, and y_6 (of around 14% on average), slight reductions for y_1 and y_2 (around 3% average), and only slight increase for y_5 (around 4%). The reduced-rank model (3.13) thus provides considerable simplification over the full-rank model fitted by LS and generally provides a better fit of model as well. In the reduced-rank model results, it might also be of interest and instructive to examine in some detail the vectors $\widehat{\ell}_j = \widetilde{\Sigma}_{\epsilon\epsilon}^{-1/2}\widehat{V}_j$, $j = 3, \ldots, 6$, which provide estimates for linear combinations $y^*_{jk} = \widehat{\ell}'_j Y_k$ of the response variables that, supposedly, are little influenced by the set of temperature variables. For example, two such linear combinations are estimated to be, approximately, $y^*_6 \approx 1.51y_1 - 1.84y_2 + 2.96y_5$ and $y^*_5 \approx -4.98y_1 + 2.40y_4 - 2.94y_6$. Interpretations for these linear combinations require further knowledge of the subject matter which goes beyond the scope of this illustrative example.

An additional application of the reduced-rank methods in the context of this example is also suggested. One of the goals of the study by Skagerberg et al. (1992) was to "monitor the performance of the process by means of multivariate control charts" over time, to enable quick detection of changes early in the process. In the case of a univariate measurement, this is typically accomplished by use of statistical process control charts. For this multivariate chemometrics data set, one might require as many as 22 charts for the predictor variables and 6 for the response variables. Because of the association established between the response and the predictor variables in the reduced-rank model, involving only two linear combinations of the temperature profile variables, a small number of linear combinations only might be used to effectively monitor the process. Skagerberg et al. (1992) suggested a similar idea based on linear combinations that result from partial least squares regression methods.

3.4 Both Regression Matrices of Lower Ranks: Model and Its Applications

The assumption that the coefficient matrix of the regressors Z_k in (3.1) is of full rank and cannot be collapsed may not hold in practice. We propose a model that assumes that there is a natural division of regressor variables into two sets, and each coefficient matrix can be of reduced rank. This model seems to have wider applications not only for analyzing cross-sectional data where observations are assumed to be independent but also for analyzing chronological (time series) data. The usefulness of the extended model will be illustrated with an application in meteorology where the total ozone measurements taken at five European stations are related to temperature measurements taken at various pressure levels. This model will be shown also to lead to some intuitively simpler models which cannot be otherwise identified as applicable to this data set. Thus, the merit of reduced-rank regression techniques will become more apparent.

The extended reduced-rank model is given as in (3.1) by

$$Y_k = CX_k + DZ_k + \epsilon_k, \qquad k = 1, \ldots, T, \tag{3.21}$$

with the assumption that

$$\begin{aligned} \text{rank}(C) &= r_1 \leq \min(m, n_1) \\ \text{rank}(D) &= r_2 \leq \min(m, n_2) \end{aligned} \tag{3.22}$$

accommodating the possibility that the dimension reduction could be different for each set. Hence we can write $C = AB$, where A is an $m \times r_1$ matrix and B is an $r_1 \times n_1$ matrix. Similarly, we can write $D = EF$ where E is an $m \times r_2$ matrix and F is an $r_2 \times n_2$ matrix.

A situation that readily fits into this setup is given in the appendix of the paper by Jöreskog and Goldberger (1975). Suppose there are two latent variables y_1^* and y_2^* each determined by its own set of exogenous causes and its own disturbances so that

$$y_{1k}^* = \beta_1' X_k + V_{1k}, \quad y_{2k}^* = \beta_2' Z_k + V_{2k}.$$

The observable indicators Y_k are assumed to be related to the latent variables linearly, giving $Y_k = \alpha_1 y_{1k}^* + \alpha_2 y_{2k}^* + U_k$. Therefore the reduced-rank form equations are

$$Y_k = \alpha_1 \beta_1' X_k + \alpha_2 \beta_2' Z_k + \epsilon_k = CX_k + DZ_k + \epsilon_k,$$

where both C and D are of unit rank. This example can be formally expanded into two blocks of latent variables each explained by their own causal set. This would result in a higher dimensional example for the extended model.

With the factorization for C and D, we can write the model (3.21) as

$$Y_k = ABX_k + EFZ_k + \epsilon_k, \qquad k = 1, \ldots, T. \tag{3.23}$$

Following Brillinger's interpretation, the n_1 component vector X_k and the n_2 component vector Z_k that arise from two different sources might be regarded as determinants of a "message" Y_k having m components, but there are restrictions on the number of channels through which the message can be transmitted. These restrictions are of different orders for X_k and Z_k. We assume that these vectors X_k and Z_k cannot be assembled at one place. This means that they cannot be collapsed jointly but have to be condensed individually as and when they arise from their sources. Thus BX_k and FZ_k act as codes from their sources of origin, and on receipt they are premultiplied by A and E, respectively, and are summed to yield Y_k, subject to error ϵ_k.

Estimation and inference aspects for the extended reduced-rank model (3.23) were considered by Velu (1991). While the focus of that paper and the remainder of this chapter is on the estimation of component matrices, the complementary feature of linear constraints, which are of different dimensions and forms for the different coefficient matrices, has been shown to be useful in multidimensional scaling. Takane, Kiers and de Leeuw (1995) consider a further extension of model (3.23), to allow for more than two sets of regressor variables and provide a convergent alternating least squares algorithm to compute the least squares estimates of the component matrices of the reduced-rank coefficient matrices.

A model similar to (3.23) can be used to represent the time lag dependency of a multiple autoregressive time series $\{Y_t\}$. For illustration, let $X_t = Y_{t-1}$ and $Z_t = Y_{t-2}$. Model (3.23) for the time series $\{Y_t\}$ can then be written as $Y_t = ABY_{t-1} + EFY_{t-2} + \epsilon_t$, which provides a framework for dimension reduction for multiple time series data. An even more condensed model follows if, further, $B = F$ when $\text{rank}(C) = \text{rank}(D) = r$. Then $Y_t = A(BY_{t-1}) + E(BY_{t-2}) + \epsilon_t$, and this model has an interpretation in terms of index variables. In this case the lower-dimensional $Y_t^* = BY_t$ acts as an index vector driving the higher-dimensional time series Y_t. These dimension reducing aspects in a multiple autoregressive time series context are discussed in Reinsel (1983), Velu, Reinsel and Wichern (1986), and Ahn and Reinsel (1988). The reduced-rank model discussed in Section 3.1 would also be sensible to apply in the multiple time series context. For example, we might consider $Y_t = DY_{t-1} + ABY_{t-2} + \epsilon_t$, where the coefficient matrix of the lag one variables Y_{t-1} may be of full rank. An interpretation is that because of the closer ordering in time to Y_t, all variables in Y_{t-1} are needed for Y_t, but some (linear combinations of) variables in the more distant lagged term Y_{t-2} are not needed. This provides an example of a "nested reduced-rank" autoregressive model, and a general class of models of this type was considered by Ahn and Reinsel (1988). Autoregressive models with reduced rank are discussed in further detail in Chapter 5.

3.5 Estimation and Inference for the Model

Before we present the efficient estimators of the coefficient matrices for model (3.23), it is instructive to note that combining X_k and Z_k through blocking, as in the conventional reduced-rank model considered in Chapter 2, may not necessarily produce a reduced-rank structure. The extended model (3.23) can be written as

$$Y_k = C^* \begin{bmatrix} X_k \\ Z_k \end{bmatrix} + \epsilon_k,$$

where $C^* = [C, D]$ is an $m \times (n_1 + n_2)$ matrix. From matrix theory, we have

$$\min(m, r_1 + r_2) \geq \text{rank}(C^*) \geq \max(r_1, r_2).$$

Treating the above model as a classical reduced-rank regression model, when $r = \text{rank}(C^*)$, would yield the factorization

$$C^* = GH = [GH_1, GH_2],$$

where G is an $m \times r$ matrix and H is an $r \times (n_1 + n_2)$ matrix with partitions H_1 and H_2 of dimensions $r \times n_1$ and $r \times n_2$, respectively. Note that both submatrices H_1 and H_2 in the partitioned matrix H have (at most) rank r, but this formulation does not yield the desired factorization. Even if the individual matrices H_1 and H_2 are of lower ranks, when put in partitioned form, the partitioned matrix H may be of full rank (i.e., $r = m$) and the model may have no particular reduced-rank structure at all. When m is assumed to be greater than $r_1 + r_2$, the above blocking amounts to overestimating the rank of each individual matrix by $(r_1 + r_2) - \max(r_1, r_2)$. There is also a conceptual difference between the above model and the model in (3.23). This is best illustrated through the simple case $r = r_1 + r_2$ which we now consider.

The "correct" factorization we are seeking under model (3.23) can be written as

$$[C, D] = [AB, EF] = [A, E] \begin{bmatrix} B & 0 \\ 0 & F \end{bmatrix},$$

where the orders of A, B, E, and F are defined as before. The factorization we would arrive at if we worked through a "blocked" version of model (3.23) is

$$[C, D] = GH = [G_1, G_2] \begin{bmatrix} H_{11} & H_{12} \\ H_{21} & H_{22} \end{bmatrix},$$

where G_1 is an $m \times r_1$ matrix, G_2 is an $m \times r_2$ matrix, H_{11} is an $r_1 \times n_1$, H_{12} is an $r_1 \times n_2$, H_{21} is an $r_2 \times n_1$, and H_{22} is an $r_2 \times n_2$. Recall that the right-hand side matrix in the factorization is used to form codes out of the regressors, and in the original formulation of the model (3.23), each regressor set, X_k and Z_k, has its own codes. This feature is violated if we work with the blocked model above.

3.5.1 *Efficient Estimator*

To identify the individual components in the extended reduced-rank model (3.23), e.g., A and B from C, it is necessary to impose certain normalization conditions on them. Normalizations similar to those used in Chapter 2 for the reduced-rank regression model with one set of regressor variables are chosen for this model. For the sample situation, these normalizations are

$$\begin{aligned} A'\Gamma A &= I_{r_1}; \quad B(\tfrac{1}{T}\mathbf{X}\mathbf{X}')B' = \Lambda_{r_1}^2 \\ E'\Gamma E &= I_{r_2}; \quad F(\tfrac{1}{T}\mathbf{Z}\mathbf{Z}')F' = \Lambda_{r_2}^{*2}, \end{aligned} \tag{3.24}$$

where $\mathbf{X} = [X_1, X_2, \ldots, X_T]$ and $\mathbf{Z} = [Z_1, Z_2, \ldots, Z_T]$ are data matrices, Γ is a positive-definite matrix, and $\Lambda_{r_1}^2$ and $\Lambda_{r_2}^{*2}$ are diagonal matrices. The Γ matrix can be specified by the analyst, whereas the diagonal elements of Λ_{r_1} and $\Lambda_{r_2}^*$ are dependent on C and D, respectively. Let $\theta = (\alpha', \beta', \gamma', \delta')'$ denote the vector of unknown regression parameters in model (3.23), where $\alpha = \text{vec}(A')$, $\beta = \text{vec}(B')$, $\gamma = \text{vec}(E')$, and $\delta = \text{vec}(F')$. To arrive at the "efficient" estimator of θ, consider the criterion

$$S_T(\theta) = \frac{1}{2T}\text{tr}\{\Gamma(\mathbf{Y} - AB\mathbf{X} - EF\mathbf{Z})(\mathbf{Y} - AB\mathbf{X} - EF\mathbf{Z})'\}, \tag{3.25}$$

where $\mathbf{Y} = [Y_1, Y_2, \ldots, Y_T]$ is the data matrix of the dependent variables. The estimates are obtained by minimizing criterion (3.25) subject to the normalization conditions (3.24). Criterion (3.25) is the likelihood criterion when the errors ϵ_k are assumed to be normal, and $\Gamma = \widehat{\Sigma}_{\epsilon\epsilon}^{-1}$ where $\widehat{\Sigma}_{\epsilon\epsilon} = \frac{1}{T}(\mathbf{Y} - \widehat{A}\widehat{B}\mathbf{X} - \widehat{E}\widehat{F}\mathbf{Z})(\mathbf{Y} - \widehat{A}\widehat{B}\mathbf{X} - \widehat{E}\widehat{F}\mathbf{Z})'$ is the maximum likelihood estimate of the error covariance matrix. For this reason, the resulting estimators can be called efficient. However, for the asymptotic theory, we shall initially take $\Gamma = \Sigma_{\epsilon\epsilon}^{-1}$ as known in criterion (3.25), for convenience. The asymptotic theory of the proposed estimator follows from Robinson (1973) or Silvey (1959) and will be presented later.

Several procedures are available for computing the efficient estimators. The first order equations that result from the criterion (3.25) refer only to the necessary conditions for the minimum to occur and so do not guarantee the global minimum. Therefore, we suggest various computational schemes to check on the optimality of a solution found by an alternate method.

The first order equations resulting from (3.25) are

$$\frac{\partial S_T(\theta)}{\partial A'} = -\frac{1}{T}B\mathbf{X}(\mathbf{Y} - AB\mathbf{X} - EF\mathbf{Z})'\Gamma \tag{3.26a}$$

$$\frac{\partial S_T(\theta)}{\partial B'} = -\frac{1}{T}\mathbf{X}(\mathbf{Y} - AB\mathbf{X} - EF\mathbf{Z})'\Gamma A \tag{3.26b}$$

$$\frac{\partial S_T(\theta)}{\partial E'} = -\frac{1}{T}F\mathbf{Z}(\mathbf{Y} - AB\mathbf{X} - EF\mathbf{Z})'\Gamma \tag{3.26c}$$

$$\frac{\partial S_T(\theta)}{\partial F'} = -\frac{1}{T}\mathbf{Z}(\mathbf{Y} - AB\mathbf{X} - EF\mathbf{Z})'\Gamma E \tag{3.26d}$$

An iterative procedure that utilizes only the first order equations is the partial least squares procedure [see Gabriel and Zamir (1979)]. The procedure is easy to use when compared with other suggested procedures and is now described. To start assume that the matrices B and F in model (3.23) are known. The estimates of A

and E are easily obtained by least squares regression of Y_k on $[BX_k, FZ_k]$. Thus, starting with estimates $\widehat{B}$ and $\widehat{F}$ that satisfy the constraints (3.24), $\widehat{A}$ and $\widehat{E}$ can be obtained by the usual multivariate regression procedure. The resulting $\widehat{A}$ and $\widehat{E}$ are then adjusted to satisfy the normalization constraints (3.24) before moving to the next step to obtain new estimates $\widehat{B}$ and $\widehat{F}$ of B and F.

The first order equations $\partial S_T(\theta)/\partial B = 0$ lead to the solution for B as

$$\widehat{B} = \widehat{A}'\Gamma\left[\frac{1}{T}(\mathbf{Y} - \widehat{E}\widehat{F}\mathbf{Z})\mathbf{X}'\right]\left[\frac{1}{T}\mathbf{X}\mathbf{X}'\right]^{-1}. \tag{3.27}$$

Substituting this $\widehat{B}$ into the first order equation $\partial S_T(\theta)/\partial F = 0$, we arrive at the equations

$$\widehat{F}\left(\frac{1}{T}\mathbf{Z}\mathbf{Z}'\right) - Q\widehat{F}R = P,$$

where $R = \frac{1}{T}\mathbf{Z}\mathbf{X}'(\frac{1}{T}\mathbf{X}\mathbf{X}')^{-1}\frac{1}{T}\mathbf{X}\mathbf{Z}'$, $Q = (\widehat{E}'\Gamma\widehat{A})(\widehat{A}'\Gamma\widehat{E})$, and

$$P = \widehat{E}'\Gamma\left(\frac{1}{T}\mathbf{Y}\mathbf{Z}'\right) - (\widehat{E}'\Gamma\widehat{A})\widehat{A}'\Gamma\left(\frac{1}{T}\mathbf{Y}\mathbf{X}'\right)\left(\frac{1}{T}\mathbf{X}\mathbf{X}'\right)^{-1}\left(\frac{1}{T}\mathbf{X}\mathbf{Z}'\right).$$

We can readily solve for $\widehat{F}$ by conveniently stacking the columns of $\widehat{F}$, using the "vec" operator and the Kronecker products, from

$$\left[\left(\frac{1}{T}\mathbf{Z}\mathbf{Z}' \otimes I\right) - (R \otimes Q)\right]\text{vec}(\widehat{F}) = \text{vec}(P). \tag{3.28}$$

Observe that with the regression estimates $\widehat{A}$ and $\widehat{E}$, using (3.28), we can solve for $\widehat{F}$ and then use all of these estimates in (3.27) to solve for $\widehat{B}$. The normalizations implied by the conditions (3.24) can be carried out using the spectral decomposition of matrices, e.g., Rao (1973, p. 39).

An alternative procedure that can be formulated is along the lines of the modification of the Newton–Raphson method due to Aitchison and Silvey (1958). The details of this procedure will be made clearer when the asymptotic inference results are presented later, but the essential steps can be described as follows. Let $h(\theta)$ be an $r^* = r_1^2 + r_2^2$ vector obtained by stacking the normalization conditions (3.24) without duplication, and let the matrix of first partial derivatives of these constraints be $H_\theta = (\partial h_j(\theta)/\partial\theta_i)$. Note that H_θ is of dimension $s \times r^*$ where $s = r_1(m + n_1) + r_2(m + n_2)$. The first order equations that are obtained by minimizing the criterion $S_T(\theta) - h(\theta)'\lambda$, where λ is a vector of Lagrange multipliers, can be expanded through Taylor's theorem to arrive at

$$\begin{bmatrix} B_\theta & -H_\theta \\ -H_\theta' & 0 \end{bmatrix}\begin{bmatrix} \sqrt{T}(\widehat{\theta} - \theta) \\ \sqrt{T}\widehat{\lambda} \end{bmatrix} = \begin{bmatrix} -\sqrt{T}\partial S_T(\theta)/\partial\theta \\ 0 \end{bmatrix} \tag{3.29}$$

In the above expression, $\partial S_T(\theta)/\partial\theta$ is essentially (3.26) arranged in a vector form and B_θ is the expected value of the matrix of second-order derivatives, $\{\partial^2 S_T(\theta)/\partial\theta_i\partial\theta_j\}$.

To obtain explicit expressions for these quantities, we write the matrix form of model (3.23), $\mathbf{Y} = C\mathbf{X} + D\mathbf{Z} + \boldsymbol{\epsilon} = AB\mathbf{X} + EF\mathbf{Z} + \boldsymbol{\epsilon}$, in vector form as

$$\begin{aligned} y = \text{vec}(\mathbf{Y}') &= (I_m \otimes \mathbf{X}')\text{vec}(C') + (I_m \otimes \mathbf{Z}')\text{vec}(D') + e \\ &= (I_m \otimes \mathbf{X}'B')\alpha + (I_m \otimes \mathbf{Z}'F')\gamma + e \\ &= (A \otimes \mathbf{X}')\beta + (E \otimes \mathbf{Z}')\delta + e, \end{aligned} \tag{3.30}$$

where $e = \text{vec}(\boldsymbol{\epsilon}')$, and we have used the relations $\text{vec}(C') = \text{vec}(B'A') = (I_m \otimes B')\text{vec}(A') = (A \otimes I_{n_1})\text{vec}(B')$ and similarly for $\text{vec}(D')$. Then the criterion in (3.25) can be written as $S_T(\theta) = \frac{1}{2T}e'(\Gamma \otimes I_T)e$ with the error vector e as in (3.30), and so we find that $\partial S_T(\theta)/\partial\theta$ can be expressed as

$$\frac{\partial S_T(\theta)}{\partial\theta} = \frac{1}{T}\frac{\partial e'}{\partial\theta}(\Gamma \otimes I_T)e = -\frac{1}{T}M'\begin{bmatrix}\Gamma \otimes \mathbf{X}\\ \Gamma \otimes \mathbf{Z}\end{bmatrix}e, \tag{3.31}$$

where $M' = \text{Diag}[M_1', M_2']$ is block diagonal with the matrices M_1' and M_2' in its main diagonals,

$$M_1 = [I_m \otimes B', A \otimes I_{n_1}] \equiv \left(\frac{\partial\text{vec}(C')}{\partial\alpha'}, \frac{\partial\text{vec}(C')}{\partial\beta'}\right)$$

and

$$M_2 = [I_m \otimes F', E \otimes I_{n_2}] \equiv \left(\frac{\partial\text{vec}(D')}{\partial\gamma'}, \frac{\partial\text{vec}(D')}{\partial\delta'}\right).$$

Notice that the vector of partial derivatives $\partial e/\partial\theta'$ that is needed in (3.31) is easily obtained from (3.30) as $\partial e/\partial\theta' = -[I_m \otimes \mathbf{X}'B', A \otimes \mathbf{X}', I_m \otimes \mathbf{Z}'F', E \otimes \mathbf{Z}']$. It also follows, noting that $E(e) = 0$, that the matrix B_θ can be written as

$$\begin{aligned} B_\theta \equiv E\left\{\frac{\partial^2 S_T(\theta)}{\partial\theta\partial\theta'}\right\} &= \frac{1}{T}E\left\{\frac{\partial e'}{\partial\theta}(\Gamma \otimes I_T)\frac{\partial e}{\partial\theta'}\right\} \\ &= M'\frac{1}{T}E\begin{bmatrix}\Gamma \otimes \mathbf{XX}' & \Gamma \otimes \mathbf{XZ}'\\ \Gamma \otimes \mathbf{ZX}' & \Gamma \otimes \mathbf{ZZ}'\end{bmatrix}M \\ &\equiv M'\Sigma^* M, \end{aligned} \tag{3.32}$$

where

$$\Sigma^* = \begin{bmatrix} \Gamma \otimes \Sigma_{xx} & \Gamma \otimes \Sigma_{xz} \\ \Gamma \otimes \Sigma_{zx} & \Gamma \otimes \Sigma_{zz} \end{bmatrix}.$$

The basic idea in this procedure is to solve for the unknown quantities using Eq. (3.29), where in practice the sample version of B_θ (that is, the sample version of Σ^*) is used in which the expected value matrices Σ_{xx}, Σ_{xz}, and Σ_{zz} are replaced by their corresponding sample quantities $\widehat{\Sigma}_{xx} = \frac{1}{T}\mathbf{X}\mathbf{X}'$, $\widehat{\Sigma}_{xz} = \frac{1}{T}\mathbf{X}\mathbf{Z}'$, and $\widehat{\Sigma}_{zz} = \frac{1}{T}\mathbf{Z}\mathbf{Z}'$, respectively. There are some essential differences between this procedure and the partial least squares procedure. This procedure uses the second-order derivatives of $S_T(\theta)$ in addition to the first-order derivatives. In addition, the estimates and also the constraints are simultaneously iterated. Because of these differences, the dimensions of the estimators are much larger. An added problem to this procedure is that we cannot use the results on inverting the partitioned matrix, because the rank of B_θ is the same as the rank of M' (which equals $r_1(m+n_1-r_1)+r_2(m+n_2-r_2)$) and therefore B_θ is singular. But the singularity can be removed using a modification suggested by Silvey (1959, p. 399); see also Neuenschwander and Flury (1997). Observe that the matrix H_θ is of rank $r_1^2 + r_2^2$ and it can be demonstrated that $B_\theta + H_\theta H_\theta'$ is of full rank $r_1(m+n_1) + r_2(m+n_2)$. Using this fact, we can rewrite the Eq. (3.29) as

$$\begin{bmatrix} B_\theta + H_\theta H_\theta' & -H_\theta \\ -H_\theta' & 0 \end{bmatrix} \begin{bmatrix} \sqrt{T}(\widehat{\theta} - \theta) \\ \sqrt{T}\widehat{\lambda} \end{bmatrix} = \begin{bmatrix} -\sqrt{T}\,\partial S_T(\theta)/\partial\theta \\ 0 \end{bmatrix}.$$

The iterated estimate of θ at the ith step is given as

$$\begin{bmatrix} \widehat{\theta}_i \\ \widehat{\lambda}_i \end{bmatrix} = \begin{bmatrix} \widehat{\theta}_{i-1} \\ 0 \end{bmatrix} + \begin{bmatrix} B_{\widehat{\theta}_{i-1}} + H_{\widehat{\theta}_{i-1}} H_{\widehat{\theta}_{i-1}}' & -H_{\widehat{\theta}_{i-1}} \\ -H_{\widehat{\theta}_{i-1}}' & 0 \end{bmatrix}^{-1} \begin{bmatrix} -\partial S_T(\widehat{\theta}_{i-1})/\partial\theta \\ 0 \end{bmatrix} \tag{3.33}$$

and using a stopping rule based on either the successive differences between the $\widehat{\theta}_i$ or the magnitudes of the Lagrange multipliers $\widehat{\lambda}_i$ or $\partial S_T(\widehat{\theta}_{i-1})/\partial\theta$, the estimates can be calculated. In general we would expect this procedure to perform better numerically than the partial least squares procedure, because it uses more information. However, in practical applications, the size of the matrix B_θ can be quite large.

A third estimation procedure, which is not (statistically) efficient relative to the above efficient estimator but intuitively appealing, is presented in the discussion below as an alternative estimator.

3.5.2 *An Alternative Estimator*

The proposed alternative estimator is similar to the estimator of the parameters in the model (3.3) of Section 3.1. It is inefficient but intuitive as mentioned earlier and it is based on the full-rank coefficient estimates. Recall that in the model (3.1) or (3.3), D was assumed to be full rank but C could be of reduced rank. It was noted at the end of Section 3.1 that, in the population, solutions for the components A and B of the matrix $C \equiv AB$ under model (3.3) can be obtained as follows: the columns of $\Sigma_{xx.z}^{1/2}B'$ are chosen to be the eigenvectors corresponding to the largest r_1 eigenvalues of $\Sigma_{xx.z}^{-1/2}\Sigma_{xy.z}\Gamma\Sigma_{yx.z}\Sigma_{xx.z}^{-1/2}$ and $A = \Sigma_{yx.z}B'(B\Sigma_{xx.z}B')^{-1}$. The matrix D then is obtained as $D = \Sigma_{yz}\Sigma_{zz}^{-1} - AB\Sigma_{xz}\Sigma_{zz}^{-1}$. The r_1 eigenvalues can be related to partial canonical correlations between Y_k and X_k eliminating Z_k and are used to estimate the rank of C (see Section 3.1). The corresponding estimates in the sample are easily computable using the output from the standard documented software packages as in the case of the reduced-rank model discussed in Chapter 2.

The alternative estimator is in the same spirit as the estimator in model (3.1) but differs only in the estimation of the matrix C. In Theorem 3.1, the criterion was decomposed into two parts:

$$\begin{aligned}&\operatorname{tr}\{E[\Gamma^{1/2}(Y - ABX^*)(Y - ABX^*)'\Gamma^{1/2}]\}\\&\quad + \operatorname{tr}\{E[\Gamma^{1/2}(Y - D^*Z)(Y - D^*Z)'\Gamma^{1/2}]\}.\end{aligned}$$

Here we modify the second part as $\operatorname{tr}\{E[\Gamma^{1/2}(Y - DZ^*)(Y - DZ^*)'\Gamma^{1/2}]\}$ where $D = EF$ and $Z^* = Z - \Sigma_{zx}\Sigma_{xx}^{-1}X$, leaving the first part as it is in the procedure of Section 3.1. This is equivalent to assuming that C is of full rank when estimating the components of the matrix D. Therefore the columns of $\Sigma_{zz.x}^{1/2}F'$ are given by the eigenvectors of $\Sigma_{zz.x}^{-1/2}\Sigma_{zy.x}\Gamma\Sigma_{yz.x}\Sigma_{zz.x}^{-1/2}$ and $E = \Sigma_{yz.x}F'(F\Sigma_{zz.x}F')^{-1}$. Stated differently, in each case (either C or D), in the alternative procedure we first compute the full-rank estimator and then project it to a lower dimension (reduced rank) individually, ignoring the simultaneous nature of restrictions involved in the reduced-rank problem. The alternative estimators of the components A, B and E, F are simply the corresponding sample versions of the above expressions.

Implicit in the solution is the fact that the matrices B and F are normalized with respect to partial covariance matrices $\Sigma_{xx.z}$ and $\Sigma_{zz.x}$ respectively, rather than the covariance matrices Σ_{xx} and Σ_{zz} as for the efficient estimator [see Eq. (3.8)]. For different choices of normalizations, it is assumed that the true parameter belongs to different regions of the parameter space. However, through the transformations from one set of true parameters to the other using the spectral decomposition of matrices, the estimators are made comparable to each other. Thus we provide yet another procedure to arrive at estimates of the extended model (3.23).

3.5.3 Asymptotic Inference

The asymptotic distribution of the maximum likelihood estimator and the alternative estimator can be derived using the theoretical results from Robinson (1973) or

Silvey (1959). Before we present the main results, we observe the following points. It must be noted that with the choice of $\Gamma = \widehat{\Sigma}_{\epsilon\epsilon}^{-1}$, the normalization conditions (3.24) imposed on θ vary with sample quantities. To apply Silvey (1959) results directly for estimators of θ, the constraints are required to be fixed and not vary with sample quantities. The modifications to account for the randomness could be accommodated by considering the asymptotic theory of Silvey (1959) for the combined set of ML estimators $\widehat{\theta}$ and $\widehat{\Sigma}_{\epsilon\epsilon}$ jointly, but explicit details on the asymptotic distribution of the ML estimator $\widehat{\theta}$ would then become more complicated. Hence, we do not consider results for this case explicitly. As in the case of the asymptotic results of Theorem 2.4 for the single regressor model, we use $\Gamma = \Sigma_{\epsilon\epsilon}^{-1}$ as known instead of $\Gamma = \widehat{\Sigma}_{\epsilon\epsilon}^{-1}$, to obtain more convenient explicit results on the asymptotic covariance matrix of the estimator of θ. It is expected that the use of $\Gamma = \Sigma_{\epsilon\epsilon}^{-1}$ instead of $\Gamma = \widehat{\Sigma}_{\epsilon\epsilon}^{-1}$ will not have much impact on the asymptotic distribution. Suppose we let (3.25) stand for the criterion with the choice $\Gamma = \widehat{\Sigma}_{\epsilon\epsilon}^{-1}$ and let the same criterion and the constraints in (3.24) for the choice of fixed $\Gamma = \Sigma_{\epsilon\epsilon}^{-1}$ be denoted as $S_T^*(\theta)$.

Theorem 3.2 *Assume the limits*

$$\lim_{T\to\infty}\left(\frac{1}{T}\Sigma_k X_k X_k'\right) = \Sigma_{xx}, \quad \lim_{T\to\infty}\left(\frac{1}{T}\Sigma_k Z_k Z_k'\right) = \Sigma_{zz},$$

and

$$\lim_{T\to\infty}\left(\frac{1}{T}\Sigma_k X_k Z_k'\right) = \Sigma_{xz} \text{ exist almost surely and } \begin{pmatrix} \Sigma_{xx} & \Sigma_{xz} \\ \Sigma_{zx} & \Sigma_{zz} \end{pmatrix}$$

and $\Sigma_{\epsilon\epsilon}$ are positive-definite. Let $\widehat{\theta}$ denote the estimator that absolutely minimizes the criterion (3.25) subject to the normalizations (3.24), with the choice $\Gamma = \Sigma_{\epsilon\epsilon}^{-1}$ known. Then as $T \to \infty$, $\widehat{\theta} \to \theta$ almost surely and $\sqrt{T}(\widehat{\theta} - \theta)$ has a limiting multivariate normal distribution with null mean vector and singular covariance matrix

$$W = (B_\theta + H_\theta H_\theta')^{-1} B_\theta (B_\theta + H_\theta H_\theta')^{-1}, \tag{3.34}$$

where $B_\theta = M'\Sigma^* M$ is given in (3.32) and $H_\theta' = \partial h(\theta)/\partial\theta'$.

Proof With assumptions stated in the theorem and with an assumption that the parameter space Θ is compact and if the true parameter is taken to be an interior point of Θ, the almost sure convergence follows from Silvey (1959) results. To prove the limiting distribution part of the theorem, we follow the standard steps. The first-order equations are

$$\frac{\partial S_T^*(\theta)}{\partial \theta} - H_\theta \lambda = 0$$
$$h(\theta) = 0. \tag{3.35}$$

Let $\widehat{\theta}$ and $\widehat{\lambda}$ be solutions of the above equations. Because the partial derivatives of $h_i(\theta)$ are continuous,

$$H_{\widehat{\theta}} = H_\theta + o(1) \tag{3.36a}$$

and

$$h(\widehat{\theta}) = [H_\theta' + o(1)](\widehat{\theta} - \theta). \tag{3.36b}$$

By Taylor's theorem,

$$\frac{\partial S_T^*(\theta)}{\partial \theta}|_{\widehat{\theta}} = \frac{\partial S_T^*(\theta)}{\partial \theta} + \left\{ \frac{\partial^2 S_T^*(\theta)}{\partial \theta \partial \theta'} + o(1) \right\} (\widehat{\theta} - \theta). \tag{3.37}$$

Observe that $\frac{\partial^2 S_T^*(\theta)}{\partial \theta \partial \theta'} \rightarrow B_\theta$ almost surely. Substituting (3.36) and (3.37) in Eq. (3.35) evaluated at $\widehat{\theta}, \widehat{\lambda}$, we arrive at (3.29), which forms the basic components of the asymptotic distribution of $\sqrt{T}(\widehat{\theta} - \theta)$.

If we let $\boldsymbol{\epsilon} = \mathbf{Y} - AB\mathbf{X} - EF\mathbf{Z}$, the first order quantities $\partial S_T^*(\theta)/\partial \theta$ can be written compactly as in (3.31). Observe that the same quantities written in the matrix form in (3.26) are

$$\frac{\partial S_T^*(\theta)}{\partial A'} = -B\left(\frac{1}{T}\mathbf{X}\boldsymbol{\epsilon}'\right)\Gamma$$
$$\frac{\partial S_T^*(\theta)}{\partial B'} = -\left(\frac{1}{T}\mathbf{X}\boldsymbol{\epsilon}'\right)\Gamma A$$
$$\frac{\partial S_T^*(\theta)}{\partial E'} = -F\left(\frac{1}{T}\mathbf{Z}\boldsymbol{\epsilon}'\right)\Gamma$$
$$\frac{\partial S_T^*(\theta)}{\partial F'} = -\left(\frac{1}{T}\mathbf{Z}\boldsymbol{\epsilon}'\right)\Gamma E.$$

The above can be reexpressed, as in (3.31), using the vec operator and writing $\text{vec}(\mathbf{X}\boldsymbol{\epsilon}'\Gamma) = (\Gamma \otimes \mathbf{X})e$ and $\text{vec}(\mathbf{Z}\boldsymbol{\epsilon}'\Gamma) = (\Gamma \otimes \mathbf{Z})e$, as

$$\sqrt{T}\frac{\partial S_T^*(\theta)}{\partial \theta} = -M'\sqrt{T}\begin{pmatrix} \text{vec}(\frac{1}{T}\mathbf{X}\boldsymbol{\epsilon}'\Gamma) \\ \text{vec}(\frac{1}{T}\mathbf{Z}\boldsymbol{\epsilon}'\Gamma) \end{pmatrix}.$$

A simple interpretation for this relation is as

$$\partial S_T^*(\theta)/\partial\theta = (\partial\phi'/\partial\theta)(\partial S_T^*(\theta)/\partial\phi),$$

where $\phi = \text{vec}(C', D') = (\text{vec}(C')', \text{vec}(D')')'$ denotes the vector of "unrestricted" regression coefficient parameters, with $\partial\phi'/\partial\theta = M'$ and

$$\partial S_T^*(\theta)/\partial\phi = -\text{vec}\left(\frac{1}{T}\mathbf{X}\epsilon'\Gamma, \frac{1}{T}\mathbf{Z}\epsilon'\Gamma\right) \equiv -\Gamma^{*\prime}\text{vec}\left(\frac{1}{T}\mathbf{X}\epsilon', \frac{1}{T}\mathbf{Z}\epsilon'\right),$$

with $\Gamma^{*\prime} = \text{Diag}[\Gamma \otimes I_{n_1}, \Gamma \otimes I_{n_2}]$. From the multivariate linear regression theory, we know that

$$\sqrt{T}\begin{pmatrix}\text{vec}(\frac{1}{T}\mathbf{X}\epsilon')\\ \text{vec}(\frac{1}{T}\mathbf{Z}\epsilon')\end{pmatrix} \xrightarrow{d} N(0, \Sigma^{**}),$$

where

$$\Sigma^{**} = \begin{bmatrix}\Sigma_{\epsilon\epsilon}\otimes\Sigma_{xx} & \Sigma_{\epsilon\epsilon}\otimes\Sigma_{xz}\\ \Sigma_{\epsilon\epsilon}\otimes\Sigma_{zx} & \Sigma_{\epsilon\epsilon}\otimes\Sigma_{zz}\end{bmatrix},$$

and, therefore, with $\Gamma = \Sigma_{\epsilon\epsilon}^{-1}$, $\sqrt{T}\frac{\partial S_T^*(\theta)}{\partial\theta} \xrightarrow{d} N(0, B_\theta)$, where $B_\theta = M'\Gamma^{*\prime}\Sigma^{**}\Gamma^*M = M'\Sigma^*M$. The asymptotic covariance matrix of $\widehat{\theta}$ is the relevant partition (upper left block) of the following matrix:

$$\begin{bmatrix}B_\theta & -H_\theta\\ -H_\theta' & 0\end{bmatrix}^{-1}\begin{bmatrix}B_\theta & 0\\ 0 & 0\end{bmatrix}\begin{bmatrix}B_\theta & -H_\theta\\ -H_\theta' & 0\end{bmatrix}^{-1}. \tag{3.38}$$

The above expression follows from (3.29) and the upper left partitioned portion of the resulting matrix product will yield the asymptotic covariance matrix of $\widehat{\theta}$. In order to arrive at the explicit analytical expression W, we can use Silvey (1959) modification, where we replace B_θ by $B_\theta + H_\theta H_\theta'$ in (3.38) in the matrices that require inverses. The standard results for finding the inverse of a partitioned matrix [e.g., see Rao (1973, p. 33)] will yield the asymptotic covariance matrix W.

For the numerical example that will be illustrated in the last section of this chapter, we compute estimates using the partial least squares procedure and their standard errors are calculated via the matrix W in the above theorem. Although the quantity H_θ does not appear to be easily written in a compact form, it can be computed using standard results on matrix differentiation of the quadratic functions of the elements of A, B, E, and F involved in the normalization conditions (3.24).

The asymptotic distribution of the alternative estimator, which was motivated by the model (3.1) of Anderson and which was discussed earlier, can also be considered. A limiting multivariate normal distribution result can be established for this estimator by following the same line of argument as in the proof of Theorem 3.2, but we omit the details. A main feature of the results is that the alternative estimator is asymptotically inefficient relative to the estimator of

Theorem 3.2, although the expression for the asymptotic covariance matrix of the alternative estimator is somewhat cumbersome and so is not easily compared with the asymptotic covariance matrix (3.34) in Theorem 3.2. The proposed alternative estimator, it should be again emphasized, does not have any overall optimality properties corresponding to the extended reduced-rank model (3.23), but is mainly motivated by the model (3.3) in Section 3.1. The usefulness of this estimator is that it provides initial values in the iterative procedures for the efficient estimator and more importantly that the associated partial canonical correlations can be used to specify (identify) the ranks of the coefficient matrices C and D in model (3.23). This topic is elaborated on in the following section.

Comments For the reduced-rank model (2.11) with a single regressor set X_k considered in Chapter 2, the asymptotic distribution results for the corresponding reduced-rank estimators $\widehat{A}$ and $\widehat{B}$ can be obtained as a special case of the approach and results in Theorem 3.2 simply by eliminating aspects related to the second regressor set Z_k. In Theorem 2.4 of Section 2.5, the asymptotic results for the reduced-rank estimators in the single regressor model (2.11) were stated in more explicit detail in terms of eigenvalues and eigenvectors of relevant matrices, and the estimators were obtained in more explicit noniterative form as compared to results for the extended model (3.23). Nevertheless, the final asymptotic distribution results are the same in both approaches. In addition, as in the single regressor model, the results of Theorem 3.2 enable the asymptotic distribution of the overall coefficient matrix estimators $\widehat{C} = \widehat{A}\widehat{B}$ and $\widehat{D} = \widehat{E}\widehat{F}$, obtained as functions of the reduced-rank estimators in the extended model, to be recovered. Specifically, from the relation

$$
\begin{aligned}
T^{1/2}\text{vec}(\widehat{C}' - C') &= T^{1/2}[I_m \otimes B', A \otimes I_{n_1}] \begin{bmatrix} \text{vec}(\widehat{A}' - A') \\ \text{vec}(\widehat{B}' - B') \end{bmatrix} + o_p(1) \\
&\equiv T^{1/2} M_1 \begin{bmatrix} \text{vec}(\widehat{A}' - A') \\ \text{vec}(\widehat{B}' - B') \end{bmatrix} + o_p(1)
\end{aligned}
$$

and a similar relation for $T^{1/2}\text{vec}(\widehat{D}' - D')$ in terms of $T^{1/2}\text{vec}(\widehat{E}' - E')$ and $T^{1/2}\text{vec}(\widehat{F}' - F')$, the asymptotic distribution result $T^{1/2}\text{vec}(\widehat{C}' - C', \widehat{D}' - D') \xrightarrow{d} N(0, MWM')$ readily follows, where M is defined below Eq. (3.31) and W is the asymptotic covariance matrix of Theorem 3.2 as defined in (3.34). Also notice that the covariance matrix W is singular, since the matrix $B_\theta = M'\Sigma^* M$ is singular. However, it can be verified that W is a (reflexive) generalized inverse of B_θ (that is, the relations $B_\theta W B_\theta = B_\theta$ and $W B_\theta W = W$ both hold), which we denote as $W = B_\theta^-$. Therefore, the above asymptotic covariance matrix for $T^{1/2}\text{vec}(\widehat{C}' - C', \widehat{D}' - D')$ can be expressed as $M B_\theta^- M' = M\{M'\Sigma^* M\}^- M'$. This form can be compared to the asymptotic covariance matrix for the full-rank least squares estimators $\widetilde{C}$ and $\widetilde{D}$ in the extended model, which ignore the reduced-rank restrictions. The asymptotic covariance matrix of this estimator, $T^{1/2}\text{vec}(\widetilde{C}' - C', \widetilde{D}' - D')$, is given by Σ^{*-1}.

3.6 Identification of Ranks of Coefficient Matrices

It has been assumed up to this point that the ranks of the coefficient matrices C and D in model (3.21) are known. Usually in practice these ranks are unknown a priori and have to be estimated from the sample data. A formal testing procedure is presented, but the properties of the procedure have not been fully investigated.

In the extended reduced-rank model (3.21) if the structure on the coefficient matrix C is ignored, then the rank of D can be estimated from partial canonical correlation analysis between Y_k and Z_k eliminating X_k (see Section 3.1). The rank of D is taken to be the number of eigenvalues $\widehat{\rho}_{2j}^2$ of $\widehat{\Sigma}_{yy.x}^{-1/2}\widehat{\Sigma}_{yz.x}\widehat{\Sigma}_{zz.x}^{-1}\widehat{\Sigma}_{zy.x}\widehat{\Sigma}_{yy.x}^{-1/2}$ that are not significantly different from zero. The likelihood ratio criterion for testing whether the number of corresponding population eigenvalues which are zero is equal to $m - r_2$ is given by the test statistic similar to (3.12), namely $-T\sum_{j=r_2+1}^{m}\log(1-\widehat{\rho}_{2j}^2)$, where the $\widehat{\rho}_{2j}$ are the sample partial canonical correlations between Y_k and Z_k eliminating X_k. For large samples, this statistic is approximately the same as $T\sum_{j=r_2+1}^{m}\widehat{\rho}_{2j}^2$. These criteria for testing the hypothesis that $\text{rank}(D) = r_2$ against the alternative that D is of full rank are shown to be asymptotically distributed as χ^2 with $(m - r_2)(n_2 - r_2)$ degrees of freedom (e.g., Anderson (1951)). We can use similar criteria to test for the rank of C assuming that D is full rank.

Convenient formal test procedures for the ranks when both coefficient matrices are assumed to be rank deficient are difficult to formulate because of the simultaneous nature of the problem. Given $X_k^* = BX_k$, model (3.23) is in the framework of the model (3.3) studied in Section 3.1, and hence the rank of D is precisely estimated by partial canonical correlation analysis between Y_k and Z_k eliminating X_k^*. Similarly, partial canonical correlation analysis between Y_k and X_k eliminating $Z_k^* = FZ_k$ can be used to test for the rank of C, if the matrix F were given. In these computations, it is assumed that B or F is known, but in practice these have to be estimated by the corresponding sample quantities. Hence, objective criteria that do not depend on the values of B or F are not immediately apparent. However, we argue that, in the population, conditioning on the total information X_k is equivalent to conditioning on the partial information $X_k^* = BX_k$.

We can write model (3.23) as $Y_k = [A, \quad 0]V_k + EFZ_k + \epsilon_k$, where $V_k = [B', \quad B_1']'X_k$ and B_1 is a matrix of dimension $(n_1 - r_1) \times n_1$. Assuming the population version of the normalization conditions (3.24), we have $B\Sigma_{xx}B' = \Lambda_{r_1}^2$ and we can complete the orthogonalization procedure by finding B_1 such that $B_1\Sigma_{xx}B_1' = \Lambda_{n_1-r_1}^2$ and $B\Sigma_{xx}B_1' = 0$. The additional variables B_1X_k included in V_k are intended to capture any information from X_k that exists beyond the linear combinations BX_k that are assumed to be relevant in the model (3.23). Hence, we have a one-to-one transformation from X_k to V_k, and for a known $[B', B_1']'$ conditioning on V_k is equivalent to conditioning on X_k. Thus, in the population, the rank of the matrix $D = EF$ can be determined as the number of nonzero partial canonical correlations between Y_k and Z_k given X_k as in the model of Section 3.1. A more general population result that throws some light on relative efficiencies of the two procedures, i.e., canonical correlation analysis between Y_k and Z_k eliminating $X_k^* = BX_k$ versus eliminating X_k, in the sample is presented below. An analogous sample version of the result indicates that using sample partial canonical correlations, eliminating X_k, to test whether the hypothesis that $\text{rank}(D) = r_2$ is true, is less powerful than using those eliminating only BX_k.

Lemma 3.1 *Let the true model be $Y = AX^* + DZ + \epsilon$ where $X^* = BX$. Also let ρ_j^{*2} and $\rho_j^2 (j = 1, 2, \ldots, m)$ be the (population) squared partial canonical correlations between Y and Z obtained by eliminating X^* and X, respectively. Suppose that* $\text{rank}(D) = r_2$*; then (a) $\rho_j^{*2} \geq \rho_j^2$, for all j, and (b) $\rho_j^* = \rho_j = 0$, $j = r_2 + 1, \ldots, m$.*

Proof Assume without loss of generality that the covariance matrices Σ_{zz}, Σ_{xx} and $\Sigma_{\epsilon\epsilon}$ are identity matrices. Then, since $\Sigma_{yz} = [A, 0]\Sigma_{xz} + D\Sigma_{zz} = [A,\ 0]\Sigma_{xz} + D$, $\Sigma_{yx} = [A,\ 0]\Sigma_{xx} + D\Sigma_{zx} = [A,\ 0] + D\Sigma_{zx}$, and $\Sigma_{yy} = AA' + DD' + I + [A,\ 0]\Sigma_{xz}D' + D\Sigma_{zx}[A,\ 0]'$, we find that the relevant partial covariance matrices needed in the partial canonical correlation analysis of Y and Z eliminating X are given by $\Sigma_{zz.x} = \Sigma_{zz} - \Sigma_{zx}\Sigma_{xx}^{-1}\Sigma_{xz} = I - \Sigma_{zx}\Sigma_{xz}$, $\Sigma_{yz.x} = \Sigma_{yz} - \Sigma_{yx}\Sigma_{xx}^{-1}\Sigma_{xz} = D(I - \Sigma_{zx}\Sigma_{xz})$, and $\Sigma_{yy.x} = \Sigma_{yy} - \Sigma_{yx}\Sigma_{xx}^{-1}\Sigma_{xy} = I + D(I - \Sigma_{zx}\Sigma_{xz})D'$. The partial covariance matrices when eliminating only X^* can be obtained by simply replacing X by X^* in these expressions. Thus it follows that the squared partial canonical correlations between Y and Z eliminating X are the roots of the determinantal equation $|D(I - \Sigma_{zx}\Sigma_{xz})D' - \rho^2/(1 - \rho^2)I| = 0$. Similarly, the squared partial canonical correlations between Y and Z eliminating X^* are the roots of the determinantal equation $|D(I - \Sigma_{zx^*}\Sigma_{x^*z})D' - \rho^{*2}/(1 - \rho^{*2})I| = 0$. Since $I - \Sigma_{zx}\Sigma_{xz} \geq 0$, the number of nonzero eigenvalues (roots) in either equation is equal to the rank of D.

To prove part (a) note that $D(I - \Sigma_{zx^*}\Sigma_{x^*z})D' \geq D(I - \Sigma_{zx}\Sigma_{xz})D'$ since the matrices in parentheses are respectively the residual covariance matrices obtained by regressing Z on X and Z on X^*, a subset of X. This implies that $\rho_j^{*2}/(1 - \rho_j^{*2}) \geq \rho_j^2/(1 - \rho_j^2)$ for all j, and hence, $\rho_j^{*2} \geq \rho_j^2$, $j = 1, 2, \ldots, m$.

The above lemma indicates that in the population, the rank of D for the extended model (3.23) can be determined using the partial canonical correlations similar to methods for the model (3.3) of Section 3.1. But in the sample, an analogous version of the above result implies that the procedure is not efficient. To emphasize that the quantity ρ^* depends on the choice of the matrix B, we write it as $\rho^*(B)$. Under the null hypothesis that $\text{rank}(D) = r_2$, both statistics $\widehat{\psi} = -T\sum_{j=r_2+1}^{m}\log(1-\widehat{\rho}_j^2)$ and $\widehat{\psi}^* = -T\sum_{j=r_2+1}^{m}\log(1 - \widehat{\rho}_j^{*2}(B))$ follow χ^2 distributions with $(m - r_2)(n_2 - r_2)$ degrees of freedom (when $\widehat{\psi}^*$ uses the true value of B). Since $\widehat{\psi}^* \geq \widehat{\psi}$, which follows from the result $\widehat{\rho}_j^{*2} \geq \widehat{\rho}_j^2$ for all $j = 1, \ldots, m$ that is implied by a sample analogue of the above lemma, use of (sample) partial canonical correlation analysis with the statistic $\widehat{\psi}$ to test the null hypothesis that $\text{rank}(D) = r_2$ is less powerful than with the statistic $\widehat{\psi}^*$.

An alternative data-based method for estimating the ranks of the coefficient matrices is the rank trace method given by Izenman (1980). This calls for examining the residual covariance matrix for every possible rank combination and so it involves a large amount of computational effort. Our limited experience suggests that the use of partial canonical correlations, which represents a much easier computation, is a good way to identify the ranks.

3.7 An Example on Ozone Data

Ozone is found in the atmosphere between about 5 km and 60 km above the Earth's surface and plays an important role in the life cycle on earth. It is well known that stratospheric ozone is related to other meteorological variables, such as stratospheric temperatures. Reinsel et al. (1981), for example, investigated the correlation between total column ozone measured from Dobson ground stations and measurements of atmospheric temperature at various pressure levels taken near the Dobson stations. A typical pattern of correlations between total ozone and temperature as a function of pressure level of the temperature reading was given by Reinsel et al. (1981) for the data from Edmonton, Canada.

Temperature measurements taken at pressure levels of 300 mbar and below are negatively correlated with total ozone measurements, and temperature measurements taken at pressure levels of 200 mbar and above are positively correlated with ozone. For the observations from Edmonton, a maximum negative correlation between total ozone and temperature occurs in the middle troposphere at 500 mbar ($\simeq$ 5 km), and a maximum positive correlation occurs in the stratosphere at 100 mbar ($\simeq$ 16 km). However, these patterns are known to change somewhat with location, especially with latitude.

We consider here the monthly averages of total ozone measurements taken from January 1968 to December 1977 for five stations in Europe. The stations are Arosa (Switzerland), Vigna Di Valle (Italy), Mont Louis (France), Hradec Kralove (Czech Republic), and Hohenpeissenberg (Germany). The ozone observations, which represent the time series Y_t, are to be related to temperature measurements taken at 10 different pressure levels. The temperature measurement at each pressure level is an average of the temperature measurements at that pressure level taken around the three stations Arosa, Vigna Di Valle, and Hohenpeissenberg. The series are seasonally adjusted using regression on sinusoidal functions with periods 12 and 6 months, before the multivariate regression analysis is carried out.

The correlations, averaged over the five ozone stations, between the ozone measurements and the temperature data for the 10 pressure levels are displayed in Fig. 3.1. As can be seen from Fig. 3.1, the average correlations confirm the pattern mentioned earlier. But, in general, negative correlations are of greater magnitude than positive correlations. The zone between the troposphere and stratosphere known as the tropopause provides a natural division between the temperature

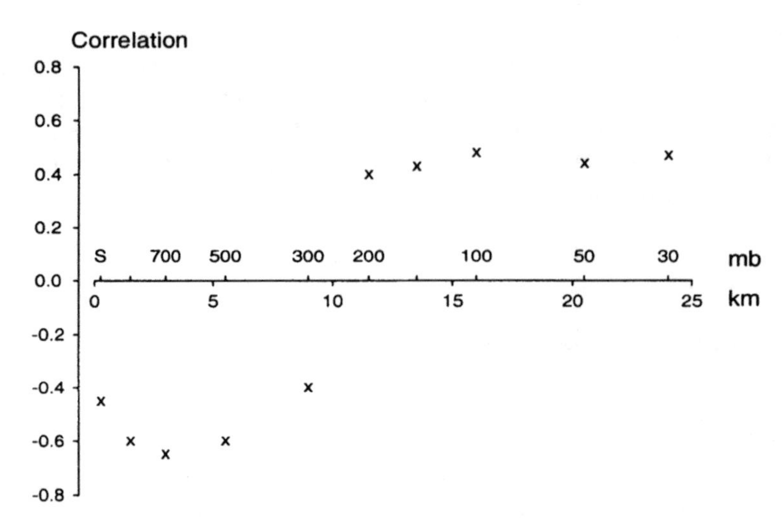

Fig. 3.1 Average correlation between ozone and temperatures at European stations versus pressure level (and average height)

measurements in influencing ozone measurements. The temperature measurements in the troposphere, at the first five pressure levels (the pressure levels below 10 km), form the first set of time series regressor variables X_t, and the temperature measurements at the next five pressure levels (above 10 km) in the stratosphere form the second set of regressors Z_t. As mentioned earlier, all series are seasonally adjusted prior to the multivariate analysis. The ozone series exhibit weak autocorrelations of magnitude about 0.2 which have been ignored in the analysis reported here. Because the ozone stations are relatively close to each other, and the temperature series are averages representing the region, with substantial correlations in neighboring pressure levels, some redundancy in the parameters can be expected and thus the extended reduced-rank multivariate regression model may be appropriate.

A standard multivariate regression model $Y_t = CX_t + DZ_t + \epsilon_t$ was initially fitted to the data, and the least squares estimates of the coefficient matrices were obtained as

$$\widetilde{C} = \begin{bmatrix} -0.99 & 1.28 & -4.51 & -5.09* & 2.66 \\ -0.95 & -0.61 & -3.70 & 0.48 & -1.87 \\ 0.33 & 1.77 & -1.51 & -2.02 & -3.35* \\ 0.59 & 0.65 & -5.74 & -7.29* & 6.15* \\ -1.12 & 2.20 & -6.14* & -5.53* & 3.95* \end{bmatrix}$$

Table 3.2 Summary of results for reduced-rank regression models fitted to ozone-temperature data

Type of regression model	Rank of C	Rank of D	Number of parameters, n^*	$\text{BIC} = \log|\widehat{\Sigma}| + (n^* \log T)/T$
–	–	–	0	$\log|\widehat{\Sigma}_{yy}| = 22.3591$
Full rank	5	5	50	22.4683
Two regressors	1	1	18	21.8511
	2	1	25	21.8466
	2	2	32	21.9704
	3	2	37	22.0670

$$\widetilde{D} = \begin{bmatrix} -0.75 & -0.48 & 1.62 & -2.55 & 4.62* \\ 0.89 & -3.30 & 4.12* & -4.14* & 3.88* \\ 1.39 & -1.45 & 2.93 & -2.39 & 3.22* \\ -3.30 & 1.11 & 1.13 & -3.48 & 5.09* \\ -1.76 & -0.05 & 1.13 & -0.65 & 3.39* \end{bmatrix}.$$

The asterisks indicate that the individual coefficients have t-ratios greater than 2. Taking that as a rule of thumb for their significance, it follows that the temperature measurements in the last three pressure levels of both the troposphere and the stratosphere have greater influence than the measurements in the first two pressure levels when all are considered together. The mere nonsignificance of the elements of the first two columns in both matrices $\widetilde{C}$ and $\widetilde{D}$ probably indicates that these coefficient matrices may be of reduced rank.

As argued in Section 3.6, formal test procedures for the ranks when both C and D are rank deficient are difficult to formulate, but we still may use partial canonical correlations to identify the ranks as in the testing procedure for the model considered in Section 3.1. Though this procedure of using partial canonical correlations is not efficient, it is a good practical tool, as illustrated by this example. The sample partial canonical correlations $\widehat{\rho}_{1j}$ between Y_t and X_t eliminating Z_t are found to be 0.71, 0.43, 0.34, 0.11, and 0.07. Using Bartlett's test statistic (i.e., the asymptotically equivalent form of the LR test statistic given in (3.12) of Section 3.1) to test the hypothesis H_0: $\text{rank}(C) = r_1$ versus H_1 : $\text{rank}(C) = 5$ for $r_1 = 1, 2$, the rejection tail probabilities (p-values) corresponding to the two null hypotheses are found to be 0.001 and 0.076. If the conventional rule of fixing the level of significance as either 0.05 or 0.01 is followed, the hypothesis that $\text{rank}(C) = 2$ cannot be rejected. Since this procedure is inefficient, the possibility that $\text{rank}(C) = 3$ (the corresponding tail probability is 0.076) may also be entertained. When a similar canonical correlation analysis between Y_t and Z_t eliminating X_t is conducted, the partial canonical correlations are 0.53, 0.35, 0.28, 0.12, and 0.08. Using Bartlett's test statistic to test the hypothesis H_0 : $\text{rank}(D) = r_2$ versus H_1 : $\text{rank}(D) = 5$ for $r_2 = 1, 2$, the rejection tail probabilities are found to be 0.04 and 0.24. Here again, the possibility that the rank of D is either 1 or 2 can be entertained.

Comparison of various model fittings is based on the information criterion (BIC) as discussed by Schwarz (1978), and a summary of the results is presented in Table 3.2. Of the two-regressor-sets reduced-rank models, the model fitted with ranks 2 and 1 for C and D, respectively, has the smallest BIC value, which suggests that the partial canonical correlations still may be useful tools for identifying the ranks of C and D. (If a stringent significance level of 0.01 had been used, the tests based on partial canonical correlations could have concluded that the ranks are 2 and 1.) The model with ranks 1 and 1 also comes close to attaining the minimum BIC value. The estimated components of the best fitting model, obtained using the partial least squares computational approach, are as follows (the entries in parentheses are the standard errors computed using Eq. (3.34) from the results in Theorem 3.2):

$$\widehat{A}' = \begin{bmatrix} 8.28 & 5.86 & 4.59 & 8.48 & 8.55 \\ (0.39) & (0.43) & (0.56) & (0.63) & (0.39) \\ & & & & \\ 1.03 & 0.21 & -5.77 & 4.12 & 1.85 \\ (1.16) & (1.10) & (0.49) & (1.35) & (1.13) \end{bmatrix}$$

$$\widehat{B} = \begin{bmatrix} -0.06 & 0.15 & -0.61 & -0.52 & 0.16 \\ (0.13) & (0.22) & (0.24) & (0.24) & (0.17) \\ & & & & \\ 0.00 & -0.38 & -0.25 & 0.46 & 0.55 \\ (0.12) & (0.21) & (0.23) & (0.19) & (0.14) \end{bmatrix}$$

$$\widehat{E}' = \begin{bmatrix} 8.07 & 5.74 & 4.65 & 9.01 & 8.25 \\ (0.84) & (0.89) & (0.85) & (1.09) & (0.84) \end{bmatrix}$$

$$\widehat{F} = \begin{bmatrix} -0.09 & -0.13 & 0.32 & -0.38 & 0.57 \\ (0.14) & (0.22) & (0.22) & (0.18) & (0.13) \end{bmatrix}.$$

In this model the temperature measurements at the first two pressure levels do not play a significant role, as can be seen from the standard errors of the corresponding coefficients in both $\widehat{B}$ and $\widehat{F}$. This aspect was also briefly noted in the discussion of the full-rank coefficient estimates. The reduced-rank model also has an interpretation that is consistent with the correlation pattern mentioned earlier. Specifically, the first linear combination of X_t has large negative values for the temperature measurements at 700 mbar and 500 mbar, and the linear combination of Z_t mainly consists of the last measurement in the stratosphere with a predominantly positive influence.

The temperature series, it may be recalled, are averages taken over three stations and thus are meant to represent the average temperature across the region, representative for all the stations. Because of the average nature of the temperature series and the proximity of the Dobson ozone stations under study, it may be expected that the most significant linear combinations of the temperature measurements will have a similar influence on ozone in all five stations. This is broadly confirmed by the first row elements of the component $\widehat{A}$ and by the elements of the component $\widehat{E}$ in the model. The first linear combination of $\widehat{B}X_t$ and $\widehat{F}Z_t$ may have the same influence on ozone measurements of all five stations. This is tested by fitting a model where individual ranks of C and D are assumed to be unity and further constraining the elements within A and within E to be equal. This is equivalent to a multiple regression procedure where the collection of all ozone measurements (combined over all stations) is regressed on the 10 temperature measurements, taking account of contemporaneous correlations across the five stations. This will be called the "constrained" model,

$$Y_t = \mathbf{1}BX_t + \mathbf{1}FZ_t + \epsilon_t \quad \text{or} \quad \mathbf{Y} = \mathbf{1}[B,\ F][\mathbf{X}', \mathbf{Z}']' + \boldsymbol{\epsilon},$$

where $\mathbf{1}$ denotes an $m \times 1$ vector of ones, and the following estimation results were obtained:

$$\widehat{B} = 7.5\begin{bmatrix}-0.06 & 0.14 & -0.58^{*} & -0.52 & 0.20\end{bmatrix}$$
$$\widehat{F} = 7.0\begin{bmatrix}-0.10 & -0.12^{*} & 0.31^{*} & -0.38^{*} & 0.58^{*}\end{bmatrix}.$$

Notice that this constrained model can be viewed as a special case of a conventional reduced-rank model of rank one in which the left factor in the regression coefficient matrix is constrained to be equal to $\mathbf{1}$.

It is interesting to see that the estimated combinations of temperature measurements using the constrained model are roughly proportional to the first rows of $\widehat{B}$ and $\widehat{F}$. The merit of the reduced-rank regression techniques is apparent when we consider that we might not have identified the constrained model structure from the data by merely looking at the full-rank coefficient estimates $\widetilde{C}$ and $\widetilde{D}$. As a contrast to the reduced-rank model and the constrained model, a "simplified" model can be fitted where a temperature measurement at one fixed pressure level from the tropospheric region and one from the stratospheric region are selected as the potential explanatory variables. The two particular levels may vary from station to station, and we select the temperature level whose estimated regression coefficient has the largest negative t-ratio in the troposphere and the level whose estimate has the largest positive t-ratio in the stratosphere, in the full-rank regression analysis, as potential variables.

The performance of the three models is compared in terms of residual variances of individual series rather than using an overall measure such as BIC. The results are given in Table 3.3. It is clear from Table 3.3 that the results for the reduced-rank model and the constrained model are fairly close to those of the full-rank model. This indicates that we do not lose too much information by reducing the dimension of the temperature measurements. Even the simplified model performs close to the other two models for the three stations Vigna Di Valle, Mont Louis, and Hohenpeissenberg. The analysis can be further improved with the models that can account for spatial correlations as well.

Table 3.3 Summary of the analysis of the ozone-temperature data residual variances

Station	Total variance	Residual variance Full rank model	Reduced rank model	Simplified model	Constrained model
Arosa	259.08	99.84	100.67	114.13	104.04
Vigna Di Valle	159.97	73.35	80.27	84.09	85.74
Mont Louis	124.10	56.85	60.72	68.95	88.92
Hradec Kralove	323.94	139.59	147.35	182.14	158.76
Hohenpeissenberg	274.78	103.28	105.99	112.36	112.36

3.8 Conclusion

The suggestion is to first estimate the regression of Y_k on a small number of the "preferred" predictor variables, say Z_k, which are expected to yield canonical correlations that are a large portion of the total correlations due to all predictor variables, in the usual ordinary least squares method. Then use the previous canonical analysis multivariate shrinkage methods applied to the residuals, where the methods can be represented in terms of partial canonical correlation analysis. That is, suppose there exists a partition of the predictor variables as $(X_k', Z_k')'$, where X_k is $n_1 \times 1$, and Z_k is $n_2 \times 1$ with n_2 "small", such that Z_k contains predictor variables that are "known" or "expected" to give a large portion of the correlation with Y_k. Accordingly, write the multivariate regression model as $Y_k = CX_k + DZ_k + \epsilon_k$. Then, in the alternate suggested procedure, obtain the usual LS estimates in the regression of Y_k on Z_k only as $\widetilde{D}^* = \widehat{\Sigma}_{yz}\widehat{\Sigma}_{zz}^{-1}$, and similarly perform the LS regression of X_k on Z_k to obtain the LS coefficient matrix $\widetilde{D}_2 = \widehat{\Sigma}_{xz}\widehat{\Sigma}_{zz}^{-1}$. Then consider the shrinkage estimator of the regression coefficient matrix C of the form

$$\widehat{C} = \widehat{H}_1^{-1}\widehat{\Delta}_1\widehat{H}_1\widetilde{C}, \tag{3.39}$$

where

$$\widetilde{C} = \widehat{\Sigma}_{yx.z}\widehat{\Sigma}_{xx.z}^{-1} \equiv (\mathbf{Y} - \widetilde{D}^*\mathbf{Z})(\mathbf{X} - \widetilde{D}_2\mathbf{Z})'\{(\mathbf{X} - \widetilde{D}_2\mathbf{Z})(\mathbf{X} - \widetilde{D}_2\mathbf{Z})'\}^{-1}$$

is the ordinary LS estimate of C, with the corresponding estimate of D given by $\widehat{D} = \widetilde{D}^* - \widehat{C}\widetilde{D}_2$. In the same spirit as above, the shrinkage involved for $\widetilde{C}$ is developed and expressible in terms of the squared sample *partial* canonical correlations $\widehat{\rho}_{1j}^2$ of the Y_k and X_k, after adjusting for Z_k. The rows of the matrix $\widehat{H}_1$ in this case correspond to the sample partial canonical vectors for the variates Y_k, and $\widehat{\Delta}_1$ in (3.39) is a diagonal matrix with elements $\widehat{d}_{1j}$ of a similar form as in (2.48) but in terms of the $\widehat{\rho}_{1j}^2$, that is,

$$\widehat{d}_{1j} = \frac{(1 - p_1)(\widehat{\rho}_{1j}^2 - p_1)}{(1 - p_1)^2 \widehat{\rho}_{1j}^2 + p_1^2(1 - \widehat{\rho}_{1j}^2)}, \qquad j = 1, \ldots, m,$$

where $p_1 = n_1/(T - n_2)$. This form is derived based on consideration of the mean squared error matrix of prediction of Y_k using $(X_k', Z_k')'$, and bias correction of the usual squared sample partial canonical correlation estimators.

Some Recent Developments In Chapter 2, we have mentioned that the reduced-rank regression problem had been studied under other names such as "redundancy" analysis. With the extended models presented in this chapter, the redundancy analysis has been extended to "partial redundancy" analysis, taking into account sparseness and other known constraints beyond what are implied by the reduced-rank set-up, see Takane and Jung (2008). In the case of linear discriminant analysis discussed in Section 3.2, one should observe that if "m" is large, the estimation of the error covariance matrix $\widetilde{\Sigma}_{\epsilon\epsilon} = \frac{1}{T} S$ is not efficient unless some structure is imposed. In practice, the so-called spiked structure where except for a few large eigenvalues, others are equal and small, is commonly observed. Note that Fisher-Rao discriminant functions are based on the eigenvectors of $S^{-1} S_B$, where S_B is the between group sum of squares. In addition to entertaining a spiked structure on "S", tuning parameter "γ" is introduced so that the eigenvalues and eigenvectors can be computed based on $S^{-1}(\gamma S_B)$. This is shown to provide a competitive classification scheme by Niu et al. (2018). Finally, the model (3.23) is extended to many subsets of predictors and is shown to have interesting practical applications. The model known as integrative reduced-rank regression is studied by Li et al. (2019) and is covered in some detail in Chapter 11.

Chapter 4
Reduced-Rank Regression Model With Autoregressive Errors

4.1 Introduction and the Model

The classical multivariate regression methods are based on the assumptions that (i) the regression coefficient matrix is of full rank and (ii) the error terms in the model are independent. In Chapters 2 and 3, we have presented regression models that describe the linear relationships between two or more large sets of variables with a fewer number of parameters than those posited by the classical model. The assumption (i) of full rank of the coefficient matrix was relaxed, and the possibility of reduced rank for the coefficient matrix has produced a rich class of models. In this chapter, we also relax the assumption (ii) that the errors are independent, to allow for possible correlation in the errors that may be likely in the regression modeling of time series data. For the ozone/temperature time series data considered in Chapter 3, the assumption of independence of errors appears to hold.

When the data are in the form of time series, as we consider in this chapter, the assumption of serial independence of errors is often not appropriate. In these circumstances, it is important to account for serial correlation in the errors. Robinson (1973) considered the reduced-rank model for time series data in the frequency domain with a general structure for errors. In particular, assuming the errors follow a general stationary process, Robinson studied the asymptotic properties of both the least squares and other more efficient estimators of the regression coefficient matrix. However, often the serially correlated disturbances will satisfy more restrictive assumptions, and, in these circumstances, it would seem possible to exploit this structure to obtain estimates that may have better small sample properties than the estimators proposed by Robinson. (For a related argument, see Gallant and Goebel (1976).) This approach was taken by Velu and Reinsel (1987), and in this chapter, we present the relevant methodology and results.

The assumption we make on the errors is that they follow a (first order) multivariate autoregressive process. We consider the estimation problem with the

G. C. Reinsel et al., *Multivariate Reduced-Rank Regression*, Lecture Notes in Statistics 225, https://doi.org/10.1007/978-1-0716-2793-8_4

autoregressive error assumption in the time domain. The asymptotic distribution theory for the proposed estimator is derived, and computational algorithms for the estimation procedure are also briefly discussed. Discussion of methods for the initial specification and testing of the rank of the coefficient matrix is presented, and an alternative estimator, which is motivated by the need for initial identification of the rank of the model, is also suggested. As an example, we present an analysis of some macroeconomic data of the United Kingdom that was previously considered by Gudmundsson (1977). This has been discussed earlier in Example 2.2 of Section 2.2.

Consider the multivariate regression model for the vector time series $\{Y_t\}$,

$$Y_t = CX_t + U_t, \qquad t = 1, 2, \dots, T, \tag{4.1}$$

where Y_t is an $m \times 1$ vector of response variables, X_t is an $n \times 1$ vector of predictor variables, and C is an $m \times n$ regression coefficient matrix. As in Chapter 2, we assume that C is of reduced rank,

$$\text{rank}(C) = r \leq \min(m, n). \tag{4.2}$$

We assume further that the error terms U_t satisfy the stochastic difference equation

$$U_t = \mathbf{\Phi} U_{t-1} + \epsilon_t, \tag{4.3}$$

where U_0 is taken to be fixed. Here the ϵ_t are independently distributed random vectors with mean zero and positive-definite covariance matrix $\Sigma_{\epsilon\epsilon}$, and $\mathbf{\Phi} = (\phi_{ij})$ is an $m \times m$ matrix of unknown parameters, with all eigenvalues of $\mathbf{\Phi}$ less than one in absolute value. Hendry (1971), Hatanaka (1976), and Reinsel (1979), among others, have considered maximum likelihood estimation of the model parameters for the specification (4.1) and (4.3). When the regression matrix C is assumed to be of lower rank r, following the discussion in Chapter 2, write $C = AB$ with normalization conditions as in (2.14),

$$A'\Gamma A = I, \quad \text{and} \quad B\Sigma_{xx}B' = \Lambda^2, \tag{4.4}$$

where $\Sigma_{xx} = \lim_{T\to\infty} \frac{1}{T} \sum_{t=1}^{T} X_t X_t'$, and Γ is a specified positive-definite matrix.

The main focus in this chapter is on the estimation of A and B based on a sample of T observations. Although we do not consider the complementary result of (4.2) that there are $(m - r)$ linear constraints (assuming $m < n$) in detail here, the implications for the extended model (4.1)–(4.3) are the same as the model discussed in Chapter 2. The linear combinations $\ell_i' Y_t$, $i = 1, 2, \dots, (m-r)$, such that $\ell_i' C = 0$, do not depend on the predictor variables X_t, but they are no longer white noise processes.

4.2 Example on the U.K. Economy: Basic Data and Their Descriptions

To illustrate the procedures to be developed in this chapter, we consider the macroeconomic data of the United Kingdom that has been previously analyzed by Klein et al. (1961) and Gudmundsson (1977) (see Section 2.2). Klein et al. (1961) present a detailed description of an econometric model based on an extensive database including the data used in our calculations. Quarterly observations, starting from the first quarter of 1948 to the last quarter of 1956, are used in the following analysis. The endogenous (response) variables are y_1 = index of industrial production, y_2 = consumption of food, drinks, and tobacco at constant prices, y_3 = total unemployment, y_4 = index of volume of total imports, and y_5 = index of volume of total exports, and the exogenous (predictor) variables are x_1 = total civilian labor force, x_2 = index of weekly wage rates, x_3 = price index of total imports, x_4 = price index of total exports, and x_5 = price index of total consumption. The relationship between these two sets of variables is taken to reflect the demand side of the macrosystem of the economy of the United Kingdom. The first three series of the set of endogenous variables are seasonally adjusted before the analysis is carried out. Time series plots of the resulting data are given in Fig. 4.1. These data are also presented in Table A.3 of the Appendix.

Initially, the least squares regression of $Y_t = (y_{1t}, y_{2t}, y_{3t}, y_{4t}, y_{5t})'$ on $X_t = (x_{1t}, x_{2t}, x_{3t}, x_{4t}, x_{5t})'$ was performed without taking into account any autocorrelation structure on the errors. In order to get a rough view on the rank of the regression coefficient matrix, canonical correlation analysis was also performed. The squared sample canonical correlations between Y_t and X_t are 0.95, 0.67, 0.29, 0.14, and 0.01, which suggest a possible deficiency in the rank of C.

However, the residuals resulting from the least squares regression indicated a strong serial correlation in the errors. While commenting on the model of Klein et al. (1961), Gudmundsson (1977, p. 51) had a similar view that some allowance should be made for the serial correlation in the residuals of many of their equations. Therefore, we consider incorporating an autoregressive errors structure in the analysis.

4.3 Maximum Likelihood Estimators for the Model

We now consider an estimation procedure that will yield an efficient estimator, in the sense that it has the same asymptotic distribution as the maximum likelihood estimator under the normality assumption for the errors. From the model (4.1) and (4.3), we have

$$Y_t - \Phi Y_{t-1} = CX_t - \Phi CX_{t-1} + \epsilon_t, \tag{4.5}$$

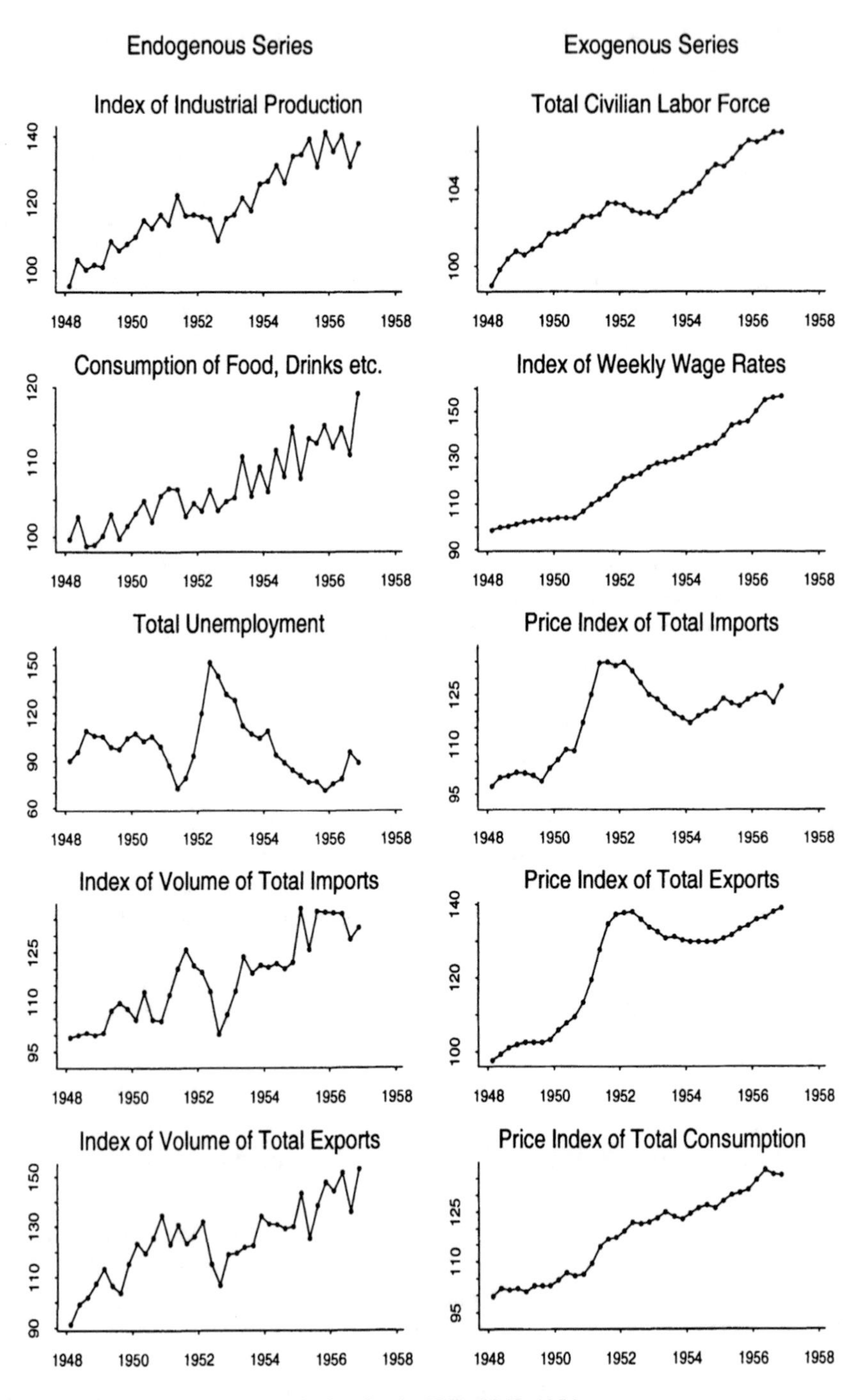

Fig. 4.1 Quarterly macroeconomic data for the U.K., 1948–1956

where the error term ϵ_t in (4.5) satisfies the usual independence assumptions. Let $\boldsymbol{\epsilon} = [\epsilon_1, \dots, \epsilon_T]$, and note that (4.5) leads to

$$\mathbf{Y} - \boldsymbol{\Phi}\mathbf{Y}_{-1} = C\mathbf{X} - \boldsymbol{\Phi}C\mathbf{X}_{-1} + \boldsymbol{\epsilon}, \tag{4.6}$$

where $\mathbf{Y} = [Y_1, \dots, Y_T]$ and $\mathbf{X} = [X_1, \dots, X_T]$ are the $m \times T$ and $n \times T$ data matrices of observations on the response and predictor variables, and $\mathbf{Y}_{-1} = [Y_0, \dots, Y_{T-1}]$ and $\mathbf{X}_{-1} = [X_0, \dots, X_{T-1}]$ are the lagged data matrices. Before we examine the reduced-rank situation, we first briefly consider estimation of parameters C and $\boldsymbol{\Phi}$ in the full-rank version of the model, that is, without the rank restriction on C. The full-rank estimator is useful, in particular, as a basis for the alternative reduced-rank estimator that is discussed in Section 4.5 and that is used in the identification procedure described in Section 4.6 to specify the rank of the coefficient matrix C.

We consider minimization of the criterion $S_T(C, \boldsymbol{\Phi}) = \frac{1}{2T}\text{tr}(\boldsymbol{\epsilon}\boldsymbol{\epsilon}'\Gamma) = \frac{1}{2T}e'(\Gamma \otimes I_T)e$, where $\boldsymbol{\epsilon} = \mathbf{Y} - \boldsymbol{\Phi}\mathbf{Y}_{-1} - C\mathbf{X} + \boldsymbol{\Phi}C\mathbf{X}_{-1}$, $e = \text{vec}(\boldsymbol{\epsilon}')$, and the choice $\Gamma = \Sigma_{\epsilon\epsilon}^{-1}$ is taken. From (4.6), we have

$$\text{vec}(\mathbf{Y}^{*\prime}) = \mathbf{X}^{*}\text{vec}(C') + \text{vec}(\boldsymbol{\epsilon}'),$$

where $\mathbf{Y}^{*} = \mathbf{Y} - \boldsymbol{\Phi}\mathbf{Y}_{-1}$ and $\mathbf{X}^{*} = (I_m \otimes \mathbf{X}') - (\boldsymbol{\Phi} \otimes \mathbf{X}'_{-1})$ is an $mT \times mn$ matrix. It follows from standard results for the linear regression model that, for given values (estimates) $\widetilde{\boldsymbol{\Phi}}$ and $\widetilde{\Sigma}_{\epsilon\epsilon}$ of $\boldsymbol{\Phi}$ and $\Sigma_{\epsilon\epsilon}$, the full-rank estimate of C is obtained as the (estimated) generalized least squares (GLS) estimate, given by

$$\text{vec}(\widetilde{C}') = \widetilde{G}^{-1}\widetilde{\mathbf{X}}^{*\prime}(\widetilde{\Sigma}_{\epsilon\epsilon}^{-1} \otimes I_T)\text{vec}(\widetilde{\mathbf{Y}}^{*\prime})/T, \tag{4.7}$$

where $\widetilde{G} = \widetilde{\mathbf{X}}^{*\prime}(\widetilde{\Sigma}_{\epsilon\epsilon}^{-1} \otimes I_T)\widetilde{\mathbf{X}}^{*}/T$, with $\widetilde{\mathbf{Y}}^{*} = \mathbf{Y} - \widetilde{\boldsymbol{\Phi}}\mathbf{Y}_{-1}$ and $\widetilde{\mathbf{X}}^{*} = (I_m \otimes \mathbf{X}') - (\widetilde{\boldsymbol{\Phi}} \otimes \mathbf{X}'_{-1})$. The covariance matrix of the full-rank estimate is estimated by $\widehat{\text{Cov}}[\text{vec}(\widetilde{C}')] = \frac{1}{T}\widetilde{G}^{-1}$. Conversely, for a given value (estimate) $\widetilde{C}$ of C, the corresponding estimate of $\boldsymbol{\Phi}$ to minimize the above criterion $S_T(C, \boldsymbol{\Phi})$ is given by

$$\widetilde{\boldsymbol{\Phi}} = \widetilde{\mathbf{U}}\widetilde{\mathbf{U}}'_{-1}(\widetilde{\mathbf{U}}_{-1}\widetilde{\mathbf{U}}'_{-1})^{-1},$$

where $\widetilde{\mathbf{U}} = \mathbf{Y} - \widetilde{C}\mathbf{X} = [\widetilde{U}_1, \dots, \widetilde{U}_T]$, $\widetilde{\mathbf{U}}_{-1} = [\widetilde{U}_0, \dots, \widetilde{U}_{T-1}]$, with $\widetilde{U}_t = Y_t - \widetilde{C}X_t$, $t = 1, \dots, T$, and $\Sigma_{\epsilon\epsilon}$ is estimated as $\widetilde{\Sigma}_{\epsilon\epsilon} = \widetilde{\boldsymbol{\epsilon}}\widetilde{\boldsymbol{\epsilon}}'/T$, where $\widetilde{\boldsymbol{\epsilon}} = \widetilde{\mathbf{U}} - \widetilde{\boldsymbol{\Phi}}\widetilde{\mathbf{U}}_{-1}$. The estimation scheme is thus performed most conveniently by an iterative procedure that alternates between estimation of C through (4.7), for given estimate $\widetilde{\boldsymbol{\Phi}}$, and estimation of $\boldsymbol{\Phi}$ as indicated for given estimate $\widetilde{C}$. The most common starting value is the least squares estimate of C, given in (1.7), that ignores the AR(1) errors structure.

Now for estimation of the reduced-rank model with $C = AB$, let $\theta = (\alpha', \beta', \phi')'$, where $\alpha = \text{vec}(A')$, $\beta = \text{vec}(B')$, and $\phi = \text{vec}(\boldsymbol{\Phi}')$. We again consider minimization of the criterion

$$S_{1T}(\theta) = \frac{1}{2T}\operatorname{tr}(\boldsymbol{\epsilon}\boldsymbol{\epsilon}'\Gamma), \tag{4.8}$$

subject to normalization conditions

$$A'\Gamma A = I_r, \qquad B\left(\frac{1}{T}\mathbf{X}\mathbf{X}'\right)B' = \Lambda^2, \tag{4.9}$$

where Λ^2 is diagonal. If the autocorrelation structure implied by (4.3) is ignored by setting $\boldsymbol{\Phi} = 0$, then one is led to the least squares estimator considered by Robinson (1973). It has been shown that the resulting least squares estimator is consistent but not efficient. To arrive at the asymptotic theory for the estimator that is obtained by minimizing (4.8) subject to (4.9), we shall again use the results of Silvey (1959) and Robinson (1972, 1973). For the normalization conditions (4.9), we shall be particularly interested in the choice of $\Gamma = \widehat{\Sigma}_{\epsilon\epsilon}^{-1}$ for the asymptotic results where $\widehat{\Sigma}_{\epsilon\epsilon}$ is the sample covariance matrix of the residuals $\widehat{\boldsymbol{\epsilon}}$ obtained from the maximum likelihood estimation of the model (4.1) and (4.3) with the rank restriction on C. Note because of this choice of Γ, the normalization conditions vary with sample quantities. As in Chapters 2 and 3, however, to provide more explicit results, the asymptotic theory presented here is for the estimator with the choice of $\Gamma = \Sigma_{\epsilon\epsilon}^{-1}$, which is a population quantity. For the sake of brevity, we omit the details, but the main result follows.

Theorem 4.1 *Consider the model (4.1) and (4.3) under the stated assumptions, with $C = AB$, where A and B satisfy the normalization conditions (4.4), and let $h(\theta) = 0$ denote the $r^2 \times 1$ vector of normalization conditions (4.4) obtained by stacking the conditions (4.4) without duplication. The first $r(r+1)/2$ components of $h(\theta)$ are of the form $\alpha_i'\Sigma_{\epsilon\epsilon}^{-1}\alpha_j - \delta_{ij}$, $i \le j$, where α_i denotes the ith column of the matrix A and $\delta_{ij} = 1$ for $i = j$ and 0 otherwise, and the remaining $r(r-1)/2$ elements are $\beta_i'\Sigma_{xx}\beta_j$, $i < j$, where β_i' denotes the ith row of B. Assume the limits*

$$\lim_{T\to\infty}\frac{1}{T}\sum_{t=1}^{T}X_tX_t' = \Sigma_{xx} = \Gamma_x(0), \qquad \lim_{T\to\infty}\frac{1}{T}\sum_{t=2}^{T}X_{t-1}X_t' = \Gamma_x(1)$$

exist almost surely, and define

$$\begin{aligned} G &= (\Sigma_{\epsilon\epsilon}^{-1}\otimes\Gamma_x(0)) - (\Sigma_{\epsilon\epsilon}^{-1}\boldsymbol{\Phi}\otimes\Gamma_x(1)') \\ &\quad -(\boldsymbol{\Phi}'\Sigma_{\epsilon\epsilon}^{-1}\otimes\Gamma_x(1)) + (\boldsymbol{\Phi}'\Sigma_{\epsilon\epsilon}^{-1}\boldsymbol{\Phi}\otimes\Gamma_x(0)) \\ &= \operatorname{Cov}[\operatorname{vec}(X_t\epsilon_t'\Sigma_{\epsilon\epsilon}^{-1} - X_{t-1}\epsilon_t'\Sigma_{\epsilon\epsilon}^{-1}\boldsymbol{\Phi})]. \end{aligned}$$

Let $\widehat{\theta}$ denote the estimator that absolutely minimizes (4.8) subject to (4.9) for the choice $\Gamma = \Sigma_{\epsilon\epsilon}^{-1}$. Then as $T \to \infty$, $\widehat{\theta} \to \theta$ almost surely and $\sqrt{T}(\widehat{\theta} - \theta)$ has a multivariate normal distribution with zero mean vector and covariance matrix given *by the leading $(r(m+n)+m^2)$-order square submatrix of the $(r(m+n+r)+m^2)$-order matrix*

$$\begin{bmatrix} B_\theta & -H_\theta \\ -H'_\theta & 0 \end{bmatrix}^{-1} \begin{bmatrix} B_\theta & 0 \\ 0 & 0 \end{bmatrix} \begin{bmatrix} B_\theta & -H_\theta \\ -H'_\theta & 0 \end{bmatrix}^{-1}, \tag{4.10}$$

where

$$B_\theta = M' \begin{bmatrix} G & 0 \\ 0 & \Sigma_{\epsilon\epsilon}^{-1} \otimes \Sigma_u \end{bmatrix} M, \qquad M = \begin{bmatrix} I_m \otimes B' & A \otimes I_n & 0 \\ 0 & 0 & I_m \otimes I_m \end{bmatrix},$$

$H_\theta = \{(\partial h_j(\theta)/\partial\theta_i)\}$ is the matrix of first derivatives of the constraints in (4.4) and is of order $r(m+n)+m^2 \times r^2$, and $\Sigma_u = \mathbf{\Phi}\Sigma_u\mathbf{\Phi}' + \Sigma_{\epsilon\epsilon}$, which can be obtained from $\text{vec}(\Sigma_u) = [I_{m^2} - \mathbf{\Phi} \otimes \mathbf{\Phi}]^{-1}\text{vec}(\Sigma_{\epsilon\epsilon})$. Specifically, it can be shown as in Theorem 3.2 of Chapter 3 that the covariance matrix of this limiting distribution is given by $W = (B_\theta + H_\theta H'_\theta)^{-1}B_\theta(B_\theta + H_\theta H'_\theta)^{-1}$. We note that the choice of normalization condition (4.4) guarantees the existence of the inverse matrix in (4.10).

Proof We shall only briefly indicate the proof here because many of the details follow along similar lines as Silvey (1959) and have been discussed in the proof of Theorem 3.2 in Chapter 3. Initially, we consider estimates that minimize $S^*_{1T}(\theta) - h(\theta)'\lambda$, where λ is the Lagrangian multiplier and $S^*_{1T}(\theta)$ denotes the criterion (4.8) with the choice $\Gamma = \Sigma_{\epsilon\epsilon}^{-1}$. Let $\widehat{\theta}$ and $\widehat{\lambda}$ be solutions of the first order equations. It can be shown that $\{\partial^2 S^*_{1T}(\theta)/\partial\theta\partial\theta'\} \to B_\theta$ almost surely. So following Silvey (1959), we have the result below (similar to (3.29)).

$$\begin{bmatrix} B_\theta + o(1) & -H_\theta + o(1) \\ -H'_\theta + o(1) & 0 \end{bmatrix} \begin{bmatrix} \sqrt{T}(\widehat{\theta} - \theta) \\ \sqrt{T}\widehat{\lambda} \end{bmatrix} = \begin{bmatrix} -\sqrt{T}\partial S^*_{1T}(\theta)/\partial\theta \\ 0 \end{bmatrix}. \tag{4.11}$$

Thus the asymptotic distribution of $\sqrt{T}(\widehat{\theta} - \theta)$ follows from the asymptotic distribution of the quantity $\sqrt{T}\partial S^*_{1T}(\theta)/\partial\theta$. For the model considered, we can express in vector form as

$$\begin{aligned} y^* &= \text{vec}(\mathbf{Y}^{*\prime}) = \mathbf{X}^*\text{vec}(C') + \text{vec}(\boldsymbol{\epsilon}') \\ &= \mathbf{X}^*(I_m \otimes B')\alpha + e = \mathbf{X}^*(A \otimes I_n)\beta + e, \end{aligned}$$

where $e = \text{vec}(\boldsymbol{\epsilon}')$ has covariance matrix $\Sigma_{\epsilon\epsilon} \otimes I_T$. The vector AR(1) error part of the model, $\mathbf{U} = \mathbf{\Phi}\mathbf{U}_{-1} + \boldsymbol{\epsilon}$, can similarly be written as $\text{vec}(\mathbf{U}') = \text{vec}(\mathbf{U}'_{-1}\mathbf{\Phi}') + \text{vec}(\boldsymbol{\epsilon}') = (I_m \otimes \mathbf{U}'_{-1})\phi + e$. The criterion in (4.8) with $\Gamma = \Sigma_{\epsilon\epsilon}^{-1}$ can also be rewritten as $S^*_{1T}(\theta) = \frac{1}{2T}e'(\Sigma_{\epsilon\epsilon}^{-1} \otimes I_T)e$, and we therefore find that

$$\frac{\partial S_{1T}^*(\theta)}{\partial \theta} = \frac{1}{T}\frac{\partial e'}{\partial \theta}(\Sigma_{\epsilon\epsilon}^{-1} \otimes I_T)e = -\frac{1}{T}\begin{bmatrix} (I_m \otimes B)\mathbf{X}^{*\prime} \\ (A' \otimes I_n)\mathbf{X}^{*\prime} \\ (I_m \otimes \mathbf{U}_{-1}) \end{bmatrix}(\Sigma_{\epsilon\epsilon}^{-1} \otimes I_T)e$$

$$= -\frac{1}{T}\begin{bmatrix} I_m \otimes B & 0 \\ A' \otimes I_n & 0 \\ 0 & I_m \otimes I_m \end{bmatrix}\begin{bmatrix} (\Sigma_{\epsilon\epsilon}^{-1} \otimes \mathbf{X})e - (\mathbf{\Phi}'\Sigma_{\epsilon\epsilon}^{-1} \otimes \mathbf{X}_{-1})e \\ (\Sigma_{\epsilon\epsilon}^{-1} \otimes \mathbf{U}_{-1})e \end{bmatrix},$$

where $\mathbf{X}^* = (I_m \otimes \mathbf{X}') - (\mathbf{\Phi} \otimes \mathbf{X}'_{-1})$.

Hence, noting that $(\mathbf{\Phi}'\Sigma_{\epsilon\epsilon}^{-1} \otimes \mathbf{X}_{-1})e = (\mathbf{\Phi}'\Sigma_{\epsilon\epsilon}^{-1} \otimes I_n)\text{vec}(\mathbf{X}_{-1}\boldsymbol{\epsilon}')$ and so on, the vector of partial derivatives can be written as $\sqrt{T}\partial S_{1T}^*(\theta)/\partial\theta = M'F'\sqrt{T}Z_T$, where $Z_T = \text{vec}\{\mathbf{X}_{-1}\boldsymbol{\epsilon}', \mathbf{X}\boldsymbol{\epsilon}', \mathbf{U}_{-1}\boldsymbol{\epsilon}'\}/T$, and

$$F' = \begin{bmatrix} \mathbf{\Phi}'\Sigma_{\epsilon\epsilon}^{-1} \otimes I_n & -(\Sigma_{\epsilon\epsilon}^{-1} \otimes I_n) & 0 \\ 0 & 0 & -(\Sigma_{\epsilon\epsilon}^{-1} \otimes I_m) \end{bmatrix}.$$

It is known (e.g., Anderson (1971, Chapter 5)) that $\sqrt{T}Z_T \xrightarrow{d} N(0, V)$ as $T \to \infty$, where

$$V = \begin{bmatrix} \Sigma_{\epsilon\epsilon} \otimes \Gamma_x(0) & \Sigma_{\epsilon\epsilon} \otimes \Gamma_x(1) & 0 \\ \Sigma_{\epsilon\epsilon} \otimes \Gamma_x(1)' & \Sigma_{\epsilon\epsilon} \otimes \Gamma_x(0) & 0 \\ 0 & 0 & \Sigma_{\epsilon\epsilon} \otimes \Sigma_u \end{bmatrix}.$$

It thus follows that $\sqrt{T}\partial S_{1T}^*(\theta)/\partial\theta \xrightarrow{d} N(0, B_\theta)$, where $B_\theta = M'F'VFM$ is as defined, and also that $\{\partial^2 S_{1T}^*(\theta)/\partial\theta\partial\theta'\} \to E\{\frac{1}{T}\frac{\partial e'}{\partial\theta}(\Sigma_{\epsilon\epsilon}^{-1} \otimes I_T)\frac{\partial e}{\partial\theta'}\} = B_\theta$. The asymptotic theory of Theorem 4.1 thus follows from (4.11).

Note that $\text{rank}(B_\theta) = \text{rank}(M) = r(m+n-r)+m^2 < r(m+n)+m^2$ because of the existence of r^2 linearly independent restrictions between the columns. Hence, the information matrix B_θ is singular, and this makes the computation of the asymptotic covariances of the estimators in Theorem 4.1 more difficult. Similar to the discussion in Section 3.5, Silvey (1959, p. 401) suggested alternatively that the asymptotic covariance matrix of $\widehat{\theta}$ is given by the leading submatrix of

$$\begin{bmatrix} B_\theta + H_\theta H_\theta' & -H_\theta \\ -H_\theta' & 0 \end{bmatrix}^{-1}\begin{bmatrix} B_\theta & 0 \\ 0 & 0 \end{bmatrix}\begin{bmatrix} B_\theta + H_\theta H_\theta' & -H_\theta \\ -H_\theta' & 0 \end{bmatrix}^{-1},$$

which can be shown to be equivalent to the matrix W given in Theorem 4.1.

The asymptotic theory for the estimators $\widehat{A}$ and $\widehat{B}$ is useful for inferences concerning whether certain components of X_t contribute in the formation of the explanatory indexes BX_t and also whether certain components of Y_t are influenced by certain of these indexes. The overall regression coefficient matrix $C = AB$ is, of course, also of interest, and the asymptotic distribution of the estimator $\widehat{C} = \widehat{A}\widehat{B}$ follows directly from results of Theorem 4.1. Based on the relation

$\widehat{C} - C = \widehat{A}\widehat{B} - AB = (\widehat{A} - A)B + A(\widehat{B} - B) + O_p(T^{-1})$, it follows directly that $\sqrt{T}\text{vec}(\widehat{C}' - C') \xrightarrow{d} N(0, M_1 W_1 M_1')$, where $M_1 = [I_m \otimes B', A \otimes I_n]$ and W_1 denotes the asymptotic covariance matrix of $\sqrt{T}\{(\widehat{\alpha} - \alpha)', (\widehat{\beta} - \beta)'\}'$, that is, the leading $r(m+n)$-order square submatrix of the matrix W in Theorem 4.1.

Remark–Asymptotic Independence of Estimators As will be discussed in more detail in the next section, we note that the asymptotic covariance matrix W in Theorem 4.1 is block diagonal of the form $W = \text{diag}(W_1, \Sigma_{\epsilon\epsilon}^{-1} \otimes \Sigma_u)$, where $W_1 = (B_\delta + H_1 H_1')^{-1} B_\delta (B_\delta + H_1 H_1')^{-1}$, with $\delta = (\alpha', \beta')'$, $B_\delta = M_1' G M_1$, and $H_1 = \partial h(\theta)'/\partial\delta$. This indicates that the estimates $\widehat{A}$ and $\widehat{B}$ of the regression components of the model (4.1) and (4.3) are asymptotically independent of the estimate $\widehat{\boldsymbol{\Phi}}$ of the AR(1) error component of the model. In addition, it follows that the asymptotic covariance matrix of $\sqrt{T}\text{vec}(\widehat{C}' - C')$, where $\widehat{C} = \widehat{A}\widehat{B}$, can be expressed more explicitly as

$$\begin{aligned} M_1 W_1 M_1' &= M_1(M_1' G M_1 + H_1 H_1')^{-1} M_1' G M_1 (M_1' G M_1 + H_1 H_1')^{-1} M_1' \\ &\equiv M_1 (M_1' G M_1)^{-} M_1'. \end{aligned}$$

4.4 Computational Algorithms for Efficient Estimators

We now briefly comment on the procedure for computing the efficient estimator. An iterative procedure can be formulated along the lines of Aitchison and Silvey (1958) modification of the Newton–Raphson method to account for constraints as discussed in Section 3.5. It is based on a modification of the first order equations in (4.11) to

$$\begin{bmatrix} B_\theta + H_\theta H_\theta' & -H_\theta \\ -H_\theta' & 0 \end{bmatrix} \begin{bmatrix} \widehat{\theta} - \theta \\ \widehat{\lambda} \end{bmatrix} = \begin{bmatrix} -\partial S_{1T}^*(\theta)/\partial\theta \\ 0 \end{bmatrix}, \tag{4.12}$$

where explicitly

$$\partial S_{1T}^*(\theta)/\partial\theta = \text{vec}(\partial S_{1T}^*/\partial A',\ \partial S_{1T}^*(\theta)/\partial B',\ \partial S_{1T}^*(\theta)/\partial\boldsymbol{\Phi}'),$$

with

$$\begin{aligned} \partial S_{1T}^*(\theta)/\partial A' &= T^{-1}(B\mathbf{X}_{-1}\boldsymbol{\epsilon}'\Sigma_{\epsilon\epsilon}^{-1}\boldsymbol{\Phi} - B\mathbf{X}\boldsymbol{\epsilon}'\Sigma_{\epsilon\epsilon}^{-1}), \\ \partial S_{1T}^*(\theta)/\partial B' &= T^{-1}(\mathbf{X}_{-1}\boldsymbol{\epsilon}'\Sigma_{\epsilon\epsilon}^{-1}\boldsymbol{\Phi}A - \mathbf{X}\boldsymbol{\epsilon}'\Sigma_{\epsilon\epsilon}^{-1}A), \\ \partial S_{1T}^*(\theta)/\partial\boldsymbol{\Phi}' &= -T^{-1}\mathbf{U}_{-1}\boldsymbol{\epsilon}'\Sigma_{\epsilon\epsilon}^{-1}, \end{aligned}$$

$\mathbf{U} = \mathbf{Y} - AB\mathbf{X}$, and $\boldsymbol{\epsilon} = \mathbf{U} - \boldsymbol{\Phi}\mathbf{U}_{-1}$. Hence, at the ith step, the iterative scheme is given by

$$\begin{bmatrix} \widehat{\theta}_i \\ \widehat{\lambda}_i \end{bmatrix} = \begin{bmatrix} \widehat{\theta}_{i-1} \\ 0 \end{bmatrix} - \begin{bmatrix} W_{i-1} & Q_{i-1} \\ Q'_{i-1} & -I_{r^2} \end{bmatrix} \begin{bmatrix} \partial S^*_{1T}(\theta)/\partial\theta|_{\widehat{\theta}_{i-1}} \\ h(\theta)|_{\widehat{\theta}_{i-1}} \end{bmatrix}, \tag{4.13}$$

where

$$\begin{bmatrix} W & Q \\ Q' & -I_{r^2} \end{bmatrix} = \begin{bmatrix} B_\theta + H_\theta H'_\theta & -H_\theta \\ -H'_\theta & 0 \end{bmatrix}^{-1},$$

so that

$$W = (B_\theta + H_\theta H'_\theta)^{-1} B_\theta (B_\theta + H_\theta H'_\theta)^{-1},$$

and

$$Q = -(B_\theta + H_\theta H'_\theta)^{-1} H_\theta.$$

The covariance matrix of $\widehat{\theta}_i$ is approximated by

$$(1/T)(B_\theta + H_\theta H'_\theta)^{-1} B_\theta (B_\theta + H_\theta H'_\theta)^{-1} = (1/T)W$$

evaluated at $\widehat{\theta}_i$. The iterations are carried out until either the successive differences between $\widehat{\theta}_i$ or the magnitudes of the Lagrangian multipliers $\widehat{\lambda}_i$ or the magnitudes of first order quantities $\partial S^*_{1T}(\widehat{\theta}_{i-1})/\partial\theta$ are reasonably small.

However, as in the full-rank estimation discussed in Section 4.3, the calculations in the iterative scheme for $\widehat{\theta}$ can be presented in a somewhat simpler form, in which one alternates between estimation of A and B (α and β), for given estimate of $\boldsymbol{\Phi}$, and estimation of $\boldsymbol{\Phi}$ for given estimates of A and B. Let $\delta = (\alpha', \beta')'$ so that $\theta = (\delta', \phi')'$. Then note that the matrix B_θ is block diagonal, having the form $B_\theta = \text{diag}(M'_1 G M_1, \Sigma^{-1}_{\epsilon\epsilon} \otimes \Sigma_u) \equiv \text{diag}(B_\delta, B_\phi)$, where $M_1 = [I_m \otimes B', A \otimes I_n]$. In addition, because the normalization constraints $h(\theta) = 0$ ($h(\delta) = 0$) do not involve $\boldsymbol{\Phi}$, the matrix of first derivatives of $h(\theta)$ with respect to θ can be partitioned as $H_\theta = \partial h(\theta)'/\partial\theta = [H'_1, 0']'$, where $H_1 = \partial h(\theta)'/\partial\delta$. Therefore, upon rearranging the terms, we see that the first order equations in (4.12) separate as

$$\begin{bmatrix} B_\delta + H_1 H'_1 & -H_1 \\ -H'_1 & 0 \end{bmatrix} \begin{bmatrix} \widehat{\delta} - \delta \\ \widehat{\lambda} \end{bmatrix} = \begin{bmatrix} -\partial S^*_{1T}(\theta)/\partial\delta \\ 0 \end{bmatrix}, \tag{4.14}$$

and $B_\phi(\widehat{\phi} - \phi) = -\partial S^*_{1T}(\theta)/\partial\phi$. These last equations are equivalent to $(\widehat{\boldsymbol{\Phi}} - \boldsymbol{\Phi})\Sigma_u = T^{-1}\boldsymbol{\epsilon}\mathbf{U}'_{-1} \equiv T^{-1}(\mathbf{U} - \boldsymbol{\Phi}\mathbf{U}_{-1})\mathbf{U}'_{-1}$. On substituting the sample estimate $\widetilde{\Sigma}_u = T^{-1}\mathbf{U}_{-1}\mathbf{U}'_{-1}$ for Σ_u, we obtain the solution $\widehat{\boldsymbol{\Phi}} = \widehat{\mathbf{U}}\widehat{\mathbf{U}}'_{-1}(\widehat{\mathbf{U}}_{-1}\widehat{\mathbf{U}}'_{-1})^{-1}$ for the estimate of $\boldsymbol{\Phi}$, where $\widehat{\mathbf{U}} = \mathbf{Y} - \widehat{C}\mathbf{X}$ with $\widehat{C} = \widehat{A}\widehat{B}$. The iteration scheme in

(4.13) is then reduced in dimension, based on the uncoupled first order equations in (4.14), by replacing $\widehat{\theta}_i$ by $\widehat{\delta}_i$ and making the additional appropriate simplifications. Specifically, because

$$\begin{bmatrix} B_\delta + H_1 H_1' & -H_1 \\ -H_1' & 0 \end{bmatrix}^{-1} = \begin{bmatrix} W_1 & Q_1 \\ Q_1' & -I_{r^2} \end{bmatrix},$$

where $Q_1 = -(M_1' G M_1 + H_1 H_1')^{-1} H_1$, this leads to the iterative step to obtain the estimates $\widehat{A}$ and $\widehat{B}$ as follows:

$$\begin{aligned} \widehat{\delta}_i &= \widehat{\delta}_{i-1} - W_{1,i-1}(\partial S_{1T}^*(\theta)/\partial\delta)|_{\widehat{\theta}_{i-1}} - Q_{1,i-1} h(\delta)|_{\widehat{\delta}_{i-1}} \\ \widehat{\lambda}_i &= -Q_{1,i-1}'(\partial S_{1T}^*(\theta)/\partial\delta)|_{\widehat{\theta}_{i-1}} + h(\delta)|_{\widehat{\delta}_{i-1}}. \end{aligned}$$

The iterations thus alternate between the above equations to update the estimates of A and B and the step given to update the estimate of $\mathbf{\Phi}$. The covariance matrices of the final estimates $\widehat{\delta}$ and $\widehat{\phi}$ are provided by $\frac{1}{T} W_1$ and $\frac{1}{T} W_2$, where $W_2 = \Sigma_{\epsilon\epsilon} \otimes \Sigma_u^{-1} \equiv B_\phi^{-1}$.

4.5 Alternative Estimators and Their Properties

Initial specification of an appropriate rank of the regression coefficient matrix is desirable before the iterative estimation procedure is employed and the initial estimates of the component matrices are important for the convergence of iterative methods. This dictates a need for an alternative preliminary estimator that is computationally simpler and that may also be useful at the initial model specification stages to identify an appropriate rank for the model. In practice, the full-rank estimator may be computed first, and if there is a structure to the true parameter matrix, we should be able to detect it by using the full-rank estimator directly. The alternative initial estimator we propose below is essentially based on this simple concept.

To motivate the alternative estimator, we first recall the solution to the reduced-rank regression problem for the single regressor case under independent errors, as presented in Chapter 2. The sample version of the criterion that led to the solution in the single regressor case reduces to finding A and B by minimizing (see Eq. (2.12′))

$$\begin{aligned} &\operatorname{tr}[\Gamma^{1/2}(\widetilde{C} - AB)\widehat{\Sigma}_{xx}(\widetilde{C} - AB)'\Gamma^{1/2}] \\ &= [\operatorname{vec}\{(\widetilde{C} - AB)'\}]'(\Gamma \otimes \widehat{\Sigma}_{xx})\operatorname{vec}\{(\widetilde{C} - AB)'\}, \end{aligned} \tag{4.15}$$

where $\widetilde{C} = \widehat{\Sigma}_{yx}\widehat{\Sigma}_{xx}^{-1}$ is the full-rank least squares estimator. Using the Householder–Young theorem, the problem is solved explicitly as in (2.17). This is not the case with an AR(1) error model since the full-rank estimate of C, as displayed in (4.7) of Section 4.3, does not have an analytically simple form. For given $\boldsymbol{\Phi}$ and $\Sigma_{\epsilon\epsilon}(\equiv \Gamma^{-1})$, minimization of the efficient criterion (4.8) reduces to the problem of finding A and B such that

$$[\text{vec}\{(\widetilde{C} - AB)'\}]'\widetilde{G}\text{vec}\{(\widetilde{C} - AB)'\} \tag{4.16}$$

is minimized, where $\widetilde{C}$ is the corresponding full-rank GLS estimate as described in Section 4.3, $\text{vec}(\widetilde{C}') = \widetilde{G}^{-1}\mathbf{X}^{*\prime}(\widetilde{\Sigma}_{\epsilon\epsilon}^{-1} \otimes I_T)\text{vec}(\mathbf{Y}^{*\prime})/T$, and $\widetilde{G} = \mathbf{X}^{*\prime}(\widetilde{\Sigma}_{\epsilon\epsilon}^{-1} \otimes I_T)\mathbf{X}^*/T$. Now comparison of this latter expression with (4.15) indicates that an explicit solution via the Householder–Young Theorem is not possible unless the matrix $\widetilde{G}$ can be written as the Kronecker product of two matrices of appropriate dimensions. This can happen only under rather special circumstances, for instance, when $\boldsymbol{\Phi} = \rho I$. Hence, the solution is not simple as in the standard independent error case.

Motivated by a desire to obtain initial estimates for the reduced-rank model (for various choices of the rank r) that are relatively easy to compute in terms of the full-rank estimator $\widetilde{C}$ of model (4.1)-(4.3), we consider an alternate preliminary estimator. This estimator essentially is chosen to minimize (4.16), but with $\widetilde{G}$ replaced by $(\Gamma \otimes \mathbf{XX}'/T)$, the weight matrix that appears in (4.15), the inefficient least squares criterion. Note since this results in a criterion that is essentially the least squares criterion (4.15) with a better estimate $\widetilde{C}$ of C in place of the least squares estimate $\widehat{\Sigma}_{yx}\widehat{\Sigma}_{xx}^{-1}$, the natural choice of Γ is $\Gamma = \widetilde{\Sigma}_u^{-1}$, where $\widetilde{\Sigma}_u = (\mathbf{Y} - \widetilde{C}\mathbf{X})(\mathbf{Y} - \widetilde{C}\mathbf{X})'/T$. This proposed estimation procedure; it should be kept in mind, does not have any overall optimality properties corresponding to our model, but is motivated by the standard model and computational convenience.

As in the single regressor case, we can obtain the alternative estimators $\widetilde{A}^*$ and $\widetilde{B}^*$ using the Householder–Young theorem, which satisfy the normalization conditions $\widetilde{A}^{*\prime}\widetilde{\Sigma}_u^{-1}\widetilde{A}^* = I_r$, $\widetilde{B}^*(\mathbf{XX}'/T)\widetilde{B}^{*\prime} = \widetilde{\Lambda}^{*2}$. From the results in Chapter 2, it follows that the columns of $\widehat{\Sigma}_{xx}^{1/2}\widetilde{B}^{*\prime}$ are chosen to be the eigenvectors corresponding to the first r eigenvalues of the matrix $\widehat{\Sigma}_{xx}^{1/2}\widetilde{C}'\widehat{\Sigma}_u^{-1}\widetilde{C}\widehat{\Sigma}_{xx}^{1/2}$, the positive diagonal elements of $\widetilde{\Lambda}^*$ are the square roots of the eigenvalues, and $\widetilde{A}^* = \widetilde{C}\widehat{\Sigma}_{xx}\widetilde{B}^{*\prime}\widetilde{\Lambda}^{*-2}$. Equivalently, we can represent the alternate estimators by taking $\widetilde{A}^* = \widetilde{\Sigma}_u^{1/2}\widetilde{V}_{(r)}$ and $\widetilde{B}^* = \widetilde{V}_{(r)}'\widetilde{\Sigma}_u^{-1/2}\widetilde{C}$, where $\widetilde{V}_{(r)} = [\widetilde{V}_1, \ldots, \widetilde{V}_r]$ and the $\widetilde{V}_j$ are normalized eigenvectors of the matrix $\widetilde{\Sigma}_u^{-1/2}\widetilde{C}\widehat{\Sigma}_{xx}\widetilde{C}'\widetilde{\Sigma}_u^{-1/2}$ corresponding to its r largest eigenvalues. Although this is primarily of interest, for preliminary model specification, and used as an initial estimator, the asymptotic distribution of the alternative estimator can also be developed along similar lines as in Theorem 4.1. We shall not present details, but if we let $\widetilde{\alpha}^* = \text{vec}(\widetilde{A}^{*\prime})$, $\widetilde{\beta}^* = \text{vec}(\widetilde{B}^{*\prime})$, $\widetilde{\delta}^* = (\widetilde{\alpha}^{*\prime}, \widetilde{\beta}^{*\prime})'$ denote the alternative estimator and $\delta^* = (\alpha^{*\prime}, \beta^{*\prime})'$, the parameter values satisfying the corresponding population normalizations, then it can be shown that $\sqrt{T}(\widetilde{\delta}^* - \delta^*)$ has a limiting multivariate normal distribution with zero mean vector and covariance matrix given by the leading $r(m+n)$-order square submatrix of the $r(m+n+r)$-order matrix

$$\begin{bmatrix} M_1^{*\prime}(\Sigma_u^{-1} \otimes \Sigma_{xx})M_1^* & -H_1^* \\ -H_1^{*\prime} & 0 \end{bmatrix}^{-1}$$

$$\times \begin{bmatrix} M_1^{*\prime}(\Sigma_u^{-1} \otimes \Sigma_{xx})G^{-1}(\Sigma_u^{-1} \otimes \Sigma_{xx})M_1^* & 0 \\ 0 & 0 \end{bmatrix}$$

$$\times \begin{bmatrix} M_1^{*\prime}(\Sigma_u^{-1} \otimes \Sigma_{xx})M_1^* & -H_1^* \\ -H_1^{*\prime} & 0 \end{bmatrix}^{-1},$$

where $M_1^* = [I_m \otimes B^{*\prime}, A^* \otimes I_n]$ and $H_1^* = \{(\partial h_j(\delta^*)/\partial \delta_i^*)\}$ is $r(m+n) \times r^2$.

4.5.1 A Comparison Between Efficient and Other Estimators

Though efficient estimators are desirable, the alternative estimator discussed above, as well as the least squares estimator, is more easily computable and may also be quite useful as initial estimators and for model specification purposes. We will briefly consider their merits in terms of asymptotic efficiencies relative to the efficient estimator.

To obtain relatively explicit results, we shall restrict attention to the rank 1 case only, so that $C = \alpha\beta'$, where α and β are column vectors that satisfy the normalizations $\alpha'\Sigma_{\epsilon\epsilon}^{-1}\alpha = 1$, $\beta'\Gamma_x(0)\beta = \lambda^2$. In addition, we assume for simplicity that $\mathbf{\Phi}$ is known and has the form $\mathbf{\Phi} = \rho I$, where $|\rho| < 1$. In this case, we can show that the asymptotic covariance matrix of the efficient estimator $(\widehat{\alpha}', \widehat{\beta}')'$ is

$$W_1 = \begin{bmatrix} \frac{1}{a}(\Sigma_{\epsilon\epsilon} - \alpha\alpha') & 0 \\ 0 & \mathbf{\Psi}^{-1} \end{bmatrix}, \tag{4.17}$$

with

$$\mathbf{\Psi} = (1+\rho^2)\Gamma_x(0) - \rho(\Gamma_x(1) + \Gamma_x(1)'),$$

and

$$a = (1+\rho^2)\lambda^2 - 2\rho(\beta'\Gamma_x(1)\beta) = \beta'\mathbf{\Psi}\beta.$$

In the classical independent error case, $\rho = 0$, W_1 reduces to the result given in Robinson (1974), $W_1 = \mathrm{diag}\{(\beta'\Gamma_x(0)\beta)^{-1}(\Sigma_{\epsilon\epsilon} - \alpha\alpha'), \Gamma_x(0)^{-1}\}$ as given in the discussion following the proof of Theorem 2.4 in Section 2.5.

For the alternative estimator, note the decomposition is $C = \alpha^*\beta^{*\prime}$, where the column vector α^* satisfies the normalization condition $\alpha^{*\prime}\Sigma_u^{-1}\alpha^* = 1$. Because of this change in the constraint on α^*, it is necessary to make a transformation of the parameter estimator so that the alternative estimator will be comparable to the efficient estimator. This transformation is easy to obtain (since $\Sigma_u = \{1/(1-\rho^2)\}\Sigma_{\epsilon\epsilon}$ for the special case $\mathbf{\Phi} = \rho I$) as follows: $\beta = \{1/(1-\rho^2)^{1/2}\}\beta^*$, $\alpha = (1-\rho^2)^{1/2}\alpha^*$, and $\lambda = \lambda^*/(1-\rho^2)^{1/2}$. Hence, the asymptotic covariance matrix of the alternative estimator when normalized in the same way as the efficient estimator is

$$R = \begin{bmatrix} \frac{a^*}{\lambda^4}(\Sigma_{\epsilon\epsilon} - \alpha\alpha') & 0 \\ 0 & \mathbf{\Psi}^{-1} \end{bmatrix}, \tag{4.18}$$

where

$$a^* = \beta'\Gamma_x(0)\mathbf{\Psi}^{-1}\Gamma_x(0)\beta.$$

Comparison of the matrix W_1 in (4.17) with R in (4.18) indicates that, asymptotically, the alternative estimator of β is efficient, but the alternative estimator of α is not. In fact, the inefficiency of the alternative estimator of α can be readily established by verifying that $a^*/\lambda^4 \geq 1/a$ through the Cauchy–Schwarz inequality. This asymptotic comparison has an interesting interpretation that indicates that an efficient estimator can be achieved indirectly by a two-step procedure in which the efficient alternative estimator $\widetilde{\beta}$ is first obtained, and then an efficient estimate of α is obtained by regression of Y_t on the index $\widetilde{\beta}'X_t$, taking into account the AR error structure. We note that one special case in which the alternate estimator $\widetilde{\alpha}$ is efficient occurs when the X_t have first order autocovariance matrix of the form $\Gamma_x(1) = \gamma\Gamma_x(0)$, since then $\mathbf{\Psi} = (1+\rho^2 - 2\gamma\rho)\Gamma_x(0)$ and hence $a^*/\lambda^4 = 1/a$, and so in particular $\widetilde{\alpha}$ is efficient when the X_t are white noise ($\gamma = 0$).

Now we shall compare the least squares (LS) estimator with the alternative estimator for the rank one case. From the asymptotic results of Robinson (1973) and using a similar parameter transformation as in the case of the alternative estimator, we obtain the covariance matrix for the least squares estimator under the normalization $\alpha'\Sigma_{\epsilon\epsilon}^{-1}\alpha = 1$ as

$$S = \begin{bmatrix} \frac{a_1}{\lambda^4(1-\rho^2)}(\Sigma_{\epsilon\epsilon} - \alpha\alpha') & 0 \\ 0 & \frac{1}{(1-\rho^2)}\Gamma_x(0)^{-1}\Gamma_x^*(0)\Gamma_x(0)^{-1} \end{bmatrix} \tag{4.19}$$

with

$$a_1 = \lambda^2 + 2\sum_{j=1}^{\infty}(\beta'\Gamma_x(j)\beta)\rho^j,$$

and

$$\Gamma_x^*(0) = \Gamma_x(0) + \sum_{j=1}^{\infty}(\Gamma_x(j) + \Gamma_x(j)')\rho^j.$$

It is not easy to analytically compare the above covariance matrix with R in (4.18) in general. However, for illustration, suppose the X_t's are first order autoregressive with coefficient matrix equal to γI. With these simplifying conditions, we have shown before that the alternative estimator is efficient. However, the LS estimator performs poorly. We shall particularly concentrate on the estimators of α and compare the elements in the first diagonal block in S and R. Thus we compare $a_1/\lambda^4(1-\rho^2)$ with a^*/λ^4, where $a_1 = \lambda^2(1+\gamma\rho)/(1-\gamma\rho)$ and $a^* = \lambda^2/(1+\rho^2-2\gamma\rho)$. For $\gamma = \rho = 0.8$, the efficiency of the LS estimator of α relative to the alternative estimator of α is 0.22. This simple comparison suggests that the alternative estimator may perform substantially better than the LS estimator, and as mentioned before, the alternative estimator is also computationally convenient.

4.6 Identification of Rank of the Regression Coefficient Matrix

It is obviously important to be able to determine the rank of the coefficient matrix C for it is a key determinant in the structure of the reduced-rank regression problem. From the asymptotic distribution theory presented earlier, we do not learn about the rank of C because the asymptotic results are derived assuming the rank conditions are true. In the single regressor model with independent errors, the exact relationship between canonical correlation analysis and reduced-rank regression allows one to check on the rank by testing if certain canonical correlations are zero (see Section 2.6). When the rank of C is r, under independence and normality assumptions, the statistic suggested by Bartlett (1947), $T\sum_{j=r+1}^{m}\log(1+\widehat{\lambda}_j^2) \equiv -T\sum_{j=r+1}^{m}\log(1-\widehat{\rho}_j^2)$, which is also the likelihood ratio statistic as discussed in Section 2.6, has a limiting $\chi^2_{(m-r)(n-r)}$ distribution (Hsu 1941). However, this test procedure is not valid under the autoregressive error model of (4.3). We need to develop other large sample quantities to test the hypothesis of interest in the framework of a correlated error structure.

Likelihood ratio test procedures have been shown to be valid for rather general autocorrelated error structures by Kohn (1979, Section 8). The likelihood ratio test statistic takes the form

$$\chi^2 = T\log(|\widehat{\Sigma}_{\epsilon\epsilon}|/|\widetilde{\Sigma}_{\epsilon\epsilon}|), \tag{4.20}$$

where $\widetilde{\Sigma}_{\epsilon\epsilon}$ denotes the unrestricted maximum likelihood estimate of $\Sigma_{\epsilon\epsilon}$ under the autoregressive error structure and $\widehat{\Sigma}_{\epsilon\epsilon}$ the restricted estimate based on the restricted maximum likelihood estimate $\widehat{C} = \widehat{A}\widehat{B}$ of C subject to the constraint that rank$(C) = r$. Hence, this statistic can be used to test the null hypothesis that rank$(C) = r$, and the statistic will follow a χ^2-distribution with $mn - \{r(m+n) - r^2\} = (m-r)(n-r)$ degrees of freedom, the number of constrained parameters, under the null hypothesis. This statistic can be used to test the validity of the assumption of a reduced-rank structure of rank r for C, once the efficient estimates $\widehat{A}$ and $\widehat{B}$ have been obtained as in Section 4.3 with $\widehat{C} = \widehat{A}\widehat{B}$. However, this procedure would not be useful at the preliminary specification stage where we are interested in selecting an appropriate rank for C prior to efficient estimation. This leads to consideration of an "approximate" likelihood ratio procedure, as a rough guide to preliminary model specification (e.g., see Robinson (1973, p. 159) for a related suggestion), in which the alternative estimator of C derived under the assumption of AR(1) errors is used to form the estimate of $\Sigma_{\epsilon\epsilon}$ in the statistic (4.20) as an approximation to the efficient constrained estimator $\widehat{\Sigma}_{\epsilon\epsilon}$, with the resulting statistic treated as having roughly a chi-squared distribution with $(m-r)(n-r)$ degrees of freedom under H_0. Thus, if $\widetilde{C}^* = \widetilde{A}^*\widetilde{B}^*$ denotes the alternative estimator of C from Section 4.5, then we use $\widehat{\Sigma}^*_{\epsilon\epsilon} = \widehat{\boldsymbol{\epsilon}}^*\widehat{\boldsymbol{\epsilon}}^{*\prime}/T$ in (4.20) in place of the restricted maximum likelihood estimate $\widehat{\Sigma}_{\epsilon\epsilon}$, where $\widehat{\boldsymbol{\epsilon}}^* = \widehat{\mathbf{U}}^* - \widehat{\boldsymbol{\Phi}}^*\widehat{\mathbf{U}}^*_{-1}$, $\widehat{\boldsymbol{\Phi}}^* = \widehat{\mathbf{U}}^*\widehat{\mathbf{U}}^{*\prime}_{-1}(\widehat{\mathbf{U}}^*_{-1}\widehat{\mathbf{U}}^{*\prime}_{-1})^{-1}$, and $\widehat{\mathbf{U}}^* = \mathbf{Y} - \widetilde{C}^*\mathbf{X}$ is obtained from the alternative estimate.

It should be emphasized that the main motivation (and advantage) of using this approximate test, with the alternative estimator approximating the exact restricted ML estimate, is that preliminary tests for models of various ranks can all be tested simultaneously based on one set of eigenvalue–eigenvector calculations for the alternative estimator. The validity of the model (rank) selected at this preliminary stage would be checked further, of course, following efficient estimation of the parameters A and B by use of the chi-squared likelihood ratio statistic (4.20) with the efficient estimates $\widehat{A}$, $\widehat{B}$, and $\widehat{\Sigma}_{\epsilon\epsilon}$. However, it should be noted that the use of the preliminary approximate likelihood ratio test would tend to be "conservative," in the sense of yielding larger values for the test statistic than the exact procedure, since the value $|\widehat{\Sigma}^*_{\epsilon\epsilon}|$ based on the preliminary alternative estimator must be larger than that of $|\widehat{\Sigma}_{\epsilon\epsilon}|$ from the final restricted ML estimate. It is possible that a computationally simple preliminary test that has an exact asymptotic chi-squared distribution may be obtainable, and an investigation of this matter is a worthwhile topic for further research.

4.7 Inference for the Numerical Example

The initial analysis of the macroeconomic data indicated that the serial correlations must be taken into account for proper analysis. Thus, having tentatively identified that the data set may be analyzed in the framework of the models described in this chapter, we first compute the full-rank estimator under an autoregressive structure for the errors. A first order autoregressive error model is considered for the purpose of illustration, although one might argue toward a second-order model. The trace of the residual covariance matrix $\widetilde{\Sigma}_{\epsilon\epsilon}$ for the AR(1) model is 133.52, and the trace under the ordinary least squares fit, i.e., the zero autocorrelation model, is 215.83, which indicates that significant improvement occurs when an AR(1) model for the errors is assumed. The full-rank estimate of the regression coefficient matrix under this assumption on the errors is given as follows:

$$\widetilde{C} = \begin{bmatrix} 5.42^* & -0.90^* & -0.06 & -0.42^* & 1.90^* \\ 1.24^* & -0.10 & 0.13 & -0.32^* & 0.56^* \\ -5.92 & 0.29 & -0.37 & 1.21 & -1.05 \\ 3.13^* & -0.51 & -0.11 & -0.13 & 1.38^* \\ 9.46^* & -0.37 & 0.71^* & -0.86^* & 0.39 \end{bmatrix}.$$

The starred entries indicate that the t-ratios of these elements are greater than 1.65. Although the elements of $\widetilde{C}$ are not too different from the elements of the least squares estimate, there are some differences in the significance of individual elements when autocorrelation in the errors is accounted for. The corresponding estimate of the coefficient matrix $\boldsymbol{\Phi}$ in the AR(1) error model (4.3) is given by

$$\widetilde{\boldsymbol{\Phi}} = \begin{bmatrix} -0.186 & -0.323 & -0.121 & -0.147 & 0.170 \\ -0.121 & -0.450 & -0.031 & -0.106 & 0.082 \\ -0.235 & -0.324 & 0.853 & 0.596 & 0.123 \\ 0.151 & 0.393 & -0.266 & -0.081 & -0.096 \\ -0.946 & 0.801 & -0.116 & -0.275 & 0.266 \end{bmatrix}$$

with eigenvalues of $\widetilde{\boldsymbol{\Phi}}$ equal to 0.649, −0.508, 0.094, and $0.083 \pm 0.240i$. These eigenvalues provide an additional indication that autocorrelation in the error term U_t in the regression model (4.1) is substantial and the error term is stationary.

In order to initially specify the rank of the regression coefficient matrix C, the approximate chi-squared likelihood ratio statistic (4.20), which is based on the full-rank estimator $\widetilde{C}$ and the alternative estimators under the AR(1) error structure, was computed for various values of the rank r, and the results are presented in Table 4.1. It is seen from this table that the rank of the matrix C may be taken to be equal to 2 when serial correlations are accounted for. For the purpose of illustration, we take the rank to be equal to 2, and the efficient estimates of the component matrices and their standard errors can be computed using the iterative algorithm given in Section 4.4. These are presented below:

Table 4.1 Summary of computations of approximate chi-squared statistics used for specification of rank of the model

r = rank of the matrix C under H_0	Chi-squared statistic	d.f.	Critical value (5%)
	78.42	25	37.70
1	31.09	16	26.30
2	13.48	9	16.90
3	3.35	4	9.49
4	0.20	1	3.84

$$\widehat{A}' = \begin{bmatrix} 2.70 & 1.09 & -1.69 & 2.54 & 2.84 \\ (0.13) & (0.07) & (1.16) & (0.34) & (0.43) \\ 0.21 & 0.31 & 0.15 & -0.86 & 3.80 \\ (0.43) & (0.21) & (1.28) & (0.49) & (0.50) \end{bmatrix},$$

$$\widehat{B} = \begin{bmatrix} 0.81 & -0.13 & 0.08 & -0.24 & 0.61 \\ (0.41) & (0.09) & (0.07) & (0.08) & (0.16) \\ 1.57 & 0.19 & 0.28 & -0.24 & -0.49 \\ (0.47) & (0.12) & (0.08) & (0.10) & (0.19) \end{bmatrix}.$$

We compute the matrix $\widehat{C} = \widehat{A}\widehat{B}$ to see how well the full-rank estimate $\widetilde{C}$ is reproduced by the product of the lower-rank matrices,

$$\widehat{C} = \begin{bmatrix} 2.51 & -0.30 & 0.27 & -0.70 & 1.54 \\ 1.38 & -0.08 & 0.17 & -0.34 & 0.51 \\ -1.14 & 0.24 & -0.09 & 0.37 & -1.10 \\ 0.72 & -0.49 & -0.05 & -0.41 & 1.97 \\ 8.26 & 0.37 & 1.30 & -1.61 & -0.12 \end{bmatrix}.$$

It can be seen that the entries of the full-rank estimate $\widetilde{C}$ are fairly well reproduced in $\widehat{C}$, and this could certainly be improved by entertaining the possibility that the rank of the matrix is 3. Alternatively, we can compare the trace of the residual covariance matrix $\widehat{\Sigma}_{\epsilon\epsilon}$, 145.42, obtained from estimation of the rank 2 model, with the trace of $\widetilde{\Sigma}_{\epsilon\epsilon}$, which we had earlier mentioned as 133.52. The 9% increase in the trace value has to be viewed in conjunction with the reduction in the number of (functionally) independent regression parameters from 25 under the full-rank model to 16 under the rank 2 model. Under the null hypothesis that $\text{rank}(C) = 2$, we also compute the exact likelihood ratio statistic (4.20) using the restricted ML estimate $\widehat{C}$ to check the validity of the reduced-rank specification. The computed value is $\chi^2 = 6.92$ with 9 degrees of freedom, which suggests adequacy in the reduced-rank specification. Subsequent calculation of the restricted ML estimate for a rank 1 model specification and of the associated likelihood ratio statistic (4.20) led to rejection of a regression model of rank 1.

It is not our intention to give a detailed interpretation of the estimates of the component matrices, but some observations are worth noting. The two linear combinations of X_t, $X_t^* = \widehat{B}X_t$, are the most useful linear combinations of the exogenous variables for predicting the endogenous variables Y_t, and these are plotted in Fig. 4.2. Based on the standard errors, note that the index of weekly wage rates does not appear to play any significant role in explaining the endogenous variables considered here. The first linear combination is (approximately) $x_{1t}^* = 0.61x_{5t} - 0.24x_{4t} + 0.81x_{1t}$, an index based on only the price index of total consumption, the price index of total exports, and the labor force. This combination is essentially a trend variable and affects all the endogenous variables except

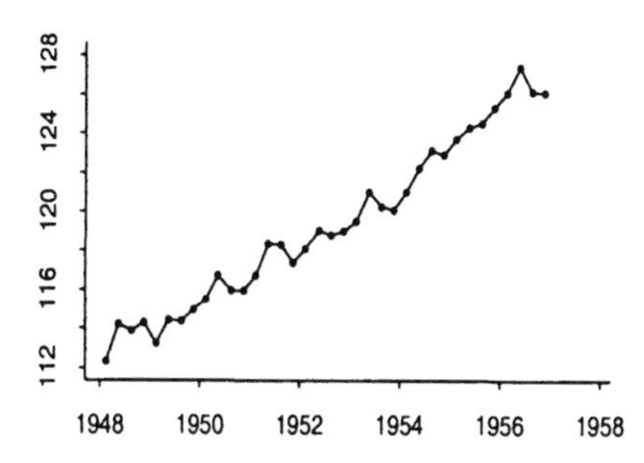

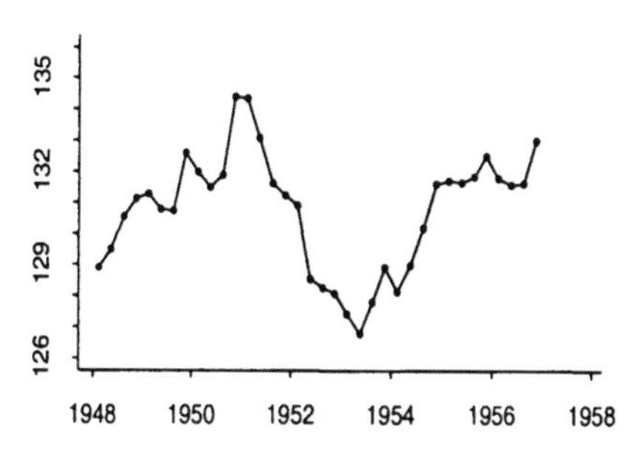

Fig. 4.2 First two exogenous "index" series for the U.K. economic data

unemployment. The second linear combination $x_{2t}^* = 1.57x_{1t} + 0.28x_{3t} - 0.24x_{4t} - 0.49x_{5t}$ seems to be especially useful only to explain y_{5t}, the index of volume of total exports. Also, it may be observed that neither combination significantly influences the unemployment series. To come up with a serious model of the U.K. economy, we may have to expand the set of exogenous variables, and the nature of that investigation would be much more detailed than that considered here.

4.8 An Alternate Estimator with Kronecker Approximation

In Section 4.5, we replaced the estimate of the covariance matrix of $\text{vec}(\widetilde{C}') = \widetilde{G}$ by $\left(\widehat{\Sigma}_u^{-1} \otimes \widetilde{\Sigma}_{xx}\right)$ to obtain the alternate estimators of A and B. As shown earlier, the Kronecker structure of separable covariance matrices leads to fast computation of the parameters and may be especially useful for dimension reduction aspects of data with a large number of variables. The suggested Kronecker structures in Section 4.5 are based on intuitive ideas. In this section, we describe the nearest Kronecker product approximation of a matrix of general form suggested by Van Loan and Pitsianis (1993). The algorithm is simple to use, and the solution preserves the positive definiteness required for the covariance matrices.

Let G be a covariance matrix of dimension $mn \times mn$. This could be a sum of several Kronecker products or simply a matrix that is not in the form of a Kronecker product. The solution to the approximation of G by $G_1 \otimes G_2$, i.e., to minimize

$$\|G - G_1 \otimes G_2\|_F = \|R(G) - (\widetilde{G}_1) \otimes (\widetilde{G}_2)\|_F, \tag{4.21}$$

where $R(G)$ is a matrix that is rearranged from the columns and rows of G so that $\|G\|_F = \|R(G)\|_F$. The rearrangement function is $R(G) = [\text{vec}'(G^{11}), \ldots \text{vec}'(G^{m1}), \text{vec}'(G^{21}), \ldots, \text{vec}'(G^{mm})]$, where G^{kl} is the k, lth $n \times n$

block of G. The rearranged matrix $R(G)$ is $m^2 \times n^2$ and can be rectangular. Here $\|\cdot\|_F$ denotes the Frobenius norm.

It is shown that minimizing (4.21) is essentially reduced to a rank one approximation of the rectangular matrix, $R(G)$. By singular value decomposition of $R(G)$, $U'R(G)V = \Delta = \text{diag}(\delta_1, \ldots, \delta_q)$, where $U \in \mathbb{R}^{n^2 \times n^2}$ and $V \in \mathbb{R}^{m^2 \times m^2}$ are orthogonal matrices, and the singular values are arranged in descending order and are positive; observe $q = \min\{m^2, n^2\}$. The solution to (4.21) is given by

$$\text{vec}(G_1) = \sqrt{\delta_1}\, U_1 \quad \text{and} \quad \text{vec}(G_2) = \sqrt{\delta_1}\, V_1, \tag{4.22}$$

where U_1 and V_1 are the first column of U and V matrices. Application of this method in various statistical contexts is given in Genton (2007) and in Guggenberger, Kleibergen and Mavroeidis (2022). Note this approximation is based on the elements of G matrix, and therefore, we expect it to perform better than other Kronecker structures suggested earlier. Genton (2007) defines an error index to measure the error in the approximation as

$$K_G(G_1, G_2) = \frac{\|G - G_1 \otimes G_2\|_F}{\|G\|_F} = \sqrt{\frac{\sum_{i=2}^{q} \delta_i^2}{\sum_{i=1}^{q} \delta_i^2}}. \tag{4.23}$$

The index is bounded below by zero and above by $\sqrt{1 - 1/q}$.

4.8.1 Computational Results

In order to study the effectiveness of the Kronecker's approximation of G in (4.21), we compute the trace and the determinant of the error covariance matrix, $\Sigma_{\epsilon\epsilon}$ for ranks $r = 1, 2, \ldots, 5$. Recall that the reduced-rank estimators of "C" are obtained from the projection of $\Gamma_1^{1/2} C \Gamma_2^{1/2}$ to the lower dimension. Results we present below are for the following alternatives; we assume $\widetilde{C}$ is the full rank GLS estimator of C under AR(1) structure for the errors, as given in (4.7).

(a) Weight matrix, G, not in Kronecker form
(b) Weight matrices: $\Gamma_1 = \widetilde{\Sigma}_u^{-1}$; $\Gamma_2 = \widetilde{\Sigma}_{xx}$
(c) Weight matrices: $\Gamma_1 = \widetilde{G}_1$; $\Gamma_2 = \widetilde{G}_2$

Note that $\widetilde{C}^{(r)} = P\widetilde{C}$, where the projection matrix $P = \Gamma_1^{1/2} \widetilde{V}_r \widetilde{V}_r' \Gamma_1^{-1/2}$ and $\widetilde{V}_{(r)}$ is the $m \times r$ matrix of eigenvectors corresponding to the first "r" eigenvalues of $\Gamma_1^{1/2} \widetilde{C} \Gamma_2 \widetilde{C}' \Gamma_!^{1/2}$. The results under (a) are given in the last section. For (b) and (c), we also compute results for $r = 2$.

The $\widetilde{G}$-matrix is of dimension 25×25 and is not reproduced here. Now the results of the Kronecker's product: the $\widetilde{G}$ matrix is decomposed into $\widetilde{G}_1 \otimes \widetilde{G}_2$ and they are given below:

$$\widetilde{G}_1 = \begin{bmatrix} 6.61 & 0.20 & 0.30 & -1.42 & -0.79 \\ 0.20 & 25.59 & -0.05 & -0.92 & -1.45 \\ 0.30 & -0.05 & 0.22 & 0.53 & -0.15 \\ -1.42 & -0.92 & 0.53 & 2.40 & 0.06 \\ -0.79 & -1.45 & -0.15 & 0.06 & 0.98 \end{bmatrix}, \quad \widetilde{G}_2 = \begin{bmatrix} 0.16 & 1.35 & 0.55 & 0.88 & 0.87 \\ 1.35 & 12.37 & 4.54 & 7.93 & 7.98 \\ 0.58 & 4.54 & 4.82 & 5.57 & 3.56 \\ 0.88 & 7.93 & 5.57 & 7.60 & 5.74 \\ 0.87 & 7.98 & 3.56 & 5.74 & 5.33 \end{bmatrix}.$$

The matrices $\widetilde{G}_1$ and $\widetilde{G}_2$ are positive definite. The resulting rank(2) approximation of C matrix from this decomposition is given below:

$$\widetilde{C}_{\text{KR}}^{(2)} = \begin{bmatrix} 4.37 & -0.62 & 0.10 & -0.55 & 1.70 \\ 1.81 & -0.24 & 0.06 & -0.23 & 0.63 \\ -6.15 & 0.23 & -0.55 & 0.89 & -0.48 \\ 2.88 & -0.65 & -0.09 & -0.32 & 1.81 \\ 8.73 & -0.43 & 0.72 & -1.24 & 0.98 \end{bmatrix}.$$

The approximation performs well compared to the rank 2 efficient coefficient matrix. The effectiveness of this method crucially depends on how well $\widetilde{G}$ is approximated by the Kronecker product $\widetilde{G}_1 \otimes \widetilde{G}_2$. The eigenvalues of $\widetilde{G}$ and the eigenvalues of $\widetilde{G}_1 \otimes \widetilde{G}_2$ are virtually indistinguishable (see Fig. 4.3); other refers to the eigenvalues of $\Sigma_u^{-1} \otimes \Sigma_{xx}$. To make the comparisons more meaningful, we can also compute the AIC criterion for various ranks. It can be shown that weighing the matrix based on the Kronecker approximation yields uniformly better results. Further details on this can be found in Velu and Herman (2017).

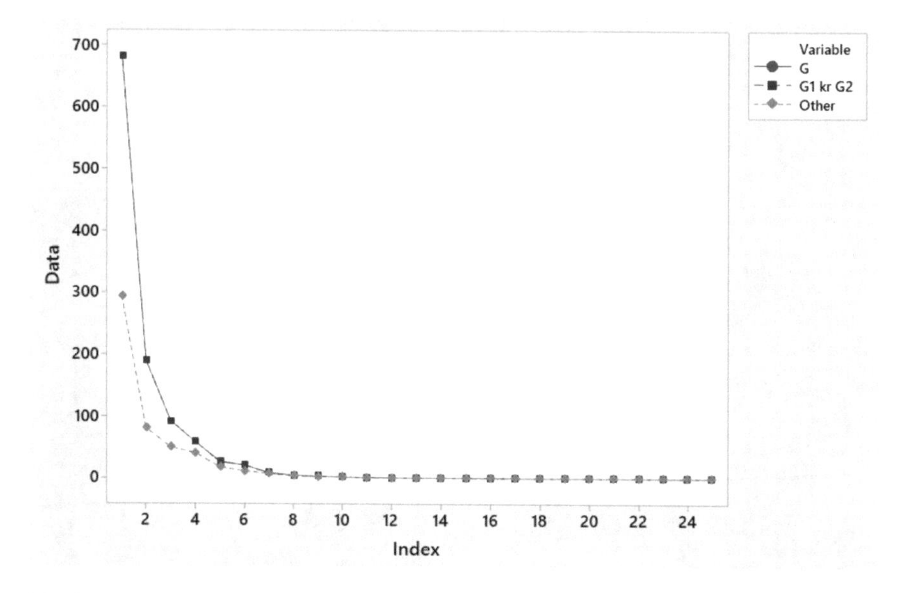

Fig. 4.3 Eigenvalue data

Chapter 5
Multiple Time Series Modeling With Reduced Ranks

5.1 Introduction and Time Series Models

There has been a growing interest in multiple time series modeling, particularly through use of vector autoregressive moving average models. The subject has found an appeal and has applications in various disciplines, including engineering, physical sciences, business and economics, and the social sciences. In general, multiple time series analysis is concerned with modeling and estimation of dynamic relationships among m related time series $y_{1t}, \ldots, y_{mt}$, based on observations on these series over T equally spaced time points $t = 1, \ldots, T$, and also between these series and potential input or exogenous time series variables $x_{1t}, \ldots, x_{nt}$, observed over or around the same time period. In this chapter, we shall explore the use of certain reduced-rank modeling techniques for analysis of multiple time series in practice. We first introduce a general model for multiple time series modeling but will specialize to multivariate autoregressive (AR) models for more detailed investigation.

Let $Y_t = (y_{1t}, \ldots, y_{mt})'$ be an $m \times 1$ multiple time series vector of response variables, and let $X_t = (x_{1t}, \ldots, x_{nt})'$ be an $n \times 1$ vector of input or predictor time series variables. Let ϵ_t denote an $m \times 1$ white noise vector of errors, independently distributed over time with $E(\epsilon_t) = 0$ and $\mathrm{Cov}(\epsilon_t) = \Sigma_{\epsilon\epsilon}$, a positive-definite matrix. We consider the multivariate time series model

$$Y_t = \sum_{s=0}^{p^*} C_s X_{t-s} + \epsilon_t, \tag{5.1}$$

where the C_s are $m \times n$ matrices of unknown parameters. In the general setting, the "input" vectors X_t could include past (lagged) values of the response series Y_t. Even for moderate values of the dimensions m and n, the number of parameters that must be estimated in model (5.1), when no constraints are imposed on the matrices C_s,

G. C. Reinsel et al., *Multivariate Reduced-Rank Regression*, Lecture Notes in Statistics 225, https://doi.org/10.1007/978-1-0716-2793-8_5

can become quite large. Because of the potential complexity of these models, it is often useful to consider canonical analyses or other dimension reduction procedures such as reduced-rank regression methods described in earlier chapters. This may also lead to more efficient models due to the reduction in the number of unknown parameters.

The multivariate time series model in (5.2) below, which places some structure on the coefficient matrices C_s in (5.1), has some useful properties and is in the spirit of reduced-rank regression models. We consider the model

$$Y_t = \sum_{s=0}^{p_2} \sum_{u=0}^{p_1} A_s B_u X_{t-s-u} + \epsilon_t, \tag{5.2}$$

where the A_s are $m \times r$ and the B_u are $r \times n$ matrices, respectively, with $1 \leq r \leq \min(m, n)$, and we shall assume $m \leq n$ for convention. Let L denote the lag operator such that $LY_t = Y_{t-1}$, and define the $m \times r$ and $r \times n$ matrix polynomial operators, of degrees p_2 and p_1, respectively, by

$$A(L) = \sum_{s=0}^{p_2} A_s L^s, \qquad B(L) = \sum_{u=0}^{p_1} B_u L^u.$$

Model (5.2) can then be written as

$$Y_t = A(L)B(L)X_t + \epsilon_t. \tag{5.3}$$

Brillinger (1969) considered a generalization of (5.3) where $A(L)$ and $B(L)$ are two-sided infinite order operators. He gives the interpretation already mentioned in Section 2.1, namely, that the $r \times 1$ series $X_t^* = B(L)X_t$ represents a reduction of the $n \times 1$ input series X_t. The reduction might be chosen, in terms of the operator $B(L)$ and, eventually, the operator $A(L)$, so that the information about Y_t contained in the series X_t is recovered as closely as possible by the factor $A(L)X_t^* = A(L)B(L)X_t$. Observe that in the standard reduced-rank regression model considered in Chapter 2, we have simply $A(L) = A$ and $B(L) = B$.

The model (5.3) can accommodate many situations that may be useful in practice. Suppose we partition X_t to explicitly allow for lagged values of the series Y_t, as $X_t = (Z_t', Y_{t-1}')'$. Partitioning the matrices B_u in (5.2) conformably, with $B(L) = [B_0(L), B_1(L)]$, allows us to write (5.3) as

$$Y_t - A(L)B_1(L)Y_{t-1} = A(L)B_0(L)Z_t + \epsilon_t. \tag{5.4}$$

The quantities $Z_t^* = B_0(L)Z_t$ have, in certain situations, interpretations as dynamic indices for the Z_t, and consequently, a model such as (5.4) is sometimes referred to as an index model (e.g., Sims 1981).

We now focus on the class of multivariate autoregressive (AR) models that results as a special case of (5.2). When $m = n$ and X_t contains only lagged values of Y_t so that $X_t = Y_{t-1}$, model (5.3) becomes

$$Y_t = A(L)B(L)Y_{t-1} + \epsilon_t, \tag{5.5}$$

and this model for Y_t represents a $p^* = p_1 + p_2 + 1$ order multivariate AR process. We mention two particular cases of special interest. First, with $p_2 = 0$ and $p_1 = p - 1$, we have

$$\begin{aligned} Y_t = A_0 B(L) Y_{t-1} + \epsilon_t &= A_0 B_0 Y_{t-1} + \cdots + A_0 B_{p-1} Y_{t-p} + \epsilon_t \\ &= A_0 B Y^*_{t-1} + \epsilon_t \equiv C Y^*_{t-1} + \epsilon_t, \end{aligned} \tag{5.6}$$

where $B = [B_0, B_1, \ldots, B_{p-1}]$, $C = A_0 B$, and $Y^*_{t-1} = (Y'_{t-1}, \ldots, Y'_{t-p})'$. Since $\text{rank}(C) = r \leq m$, model (5.6) has the same structure as the multivariate reduced-rank regression model (2.5). Various aspects of this particular reduced-rank AR model form were studied by Velu et al. (1986) and are also presented by Lütkepohl (1993, Chapter 5). Second, with $p_1 = 0$ and $p_2 = p - 1$, we obtain

$$\begin{aligned} Y_t &= A_0 B_0 Y_{t-1} + \cdots + A_{p-1} B_0 Y_{t-p} + \epsilon_t \\ &= A_0 Y^*_{t-1} + \cdots + A_{p-1} Y^*_{t-p} + \epsilon_t, \end{aligned} \tag{5.7}$$

where $Y^*_t = B_0 Y_t$ is an $r \times 1$ vector of linear combinations of Y_t, $r \leq m$. In model (5.7), the Y^*_t may be regarded as a lower-dimensional set of indices for the Y_t, whose p lagged values are sufficient to represent the past behavior of Y_t. So this model may be called an index model. Note that on multiplying (5.7) on the left by B_0, we find that the series Y^*_t also has the structure of an AR process of order p. Estimation of parameters and other aspects of model (5.7) have been considered by Reinsel (1983).

The class of models stated in (5.2) is rather flexible and provides many options for modeling multivariate time series data. The structure of coefficient matrices in (5.2) in its various forms is within a reduced-rank framework. There are other forms and extensions of reduced-rank model for multiple time series, which may not directly follow from (5.2), that are found to also be of practical importance. For example, the extended reduced-rank regression model (3.3) considered by Anderson (1951) and the model (3.23) may have utility in multivariate time series modeling. Specifically, consider the vector AR(p) model

$$Y_t = \sum_{j=1}^{p} \mathbf{\Phi}_j Y_{t-j} + \epsilon_t. \tag{5.8}$$

Suppose that for some p_1, $1 \leq p_1 < p$, the matrix $[\mathbf{\Phi}_{p_1+1}, \ldots, \mathbf{\Phi}_p]$ is of reduced rank, while $[\mathbf{\Phi}_1, \ldots, \mathbf{\Phi}_{p_1}]$ is of full rank. (For instance, a particular case occurs with $p = 2$, where $\mathbf{\Phi}_2$ is reduced rank but $\mathbf{\Phi}_1$ is full rank.) Such models with this structure fit into the framework of model (3.3), with the correspondences $Z_t =$

$(Y'_{t-1}, \ldots, Y'_{t-p_1})'$ and $X_t = (Y'_{t-p_1-1}, \ldots, Y'_{t-p})'$. This type of model might be useful if dimension reduction is possible in representing the influence on Y_t of more distant past values, whereas the recent lagged values must be retained intact. In the case where both matrices $[\mathbf{\Phi}_1, \ldots, \mathbf{\Phi}_{p_1}]$ and $[\mathbf{\Phi}_{p_1+1}, \ldots, \mathbf{\Phi}_p]$ are of reduced ranks, we have an example of the model (3.23).

A more general reduced-rank setup for vector AR models was proposed by Ahn and Reinsel (1988). In the vector AR(p) model (5.8), suppose it is assumed that the matrices $\mathbf{\Phi}_j$ have a particular reduced-rank structure, such that $\text{rank}(\mathbf{\Phi}_j) = r_j \geq \text{rank}(\mathbf{\Phi}_{j+1}) = r_{j+1}$, $j = 1, \ldots, p-1$. The $\mathbf{\Phi}_j$ then have the decomposition $\mathbf{\Phi}_j = A_j B_j$, where A_j is $m \times r_j$ and B_j is $r_j \times m$, and it is further assumed that $\text{range}(A_j) \supset \text{range}(A_{j+1})$. These models are referred to as nested reduced-rank AR models. We will discuss this more general case further in Section 5.3 (also see Reinsel (1997, Section 6.1) for more details on this topic).

5.2 Reduced-Rank Autoregressive Models

The dynamic relationships among components of the vector time series Y_t can be modeled in various ways. For example, see Brillinger (1969), Parzen (1969), Akaike (1976), Sims (1981), Jenkins and Alavi (1981), Tiao and Box (1981), Hannan and Kavalieris (1984), Tiao and Tsay (1989), and Lütkepohl (1993) for various approaches. Doan et al. (1984) considered Bayesian methods applied to time-varying coefficients vector autoregressive models, with a particular emphasis on forecasting. Some considerations to the dimension reduction aspects in modeling have been given by the authors above, as well as by Anderson (1963), Brillinger (1981), Box and Tiao (1977), Priestley et al. (1974a,b), and others. In this section, we consider the vector AR model (5.8) where the coefficient matrices $\mathbf{\Phi}_j$ are of reduced-rank structure similar to the standard reduced-rank regression model. Because the vector AR model (5.8) is in the same framework as the multivariate regression model (1.1), we can readily make use of earlier results from Chapters 1 and 2. Therefore, we will avoid unnecessary repetition of details and shall be brief in our presentation.

5.2.1 *Estimation and Inference*

We consider the vector autoregressive model (5.8) of order p, given by $Y_t = \sum_{j=1}^{p} \mathbf{\Phi}_j Y_{t-j} + \epsilon_t$. Defining $\mathbf{\Phi}(L) = I - \mathbf{\Phi}_1 L - \cdots - \mathbf{\Phi}_p L^p$, where L denotes the lag operator, we assume that $\det\{\mathbf{\Phi}(z)\} \neq 0$ for all complex numbers $|z| \leq 1$. The series Y_t is thus assumed to be stationary. Set $C = [\mathbf{\Phi}_1, \ldots, \mathbf{\Phi}_p]$ and assume that $\text{rank}(C) = r < m$, which yields the decomposition $C = AB$ as in (2.4). From model (5.6), the correspondence between the reduced-rank regression model in Chapter 2 and this particular reduced-rank AR model follows by setting $A = A_0$,

$B = [B_0, B_1, \ldots, B_{p-1}]$, and $X_t = Y^*_{t-1} = (Y'_{t-1}, \ldots, Y'_{t-p})'$ in (2.5). Estimation of the component matrices A and B follows easily from Theorem 2.2 and subsequent results presented in Section 2.3.

For the stationary vector AR(p) process $\{Y_t\}$ above, define the covariance matrix at lag j as $\Gamma(j) = E(Y_{t-j}Y'_t)$, and set $\Gamma_* = [\Gamma(1)', \ldots, \Gamma(p)']' = E(Y^*_{t-1}Y'_t)$, and $\Gamma_p = E(Y^*_{t-1}Y^{*\prime}_{t-1})$ as the $mp \times mp$ matrix that has $\Gamma(i-j)$ in the (i, j)th block. As in Chapter 2, under the assumption of reduced rank, for any choice of positive-definite matrix Γ, the population coefficient matrices A and B in the reduced-rank AR model can be expressed as

$$A = \Gamma^{-1/2}[V_1, \ldots, V_r] = \Gamma^{-1/2}V = [\alpha_1, \ldots, \alpha_r], \tag{5.9a}$$

$$B = V'\Gamma^{1/2}\Gamma'_*\Gamma_p^{-1} = [\beta_1, \ldots, \beta_r]', \tag{5.9b}$$

where the V_j are (normalized) eigenvectors of the matrix $\Gamma^{1/2}\Gamma'_*\Gamma_p^{-1}\Gamma_*\Gamma^{1/2}$ associated with the r largest eigenvalues λ_j^2. This also implies that A and B in (5.9) satisfy the normalizations $B\Gamma_p B' = \Lambda^2$ and $A'\Gamma A = I_r$, where $\Lambda^2 = \text{diag}(\lambda_1^2, \ldots, \lambda_r^2)$.

Based on a sample $Y_1, \ldots, Y_t$, let $\widehat{\Gamma}_p = (1/T)\sum_t Y^*_{t-1}Y^{*\prime}_{t-1}$ and $\widehat{\Gamma}_* = (1/T)\sum_t Y^*_{t-1}Y'_t$. Then the (conditional) ML estimates of A and B corresponding to the population quantities in (5.9) are given by

$$\widehat{A} = \widetilde{\Sigma}_{\epsilon\epsilon}^{1/2}[\widehat{V}_1, \ldots, \widehat{V}_r] = \widetilde{\Sigma}_{\epsilon\epsilon}^{1/2}\widehat{V} = [\widehat{\alpha}_1, \ldots, \widehat{\alpha}_r], \tag{5.10a}$$

$$\widehat{B} = \widehat{V}'\widetilde{\Sigma}_{\epsilon\epsilon}^{-1/2}\widehat{\Gamma}'_*\widehat{\Gamma}_p^{-1} = [\widehat{\beta}_1, \ldots, \widehat{\beta}_r]', \tag{5.10b}$$

where $\widehat{V} = [\widehat{V}_1, \ldots, \widehat{V}_r]$ and $\widehat{V}_j$ is the (normalized) eigenvector of the matrix $\widetilde{\Sigma}_{\epsilon\epsilon}^{-1/2}\widehat{\Gamma}'_*\widehat{\Gamma}_p^{-1}\widehat{\Gamma}_*\widetilde{\Sigma}_{\epsilon\epsilon}^{-1/2}$ corresponding to the jth largest eigenvalue $\widehat{\lambda}_j^2$, and

$$\widetilde{\Sigma}_{\epsilon\epsilon} = (1/T)[\mathbf{Y}\mathbf{Y}' - \mathbf{Y}\mathbf{Y}^{*\prime}_{-1}(\mathbf{Y}^*_{-1}\mathbf{Y}^{*\prime}_{-1})^{-1}\mathbf{Y}^*_{-1}\mathbf{Y}'].$$

Here, $\mathbf{Y} = [Y_1, \ldots, Y_t]$ and $\mathbf{Y}^*_{-1} = [Y^*_0, \ldots, Y^*_{T-1}]$ are the $m \times T$ and $mp \times T$ data matrices containing the observations on Y_t and Y^*_{t-1}, respectively, for $t = 1, \ldots, T$. It follows that the ML estimators $\widehat{A} = [\widehat{\alpha}_1, \ldots, \widehat{\alpha}_r]$ and $\widehat{B} = [\widehat{\beta}_1, \ldots, \widehat{\beta}_r]'$ above, with $\Gamma = \Sigma_{\epsilon\epsilon}^{-1}$ taken as fixed, possess the same form of asymptotic distribution as given in Theorem 2.4 for the multivariate reduced-rank regression model. The proof of this result is essentially the same as given for Theorem 2.4, and so to avoid repetition, we omit any further details. These asymptotic results are useful in the calculation of approximate standard errors for the ML estimates in (5.10) and will be used for the numerical example on the analysis of the U.S. hog data that will be presented later in the chapter. It also follows that the same asymptotic theory holds for the LR testing for the rank of the coefficient matrix C in the reduced-rank autoregressive model as was presented in Section 2.6 for the reduced-rank regression model (e.g., Kohn 1979). That is, under the null hypothesis H_0: rank$(C) \leq r$, the LR test statistic

$$\mathcal{M} = [T - mp + (mp - m - 1)/2] \sum_{j=r+1}^{m} \log(1 + \widehat{\lambda}_j^2)$$

has an asymptotic $\chi^2_{(m-r)(mp-r)}$ distribution.

5.2.2 Relationship to Canonical Analysis of Box and Tiao

In Section 2.4, the connection between reduced-rank regression and canonical correlation analysis was demonstrated. In the multiple time series context, Box and Tiao (1977) considered a canonical analysis of a vector time series from a prediction point of view. Consider the m-variate autoregressive process $\{Y_t\}$ defined by (5.8), with $\text{Cov}(Y_t) = \Gamma(0)$ and $\text{Cov}(\epsilon_t) = \Sigma_{\epsilon\epsilon}$, both positive definite. For an arbitrary linear combination of Y_t, $\omega_t = \ell' Y_t$, we have

$$\omega_t = \ell' Y_t = \ell' \widehat{Y}_t + \ell' \epsilon_t \equiv \widehat{\omega}_t + e_t,$$

where $\widehat{Y}_t = \mathbf{\Phi}_1 Y_{t-1} + \cdots + \mathbf{\Phi}_p Y_{t-p}$, $\widehat{\omega}_t = \ell' \widehat{Y}_t$, and $e_t = \ell' \epsilon_t$. Thus, we find that the variance of ω_t can be decomposed into the sum of a "predictor" variance and an "error" variance as $\sigma^2_\omega = \sigma^2_{\widehat{\omega}} + \sigma^2_e$. A quantity that measures the predictability of the linear combination ω_t, based on the past values of the vector series $\{Y_t\}$, is the ratio

$$\rho^2 = \frac{\text{Var}(\widehat{\omega}_t)}{\text{Var}(\omega_t)} = \frac{\sigma^2_{\widehat{\omega}}}{\sigma^2_\omega} = \frac{\ell' \, \Gamma_*(0) \, \ell}{\ell' \, \Gamma(0) \, \ell}, \tag{5.11}$$

where $\Gamma_*(0) = \text{Cov}(\widehat{Y}_t) = \sum_{j=1}^{p} \mathbf{\Phi}_j \Gamma(j) = C\Gamma_*$. It follows that, for maximum predictability, the quantity ρ^2 must be the largest eigenvalue of the matrix $\Gamma(0)^{-1}\Gamma_*(0)$ and ℓ is the corresponding eigenvector. Similarly, the eigenvector corresponding to the smallest eigenvalue of $\Gamma(0)^{-1}\Gamma_*(0)$ gives the least predictable linear combination of Y_t. In general, there will be m eigenvalues $1 \geq \rho_1^2 \geq \rho_2^2 \geq \cdots \geq \rho_m^2 \geq 0$, and let $\ell_1, \ldots, \ell_m$ denote the corresponding eigenvectors. Then define the $m \times m$ matrix L' that has ℓ_i' as its ith row, and consider the transformed process $\mathcal{W}_t = L' Y_t = (\omega_{1t}, \ldots, \omega_{mt})'$. It follows that the transformation $\mathcal{W}_t = L' Y_t$ produces m new component series that are ordered from "most predictable" to "least predictable." The developments of this canonical analysis approach are due to Box and Tiao (1977), and we will now indicate that they also correspond to the classical canonical correlation analysis originally developed by Hotelling (1935, 1936). Therefore, we see that this approach is directly related to the reduced-rank AR modeling and the ML estimates $\widehat{A}$ and $\widehat{B}$ given in (5.10) because of the previously established connection of reduced-rank regression with classical canonical correlation analysis.

For the stationary vector AR(p) model (5.8), we know that

$$C = [\boldsymbol{\Phi}_1, \dots, \boldsymbol{\Phi}_p] = \Gamma_*' \Gamma_p^{-1}$$

and that $\Gamma_*(0) = \text{Cov}(\widehat{Y}_t) = \sum_{j=1}^{p} \boldsymbol{\Phi}_j \Gamma(j) = C\Gamma_* = \Gamma_*' \Gamma_p^{-1} \Gamma_*$ (e.g., see Reinsel 1997, Sections 2.2 and 6.4). From (5.11), it follows that the choice of ℓ that gives the maximum predictability is then the eigenvector ℓ_1 satisfying

$$\Gamma(0)^{-1} \Gamma_*' \Gamma_p^{-1} \Gamma_* \ell_1 = \rho_1^2 \ell_1, \tag{5.12}$$

with ρ_1^2 being the largest eigenvalue of $\Gamma(0)^{-1} \Gamma_*' \Gamma_p^{-1} \Gamma_*$. Note that this is the same as the largest eigenvalue of the matrix $\Gamma(0)^{-1/2} \Gamma_*' \Gamma_p^{-1} \Gamma_* \Gamma(0)^{-1/2}$, which is of the same form as the matrix in (2.28) of Section 2.4. Thus we see that this is equivalent to a canonical correlation analysis between the vectors Y_t and $X_t \equiv Y_{t-1}^* = (Y_{t-1}', \dots, Y_{t-p}')'$. If $h_i' Y_t$ and $g_i' X_t$ denote the ith canonical variables, then it is easy to see that $h_i = \ell_i$ and $g_i = C' \ell_i$, for $i = 1, \dots, m$. The ρ_i are the canonical correlations between Y_t and X_t. Hence, the canonical analysis based on the predictability measure of (5.11) is the same as the classical canonical correlation analysis for the stationary series.

It is to be noted that the number of nonzero eigenvalues associated with (5.12) is the same as the rank of Γ_*, which in turn is equal to rank(C). So the preceding canonical analysis techniques for vector AR processes considered by Box and Tiao are related to the reduced-rank regression model (2.5). Specifically, from the results in Section 2.4 and from (5.9), $\ell_i = \Sigma_{\epsilon\epsilon}^{-1/2} V_i$, $i = 1, \dots, m$, where V_i denotes the (normalized) eigenvector in (5.9) for the choice of $\Gamma = \Sigma_{\epsilon\epsilon}^{-1}$ and ℓ_i is the eigenvector corresponding to the ith largest root ρ_i^2 in (5.12), normalized such that $\ell_i' \Gamma(0) \ell_i = (1 - \rho_i^2)^{-1}$. Hence, under a reduced-rank structure for the coefficient matrix C in the vector AR(p) model, the component matrices A and B in (5.9) can equivalently be expressed in terms of $L_{(r)} = [\ell_1, \dots, \ell_r]$ as $A = \Sigma_{\epsilon\epsilon} L_{(r)}$ and $B = L_{(r)}' \Gamma_*' \Gamma_p^{-1}$. With $m - r$ zero eigenvalues in (5.12), we see that $\ell_i' C = 0$ for $i = r + 1, \dots, m$. Thus, there are $m - r$ corresponding linear combinations $\omega_{it} = \ell_i' Y_t \equiv \ell_i' \epsilon_t$ that form white noise series. It is useful to identify such white noise series in the multiple time series modeling, because that will simplify the structure of the model, and this can be accomplished through the canonical correlation analysis and associated reduced-rank regression modeling.

5.3 An Extended Reduced-Rank Autoregressive Model

We discuss briefly the use of the extended reduced-rank regression model (3.3) considered by Anderson (1951), as presented in Section 3.1, within the time series framework of the vector autoregressive model (5.8). As noted in the discussion following (5.8), for the vector AR(p) model in (5.8), if for some p_1, $1 \leq p_1 < p$, the matrix $[\Phi_{p_1+1}, \ldots, \Phi_p]$ of AR coefficients for more distant lags is of reduced rank $r_1 < m$, but the matrix $[\Phi_1, \ldots, \Phi_{p_1}]$ of AR coefficients for the recent past is of full rank, then the resulting time series model is in the form of (3.3). Writing $C = [\Phi_{p_1+1}, \ldots, \Phi_p]$ and $D = [\Phi_1, \ldots, \Phi_{p_1}]$, with $Z_t = (Y'_{t-1}, \ldots, Y'_{t-p_1})'$ and $X_t = (Y'_{t-p_1-1}, \ldots, Y'_{t-p})'$, the AR($p$) model can be restated as

$$Y_t = CX_t + DZ_t + \epsilon_t, \qquad t = 1, \ldots, T, \tag{5.13}$$

where $\text{rank}(C) = r_1 < m$. Hence, we have $C = AB$, where A is an $m \times r_1$ matrix and B is an $r_1 \times m(p - p_1)$ matrix.

The ML estimation of the component parameter matrices A and B, and D in (5.13) follows from the results developed in Section 3.1. Likewise, the corresponding population results, which we discuss first, follow from Theorem 3.1. As in Section 5.2, denote $\Gamma_*(p) = [\Gamma(1)', \ldots, \Gamma(p)']' = E(Y^*_{t-1}Y'_t)$ and $\Gamma_p = E(Y^*_{t-1}Y^{*\prime}_{t-1})$. Then note that $\Sigma_{yz} \equiv E(Y_tZ'_t) = \Gamma_*(p_1)'$, $\Sigma_{zz} \equiv E(Z_tZ'_t) = \Gamma_{p_1}$, $\Sigma_{xx} \equiv E(X_tX'_t) = \Gamma_{p-p_1}$, and denote $\Sigma_{yx} \equiv E(Y_tX'_t) = [\Gamma(p_1 + 1)', \ldots, \Gamma(p)'] \equiv \Gamma_{**}(p - p_1)'$ and $\Sigma_{xz} \equiv E(X_tZ'_t) = \Gamma_{(p-p_1,p_1)}$. From the results of Theorem 3.1, and relation (3.5) in particular, it follows that the population coefficient matrices in the extended reduced-rank vector AR model (5.13) can be written as

$$A^{(r_1)} = \Sigma^{1/2}_{\epsilon\epsilon}[V_1, \ldots, V_{r_1}] = \Sigma^{1/2}_{\epsilon\epsilon}V_{(r_1)}, \tag{5.14a}$$

$$B^{(r_1)} = V'_{(r_1)}\Sigma^{-1/2}_{\epsilon\epsilon}\Sigma_{yx.z}\Sigma^{-1}_{xx.z}, \tag{5.14b}$$

and $D = (\Sigma_{yz} - A^{(r_1)}B^{(r_1)}\Sigma_{xz})\Sigma^{-1}_{zz} = (\Gamma_*(p_1)' - A^{(r_1)}B^{(r_1)}\Gamma_{(p-p_1,p_1)})\Gamma^{-1}_{p_1}$, where the V_j are (normalized) eigenvectors of $\Sigma^{-1/2}_{\epsilon\epsilon}\Sigma_{yx.z}\Sigma^{-1}_{xx.z}\Sigma_{xy.z}\Sigma^{-1/2}_{\epsilon\epsilon}$ associated with the r_1 largest eigenvalues λ^2_{1j}, and

$$\Sigma_{yx.z} = \Sigma_{yx} - \Sigma_{yz}\Sigma^{-1}_{zz}\Sigma_{zx} = \Gamma_{**}(p - p_1)' - \Gamma_*(p_1)'\Gamma^{-1}_{p_1}\Gamma'_{(p-p_1,p_1)},$$

$$\Sigma_{xx.z} = \Sigma_{xx} - \Sigma_{xz}\Sigma^{-1}_{zz}\Sigma_{zx} = \Gamma_{p-p_1} - \Gamma_{(p-p_1,p_1)}\Gamma^{-1}_{p_1}\Gamma'_{(p-p_1,p_1)}$$

are partial covariance matrices eliminating the linear effects of Z_t from both Y_t and X_t. This implies that $A^{(r_1)}$ and $B^{(r_1)}$ in (5.14) satisfy the normalizations $B^{(r_1)}\Sigma_{xx.z}B^{(r_1)\prime} = \Lambda^2_1$ and $A^{(r_1)\prime}\Sigma^{-1}_{\epsilon\epsilon}A^{(r_1)} = I_{r_1}$, where $\Lambda^2_1 = \text{diag}(\lambda^2_{11}, \ldots, \lambda^2_{1r_1})$. Also recall that the rank of C in (5.13) is equal to the number of nonzero partial canonical correlations $[\rho_{1j} = \lambda_{1j}/(1 + \lambda^2_{1j})^{1/2}]$ between Y_t and the more distant lagged values $X_t = (Y'_{t-p_1-1}, \ldots, Y'_{t-p})'$, eliminating the influence of the closer lagged values $Z_t = (Y'_{t-1}, \ldots, Y'_{t-p_1})'$.

For the sample $Y_1, \ldots, Y_t$, let

$$\hat{\Gamma}_*(p_1)' = \frac{1}{T}\sum_t Y_tZ'_t, \quad \hat{\Gamma}_{p_1} = \frac{1}{T}\sum_t Z_tZ'_t, \quad \hat{\Gamma}_{**}(p - p_1)' = \frac{1}{T}\sum_t Y_tX'_t,$$

and so on, denote the sample covariance matrices corresponding to the population quantities above, with $\widehat{\Sigma}_{yx.z} = \widehat{\Gamma}_{**}(p - p_1)' - \widehat{\Gamma}_*(p_1)'\widehat{\Gamma}_{p_1}^{-1}\widehat{\Gamma}'_{(p-p_1,p_1)}$ and $\widehat{\Sigma}_{xx.z} = \widehat{\Gamma}_{p-p_1} - \widehat{\Gamma}_{(p-p_1,p_1)}\widehat{\Gamma}_{p_1}^{-1}\widehat{\Gamma}'_{(p-p_1,p_1)}$ the associated sample partial covariance matrices. Then it follows directly from the derivation in Section 3.1, leading to (3.11), that the (conditional) ML estimates are

$$\widehat{A}^{(r_1)} = \widetilde{\Sigma}_{\epsilon\epsilon}^{1/2}\,[\widehat{V}_1, \ldots, \widehat{V}_{r_1}] = \widetilde{\Sigma}_{\epsilon\epsilon}^{1/2}\,\widehat{V}_{(r_1)}, \tag{5.15a}$$

$$\widehat{B}^{(r_1)} = \widehat{V}'_{(r_1)}\,\widetilde{\Sigma}_{\epsilon\epsilon}^{-1/2}\,\widehat{\Sigma}_{yx.z}\,\widehat{\Sigma}_{xx.z}^{-1}, \tag{5.15b}$$

and $\widehat{D} = (\widehat{\Sigma}_{yz} - \widehat{A}^{(r_1)}\widehat{B}^{(r_1)}\widehat{\Sigma}_{xz})\widehat{\Sigma}_{zz}^{-1} = (\widehat{\Gamma}_*(p_1)' - \widehat{A}^{(r_1)}\widehat{B}^{(r_1)}\widehat{\Gamma}_{(p-p_1,p_1)})\widehat{\Gamma}_{p_1}^{-1}$, where the $\widehat{V}_j$ are (normalized) eigenvectors of $\widetilde{\Sigma}_{\epsilon\epsilon}^{-1/2}\widehat{\Sigma}_{yx.z}\widehat{\Sigma}_{xx.z}^{-1}\widehat{\Sigma}_{xy.z}\widetilde{\Sigma}_{\epsilon\epsilon}^{-1/2}$ associated with the r_1 largest eigenvalues $\widehat{\lambda}_{1j}^2$. As with the ML estimators for the regression model (3.3) in Section 3.1, the asymptotic distribution theory for the ML estimators $\widehat{A}^{(r_1)}$, $\widehat{B}^{(r_1)}$, and $\widehat{D}$ in (5.15) above follows easily from the results in Theorem 2.4. It also follows that LR testing for the appropriate rank of the coefficient matrix C in the extended reduced-rank AR model (5.13) can be carried out in the same manner as described in Section 3.1. In particular, for testing the null hypothesis $H_0\colon \text{rank}(C) \leq r_1$, we can use the LR test statistic

$$\begin{aligned}\mathcal{M} &= [T - mp + (m(p - p_1) - m - 1)/2] \sum_{j=r_1+1}^{m} \log(1 + \widehat{\lambda}_{1j}^2) \\ &= -[T - mp + (m(p - p_1) - m - 1)/2] \sum_{j=r_1+1}^{m} \log(1 - \widehat{\rho}_{1j}^2),\end{aligned}$$

which has an asymptotic $\chi^2_{(m-r_1)(m(p-p_1)-r_1)}$ null distribution, where $\widehat{\rho}_{1j}^2 = \widehat{\lambda}_{1j}^2/(1 + \widehat{\lambda}_{1j}^2)$ are the squared sample partial canonical correlations between Y_t and X_t eliminating Z_t.

To conclude, we observe that the extended reduced-rank vector autoregressive model (5.13) has the feature that $m - r_1$ linear combinations of Y_t exist such that their dependence on the past can be represented in terms of only the recent past lags $Z_t = (Y'_{t-1}, \ldots, Y'_{t-p_1})'$. This type of feature may be useful in modeling and interpreting time series data with several component variables. The specification of the appropriate reduced rank r_1 and the corresponding ML estimation of the model parameters can be performed using the sample partial canonical correlation analysis and associated methods as highlighted above and as described in more detail in Section 3.1.

5.4 Nested Reduced-Rank Autoregressive Models

We now briefly discuss analysis of vector AR(p) models that have simplifying nested reduced-rank structures in their coefficient matrices $\mathbf{\Phi}_j$. Specifically, we consider the vector AR(p) model in (5.8), in which it is assumed that the matrices $\mathbf{\Phi}_j$ have a particular reduced-rank structure, such that $\text{rank}(\mathbf{\Phi}_j) = r_j \geq \text{rank}(\mathbf{\Phi}_{j+1}) = r_{j+1}$, $j = 1, 2, \ldots, p-1$. Then, the $\mathbf{\Phi}_j$ can be represented in the form $\mathbf{\Phi}_j = A_j B_j$, where A_j is $m \times r_j$ and B_j is $r_j \times m$, and we further assume that $\text{range}(A_j) \supset \text{range}(A_{j+1})$. Thus, we can write (5.8) as

$$Y_t = \sum_{j=1}^{p} A_j B_j Y_{t-j} + \epsilon_t. \tag{5.16}$$

We refer to such models as *nested reduced-rank* AR models, and these models, which generalize the earlier work by Velu et al. (1986), were studied by Ahn and Reinsel (1988). These models can result in more parsimonious parameterization, provide more details about structure possibly with simplifying features, and offer more interesting and useful interpretations concerning the interrelations among the m time series.

5.4.1 Specification of Ranks

To obtain information on the ranks of the matrices $\mathbf{\Phi}_j$ in the nested reduced-rank model, a partial canonical correlation analysis approach can be used. One fundamental consequence of the model above is that there will exist at least $m - r_j$ zero partial canonical correlations between Y_t and Y_{t-j}, given $Y_{t-1}, \ldots, Y_{t-j+1}$. This follows because we can find a $(m - r_j) \times m$ matrix F_j', whose rows are linearly independent, such that $F_j' A_j = 0$ and, hence, $F_j' A_i = 0$ for all $i \geq j$ because of the nested structure of the A_j. Therefore,

$$F_j'(Y_t - \sum_{i=1}^{j-1} \mathbf{\Phi}_i Y_{t-i}) = F_j'(Y_t - \sum_{i=1}^{p} \mathbf{\Phi}_i Y_{t-i}) \equiv F_j' \epsilon_t \tag{5.17}$$

is independent of $(Y_{t-1}', \ldots, Y_{t-j}')'$ and consists of $m - r_j$ linear combinations of $(Y_t', \ldots, Y_{t-j+1}')'$. Thus, $m - r_j$ zero partial canonical correlations between Y_t and Y_{t-j} occur. Hence, performing a (partial) canonical correlation analysis for various values of $j = 1, 2, \ldots$ will identify the rank structure of the nested reduced-rank model, as well as the overall order p of the AR model.

The sample test statistic that can be used to (tentatively) specify the ranks is

$$\mathcal{M}(j,s) = -(T-j-jm-1-1/2) \sum_{i=(m-s)+1}^{m} \log(1-\widehat{\rho}_i^2(j)) \tag{5.18}$$

for $s = 1, 2, \ldots, m$, where $1 \geq \widehat{\rho}_1(j) \geq \cdots \geq \widehat{\rho}_m(j) \geq 0$ are the *sample partial canonical correlations* between Y_t and Y_{t-j}, given $Y_{t-1}, \ldots, Y_{t-j+1}$. Under the null hypothesis that $\text{rank}(\mathbf{\Phi}_j) \leq m-s$ within the nested reduced-rank model framework, the statistic $\mathcal{M}(j,s)$ is asymptotically distributed as chi-squared with s^2 degrees of freedom (e.g., Tiao and Tsay 1989). Hence, if the value of the test statistic is not "unusually large," we would not reject the null hypothesis and might conclude that $\mathbf{\Phi}_j$ has reduced rank equal to the smallest value $r_j(\equiv m-s_j)$ for which the test does not reject the null hypothesis. Note, in particular, that when $s = m$ the statistic in (5.18) is the same as the LR test statistic for testing $H_0 : \mathbf{\Phi}_j = 0$ in an AR(j) model, since $\log(U_j) = \log\{|S_j|/|S_{j-1}|\} = \sum_{i=1}^{m} \log(1-\widehat{\rho}_i^2(j))$, where S_j denotes the residual sum of squares matrix after LS estimation of the AR model of order j.

5.4.2 A Canonical Form

of the matrices $\mathbf{\Phi}_j = A_j B_j$ that one can construct a nonsingular $m \times m$ matrix P such that the transformed series $\mathcal{Z}_t = PY_t$ will have the following simplified model structure. The model for $\mathcal{Z}_t$ is AR(p),

$$\begin{aligned}\mathcal{Z}_t = PY_t &= \sum_{i=1}^{p} P(A_i B_i)P^{-1} PY_{t-i} + P\epsilon_t \\ &= \sum_{i=1}^{p} A_i^* B_i^* \mathcal{Z}_{t-i} + e_t \equiv \sum_{i=1}^{p} \mathbf{\Phi}_i^* \mathcal{Z}_{t-i} + e_t,\end{aligned} \tag{5.19}$$

where $A_i^* = PA_i$, $B_i^* = B_i P^{-1}$, $\mathbf{\Phi}_i^* = A_i^* B_i^*$, and $e_t = P\epsilon_t$. The rows of P will consist of a basis of the rows of the various matrices F_j' indicated in (5.17). Thus, the matrix P can be chosen so that its last $m - r_j$ rows are orthogonal to the columns of A_j, and, hence, $A_i^* = [A_{i1}^{*\prime}, 0']'$ has its last $m - r_i$ rows equal to zero. This implies that $\mathbf{\Phi}_i^* = A_i^* B_i^*$ also has its last $m - r_i$ rows being zero. Hence, the "canonical form" of the model has

$$\mathcal{Z}_t = \sum_{i=1}^{p} \begin{bmatrix} \mathbf{\Phi}_{i1}^* \\ 0 \end{bmatrix} \mathcal{Z}_{t-i} + e_t,$$

with the number $m - r_i$ of zero rows for $\mathbf{\Phi}_i^* = [\mathbf{\Phi}_{i1}^{*\prime}, 0']'$ increasing as i increases. Therefore, the components z_{jt} of $\mathcal{Z}_t = (z_{1t}, z_{2t}, \ldots, z_{mt})'$ are represented in an

AR(p) model by fewer and fewer past lagged variables $\mathcal{Z}_{t-i}$ as j increases from 1 to m.

For the nested reduced-rank AR model with parameters $\boldsymbol{\Phi}_j = A_j B_j$, for any $r_j \times r_j$ nonsingular matrix Q, we can write $\boldsymbol{\Phi}_j = A_j Q^{-1} Q B_j = \bar{A}_j \bar{B}_j$, where $\bar{A}_j = A_j Q^{-1}$ and $\bar{B}_j = Q B_j$. Therefore, some normalization conditions on the A_j and B_j are required to ensure a unique set of parameters. Assuming the components of Y_t have been arranged so that the upper $r_j \times r_j$ block matrix of each A_j is full rank, this parameterization can be obtained by expressing the $\boldsymbol{\Phi}_j$ as $\boldsymbol{\Phi}_j = A_j B_j = A_1 D_j B_j$, where A_1 is $m \times r_1$ and has certain elements "normalized" to fixed values of ones and zeros, and $D_j = [I_{r_j}, 0]'$ is $r_1 \times r_j$. More specifically, the matrix A_1 can always be formed to be lower triangular with ones on the main diagonal, and when augmented with the last $m - r_1$ columns of the identity matrix, the inverse of the resulting matrix A can form the necessary matrix P of the canonical transformation in (5.19). (See Ahn and Reinsel 1988 for further details concerning the normalization.) Thus, with these conventions, we can write the model as $Y_t = A_1 \sum_{i=1}^{p} D_i B_i Y_{t-i} + \epsilon_t$. In addition, if we set $\boldsymbol{\Phi}_0^{\#} = A^{-1} = P$ and define the $m \times m$ coefficient matrices $\boldsymbol{\Phi}_i^{\#} = [B_i', 0]'$, $i = 1, \ldots, p$, the nested reduced-rank model can be written in an equivalent form as $\boldsymbol{\Phi}_0^{\#} Y_t = \sum_{i=1}^{p} \boldsymbol{\Phi}_i^{\#} Y_{t-i} + \boldsymbol{\Phi}_0^{\#} \epsilon_t$, which is sometimes referred to as an "echelon canonical form" for the vector AR(p) model (e.g., see Reinsel 1997, Sec. 3.1).

For example, consider the special case where the ranks of all $\boldsymbol{\Phi}_i$ are equal, $r_1 = \cdots = r_p \equiv r$, and notice that this special nested model then reduces to the reduced-rank AR model discussed in Section 5.2 where rank$[\boldsymbol{\Phi}_1, \ldots, \boldsymbol{\Phi}_p] = r$. The normalization for the matrix A_1 discussed above in this case becomes $A_1 = [I_r, A_{10}']'$, where A_{10} is an $(m - r) \times r$ matrix. The augmented matrix A and the matrix $P = A^{-1}$ in the canonical form (5.19) can be taken as

$$A = \begin{bmatrix} I_r & 0 \\ A_{10} & I_{m-r} \end{bmatrix} \quad \text{and} \quad P = A^{-1} = \begin{bmatrix} I_r & 0 \\ -A_{10} & I_{m-r} \end{bmatrix},$$

where the last $m - r$ rows of P consist of the (linearly independent) vectors F_j' from (5.17) such that $F_j' \boldsymbol{\Phi}_i = 0$ for all $i = 1, \ldots, p$. Thus, for this special model, the last $m - r$ components $z_{jt} = F_j' Y_t$ in the transformed canonical form (5.19) will be white noise series.

5.4.3 Maximum Likelihood Estimation

We let β_0 denote the $a \times 1$ vector of unknown parameters in the matrix A_1, where it follows that $a = \sum_{j=1}^{p} (m - r_j)(r_j - r_{j+1})$. Also, let $\beta_j = \text{vec}(B_j')$, a $mr_j \times 1$ vector, for $j = 1, \ldots, p$, and set $\theta = (\beta_0', \beta_1', \ldots, \beta_p')'$. The unknown parameters θ are estimated by (conditional) maximum likelihood, using a Newton–Raphson iterative procedure. The (conditional) log-likelihood for T observations $Y_1, \ldots, Y_t$ (given $Y_0, \ldots, Y_{1-p}$) is

$$l(\theta, \Sigma_{\epsilon\epsilon}) = -(T/2)\log|\Sigma_{\epsilon\epsilon}| - (1/2)\sum_{t=1}^{T}\epsilon_t'\Sigma_{\epsilon\epsilon}^{-1}\epsilon_t.$$

It can be shown that an approximate Newton–Raphson iterative procedure to obtain the ML estimate of θ is given by

$$\begin{aligned}\widehat{\theta}^{(i+1)} &= \widehat{\theta}^{(i)} - \{\partial^2 l/\partial\theta\partial\theta'\}^{-1}_{\widehat{\theta}^{(i)}}\{\partial l/\partial\theta\}_{\widehat{\theta}^{(i)}} \\ &= \widehat{\theta}^{(i)} + \left\{M'\left(\sum_{t=1}^{T}\mathcal{U}_t\Sigma_{\epsilon\epsilon}^{-1}\mathcal{U}_t'\right)M\right\}^{-1}_{\widehat{\theta}^{(i)}}\left\{M'\sum_{t=1}^{T}\mathcal{U}_t\Sigma_{\epsilon\epsilon}^{-1}\epsilon_t\right\}_{\widehat{\theta}^{(i)}},\end{aligned}$$

where $\widehat{\theta}^{(i)}$ denotes the estimate at the ith iteration, $\partial\epsilon_t'/\partial\theta = M'\mathcal{U}_t$, $\mathcal{U}_t = [I_m \otimes Y_{t-1}', \ldots, I_m \otimes Y_{t-p}']'$, and M' is a specific matrix of dimension $(a + m\sum_{j=1}^{p} r_j) \times m^2 p$ whose elements are given functions of the parameters θ (but not the observations).

It can be shown that the ML estimate $\widehat{\theta}$ has an asymptotic distribution such that $T^{1/2}(\widehat{\theta} - \theta)$ converges in distribution to $N(0, V^{-1})$ as $T \to \infty$, where $V = M'E(\mathcal{U}_t\Sigma_{\epsilon\epsilon}^{-1}\mathcal{U}_t')M \equiv M'GM$ and $G = E(\mathcal{U}_t\Sigma_{\epsilon\epsilon}^{-1}\mathcal{U}_t')$ has (i, j)th block element equal to $\Sigma_{\epsilon\epsilon}^{-1} \otimes \Gamma(i-j)$. Thus for large T, the ML estimate $\widehat{\theta}$ is approximately distributed as $N(\theta, T^{-1}\widehat{V}_t^{-1})$, where $V_t = M'(T^{-1}\sum_{t=1}^{T}\mathcal{U}_t\Sigma_{\epsilon\epsilon}^{-1}\mathcal{U}_t')M$. In addition, the asymptotic distribution of the (reduced-rank) ML estimates $\widehat{\mathbf{\Phi}}_j = \widehat{A}_j\widehat{B}_j$ of the AR parameters $\mathbf{\Phi}_j$ follows directly from the result above. It is determined from the relations

$$\begin{aligned}\widehat{\mathbf{\Phi}}_j - \mathbf{\Phi}_j &= \widehat{A}_1 D_j \widehat{B}_j - A_1 D_j B_j \\ &= (\widehat{A}_1 - A_1)D_j B_j + A_1 D_j(\widehat{B}_j - B_j) + (\widehat{A}_1 - A_1)D_j(\widehat{B}_j - B_j) \\ &= (\widehat{A}_1 - A_1)D_j B_j + A_1 D_j(\widehat{B}_j - B_j) + O_p(T^{-1}).\end{aligned}$$

From this, we find that

$$\widehat{\phi} - \phi = M(\widehat{\theta} - \theta) + O_p(T^{-1}),$$

where $\widehat{\phi} = \text{vec}\{\widehat{\mathbf{\Phi}}_1', \ldots, \widehat{\mathbf{\Phi}}_p'\}$, $\phi = \text{vec}\{\mathbf{\Phi}_1', \ldots, \mathbf{\Phi}_p'\}$, and $M = \partial\phi/\partial\theta'$, in fact, so that

$$T^{1/2}(\widehat{\phi} - \phi) \xrightarrow{d} N(0, MV^{-1}M') \quad \text{as } T \to \infty.$$

In particular, it is noted that the collection of resulting reduced-rank estimates $\widehat{\mathbf{\Phi}}_j = \widehat{A}_j\widehat{B}_j$ of the $\mathbf{\Phi}_j$ has a smaller asymptotic covariance matrix than the corresponding full-rank least squares estimates, since

$$MV^{-1}M' = M\{M'GM\}^{-1}M' < G^{-1}.$$

Note that the preceding developments and estimation results have similarities to the results presented in Section 3.5 concerning the estimation of multivariate reduced-rank regression models with two sets of regressors, except that the reduced-rank normalizations imposed on the vector of autoregressive parameters ϕ are now in the more explicit form $\mathbf{\Phi}_j = A_1 D_j B_j$ with constraints that certain elements of A_1 be fixed at zero or one. Also, it follows that LR tests of various hypotheses concerning the ranks and other restrictions on the matrices $\mathbf{\Phi}_j$ can be performed in the usual manner based on the ratio of determinants of the residual covariance matrix estimators, $\widehat{\Sigma}_{\epsilon\epsilon} = (1/T)\sum_{t=1}^{T} \widehat{\epsilon}_t \widehat{\epsilon}_t'$, in the "full" and "restricted" models, respectively. More details concerning the nested reduced-rank AR models are given by Ahn and Reinsel (1988) and Reinsel (1997, Section 6.1).

5.5 Numerical Example: U.S. Hog Data

To illustrate the reduced-rank modeling of the previous sections, we consider the U.S. hog, corn, and wage time series previously analyzed by several authors, including Quenouille (1968, Chapter 8), Box and Tiao (1977), Reinsel (1983), Velu et al. (1986), and Tiao and Tsay (1983, 1989). Annual observations from 1867 to 1948 on five variables, farm wage rate (FW), hog supply (HS), hog price (HP), corn price (CP), and corn supply (CS), will be analyzed. These measurements were logarithmically transformed and linearly coded by Quenouille, and, following Box and Tiao, the wage rate and hog price time series were shifted backward by one period. We consider this modified set of series as our basic set for analysis. The wage rate was originally included in the set as a feasible trend variable to reflect the general economic level. The other series are almost exclusively related to the hog industry. After corrections for means and recording the data in hundreds, we denote the basic 5-variate time series as Y_t. Time series plots of the five series, unadjusted for means, are shown in Fig. 5.1.

The series Y_t was first subjected to a multivariate stepwise autoregressive analysis to identify an appropriate order of an AR model. The LR chi-squared statistic (Tiao and Box 1981; Reinsel 1997, Section 4.3) for testing the significance of the coefficient matrix at lag j (i.e., testing $H_0 : \mathbf{\Phi}_j = 0$ in the AR model of order j) is given by (5.18) with $s = m$. For the first three lags (orders) $j = 1, 2, 3$, the LR statistic values are obtained as 448.51, 98.13, and 31.93, respectively. Note that the upper 5% critical value for the χ^2_{25} distribution is 37.7. Thus the LR test results suggest that an AR model of order 2 may be acceptable and that it fits much better than, for example, an AR model of order 1. Using more elaborate initial model specification techniques and criterion, Tiao and Tsay (1989) suggested an ARMA(2,1) model for the series.

The AR(2) model $Y_t = \mathbf{\Phi}_1 Y_{t-1} + \mathbf{\Phi}_2 Y_{t-2} + \epsilon_t$, when fitted without imposing any rank or other constraints, gives (conditional) ML or LS estimates

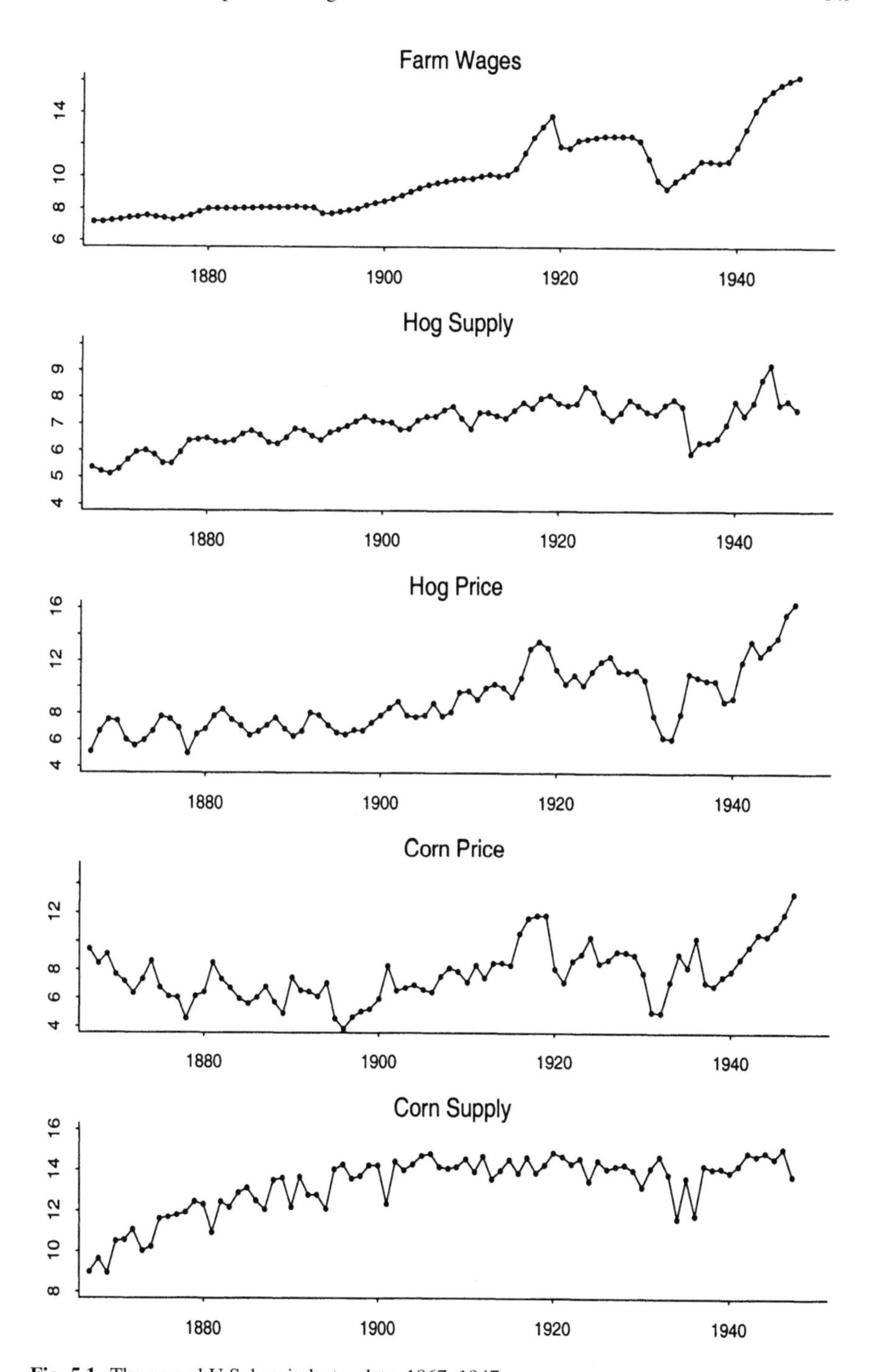

Fig. 5.1 The annual U.S. hog industry data, 1867–1947

$$\widetilde{\boldsymbol{\Phi}}_1 = \begin{bmatrix} 1.49^* & -0.09 & -0.03 & 0.01 & 0.12 \\ 0.12 & 0.43^* & -0.08 & 0.11^* & 0.29^* \\ 1.21^* & -0.42 & 0.34^* & 0.20 & 0.02 \\ 1.05^* & -0.83 & -0.27 & 0.75^* & 0.49 \\ 0.00 & 0.28 & 0.12 & -0.02 & 0.23 \end{bmatrix}$$

$$\widetilde{\boldsymbol{\Phi}}_2 = \begin{bmatrix} -0.35^* & -0.17 & -0.10 & -0.02 & 0.02 \\ -0.16 & 0.04 & 0.28^* & -0.22^* & -0.16^* \\ -0.47 & 0.14 & -0.53^* & 0.31^* & 0.39^* \\ -0.25 & -0.12 & -0.42^* & 0.07 & 0.07 \\ -0.39 & 0.34 & 0.37^* & -0.11 & 0.24 \end{bmatrix}$$

with residual ML covariance matrix estimate $\widetilde{\Sigma}_{\epsilon\epsilon}$ such that $|\widetilde{\Sigma}_{\epsilon\epsilon}| = 0.1444 \times 10^{-3}$. The asterisks indicate that the individual coefficient estimates have t-ratios greater than two in absolute value.

Two interesting points emerge from these estimation results. First, the wage rate (y_{1t}) is clearly nonstationary, and it is not influenced by any other variables; hence, it may be regarded as exogenous. This is consistent with our initial remark that the wage rate was included to reflect the general economic level. Second, the corn supply (y_{5t}) is influenced mainly by hog price (y_{3t}) in the previous year. Historically, the ratios between total feed consumption and hog production have not changed much over the years. This is true regardless of improvements in feeding efficiency; much of the corn produced is intended for hog feed, and so the price of hogs influences the corn supply.

One can attempt to simplify the estimated full-rank AR(2) model by sequentially setting to zero those coefficients whose estimates are quite small compared to their standard errors and reestimating the remaining coefficients. However, in general, this can be time-consuming and may not lead to any simplifications or improved interpretation in the structure of the model for the data, while such a simplification in structure may be uncovered by a reduced-rank analysis. Hence, we now investigate the possibility of reduced-rank autoregression models for these data. As discussed previously, the relationship between canonical correlation analysis and reduced-rank regression allows one to check on the rank of the regression coefficient matrix by testing if certain canonical correlations are zero.

The squared sample canonical correlations $\widehat{\rho}_i^2$ between Y_t and $X_t = (Y'_{t-1}, Y'_{t-2})'$, which are the same as the (estimated) predictability ratios from (5.11), are equal to 0.981, 0.866, 0.694, 0.193, and 0.084. We test the null hypothesis H_0 : rank$(C) = r$ against the alternative H_1 : rank$(C) = 5$ for $r = 1, 2, 3, 4$, where $C = [\boldsymbol{\Phi}_1, \boldsymbol{\Phi}_2]$, using the LR test statistic $\mathcal{M}$ given in Section 5.2. The LR test statistic values corresponding to the four null hypotheses are obtained as $\mathcal{M} = 244.91$, 104.07, 21.16, and 6.14, with 36, 24, 14, and 6 degrees of freedom, and the associated tail probabilities are found to be 0.00, 0.00, 0.10, and 0.41, respectively. That is, rank$(C) = 3$ is consistent with the data. (Strictly, the asymptotic chi-squared distribution theory for the LR statistics in this setting is developed for the stationary vector AR case. There are indications that the series Y_t in this example is at least "partially nonstationary"; nevertheless, it is felt that the use of the LR test procedures here is informative for rank determination.) The (conditional) ML estimates of the component matrices obtained from (5.10) for the AR(2) model of reduced rank $r = 3$ are

$$\widehat{A} = \begin{bmatrix} 0.33 & 0.06 & 0.04 \\ 0.09 & -0.15 & -0.08 \\ 0.32 & 0.15 & 0.47 \\ 0.20 & 0.17 & 0.42 \\ 0.12 & -0.28 & -0.02 \end{bmatrix}$$

$$\widehat{B} = \begin{bmatrix} 4.07 & 0.60 & 0.13 & -0.15 & 0.35 & -1.13 & -0.33 & -0.03 & -0.19 & -0.09 \\ 1.88 & -1.54 & 0.21 & -0.98 & -1.58 & 0.73 & -1.03 & -1.43 & 0.84 & 0.13 \\ -0.60 & -1.14 & 0.32 & 1.05 & 0.48 & -0.47 & 0.76 & -0.63 & 0.46 & 0.83 \end{bmatrix}$$

with ML error covariance matrix estimate $\widehat{\Sigma}_{\epsilon\epsilon}$ such that $|\widehat{\Sigma}_{\epsilon\epsilon}| = 0.1954 \times 10^{-3}$.

The standard errors of the ML estimates above can be computed using the previously mentioned asymptotic results similar to those in Theorem 2.4. All the elements in $\widehat{A}$, except for the coefficient in the (5,3) position, are found to be significant when "significance" is taken to be greater than twice the asymptotic standard error. The estimated linear combinations

$$\widetilde{Y}^*_{t-1} = \widehat{B}(Y'_{t-1}, Y'_{t-2})'$$

can be regarded as indices constructed from the past data values. For illustrative purposes, we present these linear combinations with significant terms only, together with standard errors. These are

$$\begin{aligned} \widetilde{y}^*_{1,t-1} &= 4.07(\pm 0.42) y_{1,t-1} - 1.13(\pm 0.43) y_{1,t-2}, \\ \widetilde{y}^*_{2,t-1} &= 1.88(\pm 0.42) y_{1,t-1} - 1.54(\pm 0.51) y_{2,t-1} - 0.98(\pm 0.25) y_{4,t-1} \\ &\quad -1.58(\pm 0.26) y_{5,t-1} - 1.03(\pm 0.44) y_{2,t-2} \\ &\quad -1.43(\pm 0.22) y_{3,t-2} + 0.84(\pm 0.23) y_{4,t-2}, \\ \widetilde{y}^*_{3,t-1} &= -1.14(\pm 0.50) y_{2,t-1} + 1.05(\pm 0.22) y_{4,t-1} - 0.63(\pm 0.21) y_{3,t-2} \\ &\quad +0.46(\pm 0.22) y_{4,t-2} + 0.83(\pm 0.28) y_{5,t-2}. \end{aligned}$$

The first linear combination mainly consists of farm wages at lag 1 and lag 2, whereas the third linear combination does not involve farm wages.

The total number of functionally independent parameter estimates in $\widehat{A}$ and $\widehat{B}$ is only 36, because of the normalization conditions, compared with 50 in the full-rank model. One could also consider a further simplification of the reduced-rank model by use of the procedure that sets to zero those coefficients whose estimates are quite small compared to their standard errors and reestimates the remaining coefficients. However, in this case, the explicit eigenvalue–eigenvector solution given by (5.10) would no longer be applicable, and nonlinear iterative estimation procedures must be employed, such as those described by Reinsel (1997, Sections 5.1 and 6.1). Such a procedure was, in fact, applied to the above AR(2) reduced-rank model example and resulted in a quite adequate model containing only 22 parameters, but we omit further details.

In view of the discussion in Section 5.2, the above reduced-rank regression analysis is easy to relate to the canonical analysis. As noted, the ML estimates $\widehat{A}$ and $\widehat{B}$, specifically the eigenvectors $\widehat{V}_i$ in (5.10), are related to the eigenvectors $\widehat{\ell}_i$ that are obtained using the sample version of (5.12) by

$$\widehat{\ell}_i = \widetilde{\Sigma}_{\epsilon\epsilon}^{-1/2}\widehat{V}_i, \quad i = 1, \ldots, m.$$

Since $\widehat{\ell}_i = \widetilde{\Sigma}_{\epsilon\epsilon}^{-1}\widehat{\alpha}_i$, the asymptotic distribution theory for the $\widehat{\ell}_i$ follows easily from that for $\widehat{A}$, for $i = 1, \ldots, r$, thus providing a useful inferential tool for the canonical analysis of Box and Tiao (1977). The $\widehat{\ell}_i$ can be used for the canonical analysis that transforms the series Y_t into 5 new series

$$\mathcal{W}_t = \widehat{L}'Y_t,$$

where $\widehat{L} = [\widehat{\ell}_1, \ldots, \widehat{\ell}_m]$, and the accompanying standard errors of the $\widehat{\ell}_i$ can be helpful for interpreting the extent of contribution of each original component y_{it} in the canonical variates. The matrix $\widehat{L}'$ is obtained as

$$\widehat{L}' = \begin{bmatrix} 2.561 & 1.548 & 0.610 & -0.627 & -0.388 \\ -2.286 & 3.747 & 0.180 & 0.805 & 1.627 \\ 2.219 & 0.372 & -1.235 & -1.176 & -1.104 \\ -0.480 & 0.813 & 1.366 & -1.326 & -0.646 \\ 0.869 & -2.978 & -1.115 & 0.634 & 1.647 \end{bmatrix},$$

with nearly all of the estimated coefficients being "significantly" nonzero. The components of $\mathcal{W}_t$ are contemporaneously uncorrelated and are ordered from most predictable to least predictable. Also recall from Section 5.2 that the components $\omega_{it} = \widehat{\ell}_i'Y_t$ are normalized such that they have sample variance equal to $(1 - \widehat{\rho}_i^2)^{-1}$ (i.e., $\widehat{\ell}_i'\widehat{\Gamma}(0)\widehat{\ell}_i = (1 - \widehat{\rho}_i^2)^{-1}$). The transformed series $\mathcal{W}_t$ are plotted in Fig. 5.2. This figure indicates that the first two series are nearly nonstationary, and the last two fluctuate randomly around constants and hence (on further examination) tend to behave like white noise series, consistent with the reduced rank 3 structure.

The first linear combination $\omega_{1t} = \widehat{\ell}_1'Y_t$ associated with the largest canonical correlation, that is, the largest predictability ratio, appears to have contributions from all the series but is dominated by farm wages. The nearly random (white noise) components associated with the two smallest canonical correlations may be of interest due to their relative stability over time. For instance, the sample variance of

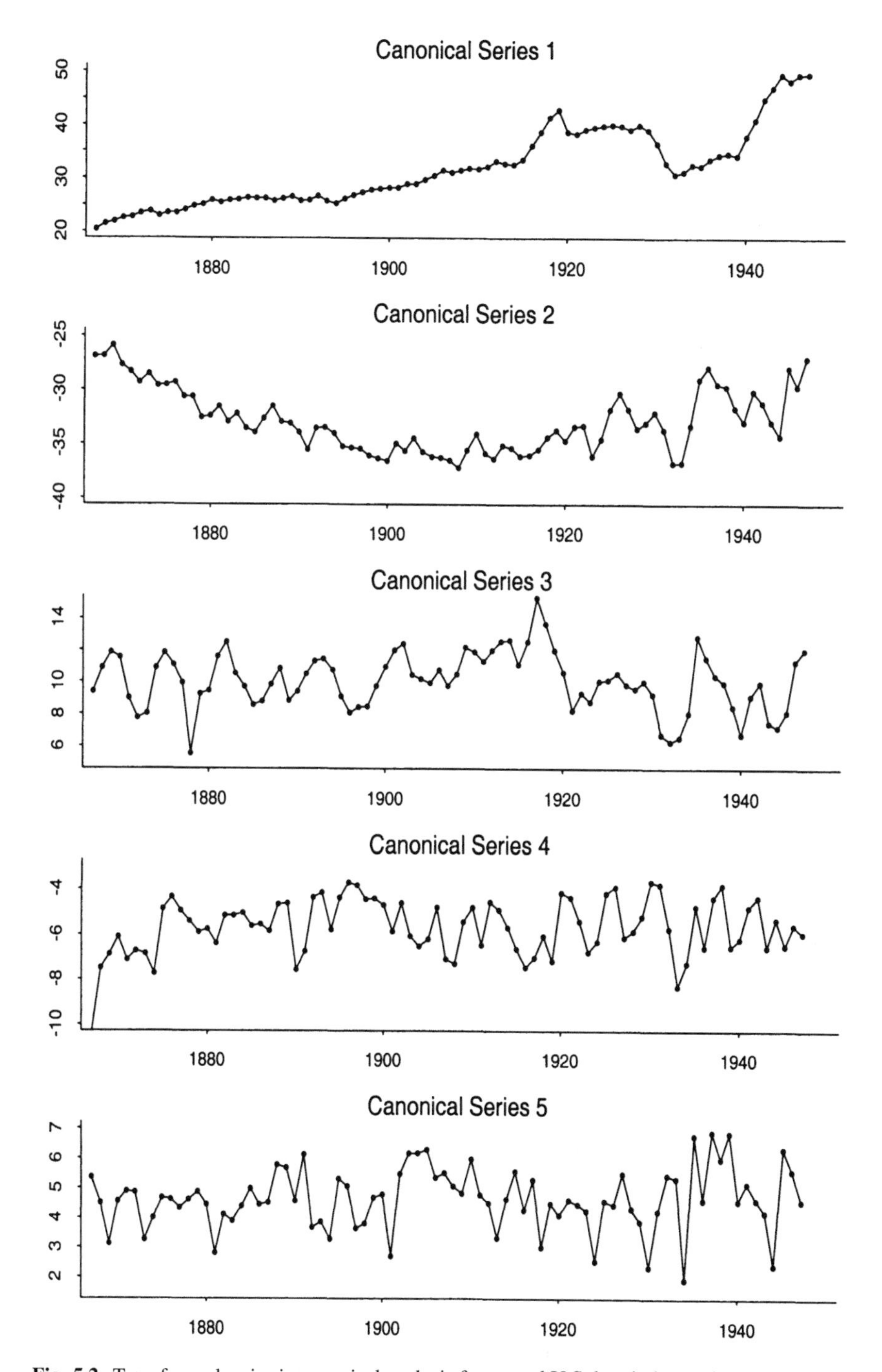

Fig. 5.2 Transformed series in canonical analysis for annual U.S. hog industry data, 1867–1947

the final white noise canonical component, $\omega_{5t}/3.739$, when normalized so that the vector of coefficients of the y_{it} has unit norm, is equal to only 0.078. It is instructive to observe that this variance is very small compared to the average variance (equal to 3.58) of the original component series y_{it}, $i = 1, \ldots, 5$.

To arrive at some additional meaningful interpretations involving the last two (white noise) components, we look at certain linear combinations of ω_{4t} and ω_{5t}. By fixing the coefficients of HS_t (y_{2t}) and HP_t (y_{3t}) to be equal to 1, we obtain the linear combination

$$0.589\omega_{4t} - 0.175\omega_{5t} = -0.435y_{1t} + y_{2t} + y_{3t} - 0.892y_{4t} - 0.668y_{5t}.$$

On taking antilogs, this implies, approximately, that the series can be related to a Cobb–Douglas production function,

$$\text{I}_t = \text{HP}_t\text{HS}_t/\{(\text{CP}_t\text{CS}_t)^{0.78}(\text{FW}_t)^{0.43}\},$$

and is a white noise series. (See Box and Tiao 1977 for similar interpretations.) The sample variance of the (essentially) white noise index above, $\omega_t^* = (0.589\omega_{4t} - 0.175\omega_{5t})/1.853$, normalized in a similar way as $\omega_{5t}/3.739$, is 0.135. Hence, we have obtained a quite stable contemporaneous linear relationship among the original 5 variables.

A nested reduced-rank AR model (5.16) can also be considered for these data. The testing procedure (5.18) for investigation of the rank structure of a possible nested reduced-rank model was used. The results of a partial canonical correlation analysis of the data, in terms of the (squared) sample partial canonical correlations $\widehat{\rho}_i^2(j)$ for lags $j = 1, 2, 3$, and the associated test statistic values $\mathcal{M}(j, s)$ from (5.18) are displayed in Table 5.1. From the statistics in this table, we reconfirm that an AR model of order 2 may be appropriate, but we also find that the hypothesis that $\text{rank}(\boldsymbol{\Phi}_2) = 3$ is clearly not rejected and that the hypothesis that $\text{rank}(\boldsymbol{\Phi}_1) = 4$ is clearly not rejected. Hence, these results suggest an AR(2) model with a nested reduced-rank structure such that $\text{rank}(\boldsymbol{\Phi}_1) = 4$ and $\text{rank}(\boldsymbol{\Phi}_2) = 3$, that is, $Y_t = A_1(B_1Y_{t-1} + D_2B_2Y_{t-2}) + \epsilon_t$ with A_1 a 5×4 matrix, B_1 a 4×5 matrix, B_2 a 3×5 matrix, and $D_2 = [I_3, 0]'$. Subsequent estimation of this model, by (conditional) ML methods, as outlined in a subsection of Section 5.3, resulted in a fitted model with the following estimates:

$$\widehat{A}_1 = \begin{bmatrix} 1.000 & 0.000 & 0.000 & 0.000 \\ 0.000 & 1.000 & 0.000 & 0.000 \\ 0.000 & 0.000 & 1.000 & 0.000 \\ 0.437 & 0.103 & 0.647 & 1.000 \\ -0.701 & 1.765 & 0.430 & -0.394 \end{bmatrix}$$

Table 5.1 Summary of partial canonical correlation analysis between Y_t and Y_{t-j} and associated chi-squared test statistics $\mathcal{M}(j, s)$ from (5.18) for the U.S. hog data. (Test statistics are used to determine the appropriate AR order and ranks $r = m - s$ of coefficient matrices $\mathbf{\Phi}_j$)

		$\mathcal{M}(j, s)$				
j	Partial correlations	$s = 5$	$s = 4$	$s = 3$	$s = 2$	$s = 1$
1	0.987, 0.876, 0.725, 0.413, 0.153	448.5	177.5	70.4	15.5	1.75
2	0.741, 0.561, 0.473, 0.140, 0.068	98.1	44.3	18.8	1.7	0.31
3	0.471, 0.386, 0.310, 0.073, 0.023	31.9	16.5	6.6	0.4	0.03
	Critical value $= \chi^2_{s^2}(.05)$	37.7	26.3	16.9	9.5	3.84

$$\widehat{B}_1 = \begin{bmatrix} 1.482 & -0.032 & -0.022 & 0.012 & 0.116 \\ 0.152 & 0.332 & -0.098 & 0.096 & 0.278 \\ 1.195 & -0.449 & 0.337 & 0.209 & 0.020 \\ -0.152 & -0.919 & -0.551 & 0.526 & 0.331 \end{bmatrix}$$

$$\widehat{B}_2 = \begin{bmatrix} -0.362 & -0.160 & -0.097 & -0.020 & -0.007 \\ -0.180 & 0.054 & 0.281 & -0.200 & -0.106 \\ -0.442 & 0.120 & -0.542 & 0.307 & 0.401 \end{bmatrix}$$

with ML error covariance matrix estimate $\widehat{\Sigma}_{\epsilon\epsilon}$ such that $|\widehat{\Sigma}_{\epsilon\epsilon}| = 0.1584 \times 10^{-3}$. In the results above, the elements in $\widehat{A}_1$ equal to 0 and 1 are fixed for normalization purposes.

Notice that when $\widehat{A}_1$ is augmented with the last column of the identity matrix, then P equal to the inverse of this matrix provides the canonical transformation $\mathcal{Z}_t = PY_t$ in (5.19) for this nested reduced-rank model. From this, for example, we find that the series

$$z_{5t} = 0.529y_{1t} - 1.805y_{2t} - 0.685y_{3t} + 0.394y_{4t} + y_{5t}$$

forms a white noise process, while the series

$$z_{4t} = -0.437y_{1t} - 0.103y_{2t} - 0.647y_{3t} + y_{4t}$$

has dependence on the past of $\{Y_t\}$ that can be represented by only one lag (an AR(1) scalar component in the terminology of Tiao and Tsay 1989). We note that the estimated white noise series z_{5t} under this nested model is almost the same (except for a scale factor) as the estimated "least predictable" canonical series $\omega_{5t} = \widehat{\ell}'_5 Y_t$ obtained from the earlier reduced-rank model and associated canonical correlation analysis. Again, further simplification of the nested reduced-rank model above could be considered by setting to zero those coefficients whose estimates are quite small compared to their standard errors and reestimating the remaining coefficients, but we will not go into any further details here.

Table 5.2 Summary of various analyses of the U.S. hog data

	ML residual variances				
Variable	Full-rank model	Reduced-rank model ($r = 3$)	Nested reduced rank ($r_1, r_2 = 4, 3$)	Index model ($r = 3$)	Univariate model
Wage rate	0.1332	0.1374	0.1336	0.1449	0.1456
Hog supply	0.0544	0.0566	0.0554	0.0710	0.1379
Hog price	0.3665	0.3729	0.3668	0.5632	0.8165
Corn price	0.9448	1.0916	0.9580	1.0776	1.4270
Corn supply	0.4623	0.5130	0.4853	0.4845	0.4594
$\lvert\widehat{\Sigma}_{\epsilon\epsilon}\rvert(\times 10^{-3})$	0.1444	0.1954	0.1584		
n^* = number of parameters	50	36	42		
AIC	−7.577	−7.629	−7.687		

The empirical results of several analyses of the hog data are summarized in Table 5.2 in terms of the variances of the individual residual series, as well as the determinant of the ML residual covariance matrix $\widehat{\Sigma}_{\epsilon\epsilon}$ and the AIC criterion,

$$\text{AIC} = \log(|\widehat{\Sigma}_{\epsilon\epsilon}|) + 2n^*/T,$$

where n^* denotes the number of estimated parameters in the model. Included in Table 5.2 are results for both the full-rank and reduced-rank second-order autoregressive models discussed earlier, as well as results from the nested reduced-rank AR(2) model and a second-order index model of rank 3 considered by Reinsel (1983). As mentioned before, various authors have analyzed the U.S. hog data, and in particular, Tiao and Tsay (1983) used the hog data to illustrate specification procedures for multivariate models. For comparative purposes, they also fit univariate autoregressive moving average models. They observed that the multivariate models clearly outperform the univariate models in terms of residual variance, mainly for the hog supply, hog price, and corn price series. Table 5.2 shows that the results for the reduced-rank models (especially the nested model) are fairly close to those of the full-rank model when compared in terms of residual variances, but the reduced-rank models involve far fewer estimated parameters, and results for both these models are slightly better than those for the index model. Between the two reduced-rank models, although the nested model involves a few more parameters, the overall measures such as the AIC criterion indicate that it represents a somewhat better model fit. The results clearly indicate that little useful information is lost by reducing the dimension of the lagged series.

5.6 Relationship Between Nonstationarity and Canonical Correlations

The reduced-rank models for vector autoregressive time series discussed in the previous sections can be viewed generally in the framework of models related to multivariate linear regression with constraints on the autoregressive coefficient matrices. The associated canonical analysis of Box and Tiao (1977) has been seen to be the same as the classical canonical correlation analysis originally introduced by Hotelling (1935, 1936). Only a few concepts specific to time series analysis, such as the notion of the degree of predictability of linear combinations based on the part of the series $\{Y_t\}$, have thus far entered into the presentation of the reduced-rank AR models and canonical analysis. In this section, we briefly address one additional issue unique to time series analysis. This is the aspect of nonstationarity of the vector time series process $\{Y_t\}$ and the relation of nonstationarity to the canonical analysis. Here we also demonstrate how the reduced-rank autoregressive model can be formulated to test for non-stationarity.

Consider again the vector AR(p) model (5.8), which we express as $\mathbf{\Phi}(L)Y_t = \epsilon_t$, where $\mathbf{\Phi}(L) = I - \mathbf{\Phi}_1 L - \cdots - \mathbf{\Phi}_p L^p$ is the AR(p) operator. Recall that we have assumed the condition that $\det\{\mathbf{\Phi}(z)\} \neq 0$ for all complex numbers $|z| \leq 1$ (i.e., all roots of $\det\{\mathbf{\Phi}(z)\} = 0$ must lie outside the unit circle) and that this condition ensures the stationarity of the process $\{Y_t\}$. In particular, for the AR(1) process $Y_t = \mathbf{\Phi} Y_{t-1} + \epsilon_t$ [$p = 1$ in (5.8)], we have $\mathbf{\Phi}(L) = I - \mathbf{\Phi} L$, and the condition for stationarity above is equivalent to the condition that all eigenvalues of the matrix $\mathbf{\Phi}$ lie inside the unit circle.

One interesting result in the work of Box and Tiao (1977) is the development of the correspondence between nonstationarity in the vector AR(1) process, as reflected by the roots of $\det\{I - \mathbf{\Phi} z\} = 0$ (equivalently, the eigenvalues of $\mathbf{\Phi}$) and the (squared) canonical correlations between Y_t and Y_{t-1} in the canonical analysis, as reflected in the eigenvalues of the matrix

$$R_* = \Gamma(0)^{-1/2}\Gamma(1)'\Gamma(0)^{-1}\Gamma(1)\Gamma(0)^{-1/2},$$

for example, see (5.12) and the discussion following it. To develop the correspondence more easily, note that we can write the matrix R_* as

$$R_* = \Gamma(0)^{-1/2}\mathbf{\Phi}\Gamma(0)\mathbf{\Phi}'\Gamma(0)^{-1/2},$$

since $\mathbf{\Phi} = \Gamma(1)'\Gamma(0)^{-1}$. The specific result to be established is that the number of eigenvalues of $\mathbf{\Phi}$ approaching the unit circle (the nonstationary boundary) is the same as the number of canonical correlations between Y_t and Y_{t-1} that approach unity. This correspondence was presented as a result in the appendix of the paper by Box and Tiao (1977). Velu et al. (1987) gave an alternative more direct proof of the result, which we now discuss.

As a preliminary, we begin with a statement of the following useful lemma.

Lemma 5.1 *For any $m \times m$ complex-valued matrix $\mathbf{\Psi}$, with ordered eigenvalues $|\alpha_1| \geq \cdots \geq |\alpha_m|$,*

$$\prod_{j=1}^{k} |\alpha_j| \leq \prod_{j=1}^{k} \rho_j, \qquad k = 1, \ldots, m-1,$$

and

$$\prod_{j=1}^{m} |\alpha_j| = \prod_{j=1}^{m} \rho_j,$$

where ρ_j^2 is the jth largest eigenvalue of $\mathbf{\Psi}\mathbf{\Psi}^*$ and $\mathbf{\Psi}^*$ denotes the conjugate transpose of $\mathbf{\Psi}$.

Proof See Marshall and Olkin (1979, p. 231).

We now give the main result as follows.

Theorem 5.1 *Suppose Y_t follows the stationary vector AR(1) model $Y_t = \mathbf{\Phi} Y_{t-1} + \epsilon_t$, where $Cov(\epsilon_t) = \Sigma_{\epsilon\epsilon}$ is assumed to be positive definite and fixed. With respect to variation of the matrix $\mathbf{\Phi}$, a necessary and sufficient condition for $d \leq m$ of the eigenvalues of $\Gamma(0)^{-1/2}\mathbf{\Phi}\Gamma(0)\mathbf{\Phi}'\Gamma(0)^{-1/2}$ to tend to unity is that d of the eigenvalues of $\mathbf{\Phi}$ approach values on the unit circle.*

Proof (Sufficiency) Note that the eigenvalues of $\mathbf{\Phi}$ are the same as the eigenvalues of $\mathbf{\Psi} = \Gamma(0)^{-1/2}\mathbf{\Phi}\Gamma(0)^{1/2}$. Denote these eigenvalues by α_j, $j = 1, \ldots, m$. The squared canonical correlations, denoted by ρ_j^2, between Y_t and Y_{t-1} ($\equiv X_t$) are the eigenvalues of $\Gamma(0)^{-1/2}\mathbf{\Phi}\Gamma(0)\mathbf{\Phi}'\Gamma(0)^{-1/2} = \mathbf{\Psi}\mathbf{\Psi}^*$ (see Section 2.4 and also (5.12)). Because $\det\{I - \mathbf{\Phi} z\} = \prod_{j=1}^{m}(1 - \alpha_j z)$, from the stationarity assumption, it follows that $|\alpha_j| < 1$, for $j = 1, \ldots, m$. The canonical correlations ρ_j all lie between zero and one. From Lemma 5.1 with $k = d$, we have

$$\prod_{j=1}^{d} |\alpha_j| \leq \prod_{j=1}^{d} \rho_j,$$

and when $|\alpha_j| \to 1$, $j = 1, \ldots, d$, then $\prod_{j=1}^{d} \rho_j \to 1$ and hence at least the first d of the canonical correlations $\rho_j \to 1$.
(**Necessity**) See Velu et al. (1987).

Box and Tiao (1977) developed the sufficiency part of the proof with considerable detail, but in the presentation above a simple proof follows with the use of Lemma 5.1. Some extensions of the results in Theorem 5.1 to higher order AR(p) processes were presented by Velu et al. (1987), involving partial canonical correlations. However, a complete correspondence between the number of roots α_j of $\det\{\mathbf{\Phi}(z)\} = 0$ that approach the unit circle and the number of various canonical correlations and partial canonical correlations of the AR(p) process $\{Y_t\}$ has not been established.

The main implication of Theorem 5.1 is that the existence of d canonical correlations ρ_j close to one in the vector AR(1) process implies that the associated canonical components will be (close to) nonstationary series, and these may reflect the nonstationary trending or dynamic growth behavior in the multivariate system, while the remaining $r = m - d$ canonical variates possess stationary behavior. Hence, this canonical analysis may be useful for exploring the nature of nonstationarity of the multiple autoregressive time series. Specifically, the nature of nonstationarity can be examined by considering the canonical correlations between Y_t and Y_{t-1} and the associated eigenvectors (canonical variates). This approach to investigation of the nonstationary aspects of the vector AR process has been explored by Bossaerts (1988) and Bewley and Yang (1995).

A particularly important instance of nonstationarity in the vector AR process is the case where the only roots of $\det\{\mathbf{\Phi}(z)\} = 0$ on the unit circle are unity. This corresponds typically to the situation where the nonstationary vector process Y_t becomes stationary upon first differencing, that is, $W_t = (1 - L)Y_t$ is stationary. In this case, for the vector AR(1) model, an alternate canonical analysis is also extremely useful to examine the nature of nonstationarity, in particular, the number of unit roots in the system. The associated methodology is the basis for ML estimation and LR testing procedures concerning processes with unit roots, and it will be discussed and developed in much more detail in the next section. For the present, we will establish a connection between the number of unit eigenvalues of $\mathbf{\Phi}$ in the vector AR(1) process and the number of zero canonical correlations in a certain canonical analysis.

For this, we first note that the vector AR(1) model, $Y_t = \mathbf{\Phi} Y_{t-1} + \epsilon_t$, can be expressed in an equivalent form as

$$W_t \equiv Y_t - Y_{t-1} = -(I - \mathbf{\Phi})Y_{t-1} + \epsilon_t, \tag{5.20}$$

which is referred to as an error-correction form. Notice that if $\mathbf{\Phi}$ has eigenvalues approaching unity, then $I - \mathbf{\Phi}$ will have eigenvalues approaching zero (and will become of reduced rank). The methodology of interest for exploring the number of unit roots in the AR(1) model is motivated by the model form (5.20) and consists of a canonical correlation analysis between $W_t = Y_t - Y_{t-1}$ and Y_{t-1} ($\equiv X_t$). Under stationarity of the process Y_t, from (5.20), note that the covariance matrix between W_t and $Y_{t-1} \equiv X_t$ is equal to $\Sigma_{wx} = -(I - \mathbf{\Phi})\Gamma(0)$. Thus, the squared canonical correlations between W_t and Y_{t-1} are the eigenvalues of

$$\Sigma_{ww}^{-1}\Sigma_{wx}\Sigma_{xx}^{-1}\Sigma_{xw} = \Sigma_{ww}^{-1}(I - \mathbf{\Phi})\Gamma(0)(I - \mathbf{\Phi})', \tag{5.21}$$

where

$$\Sigma_{ww} = \text{Cov}(W_t) = 2\Gamma(0) - \mathbf{\Phi}\Gamma(0) - \Gamma(0)\mathbf{\Phi}' = (I - \mathbf{\Phi})\Gamma(0) + \Gamma(0)(I - \mathbf{\Phi})'.$$

We now present the basic result.

Theorem 5.2 *Suppose Y_t follows the stationary vector AR(1) model, with error-correction form given in (5.20), where Cov(ϵ_t) = $\Sigma_{\epsilon\epsilon}$ is assumed to be positive definite and fixed. With respect to variation of the matrix $\mathbf{\Phi}$, the condition that $d \leq m$ of the eigenvalues of $\Sigma_{ww}^{-1}(I - \mathbf{\Phi})\Gamma(0)(I - \mathbf{\Phi})'$ (i.e., squared canonical correlations between W_t and Y_{t-1}) tend to zero implies that at least d of the eigenvalues of $\mathbf{\Phi}$ approach unity.*

Proof Denote the squared canonical correlations between W_t and Y_{t-1}, that is, the eigenvalues of (5.21), by σ_j^2, $j = 1, \ldots, m$. Now notice that

$$\Gamma(0)^{-1/2}\Sigma_{ww}\Gamma(0)^{-1/2} = \Gamma(0)^{-1/2}(I - \mathbf{\Phi})\Gamma(0)^{1/2} + \Gamma(0)^{1/2}(I - \mathbf{\Phi})'\Gamma(0)^{-1/2}$$
$$\equiv U + U',$$

where $U = \Gamma(0)^{-1/2}(I - \mathbf{\Phi})\Gamma(0)^{1/2}$. In addition, then, the matrix in (5.21) can be reexpressed as

$$\Gamma(0)^{-1/2}\{\Gamma(0)^{-1/2}\Sigma_{ww}\Gamma(0)^{-1/2}\}^{-1}$$
$$\times\, \Gamma(0)^{-1/2}(I - \mathbf{\Phi})\Gamma(0)(I - \mathbf{\Phi})'\Gamma(0)^{-1/2}\Gamma(0)^{1/2}$$
$$\equiv \Gamma(0)^{-1/2}\{U + U'\}^{-1}UU'\Gamma(0)^{1/2},$$

and this matrix has the same eigenvalues as the matrix

$$\{U + U'\}^{-1/2}UU'\{U + U'\}^{-1/2}.$$

Let $|\alpha_1^*| \geq \cdots \geq |\alpha_m^*|$ denote the ordered eigenvalues of the matrix $\{U + U'\}^{-1/2}U$. Now suppose that $d = m - r$ canonical correlations σ_j approach 0, so that $\sigma_{r+1} \to 0$ in particular. Then, by Lemma 5.1, $\prod_{j=1}^{r+1} |\alpha_j^*| \leq \prod_{j=1}^{r+1} \sigma_j \to 0$. This implies that $|\alpha_{r+1}^*| \to 0$ (so also $|\alpha_j^*| \to 0$ for $j = r + 1, \ldots, m$) and hence that at least $d = m - r$ eigenvalues of $\{U + U'\}^{-1/2}U$ approach 0. This, in turn, implies that at least d eigenvalues of U approach 0. Since the eigenvalues of $U = \Gamma(0)^{-1/2}(I - \mathbf{\Phi})\Gamma(0)^{1/2}$ are the same as the eigenvalues of $I - \mathbf{\Phi}$, and the latter eigenvalues are equal to $1 - \alpha_j$, $j = 1, \ldots, m$, where α_j are the eigenvalues of $\mathbf{\Phi}$, it follows that at least d values $(1 - \alpha_j) \to 0$, equivalently, at least d eigenvalues $\alpha_j \to 1$.

To consider the (partial) converse result, suppose that $d = m - r$ eigenvalues α_j of $\mathbf{\Phi}$ approach unity, and all other $|\alpha_j|$ remain bounded in the limit by a constant strictly less than one. Hence, d eigenvalues $1 - \alpha_j$ of $I - \mathbf{\Phi}$ approach zero, and, in addition, we will assume that the *rank* of $I - \mathbf{\Phi}$ approaches $r = m - d$ in the limit. Then it will follow that at least d squared canonical correlations σ_j^2 (i.e., eigenvalues of (5.21)) will tend to zero. Finally, then, we notice that the results of Theorems 5.1 and 5.2 combine to establish the following relation. Namely, if $d \leq m$ canonical correlations between $W_t = Y_t - Y_{t-1}$ and Y_{t-1} tend to zero, then at least d of the canonical correlations between Y_t and Y_{t-1} approach unity.

To discuss certain relationships concerning nonstationarity in more specific detail, we consider further the particular case of a vector AR(1) model $Y_t = \mathbf{\Phi} Y_{t-1} + \epsilon_t$ in which $d = m - r$ eigenvalues α_i of $\mathbf{\Phi}$ approach unity, and all other $|\alpha_i|$ remain bounded in the limit by a constant strictly less than one. We will also assume that $\mathbf{\Phi}$ approaches its limit in such a way that the rank of $I - \mathbf{\Phi}$ is $r = m - d$ in the limit, that is, there are d linearly independent left eigenvectors $Q'_1, \ldots, Q'_d$ associated with the unit eigenvalues of $\mathbf{\Phi}$. Informally, then, in a canonical correlation analysis between $W_t = Y_t - Y_{t-1}$ and Y_{t-1}, the d linear combinations $w^*_{1t} = Q'_1 W_t, \ldots, w^*_{dt} = Q'_d W_t$ tend to provide the canonical variates having associated canonical correlations that approach zero. The corresponding canonical variates for Y_{t-1} are $-Q'_i(I - \mathbf{\Phi})Y_{t-1}$, with the squared canonical correlations

$$\begin{aligned}\sigma_i^2 &= \text{Corr}^2(Q'_i W_t, -Q'_i(I - \mathbf{\Phi})Y_{t-1})\\ &= \frac{(Q'_i(I - \mathbf{\Phi})\Gamma(0)(I - \mathbf{\Phi})' Q_i)^2}{(Q'_i \Sigma_{ww} Q_i)(Q'_i(I - \mathbf{\Phi})\Gamma(0)(I - \mathbf{\Phi})' Q_i)}\\ &= \frac{Q'_i(I - \mathbf{\Phi})\Gamma(0)(I - \mathbf{\Phi})' Q_i}{Q'_i(I - \mathbf{\Phi})\Gamma(0) Q_i + Q'_i \Gamma(0)(I - \mathbf{\Phi})' Q_i} \to 0,\end{aligned}$$

since $Q'_i(I - \mathbf{\Phi})\Gamma(0)(I - \mathbf{\Phi})' Q_i / [Q'_i(I - \mathbf{\Phi})\Gamma(0) Q_i + Q'_i \Gamma(0)(I - \mathbf{\Phi})' Q_i] \sim (1 - \alpha_i)^2 Q'_i \Gamma(0) Q_i / [2(1 - \alpha_i) Q'_i \Gamma(0) Q_i] = (1 - \alpha_i)/2 \to 0$ in the limit.

5.7 Cointegration for Nonstationary Series—Reduced Rank in Long Term

In Sections 5.2 and 5.3, we considered certain reduced-rank models for vector autoregressive time series, and in Section 5.6, we presented a relationship between nonstationarity of the process and canonical correlation analyses. In this section, reduced-rank regression methods will be shown to be useful in recognizing and explicitly modeling nonstationary aspects of a vector process. While the computational procedures of the reduced-rank methods used in modeling relationships in this nonstationary setting are similar to those presented for other reduced-rank models in previous chapters and sections, the asymptotic theory of the estimation and testing procedures differs from the usual asymptotic normal theory because of the presence of nonstationary characteristics of the series involved in the analysis.

Many time series in practice exhibit nonstationary behavior, often of a homogeneous nature, that is, drifting or trending behavior such that apart from a shifting local level or local trend the series has homogeneous patterns with respect to shifts over time. Often, in univariate integrated ARMA time series models, a nonstationary time series can be reduced to stationarity through differencing of the series. For the vector AR model (5.8), $\mathbf{\Phi}(L)Y_t = \epsilon_t$, where $\mathbf{\Phi}(L) = I - \mathbf{\Phi}_1 L - \cdots - \mathbf{\Phi}_p L^p$, the condition for stationarity of the vector process Y_t is that all roots of $\det\{\mathbf{\Phi}(z)\} = 0$ be greater than one in absolute value. To generalize this to nonstationary but nonexplosive processes, one can consider a general form of the vector AR model where some of the roots of $\det\{\mathbf{\Phi}(z)\} = 0$ are allowed to have absolute value equal to one. More specifically, because of the prominent role of the differencing operator $(1 - L)$ in univariate models, we might only allow some roots to equal one (unit roots), while the remaining roots are all greater than one in absolute value.

The nonstationary (unit-root) aspects of a vector process Y_t become more complicated in multivariate time series modeling compared to the univariate case, due in part to the possibility of cointegration among the component series y_{it} of a nonstationary vector process Y_t. For instance, the possibility exists for each component series y_{it} to be nonstationary with its first difference $(1-L)y_{it}$ stationary (in which case y_{it} is said to be integrated of order one), but such that certain linear combinations $z_{it} = b_i' Y_t$ of Y_t will be stationary. In such circumstances, the process Y_t is said to be *cointegrated* with cointegrating vectors b_i (e.g., Engle and Granger 1987). An interpretation of cointegrated vector series Y_t, particularly related to economics, is that the individual components y_{it} share some common nonstationary components or "common trends," and hence, they tend to have certain similar movements in their long-term behavior. A related interpretation is that the component series y_{it}, although they may exhibit nonstationary behavior, satisfy (approximately) a long-run equilibrium relation $b_i' Y_t \approx 0$ such that the process $z_{it} = b_i' Y_t$, which represents the deviations from the equilibrium, exhibits stable behavior and so is a stationary process. A specific nonstationary AR structure for which cointegration occurs is the model where $\det\{\mathbf{\Phi}(z)\} = 0$ has $d < m$ roots equal to one and all other roots are greater than one in absolute value, and also the matrix $\mathbf{\Phi}(1) = I - \mathbf{\Phi}_1 - \cdots - \mathbf{\Phi}_p$ has rank equal to $r = m - d$. Then, for such a process, it can be established that r linearly independent vectors b_i exist such that $b_i' Y_t$ is stationary, and Y_t is said to have cointegrating rank r. Properties of nonstationary cointegrated systems have been investigated by Engle and Granger (1987), among others, and basic properties and estimation of cointegrated vector AR models will be considered in some detail in the remainder of this section. In particular, we will illustrate the usefulness of reduced-rank estimation methods in the analysis of such models.

Thus, we consider multivariate AR models for nonstationary processes $\{Y_t\}$. We focus on situations where the only "nonstationary roots" in the AR operator $\mathbf{\Phi}(L)$ are roots equal to one (unit roots), and we assume there are $d \leq m$ such unit roots, with all other roots of $\det\{\mathbf{\Phi}(z)\} = 0$ outside the unit circle. We note immediately that this implies that $\det\{\mathbf{\Phi}(1)\} = 0$ so that the matrix $\mathbf{\Phi}(1) = I - \sum_{j=1}^{p} \mathbf{\Phi}_j$ does not have full rank. It is also assumed that $\text{rank}\{\mathbf{\Phi}(1)\} = r$, with $r = m - d$, and that $\mathbf{\Phi}(1)$ has (exactly) d zero eigenvalues. It is further noted that this last condition implies that each component of the first differences $W_t = Y_t - Y_{t-1}$ will be stationary (rather than any component of Y_t being integrated of order higher than one). The AR(p)

model, $\mathbf{\Phi}(L)Y_t = Y_t - \sum_{j=1}^{p} \mathbf{\Phi}_j Y_{t-j} = \epsilon_t$, can also be represented in the *error-correction form* (Engle and Granger 1987) as $\mathbf{\Phi}^*(L)(1-L)Y_t = -\mathbf{\Phi}(1)Y_{t-1} + \epsilon_t$, that is,

$$W_t = CY_{t-1} + \sum_{j=1}^{p-1} \mathbf{\Phi}_j^* W_{t-j} + \epsilon_t, \tag{5.22}$$

where $W_t = (1-L)Y_t = Y_t - Y_{t-1}$,

$$\mathbf{\Phi}^*(L) = I - \sum_{j=1}^{p-1} \mathbf{\Phi}_j^* L^j,$$

with $\mathbf{\Phi}_j^* = -\sum_{i=j+1}^{p} \mathbf{\Phi}_i$, and $C = -\mathbf{\Phi}(1) = -(I - \sum_{j=1}^{p} \mathbf{\Phi}_j)$.

The form (5.22) is obtained from direct matrix algebra, since $\mathbf{\Phi}(L)$ can always be reexpressed as $\mathbf{\Phi}(L) = \mathbf{\Phi}^*(L)(1-L) + \mathbf{\Phi}(1)L$ with $\mathbf{\Phi}^*(L)$ as defined. For example, in the AR(2) model, we have

$$\begin{aligned}\mathbf{\Phi}(L) = I - \mathbf{\Phi}_1 L - \mathbf{\Phi}_2 L^2 &= (I - IL) + (I - \mathbf{\Phi}_1 - \mathbf{\Phi}_2)L + \mathbf{\Phi}_2(I - IL)L \\ &= (I + \mathbf{\Phi}_2 L)(I - IL) + (I - \mathbf{\Phi}_1 - \mathbf{\Phi}_2)L \equiv \mathbf{\Phi}^*(L)(1-L) + \mathbf{\Phi}(1)L,\end{aligned}$$

where $\mathbf{\Phi}^*(L) = I + \mathbf{\Phi}_2 L \equiv I - \mathbf{\Phi}_1^* L$ with $\mathbf{\Phi}_1^* = -\mathbf{\Phi}_2$. Under the assumptions on $\mathbf{\Phi}(L)$, it can also be shown that $\mathbf{\Phi}^*(L)$ is a stationary AR operator having all roots of $\det\{\mathbf{\Phi}^*(z)\} = 0$ outside the unit circle. The error-correction form (5.22) is particularly useful because the number of unit roots in the AR operator $\mathbf{\Phi}(L)$ can conveniently be incorporated through the "error-correction" term CY_{t-1}, so that the nature of nonstationarity of the model is conveniently concentrated in the behavior of the single coefficient matrix C in this form.

From the assumptions, the matrix $I - \mathbf{\Phi}(1) = \sum_{j=1}^{p} \mathbf{\Phi}_j$ has d linearly independent eigenvectors associated with its d unit eigenvalues, while its remaining eigenvalues are less than one in absolute value. Let P and $Q = P^{-1}$ be $m \times m$ matrices such that $Q(\sum_{j=1}^{p} \mathbf{\Phi}_j)P = \text{diag}[I_d, \Lambda_r] = J$, where J is the Jordan canonical form of $\sum_{j=1}^{p} \mathbf{\Phi}_j$, so that $QCP = J - I = \text{diag}[0, \Lambda_r - I_r]$. Hence, $C = P(J-I)Q = P_2(\Lambda_r - I_r)Q_2'$, where $P = [P_1, P_2]$ and $Q' = [Q_1, Q_2]$, with P_1 and Q_1 being $m \times d$ matrices, so that C is of *reduced rank* $r < m$. Therefore, the error-correction form (5.22) can be written as

$$W_t = AQ_2'Y_{t-1} + \sum_{j=1}^{p-1} \mathbf{\Phi}_j^* W_{t-j} + \epsilon_t \equiv AZ_{2t-1} + \sum_{j=1}^{p-1} \mathbf{\Phi}_j^* W_{t-j} + \epsilon_t, \tag{5.22$'$}$$

where $A = P_2(\Lambda_r - I_r)$ is $m \times r$ of rank r and $Z_{2t} = Q_2'Y_t$. We also define $Z_{1t} = Q_1'Y_t$, and let $Z_t = (Z_{1t}', Z_{2t}')' = QY_t$.

It follows from the above that although Y_t is nonstationary (with first differences that are stationary), the r linear combinations $Z_{2t} = Q_2' Y_t$ are stationary (e.g., since through (5.22) the variables Z_{2t} are such that $Z_{2t} - \Lambda_r Z_{2t-1}$ is linearly expressible in terms of the stationary series $\{W_t\}$ and $\{\epsilon_t\}$ only and Λ_r is stable). In this situation, Y_t is *cointegrated* of rank r, and the rows of Q_2' are *cointegrating vectors*. The stationary linear combinations $Z_{2t} = Q_2' Y_t$ may be interpreted as long-term stable equilibrium relations among the variables Y_t, and the error-correction model (5.22), as written in (5.22′), formulates that the changes $W_t = Y_t - Y_{t-1}$ in the variables Y_t depend on the deviations from the equilibrium relations in the previous time period (as well as on previous changes W_{t-j}). These deviations, Z_{2t-1}, are thus viewed as being useful explanatory variables for the next change in the process Y_t, and in the context of (5.22′), they are referred to as an *error-correction mechanism*.

Conversely, the d-dimensional series $Z_{1t} = Q_1' Y_t$ is purely nonstationary, with the unit-root nonstationary behavior of the series Y_t generated by Z_{1t}. That is, from $Z_t = (Z_{1t}', Z_{2t}')' = QY_t$, we obtain the relation $Y_t = PZ_t = P_1 Z_{1t} + P_2 Z_{2t}$. We then see that Y_t is a linear combination of the d-dimensional purely nonstationary component Z_{1t} and the r-dimensional stationary component Z_{2t}. The purely nonstationary component Z_{1t} may be viewed as the common nonstationary component or the d "common trends" among the Y_t, with the interpretation that the nonstationary behavior in each of the m component series Y_{it} of Y_t is actually driven by $d(< m)$ common underlying stochastic trends (Z_{1t}).

5.7.1 LS and ML Estimation and Inference

Least squares estimates $\widetilde{C}, \widetilde{\mathbf{\Phi}}_1^*, \ldots, \widetilde{\mathbf{\Phi}}_{p-1}^*$ for the model (5.22) in error-correction form can be obtained, and the limiting distribution theory for these estimators has been derived by Ahn and Reinsel (1990). To describe the asymptotic results, note that the model (5.22) can be expressed as

$$
\begin{aligned}
W_t &= CPZ_{t-1} + \sum_{j=1}^{p-1} \mathbf{\Phi}_j^* W_{t-j} + \epsilon_t \\
&= CP_1 Z_{1t-1} + CP_2 Z_{2t-1} + \sum_{j=1}^{p-1} \mathbf{\Phi}_j^* W_{t-j} + \epsilon_t,
\end{aligned}
$$

where $Z_t = QY_t = (Z_{1t}', Z_{2t}')'$, with $Z_{1t} = Q_1' Y_t$ purely nonstationary, whereas $Z_{2t} = Q_2' Y_t$ is stationary, and note also that $CP_1 = 0$. Because of the presence of the nonstationary "regressor" variables (Z_{1t}) in the model above, the least squares and maximum likelihood estimation theory is non-standard compared to the situation with stationary regressors. In addition, from (5.22), the process $u_t = Q(W_t - CY_{t-1}) = Z_t - JZ_{t-1}$, where $J = \operatorname{diag}[I_d, \Lambda_r]$, is equal to

$u_t = Q(\sum_{j=1}^{p-1} \mathbf{\Phi}_j^* W_{t-j} + \epsilon_t)$ and, hence, u_t is stationary. Therefore, u_t has the (stationary) infinite MA representation of the form $u_t = \mathbf{\Psi}(L)a_t = \sum_{j=0}^{\infty} \mathbf{\Psi}_j a_{t-j}$, with $a_t = Q\epsilon_t$, and we let $\mathbf{\Psi} = \mathbf{\Psi}(1) = \sum_{j=0}^{\infty} \mathbf{\Psi}_j$.

Now let $F = [C, \mathbf{\Phi}_1^*, \ldots, \mathbf{\Phi}_{p-1}^*]$, $X_{t-1} = (Y_{t-1}', W_{t-1}', \ldots, W_{t-p+1}')'$, and $X_{t-1}^* = (Z_{t-1}', W_{t-1}', \ldots, W_{t-p+1}')' \equiv Q^* X_{t-1}$, where $Q^* = \text{diag}[Q, I_{k(p-1)}]$, and assume that observations $Y_{1-p}, \ldots, Y_0, Y_1, \ldots, Y_t$ are available (for convenience of notation), with $W_t = Y_t - Y_{t-1}$. Then the error-correction model (5.22) can be written as

$$W_t = FX_{t-1} + \epsilon_t = FP^* X_{t-1}^* + \epsilon_t,$$

where $P^* = \text{diag}[P, I_{k(p-1)}] = Q^{*-1}$. Thus, the least squares (LS) estimator of F can be represented by

$$\begin{aligned} \widetilde{F} &= \left(\sum_{t=1}^{T} W_t X_{t-1}'\right)\left(\sum_{t=1}^{T} X_{t-1} X_{t-1}'\right)^{-1} \\ &= \left(\sum_{t=1}^{T} W_t X_{t-1}^{*\prime}\right)\left(\sum_{t=1}^{T} X_{t-1}^* X_{t-1}^{*\prime}\right)^{-1} Q^*, \end{aligned}$$

so that

$$(\widetilde{F} - F)P^* = \left(\sum_{t=1}^{T} \epsilon_t X_{t-1}^{*\prime}\right)\left(\sum_{t=1}^{T} X_{t-1}^* X_{t-1}^{*\prime}\right)^{-1}. \tag{5.23}$$

With $U_{t-1}^* = (Z_{2t-1}', W_{t-1}', \ldots, W_{t-p+1}')'$, which is stationary, and so $X_{t-1}^* = (Z_{1t-1}', U_{t-1}^{*\prime})'$, it has been shown that $T^{-3/2} \sum_{t=1}^{T} U_{t-1}^* Z_{1t-1}' \xrightarrow{p} 0$ as $T \to \infty$. Hence, it follows that the LS estimator for the model (5.22) has the representation

$$\begin{aligned} &[T\widetilde{C}P_1, \ T^{1/2}(\widetilde{C}P_2 - CP_2, \widetilde{\mathbf{\Phi}}_1^* - \mathbf{\Phi}_1^*, \ldots, \widetilde{\mathbf{\Phi}}_{p-1}^* - \mathbf{\Phi}_{p-1}^*)] \\ &= \Bigg[P\left(T^{-1}\sum a_t Z_{1t-1}'\right)\left(T^{-2}\sum Z_{1t-1} Z_{1t-1}'\right)^{-1}, \\ &\qquad \left(T^{-1/2}\sum \epsilon_t U_{t-1}^{*\prime}\right)\left(T^{-1}\sum U_{t-1}^* U_{t-1}^{*\prime}\right)^{-1}\Bigg] + o_p(1). \end{aligned}$$

Then, the distribution theory (see Lemma 1 and Theorem 1 in Ahn and Reinsel 1990) for the least squares estimator is such that $T\widetilde{C}P_1 \xrightarrow{d} PM$, where

$$M = \Sigma_a^{1/2} \left\{ \int_0^1 B_d(u)\, dB_m(u)' \right\}' \left\{ \int_0^1 B_d(u) B_d(u)'\, du \right\}^{-1} \Sigma_{a_1}^{-1/2} \boldsymbol{\Psi}_{11}^{-1}, \tag{5.24}$$

where $\Sigma_a = Q\Sigma_{\epsilon\epsilon}Q' = \text{Cov}(a_t)$ with $\Sigma_{\epsilon\epsilon} = \text{Cov}(\epsilon_t)$, $\Sigma_{a_1} = [I_d, 0]\Sigma_a[I_d, 0]'$ is the upper-left $d \times d$ block of the matrix Σ_a, and $\boldsymbol{\Psi}_{11}$ is the upper-left $d \times d$ block of the matrix $\boldsymbol{\Psi}$, $B_m(u)$ is a m-dimensional standard Brownian motion process, and $B_d(u) = \Sigma_{a_1}^{-1/2}[I_d, 0]\Sigma_a^{1/2} B_m(u)$ is a d-dimensional standard Brownian motion. It can be shown (Reinsel 1997, Section 6.3.8) that $\boldsymbol{\Psi}_{11}$ is expressible more explicitly as $\boldsymbol{\Psi}_{11} = \{Q_1'\boldsymbol{\Phi}^*(1)P_1\}^{-1}$. It also holds that $T^{1/2}(\widetilde{C}P_2 - CP_2)'$ and $T^{1/2}(\widetilde{\boldsymbol{\Phi}}_j^* - \boldsymbol{\Phi}_j^*)'$, $j = 1, \ldots, p-1$, have a joint limiting multivariate normal distribution as in stationary model situations, such that the "vec" of these terms has limiting distribution $N(0, \Sigma_{\epsilon\epsilon} \otimes \Gamma_{U^*}^{-1})$, where $\Gamma_{U^*} = \text{Cov}(U_t^*)$. Also, the LS estimates $\widetilde{\boldsymbol{\Phi}}_1^*, \ldots, \widetilde{\boldsymbol{\Phi}}_{p-1}^*$ have the same asymptotic distribution as in the "stationary case," where Q_2 is known and one regresses W_t on the stationary variables $Z_{2t-1} = Q_2'Y_{t-1}$ and $W_{t-1}, \ldots, W_{t-p+1}$.

When maximum likelihood estimation of the parameter matrix C in (5.22) is considered subject to the reduced-rank restriction that $\text{rank}(C) = r$, it is convenient to express C as $C = AB$, where A and B are $m \times r$ and $r \times m$ full-rank matrices, respectively. As usual, some normalization conditions on A and B are required for uniqueness. One possibility is to have B normalized so that $B = [I_r, B_0]$, where B_0 is an $r \times (m - r)$ matrix of unknown parameters. The elements of the vector Y_t can always be arranged so that this normalization is possible. Implicit in this arrangement of Y_t is that if we write $Y_t = (Y_{1t}', Y_{2t}')'$, where Y_{1t} is $r \times 1$ and Y_{2t} is $d \times 1$, then there is no cointegration among the components of Y_{2t}. It is emphasized that the estimation of the model with the reduced-rank constraint imposed on C is equivalent to the estimation of the AR model with d unit roots explicitly imposed in the model. Hence, this is an alternative to (arbitrarily) differencing each component of the series Y_t prior to fitting a model in situations where the individual components tend to exhibit nonstationary behavior. This explicit form of modeling the nonstationarity may lead to a better understanding of the nature of nonstationarity among the different component series and may also improve longer-term forecasting over unconstrained model fits that do not explicitly incorporate unit roots in the model. Hence, there may be many desirable reasons to formulate and estimate the AR model with an appropriate number of unit roots explicitly incorporated in the model, and it is found that a particularly convenient way to do this is through the use of model (5.22) with the constraint $\text{rank}(C) = r$ imposed.

Maximum likelihood estimation of $C = AB = P_2(\Lambda_r - I_r)Q_2'$ and the $\boldsymbol{\Phi}_j^*$ under the constraint that $\text{rank}(C) = r$, using an iterative Newton–Raphson procedure with the normalization $B = [I_r, B_0]$, is presented by Ahn and Reinsel (1990), and the limiting distribution theory for these estimators is derived. Specifically, note that the model can be written as

$$W_t = ABY_{t-1} + \sum_{j=1}^{p-1} \mathbf{\Phi}_j^* W_{t-j} + \epsilon_t \equiv ABY_{t-1} + DW_{t-1}^* + \epsilon_t, \tag{5.25}$$

where $D = [\mathbf{\Phi}_1^*, \ldots, \mathbf{\Phi}_{p-1}^*]$ and $W_{t-1}^* = (W_{t-1}', \ldots, W_{t-p+1}')'$. It is recognized that this takes the same form as the extended reduced-rank model (3.3) considered in Section 3.1. We define the vector of unknown parameters as $\theta = (\beta_0', \alpha')'$, where $\beta_0 = \text{vec}(B_0')$, and $\alpha = \text{vec}\{[A, D]'\} = \text{vec}\{[A, \mathbf{\Phi}_1^*, \ldots, \mathbf{\Phi}_{p-1}^*]'\}$. With $b = r(m - r) + rm + m^2(p-1)$ representing the number of unknown parameters in θ, define the $b \times m$ matrices

$$\mathcal{U}_{t-1} = [(A' \otimes H'Y_{t-1})', I_m \otimes \widetilde{U}_{t-1}']',$$

where $\widetilde{U}_{t-1} = [(BY_{t-1})', W_{t-1}', \ldots, W_{t-p+1}']'$ is stationary, and the matrix $H' = [0, I_{m-r}]$ is $(m-r) \times m$ such that $Y_{2t} = H'Y_t$ is taken to be purely nonstationary by assumption. Then, based on T observations $Y_1, \ldots, Y_t$, the Gaussian estimator of θ is obtained by the iterative approximate Newton–Raphson relations

$$\widehat{\theta}^{(i+1)} = \widehat{\theta}^{(i)} + \left\{ \sum_{t=1}^{T} \mathcal{U}_{t-1} \Sigma_{\epsilon\epsilon}^{-1} \mathcal{U}_{t-1}' \right\}_{\widehat{\theta}^{(i)}}^{-1} \left\{ \sum_{t=1}^{T} \mathcal{U}_{t-1} \Sigma_{\epsilon\epsilon}^{-1} \epsilon_t \right\}_{\widehat{\theta}^{(i)}}, \tag{5.26}$$

where $\widehat{\theta}^{(i)}$ denotes the estimate at the ith iteration.

Concerning the asymptotic distribution theory of the Gaussian estimators under the model where the unit roots are imposed, it is assumed that the iterations in (5.26) are started from an initial consistent estimator $\widehat{\theta}^{(0)}$ such that $D^*(\widehat{\theta}^{(0)} - \theta) = \{T(\widehat{\beta}_0^{(0)} - \beta_0)', T^{1/2}(\widehat{\alpha}^{(0)} - \alpha)'\}'$ is $O_p(1)$, where $D^* = \text{diag}[TI_{rd}, T^{1/2}I_{(b-rd)}]$. Then, it has been established (Ahn and Reinsel 1990) that

$$T(\widehat{B}_0 - B_0) \xrightarrow{d} (A'\Sigma_{\epsilon\epsilon}^{-1}A)^{-1}A'\Sigma_{\epsilon\epsilon}^{-1}PMP_{21}^{-1},$$

where the distribution of M is specified in (5.24), and P_{21} is the $d \times d$ lower submatrix of P_1. For the remaining parameters, α, in the model, it follows that

$$T^{1/2}(\widehat{\alpha} - \alpha) = \left[I_m \otimes \left(\frac{1}{T} \sum_{t=1}^{T} \widetilde{U}_{t-1}\widetilde{U}_{t-1}' \right)^{-1} \right] \frac{1}{\sqrt{T}} \sum_{t=1}^{T} (I_m \otimes \widetilde{U}_{t-1})\epsilon_t + o_p(1).$$

Hence, it is also shown that $T^{1/2}(\widehat{\alpha} - \alpha) \xrightarrow{d} N(0, \Sigma_{\epsilon\epsilon} \otimes \Gamma_{\widetilde{U}}^{-1})$, where $\Gamma_{\widetilde{U}} = \text{Cov}(\widetilde{U}_t)$, similar to results in stationary situations.

The iterative procedure (5.26) to obtain the Gaussian estimator $\widehat{\theta}$ can readily be modified to incorporate additional constraints, such as zero constraints or nested reduced-rank constraints on the stationary parameters $\mathbf{\Phi}_j^*$ in (5.22), beyond the cointegrating constraint that $\text{rank}(C) = r$. However, when there are no constraints imposed other than rank$(C) = r$, the Gaussian reduced-rank estimator in model (5.22) can also be obtained explicitly through the *partial canonical correlation analysis*, based on the previous work as discussed in Section 3.1, since the model (5.25) has the same form as (3.3). This approach was presented by Johansen (1988, 1991). To describe the results, let $\widetilde{W}_t$ and $\widetilde{Y}_{t-1}$ denote the "adjusted" or residual vectors from the least squares regressions of W_t and Y_{t-1}, respectively, on the lagged values $W_{t-1}^* = (W_{t-1}', \ldots, W_{t-p+1}')'$, and let

$$S_{\widetilde{w}\widetilde{w}} = \frac{1}{T}\sum_{t=1}^{T} \widetilde{W}_t \widetilde{W}_t', \quad S_{\widetilde{w}\widetilde{y}} = \frac{1}{T}\sum_{t=1}^{T} \widetilde{W}_t \widetilde{Y}_{t-1}', \quad S_{\widetilde{y}\widetilde{y}} = \frac{1}{T}\sum_{t=1}^{T} \widetilde{Y}_{t-1} \widetilde{Y}_{t-1}'.$$

Then, the sample partial canonical correlations $\widehat{\rho}_i$ between W_t and Y_{t-1}, given $W_{t-1}, \ldots, W_{t-p+1}$, and the corresponding vectors $\widehat{V}_i^*$ are the solutions to

$$(\widehat{\rho}_i^2 I_m - S_{\widetilde{w}\widetilde{w}}^{-1/2} S_{\widetilde{w}\widetilde{y}} S_{\widetilde{y}\widetilde{y}}^{-1} S_{\widetilde{y}\widetilde{w}} S_{\widetilde{w}\widetilde{w}}^{-1/2}) \widehat{V}_i^* = 0, \qquad i = 1, \ldots, m. \tag{5.27}$$

The *reduced-rank Gaussian estimator* of $C = AB$ can then be obtained from results of Sections 2.3–2.4 and 3.1. It can be expressed explicitly as $\widehat{C} = S_{\widetilde{w}\widetilde{w}}^{1/2} \widehat{V}_* \widehat{V}_*' S_{\widetilde{w}\widetilde{w}}^{-1/2} \widetilde{C}$, where $\widetilde{C} = S_{\widetilde{w}\widetilde{y}} S_{\widetilde{y}\widetilde{y}}^{-1}$ is the full-rank LS estimator, and $\widehat{V}_* = [\widehat{V}_1^*, \ldots, \widehat{V}_r^*]$ are the (normalized) vectors corresponding to the r largest partial canonical correlations $\widehat{\rho}_i$, $i = 1, \ldots, r$. This form of the estimator provides the reduced-rank factorization as

$$\widehat{C} = (S_{\widetilde{w}\widetilde{w}}^{1/2} \widehat{V}_*)(\widehat{V}_*' S_{\widetilde{w}\widetilde{w}}^{-1/2} \widetilde{C}) \equiv \widehat{A}\widehat{B},$$

with $\widehat{A} = S_{\widetilde{w}\widetilde{w}}^{1/2} \widehat{V}_*$ satisfying the normalization $\widehat{A}' S_{\widetilde{w}\widetilde{w}}^{-1} \widehat{A} = I_r$. The Gaussian estimator for the other parameters $\mathbf{\Phi}_1^*, \ldots, \mathbf{\Phi}_{p-1}^*$ can be obtained by ordinary least squares regression of $W_t - \widehat{C} Y_{t-1}$ on the lagged variables $W_{t-1}, \ldots, W_{t-p+1}$. As an alternate form of the reduced-rank estimator, let $\widehat{V}_i^{**} = S_{\widetilde{w}\widetilde{w}}^{-1/2} \widehat{V}_i^* (1 - \widehat{\rho}_i^2)^{-1/2}$, $i = 1, \ldots, m$, and define the $m \times r$ matrix $\widehat{V}_{**} = S_{\widetilde{w}\widetilde{w}}^{-1/2} \widehat{V}_* (I_r - \widehat{\Lambda}_*^2)^{-1/2}$, where $\widehat{\Lambda}_* = \mathrm{diag}(\widehat{\rho}_1, \ldots, \widehat{\rho}_r)$. Then the reduced-rank estimator can be expressed equivalently as $\widehat{C} = \widetilde{\Sigma}_{\epsilon\epsilon} \widehat{V}_{**} \widehat{V}_{**}' \widetilde{C}$, where

$$\widetilde{\Sigma}_{\epsilon\epsilon} = T^{-1} \sum_{t=1}^{T} \widetilde{\epsilon}_t \widetilde{\epsilon}_t' = S_{\widetilde{w}\widetilde{w}} - S_{\widetilde{w}\widetilde{y}} S_{\widetilde{y}\widetilde{y}}^{-1} S_{\widetilde{y}\widetilde{w}}$$

is the error covariance matrix estimate of $\Sigma_{\epsilon\epsilon}$ from the full-rank LS estimation. The vectors $\widehat{V}_i^{**}$ in $\widehat{V}_{**}$ are normalized by $\widehat{V}_i^{**\prime} \widetilde{\Sigma}_{\epsilon\epsilon} \widehat{V}_i^{**} = 1$ (so that $\widehat{V}_{**}' \widetilde{\Sigma}_{\epsilon\epsilon} \widehat{V}_{**} = I_r$).

Note from the discussion in Section 2.4 (e.g., see Theorem 2.3) that the expression $\widehat{B} = \widehat{V}_*' S_{\widetilde{w}\widetilde{w}}^{-1/2} \widetilde{C}$ shows that the estimated cointegrating vectors (rows of $\widehat{B}$) are obtained from the vectors that determine the first r canonical variates of Y_{t-1} in the partial canonical correlation analysis. In addition, it is seen from the form of the reduced-rank estimator $\widehat{C} = S_{\widetilde{w}\widetilde{w}}^{1/2} \widehat{V}_* \widehat{V}_*' S_{\widetilde{w}\widetilde{w}}^{-1/2} \widetilde{C}$ given above that the Gaussian estimates of the "common trends" components $Z_{1t} = Q_1' Y_t$ can be obtained with

$$\widehat{Q}_1 = S_{\widetilde{w}\widetilde{w}}^{-1/2} [\widehat{V}_m^*, \ldots, \widehat{V}_{r+1}^*],$$

so that the rows of $\widehat{Q}_1'$ are the vectors $\widehat{V}_i^{*\prime} S_{\widetilde{w}\widetilde{w}}^{-1/2}$, $i = r+1, \ldots, m$, corresponding to the d smallest partial canonical correlations, with the property that $\widehat{Q}_1' \widehat{C} = 0$ since $\widehat{Q}_1' \widehat{A} = [\widehat{V}_m^*, \ldots, \widehat{V}_{r+1}^*]' \widehat{V}_* = 0$. This method of estimation of the common "trend" or "long-memory" components was explored by Gonzalo and Granger (1995).

5.7.2 *Likelihood Ratio Test for the Number of Cointegrating Relations*

Within the context of model (5.22), it is necessary to specify or determine the rank r of cointegration or the number d of unit roots in the model. Thus, it is also of interest to test the hypothesis H_0: $\text{rank}(C) \leq r$, which is equivalent to the hypothesis that the number of unit roots in the AR model is greater than or equal to d $(d = m - r)$, against the general alternative that $\text{rank}(C) = m$. The *likelihood ratio test* for this hypothesis is considered by Johansen (1988) and by Reinsel and Ahn (1992). The likelihood ratio (LR) test statistic is given by

$$-T \log(U) = -T \log(|S|/|S_0|),$$

where $S = \sum_{t=1}^{T} \widehat{\epsilon}_t \widehat{\epsilon}_t'$ denotes the residual sum of squares matrix in the full or unconstrained model, while S_0 is the residual sum of squares matrix obtained under the reduced-rank restriction on C that $\text{rank}(C) = r$. It follows from work of Anderson (1951) in the multivariate linear model with reduced-rank structure (see Section 3.1) that the LR statistic can be expressed equivalently as

$$-T \log(U) = -T \sum_{i=r+1}^{m} \log(1 - \widehat{\rho}_i^2), \tag{5.28}$$

where the $\widehat{\rho}_i$ are the $d = m - r$ smallest sample partial canonical correlations between $W_t = Y_t - Y_{t-1}$ and Y_{t-1}, given $W_{t-1}, \ldots, W_{t-p+1}$. The limiting distribution for the likelihood ratio statistic has been derived by Johansen (1988), and by Reinsel and Ahn (1992) using the distribution theory for the full-rank LS and Gaussian reduced-rank estimators. Specifically, it is given by

$$-T \log(U) \xrightarrow{d} \text{tr}\left\{ \left(\int_0^1 B_d(u)\, dB_d(u)' \right)' \left(\int_0^1 B_d(u) B_d(u)'\, du \right)^{-1} \right.$$

$$\times \left(\int_0^1 B_d(u) \, dB_d(u)' \right) \Bigg\}, \tag{5.29}$$

where $B_d(u)$ is a d-dimensional standard Brownian motion process. Note that the asymptotic distribution of the likelihood ratio test statistic under H_0 depends only on d and not on any parameters or the order p of the AR model.

Critical values of the asymptotic distribution of (5.29) have been obtained by simulation by Johansen (1988) and Reinsel and Ahn (1992) and can be used in the test of H_0. Some other approaches to testing for unit roots in nonstationary multivariate systems have been examined by Engle and Granger (1987), Stock and Watson (1988), Fountis and Dickey (1989), Phillips and Ouliaris (1986), and Bewley and Yang (1995). It is known that inclusion of a constant term in the estimation of the nonstationary AR model affects the limiting distribution of the estimators and test statistics. Modification of the asymptotic distribution theory for the LS and Gaussian estimators and for LR testing, to allow for inclusion of constant terms in the estimation, has been considered by Johansen (1991), Johansen and Juselius (1990), and Reinsel and Ahn (1992), and critical values for the corresponding limiting distribution of the LR test statistic have been obtained by Monte Carlo simulation.

5.8 Unit Root and Cointegration Aspects for the U.S. Hog Data Example

We illustrate the modeling and analysis procedures for nonstationary vector AR processes discussed in the preceding section by further considering the annual U.S. hog data from Section 5.5. The previous analyses for these data in Section 5.5 indicate that an AR(2) model may be adequate, and the basic time series plots in Fig. 5.1 suggest nonstationarity in the component series. Hence, we investigate the nature of the nonstationarity in terms of (the number of) unit roots in the AR(2) model operator, equivalently, the number of cointegrating relations among the components. As discussed, the AR(2) model can be expressed in the equivalent error-correction model form (5.22),

$$W_t = CY_{t-1} + \mathbf{\Phi}_1^* W_{t-1} + \epsilon_t,$$

where $W_t = Y_t - Y_{t-1}$, $C = -\mathbf{\Phi}(1) = -(I - \mathbf{\Phi}_1 - \mathbf{\Phi}_2)$, and $\mathbf{\Phi}_1^* = -\mathbf{\Phi}_2$.

Within the context of a vector AR model, we consider LR tests for the number of unit roots d or the rank of cointegration $r = m - d$. Thus, we perform the partial canonical correlation analysis between W_t and Y_{t-1}, given W_{t-j}, $j = 1, \dots, k-1$. For greater generality and to illustrate the effect of the choice of AR order k on the LR test results, we provide results for various values of $k = 1, \dots, 5$. The sample partial canonical correlations $\widehat{\rho}_i$ are shown in Table 5.3, as well as the LR

Table 5.3 Summary of partial canonical correlation analysis between $W_t = Y_t - Y_{t-1}$ and Y_{t-1} and associated LR test statistics from (5.29) for the U.S. hog data. (Test statistics are used to determine the appropriate rank $r = m - d$, the number of cointegrating relations, of matrix C)

		LR Statistic				
k	Partial correlations	$r = 0$	$r = 1$	$r = 2$	$r = 3$	$r = 4$
1	0.718, 0.708, 0.473, 0.340, 0.063	142.2	84.9	30.0	10.0	0.32
2	0.804, 0.576, 0.473, 0.354, 0.022	142.9	61.7	30.2	10.5	0.04
3	0.593, 0.530, 0.442, 0.369, 0.003	86.7	53.4	28.0	11.3	0.00
4	0.560, 0.490, 0.386, 0.344, 0.052	71.5	42.9	22.0	9.8	0.21
5	0.677, 0.569, 0.393, 0.372, 0.026	99.2	53.1	23.8	11.2	0.05
	Critical values at 0.05%	71.1	49.4	31.7	18.0	8.16

test statistic values for H_0: rank$(C) = r$ for $r = 0, \ldots, 4$. Since an AR(p) model of order $p = 2$ has been chosen as appropriate, we concentrate on the LR test statistics for $k = 2$ in Table 5.3. By referring to the upper percentiles of the limiting distribution of the LR test statistic (e.g., see Reinsel 1997, p. 205), we find that the hypotheses of $d \geq 1$ $(r \leq 4)$ or $d \geq 2$ $(r \leq 3)$ are clearly not rejected, while $d \geq 4$ $(r \leq 1)$ or $d \geq 5$ $(r \leq 0)$ are strongly rejected. There is also moderate evidence for rejection of $d \geq 3$ $(r \leq 2)$, and so the LR testing procedures suggest that there are $d = 2$ unit roots and $r = 3$ cointegrating vectors. (Note that, with the exception of the value for $r = 0$ at $k = 2$, the LR test statistic values in Table 5.3 remain relatively stable for values of $k \geq 2$, which is consistent with an AR model of order 2.)

The full-rank LS estimates $\tilde{\mathbf{\Phi}}_1$ and $\tilde{\mathbf{\Phi}}_2$ of the AR(2) model were given previously in Section 5.5, and from these, the LS estimates $\tilde{C} = -(I - \tilde{\mathbf{\Phi}}_1 - \tilde{\mathbf{\Phi}}_2)$ and $\tilde{\mathbf{\Phi}}_1^* = -\tilde{\mathbf{\Phi}}_2$ in the error-correction form can easily be deduced. The Gaussian reduced-rank estimates with $r = \text{rank}(C) = 3$ imposed, which incorporates the $d = 2$ unit roots in the model, can be obtained from the partial canonical correlation results for AR order $k = 2 \equiv p$. As described in the previous section, we obtain $\hat{C} = \hat{A}\hat{B}$, with $\hat{A} = S_{\tilde{w}\tilde{w}}^{1/2} \hat{V}_*$ and $\hat{B} = \hat{V}_*' S_{\tilde{w}\tilde{w}}^{-1/2} \tilde{C}$, as

$$\hat{C} = \hat{A}\hat{B} = \begin{pmatrix} 0.027 & -0.180 & 0.023 \\ -0.286 & -0.162 & -0.100 \\ 0.781 & -0.252 & -0.126 \\ 0.288 & -1.013 & 0.463 \\ -0.214 & 0.464 & -0.109 \end{pmatrix}$$

$$\times \begin{pmatrix} 0.639 & 0.367 & -1.240 & 0.617 & 0.262 \\ -0.747 & 1.671 & 0.609 & 0.073 & -0.735 \\ -0.214 & 1.465 & 0.613 & -0.741 & -0.812 \end{pmatrix}$$

$$= \begin{pmatrix} 0.147 & -0.258 & -0.130 & -0.013 & 0.121 \\ -0.041 & -0.522 & 0.195 & -0.115 & 0.125 \\ 0.715 & -0.319 & -1.199 & 0.557 & 0.492 \\ 0.843 & -0.910 & -0.691 & -0.240 & 0.444 \\ -0.460 & 0.537 & 0.482 & -0.018 & -0.309 \end{pmatrix}$$

with ML error covariance matrix estimate $\widehat{\Sigma}_{\epsilon\epsilon}$ such that $|\widehat{\Sigma}_{\epsilon\epsilon}| = 0.1653 \times 10^{-3}$. There is close agreement between the reduced-rank estimate $\widehat{C}$ and the (implied) LS estimate $\widetilde{C}$, and the corresponding ML estimate (not shown) $\widehat{\mathbf{\Phi}}_1^*$ of $\mathbf{\Phi}_1^*$ is found to be very close to the LS estimate $\widetilde{\mathbf{\Phi}}_1^* = -\widetilde{\mathbf{\Phi}}_2$.

From the estimation results above, estimates of three cointegrating linear combinations are given by

$$Z_{2t} = (z_{3t}, z_{4t}, z_{5t})' = \widehat{\Lambda}_*^{-1} \widehat{B} Y_t \equiv \widehat{Q}_2' Y_t,$$

which are normalized such that the variances of the corresponding adjusted series $\widetilde{Z}_{2t} = \widehat{\Lambda}_*^{-1} \widehat{B} \widetilde{Y}_t$ have unit variances (i.e., $\widehat{\Lambda}_*^{-1} \widehat{B} S_{\widetilde{y}\widetilde{y}} \widehat{B}' \widehat{\Lambda}_*^{-1} = I_r$). These three series are displayed in Fig. 5.3 and give the rather clear appearance of being stationary. Also shown in Fig. 5.3 are the two linear combinations

$$Z_{1t} = (z_{1t}, z_{2t})' = \widehat{Q}_1' Y_t,$$

where $\widehat{Q}_1 = S_{\widetilde{w}\widetilde{w}}^{-1/2} [\widehat{V}_m^*, \ldots, \widehat{V}_{r+1}^*]$ and is given by

$$\widehat{Q}_1' = \begin{pmatrix} 3.237 & -0.744 & -0.228 & -0.362 & 0.085 \\ 2.346 & -1.096 & -0.673 & -1.017 & -2.059 \end{pmatrix}.$$

The series $Z_{1t} = \widehat{Q}_1' Y_t$ are normalized such that the variances of the adjusted first differences of these series, $\widehat{Q}_1' \widetilde{W}_t$, have unit variances (i.e., $\widehat{Q}_1' S_{\widetilde{w}\widetilde{w}} \widehat{Q}_1 = I_d$). These linear combinations can be interpreted as the "common trend" series among the five original series. Notice that the first of the common trend series consists predominantly of the farm wage rates series (y_{1t}).

From a model building point of view, we recall from Section 5.5 that a reduced rank of 3 was also indicated for the lag 2 coefficient matrix $\mathbf{\Phi}_2$, and so this would also be true for the matrix $\mathbf{\Phi}_1^*$ in the error-correction form. Thus, for the final model estimation results, we fit a combined reduced-rank AR(2) model in error-correction form, $W_t = CY_{t-1} + \mathbf{\Phi}_1^* W_{t-1} + \epsilon_t$, with the constraints that $\text{rank}(C) = 3$ and $\text{rank}(\mathbf{\Phi}_1^*) = 3$, the latter constraint as suggested from the LR test statistic results of Table 5.1 in Section 5.5. Notice that there is no "nesting" of the two coefficient matrices in this form of reduced-rank model, and hence, the model has the form of the model (3.23) in Section 3.4. We write $C = AB$ and $\mathbf{\Phi}_1^* = EF$ and use the normalizations $B = [I_3, B_0]$ and $E = [I_3, E_0']'$. The final Gaussian reduced-rank

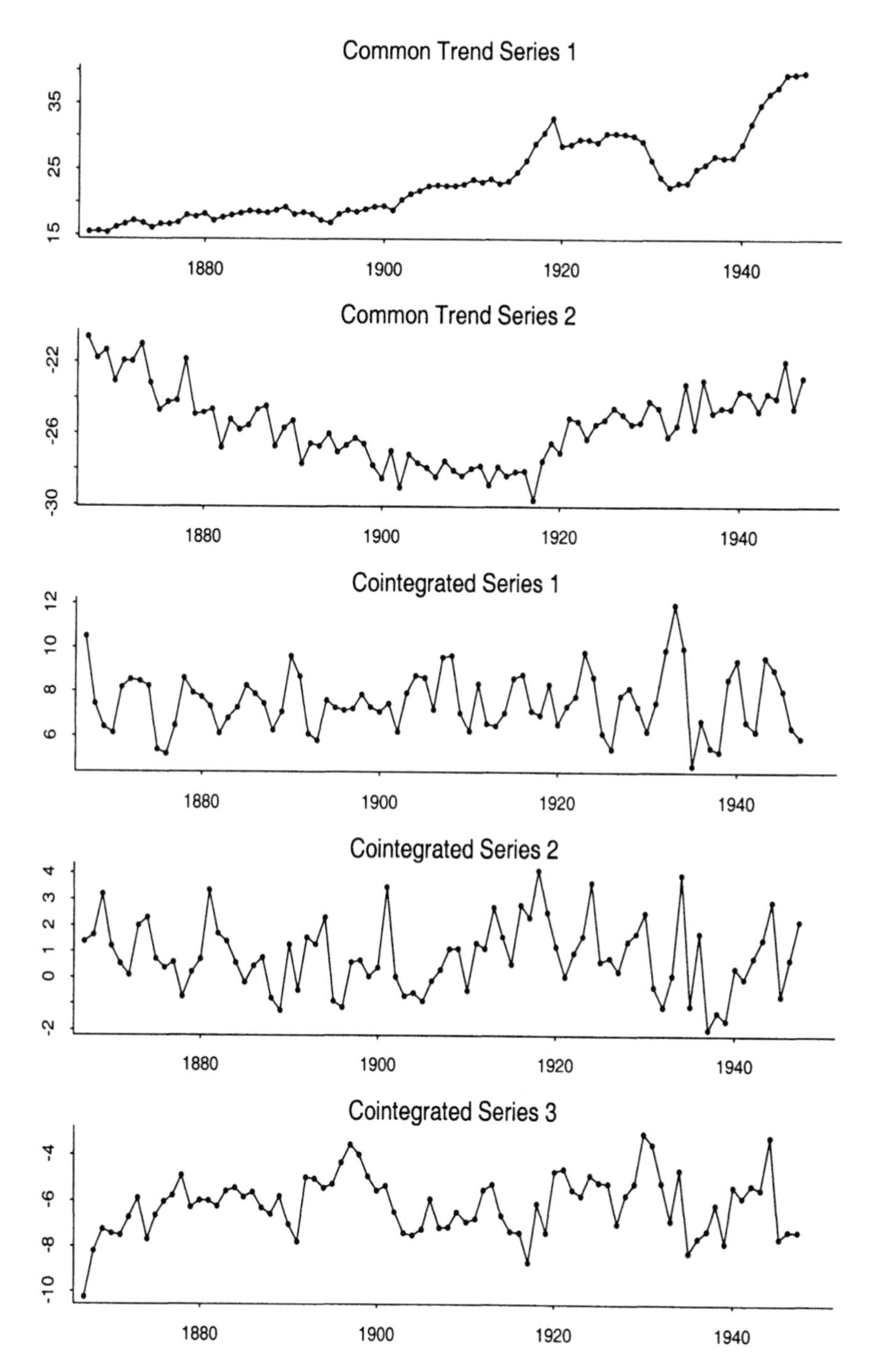

Fig. 5.3 Common trends and cointegrating series for the annual U.S. hog industry data, 1867–1947

estimates are obtained by Newton–Raphson iterative procedures, similar to those described in Sections 5.3 and 5.7 and are given by

$$\widehat{C} = \widehat{A}\widehat{B} = \begin{pmatrix} 0.136 & -0.220 & -0.118 \\ -0.036 & -0.555 & 0.184 \\ 0.714 & -0.333 & -1.196 \\ 0.849 & -1.003 & -0.734 \\ -0.493 & 0.711 & 0.548 \end{pmatrix} \begin{pmatrix} 1 & 0 & 0 & -1.6129 & -0.3114 \\ 0 & 1 & 0 & -0.1694 & -0.4095 \\ 0 & 0 & 1 & -1.3807 & -0.4947 \end{pmatrix}$$

$$= \begin{pmatrix} 0.136 & -0.220 & -0.118 & -0.020 & 0.106 \\ -0.036 & -0.555 & 0.184 & -0.103 & 0.147 \\ 0.714 & -0.333 & -1.196 & 0.557 & 0.506 \\ 0.849 & -1.003 & -0.734 & -0.187 & 0.509 \\ -0.493 & 0.711 & 0.548 & -0.082 & -0.409 \end{pmatrix}$$

$$\widehat{\boldsymbol{\Phi}}_1^* = \widehat{E}\widehat{F} = \begin{pmatrix} 1 & 0 & 0 \\ 0 & 1 & 0 \\ 0 & 0 & 1 \\ 1.511 & -0.986 & 0.072 \\ -2.728 & 3.922 & 1.572 \end{pmatrix}$$

$$\times \begin{pmatrix} 0.364 & 0.117 & 0.084 & 0.033 & 0.008 \\ 0.161 & -0.005 & -0.265 & 0.199 & 0.131 \\ 0.392 & -0.060 & 0.557 & -0.325 & -0.434 \end{pmatrix}$$

$$= \begin{pmatrix} 0.364 & 0.117 & 0.084 & 0.033 & 0.008 \\ 0.161 & -0.005 & -0.265 & 0.199 & 0.131 \\ 0.392 & -0.060 & 0.557 & -0.325 & -0.434 \\ 0.419 & 0.178 & 0.428 & -0.169 & -0.148 \\ 0.256 & -0.433 & -0.392 & 0.179 & -0.193 \end{pmatrix}$$

with ML error covariance matrix estimate $\widehat{\Sigma}_{\epsilon\epsilon}$ such that $|\widehat{\Sigma}_{\epsilon\epsilon}| = 0.1696 \times 10^{-3}$. Because the ML estimates for $\boldsymbol{\Phi}_1^*$ follow the usual normal distribution theory asymptotics, a LR test statistic for the validity of the constraint that $\text{rank}(\boldsymbol{\Phi}_1^*) = 3$ has the usual asymptotic χ_4^2 distribution. The value of this LR statistic is found to be $-(79-2\cdot5-3/2)\log(0.1653/0.1696) = 1.708$, and so the reduced-rank restriction on $\boldsymbol{\Phi}_1^*$ is quite reasonable, very similar to the LR test finding from Table 5.1 in Section 5.5. Also notice that the final reduced-rank estimate $\widehat{C} = \widehat{A}\widehat{B}$ is quite similar to the reduced-rank estimate obtained when the rank of $\boldsymbol{\Phi}_1^*$ is not constrained.

Finally, it may be of some interest to relate and compare the decomposition of Y_t into trend and stationary components, associated with the above analysis in the error-correction model form, to the canonical analysis associated with the reduced-rank modeling previously considered in Section 5.5. The transformed vectors of variates associated with the reduced-rank and error-correction form analyses, respectively, are the canonical variates $\mathcal{W}_t = \widehat{L}'Y_t$ and the common trend and cointegrating variates $Z_t = (Z_{1t}', Z_{2t}')' = \widehat{Q}Y_t$, where $\widehat{Q} = [\widehat{Q}_1, \widehat{Q}_2]'$. Hence, these two sets of transformed variates are related by $\mathcal{W}_t = \widehat{L}'\widehat{Q}^{-1}Z_t$, with the matrix $\widehat{L}'\widehat{Q}^{-1}$ given by

$$\widehat{L}'\widehat{Q}^{-1} = \begin{pmatrix} 1.1395 & -0.2594 & 0.0794 & 0.3199 & 0.3702 \\ -0.0097 & -1.1913 & 0.9551 & -0.0831 & 0.7225 \\ 0.0087 & 0.4240 & 0.8146 & -0.7074 & 0.8143 \\ 0.0114 & -0.0864 & -0.5534 & -0.3071 & 0.6034 \\ 0.0276 & -0.0792 & -0.0096 & -0.6063 & -0.4135 \end{pmatrix}.$$

In the error-correction form analysis above, the original series Y_t has been decomposed into a "purely nonstationary" component Z_{1t} and a stationary component Z_{2t}. Of course, any other series formed as linear combinations of the elements of Z_{2t} will also be stationary, whereas linear combinations that include elements of Z_{1t} will be nonstationary. The canonical analysis of Box and Tiao (1977) does not have exactly the same purpose as the analysis above, but there can be some similarities as noted previously. In particular, in the canonical analysis, components associated with canonical correlations close to one will tend to be nonstationary, while those associated with the smaller canonical correlations will tend to be stationary; canonical correlations very close to zero give rise to series with (approximate) white noise behavior. From examination of the structure of the matrix $\widehat{L}'\widehat{Q}^{-1}$ in the relation $\mathcal{W}_t = \widehat{L}'\widehat{Q}^{-1}Z_t$, it is found that the approximate white noise canonical variate components $\omega_{4t} = \widehat{\ell}_4'Y_t$ and $\omega_{5t} = \widehat{\ell}_5'Y_t$ in $\mathcal{W}_t$ are related to the components of Z_t roughly as

$$\omega_{4t} \approx -0.55z_{3t} - 0.31z_{4t} + 0.60z_{5t} \quad \text{and} \quad \omega_{5t} \approx -0.61z_{4t} - 0.41z_{5t}.$$

Thus, not surprisingly, we see that the (stationary) nearly white noise variates in the canonical correlation analysis are essentially represented as certain linear combinations of only the *stationary* components z_{3t}, z_{4t}, z_{5t} resulting from the cointegration-common trends analysis associated with the error-correction model form. By contrast, in the linear relations that connect the nearly nonstationary components, ω_{1t} and ω_{2t}, of the canonical correlation analysis with the transformed variates Z_t, it is seen that ω_{1t} has a large contribution from the first common trend component z_{1t} and ω_{2t} has a large contribution from z_{2t}.

Also notice that the final error-correction form of the model with the additional reduced-rank structure on $\mathbf{\Phi}_1^*$ imposed, that is, $W_t = ABY_{t-1}+EFW_{t-1}+\epsilon_t$, can be rewritten in the original AR(2) form as $Y_t = (I + AB + EF)Y_{t-1} - EFY_{t-2} + \epsilon_t$.

Hence, in this model, any possible white noise structure could be revealed as a linear combination $\ell' Y_t \approx \ell' \epsilon_t$ for which the relations $\ell'(I + AB + EF) \approx 0$ and $\ell'(EF) \approx 0$ hold. However, although the analysis based on the error-correction model form is not particularly designed to identify possible white noise component structure, we see from the discussion in the preceding paragraph that such white noise components (if present) would come from linear combinations of the cointegrating stationary component vector $Z_{2t} = Q_2' Y_t$ in the error-correction model analysis.

Chapter 6
The Growth Curve Model and Reduced-Rank Regression Methods

6.1 Introduction and the Growth Curve Model

One additional general model class that has aspects of reduced-rank regression, especially in its mathematical structure, is that of the growth curve or generalized multivariate ANOVA (GMANOVA) model. This type of model is often applied in the analysis of longitudinal or repeated measures data arising in biomedical and other areas. For these models, the components y_{jk} of the m-dimensional response vector $Y_k = (y_{1k}, \ldots, y_{mk})'$ usually correspond to responses on a single characteristic of a subject made over m distinct times or occasions, with interest in studying features of the mean response over time, and a sample of T independent subjects is available. Corresponding to each subject is an n-dimensional set of time-invariant explanatory variables X_k, which may include factors for treatment assignments held fixed over all occasions. In addition, it may be postulated that the pattern of mean response over time for any subject can be represented by some parametric function, e.g., polynomial function, of time whose coefficients are specified to be linear functions of the time-invariant explanatory variables X_k. That is, it is assumed that $E(y_{jk}) = a_j' \beta_k$, where a_j' is a known (specified) $1 \times r$ vector whose elements are functions of the jth time, for $j = 1, \ldots, m$, and the coefficients in β_k are unknown linear functions of X_k. Let $A = [a_1, \ldots, a_m]'$ denote the known $m \times r$ matrix, $r < \min(m, n)$, whose columns contain values of the functions of time specified in the parametric form of the mean response function over time. Then the model (2.5),

$$Y_k = A(BX_k) + \epsilon_k, \qquad k = 1, \ldots, T, \tag{6.1}$$

with unknown $r \times n$ matrix B, corresponds to this situation with $\beta_k \equiv BX_k$ interpretable as the coefficients in the mean response function over time for the kth subject, represented as linear functions of the X_k. Another interpretation for the

G. C. Reinsel et al., *Multivariate Reduced-Rank Regression*, Lecture Notes in Statistics 225, https://doi.org/10.1007/978-1-0716-2793-8_6

model is that in the standard multivariate regression model, $Y_k = CX_k + \epsilon_k$, the rows of elements of the regression coefficient matrix C are specified to be linear functions of a known basis functions of time, such as (orthogonal) polynomials up to degree $r-1$. That is, the jth row of C is specified as $C'_{(j)} = a'_j B$, where a'_j is a known $1 \times r$ vector whose elements are functions of the jth time, for $j = 1, \ldots, m$, and B is an $r \times n$ matrix of unknown parameters. Thus $C = AB$, with A specified (known), and hence, the model is again $Y_k = ABX_k + \epsilon_k$.

To briefly illustrate a situation more specifically, suppose the T individuals are from n different experimental or treatment groups, and there are no other explanatory variables. Then the $n \times T$ overall across-individual design matrix $\mathbf{X}$ is simply a matrix of indicator variables as in the one-way ANOVA situation. Assume T_i vector observations Y_{il} are available for the ith group, for $l = 1, \ldots, T_i$, with $T = T_1 + \cdots + T_n$, and observations in each group are measured at the same m time points, $t_1 < t_2 < \cdots < t_m$. In addition, we assume a model for the growth curves over time for each group to be of the form of a polynomial of degree $r-1$, so that the mean value at the jth time point for the lth individual in the ith group is given by

$$E(y_{jil}) = \beta_{1i} + \beta_{2i} t_j + \beta_{3i} t_j^2 + \cdots + \beta_{ri} t_j^{r-1} \equiv a'_j \beta_i,$$

with $a'_j = (1, t_j, \ldots, t_j^{r-1})$ and $\beta_i = (\beta_{1i}, \beta_{2i}, \ldots, \beta_{ri})'$. Then the basic (polynomial) growth curve model for all individuals can be written in matrix form as $\mathbf{Y} = AB\mathbf{X} + \boldsymbol{\epsilon}$, where $\mathbf{Y} = [Y_{11}, \ldots, Y_{1T_1}, \ldots, Y_{n1}, \ldots, Y_{nT_n}]$,

$$A = \begin{bmatrix} 1 & t_1 & \cdot & \cdot & t_1^{r-1} \\ 1 & t_2 & \cdot & \cdot & t_2^{r-1} \\ \cdot & \cdot & \cdot & \cdot & \cdot \\ \cdot & \cdot & \cdot & \cdot & \cdot \\ 1 & t_m & \cdot & \cdot & t_m^{r-1} \end{bmatrix}, \qquad B = \begin{bmatrix} \beta_{11} & \cdot & \cdot & \cdot & \beta_{1n} \\ \beta_{21} & \cdot & \cdot & \cdot & \beta_{2n} \\ \cdot & \cdot & \cdot & \cdot & \cdot \\ \cdot & \cdot & \cdot & \cdot & \cdot \\ \beta_{r1} & \cdot & \cdot & \cdot & \beta_{rn} \end{bmatrix},$$

$$\mathbf{X} = \begin{bmatrix} \mathbf{1}'_{T_1} & 0' & 0' & \cdot\cdot & 0' \\ 0' & \mathbf{1}'_{T_2} & 0' & \cdot\cdot & 0' \\ \cdot & \cdot & \cdot & \cdot\cdot & \cdot \\ \cdot & \cdot & \cdot & \cdot\cdot & \cdot \\ 0' & 0' & 0' & \cdot\cdot & \mathbf{1}'_{T_n} \end{bmatrix},$$

and $\mathbf{1}_T$ denotes a $T \times 1$ vector of ones.

Techniques for analysis of the basic growth curve model (6.1), and for various extensions of this model, will be described in the remainder of this chapter. Analysis for the growth curve model (6.1) is somewhat simpler than that of the general reduced-rank model (2.5) of Chapter 2, in that the matrix A is specified as known in the growth curve model, but the two cases still share some mathematical similarities. Moreover, application of the general model (2.5), with A *unknown* or unspecified, to the growth curve or longitudinal data setting might be useful or desirable in some situations, especially when m and n are relatively large. The interpretation could be that the columns of A represent the (unknown) basis functions for the "parametric" form of the mean response function over time, which are not explicitly specified. The growth curve model was introduced by Potthoff and Roy (1964) and was also studied extensively in its early stages by Rao (1965, 1966, 1967), Khatri (1966), and Grizzle and Allen (1969).

In some uses of the model (6.1), the covariance matrix $\Sigma_{\epsilon\epsilon}$ of the errors might be assumed to possess some special structure, rather than a general $m \times m$ covariance matrix. A particular case results from the random effects model assumption for the kth individual's regression coefficients (e.g., Laird and Ware 1982; Reinsel 1985). For this case, it is assumed that $Y_k = A\beta_k + u_k$, where β_k is a random vector with $E(\beta_k) = BX_k$ and $\text{Cov}(\beta_k) = \Sigma_\beta$, and the error vector u_k has $\text{Cov}(u_k) = \sigma^2 I_m$. Hence, with $\tau_k = \beta_k - BX_k$, we have $Y_k = ABX_k + A\tau_k + u_k \equiv ABX_k + \epsilon_k$, where $\epsilon_k = A\tau_k + u_k$ has covariance structure $\Sigma_{\epsilon\epsilon} = \text{Cov}(A\tau_k + u_k) = A\Sigma_\beta A' + \sigma^2 I_m$. Sometimes in the above model, only a subset of β_k is assumed to be random. Other types of special covariance structure, including AR and MA time series structures, are discussed by Jennrich and Schluchter (1986), Diggle (1988), Chi and Reinsel (1989), Rochon and Helms (1989), and Jones (1986, 1990).

6.2 Estimation of Parameters in the Growth Curve Model

Some basic estimation results for the growth curve model written in matrix form, $\mathbf{Y} = AB\mathbf{X} + \boldsymbol{\epsilon}$, will now be presented. Under normality and independence of the errors ϵ_k, we know from Section 2.3 that the ML estimate of B is the value that minimizes $|(\mathbf{Y} - AB\mathbf{X})(\mathbf{Y} - AB\mathbf{X})'|$ with respect to variation of B. It will be established shortly that the ML estimate of B is given by

$$\widehat{B} = (A'\widetilde{\Sigma}_{\epsilon\epsilon}^{-1}A)^{-1}A'\widetilde{\Sigma}_{\epsilon\epsilon}^{-1}\mathbf{Y}\mathbf{X}'(\mathbf{X}\mathbf{X}')^{-1} \equiv (A'\widetilde{\Sigma}_{\epsilon\epsilon}^{-1}A)^{-1}A'\widetilde{\Sigma}_{\epsilon\epsilon}^{-1}\widetilde{C}, \tag{6.2}$$

where $\widetilde{C} = \mathbf{Y}\mathbf{X}'(\mathbf{X}\mathbf{X}')^{-1}$ and $\widetilde{\Sigma}_{\epsilon\epsilon} = (1/T)(\mathbf{Y} - \widetilde{C}\mathbf{X})(\mathbf{Y} - \widetilde{C}\mathbf{X})' \equiv (1/T)S$. The corresponding ML estimate of the general error covariance matrix $\Sigma_{\epsilon\epsilon}$ is thus given by

$$\begin{aligned}\widehat{\Sigma}_{\epsilon\epsilon} &= \frac{1}{T}(\mathbf{Y} - A\widehat{B}\mathbf{X})(\mathbf{Y} - A\widehat{B}\mathbf{X})' = \widetilde{\Sigma}_{\epsilon\epsilon} + (\widetilde{C} - A\widehat{B})\widehat{\Sigma}_{xx}(\widetilde{C} - A\widehat{B})' \\ &= \widetilde{\Sigma}_{\epsilon\epsilon} + (I_m - A(A'\widetilde{\Sigma}_{\epsilon\epsilon}^{-1}A)^{-1}A'\widetilde{\Sigma}_{\epsilon\epsilon}^{-1})\widetilde{C}\widehat{\Sigma}_{xx}\widetilde{C}' \\ &\quad \times (I_m - A(A'\widetilde{\Sigma}_{\epsilon\epsilon}^{-1}A)^{-1}A'\widetilde{\Sigma}_{\epsilon\epsilon}^{-1})'.\end{aligned} \tag{6.3}$$

For derivation of the ML estimate of B, obtained by minimizing $|(\mathbf{Y}-AB\mathbf{X})(\mathbf{Y}-AB\mathbf{X})'|$, we first have the decomposition

$$
\begin{aligned}
(\mathbf{Y} - AB\mathbf{X})(\mathbf{Y} - AB\mathbf{X})' &= (\mathbf{Y} - AB\mathbf{X})[I - \mathbf{X}'(\mathbf{X}\mathbf{X}')^{-1}\mathbf{X}](\mathbf{Y} - AB\mathbf{X})' \\
&\quad + (\mathbf{Y} - AB\mathbf{X})\mathbf{X}'(\mathbf{X}\mathbf{X}')^{-1}\mathbf{X}(\mathbf{Y} - AB\mathbf{X})' \\
&= (\mathbf{Y} - \widetilde{C}\mathbf{X})(\mathbf{Y} - \widetilde{C}\mathbf{X})' \\
&\quad + (\widetilde{C} - AB)\mathbf{X}\mathbf{X}'(\widetilde{C} - AB)' \\
&= S + (\widetilde{C} - AB)\mathbf{X}\mathbf{X}'(\widetilde{C} - AB)',
\end{aligned}
$$

where $S = \mathbf{Y}[I - \mathbf{X}'(\mathbf{X}\mathbf{X}')^{-1}\mathbf{X}]\mathbf{Y}' = (\mathbf{Y} - \widetilde{C}\mathbf{X})(\mathbf{Y} - \widetilde{C}\mathbf{X})'$. So we obtain

$$
\begin{aligned}
|(\mathbf{Y} - AB\mathbf{X})(\mathbf{Y} - AB\mathbf{X})'| &= |S|\,|I_m + S^{-1}(\widetilde{C} - AB)\mathbf{X}\mathbf{X}'(\widetilde{C} - AB)'| \\
&= |S|\,|I_n + \mathbf{X}\mathbf{X}'(\widetilde{C} - AB)'S^{-1}(\widetilde{C} - AB)| \\
&= |S|\,|I_n + \widehat{\Sigma}_{xx}(\widetilde{C} - AB)'\widetilde{\Sigma}_{\epsilon\epsilon}^{-1}(\widetilde{C} - AB)|,
\end{aligned}
$$

where $\widehat{\Sigma}_{xx} = (1/T)\mathbf{X}\mathbf{X}'$. We can also write

$$
\begin{aligned}
(\widetilde{C} - AB)'\widetilde{\Sigma}_{\epsilon\epsilon}^{-1}(\widetilde{C} - AB) &= \widetilde{C}'[\widetilde{\Sigma}_{\epsilon\epsilon}^{-1} - \widetilde{\Sigma}_{\epsilon\epsilon}^{-1}A(A'\widetilde{\Sigma}_{\epsilon\epsilon}^{-1}A)^{-1}A'\widetilde{\Sigma}_{\epsilon\epsilon}^{-1}]\widetilde{C} \\
&\quad + (\widetilde{C} - AB)'\widetilde{\Sigma}_{\epsilon\epsilon}^{-1}A(A'\widetilde{\Sigma}_{\epsilon\epsilon}^{-1}A)^{-1}A'\widetilde{\Sigma}_{\epsilon\epsilon}^{-1}(\widetilde{C} - AB) \\
&= (\widetilde{C} - A\widehat{B})'\widetilde{\Sigma}_{\epsilon\epsilon}^{-1}(\widetilde{C} - A\widehat{B}) \\
&\quad + (\widehat{B} - B)'(A'\widetilde{\Sigma}_{\epsilon\epsilon}^{-1}A)(\widehat{B} - B) \\
&\equiv \mathbf{\Psi} + (\widehat{B} - B)'(A'\widetilde{\Sigma}_{\epsilon\epsilon}^{-1}A)(\widehat{B} - B),
\end{aligned}
$$

where $\widehat{B} = (A'\widetilde{\Sigma}_{\epsilon\epsilon}^{-1}A)^{-1}A'\widetilde{\Sigma}_{\epsilon\epsilon}^{-1}\widetilde{C}$, and $\mathbf{\Psi} = (\widetilde{C} - A\widehat{B})'\widetilde{\Sigma}_{\epsilon\epsilon}^{-1}(\widetilde{C} - A\widehat{B})$. Thus we find that

$$
\begin{aligned}
|(\mathbf{Y} - AB\mathbf{X})(\mathbf{Y} - AB\mathbf{X})'| &= |S|\,|I_n + \widehat{\Sigma}_{xx}\mathbf{\Psi}|\,|I_n + (I_n + \widehat{\Sigma}_{xx}\mathbf{\Psi})^{-1}\widehat{\Sigma}_{xx} \\
&\quad \times (\widehat{B} - B)'(A'\widetilde{\Sigma}_{\epsilon\epsilon}^{-1}A)(\widehat{B} - B)| \\
&= |S|\,|I_n + \widehat{\Sigma}_{xx}\mathbf{\Psi}|\,|I_n + (\widehat{\Sigma}_{xx}^{-1} + \mathbf{\Psi})^{-1} \\
&\quad \times (\widehat{B} - B)'(A'\widetilde{\Sigma}_{\epsilon\epsilon}^{-1}A)(\widehat{B} - B)|.
\end{aligned}
$$

Let $\delta_i^2 \geq 0$, $i = 1, \ldots, n$, be the eigenvalues of

$$
(\widehat{\Sigma}_{xx}^{-1} + \mathbf{\Psi})^{-1}(\widehat{B} - B)'(A'\widetilde{\Sigma}_{\epsilon\epsilon}^{-1}A)(\widehat{B} - B),
$$

so

$$|(\mathbf{Y} - AB\mathbf{X})(\mathbf{Y} - AB\mathbf{X})'| = |S|\,|I_n + \widehat{\Sigma}_{xx}\boldsymbol{\Psi}|\prod_{i=1}^{n}(1+\delta_i^2) \geq |S|\,|I_n + \widehat{\Sigma}_{xx}\boldsymbol{\Psi}|$$

with equality (under variation of B) if and only if $\delta_1^2 = \cdots = \delta_n^2 = 0$, that is, if and only if $A(\widehat{B} - B) = 0$. Since A is a full-rank matrix, this implies that the value of B given by $\widehat{B}$ above yields the minimum, so $\widehat{B} = (A'\widetilde{\Sigma}_{\epsilon\epsilon}^{-1}A)^{-1}A'\widetilde{\Sigma}_{\epsilon\epsilon}^{-1}\widetilde{C} \equiv (A'S^{-1}A)^{-1}A'S^{-1}\widetilde{C}$ is the ML estimator of B.

We briefly indicate a few simplifications that occur in the estimation for the special case mentioned in Section 6.1 where the across-individual design matrix $\mathbf{X}$ is a matrix of indicator variables corresponding to n different treatment groups, that is, $\mathbf{X} = \operatorname{diag}(\mathbf{1}'_{T_1}, \mathbf{1}'_{T_2}, \ldots, \mathbf{1}'_{T_n})$. Then we have $\mathbf{X}\mathbf{X}' = \operatorname{diag}(T_1, \ldots, T_n)$ and

$$\widetilde{C} = \mathbf{Y}\mathbf{X}'(\mathbf{X}\mathbf{X}')^{-1} = [\bar{Y}_1, \ldots, \bar{Y}_n] \equiv \bar{Y},$$

where $\bar{Y}_i = \sum_{j=1}^{T_i} Y_{ij}/T_i$ is the sample mean vector for the ith group, $i = 1, \ldots, n$. Hence, the ML estimator of B is $\widehat{B} = (A'\widetilde{\Sigma}_{\epsilon\epsilon}^{-1}A)^{-1}A'\widetilde{\Sigma}_{\epsilon\epsilon}^{-1}\bar{Y}$, where $\widetilde{\Sigma}_{\epsilon\epsilon} = (1/T)S$. In addition,

$$S = \sum_{i=1}^{n}\sum_{j=1}^{T_i}(Y_{ij} - \bar{Y}_i)(Y_{ij} - \bar{Y}_i)'$$

is the usual "within-group" error sum of squares matrix, and from (6.3), the ML estimate of the error covariance matrix $\Sigma_{\epsilon\epsilon}$ is $\widehat{\Sigma}_{\epsilon\epsilon} = (1/T)S + (1/T)\sum_{i=1}^{n} T_i(\bar{Y}_i - A\widehat{\beta}_i)(\bar{Y}_i - A\widehat{\beta}_i)'$, where $\widehat{\beta}_i = (A'\widetilde{\Sigma}_{\epsilon\epsilon}^{-1}A)^{-1}A'\widetilde{\Sigma}_{\epsilon\epsilon}^{-1}\bar{Y}_i$ is the ith column of $\widehat{B}$.

In the original study of the growth curve model (6.1) by Potthoff and Roy (1964), the following reduction of the model was adopted for the analysis. For an arbitrary $m \times m$ positive-definite matrix Γ, the equation in (6.1) is multiplied by $(A'\Gamma A)^{-1}A'\Gamma$ to obtain

$$(A'\Gamma A)^{-1}A'\Gamma Y_k = BX_k + (A'\Gamma A)^{-1}A'\Gamma\epsilon_k.$$

This can be written as

$$Y_k^* = BX_k + \epsilon_k^*, \qquad k = 1, \ldots, T, \tag{6.4}$$

which is in the form of the classical multivariate regression model. Therefore, from the classical regression framework, the LS estimator of B in (6.4) is $\widetilde{B} = \mathbf{Y}^*\mathbf{X}'(\mathbf{X}\mathbf{X}')^{-1} = (A'\Gamma A)^{-1}A'\Gamma\widetilde{C}$, where $\mathbf{Y}^* = (A'\Gamma A)^{-1}A'\Gamma\mathbf{Y}$. The optimum choice for Γ is $\Gamma = \Sigma_{\epsilon\epsilon}^{-1}$, which will yield the best linear unbiased estimator of B in model (6.1). With $\Sigma_{\epsilon\epsilon}$ unknown, we see from the previous derivations that the ML estimator $\widehat{B}$ in (6.2) has the same form as $\widetilde{B}$ using $\Gamma = \widetilde{\Sigma}_{\epsilon\epsilon}^{-1}$. Notice also that if the known matrix A was "normalized" so that the columns of $V = \Gamma^{1/2}A$ are orthonormal, that is, $A'\Gamma A = V'V = I_r$, then the above estimator $\widetilde{B}$ takes the similar form as in (2.17) of Section 2.3 for the reduced-rank model, $\widetilde{B} = V'\Gamma^{1/2}\widetilde{C}$, but with the known matrix $V = \Gamma^{1/2}A$ in place of an estimated matrix $\widehat{V}$. In addition, for both the reduced-rank and the growth curve model, we have found that the ML estimator is of the same form and is obtained with the choice $\Gamma = \widetilde{\Sigma}_{\epsilon\epsilon}^{-1}$. Finally, we mention that if the error covariance matrix $\Sigma_{\epsilon\epsilon}$ has a special "random effects" structure such as $\Sigma_{\epsilon\epsilon} = A\Sigma_\beta A' + \sigma^2 I_m$, as discussed in Section 6.1, then it follows (e.g., see Reinsel 1985) that the ML estimator of B under this model reduces to the LS estimator $\widehat{B} = (A'A)^{-1}A'\mathbf{YX}'(\mathbf{XX}')^{-1}$.

To further interpret the ML estimator of B and examine its statistical properties, we can also consider a transformation of the model (6.1) to a conditional model. Let $A_1 = A(A'A)^{-1}$ and A_2 be $m \times r$ and $m \times (m-r)$ matrices, respectively, of full ranks such that $A_1'A = I_r$ and $A_2'A = 0$, which also imply that $A_2'A_1 = 0$. The choice of A_2 as the column basis of $I_m - A(A'A)^{-1}A'$ will satisfy the above conditions. Then let $Y_{1k} = A_1'Y_k$ and $Y_{2k} = A_2'Y_k$, so that the original model (6.1) can be transformed into

$$\begin{bmatrix} Y_{1k} \\ Y_{2k} \end{bmatrix} = \begin{bmatrix} BX_k \\ 0 \end{bmatrix} + \begin{bmatrix} A_1'\epsilon_k \\ A_2'\epsilon_k \end{bmatrix}. \tag{6.5}$$

The form (6.5) indicates that the information on B is contained within the relationship between the variables Y_{1k} and X_k and the transformed responses Y_{2k}, which are referred to as the set of covariates, have mean values that do not involve B, but Y_{2k} may be correlated with Y_{1k}. The conditional model for Y_{1k}, given Y_{2k}, which follows from (6.5), can be written as

$$Y_{1k} = BX_k + \Theta Y_{2k} + \epsilon_{1k}^*, \tag{6.6}$$

where $\Theta = \mathrm{Cov}(Y_{1k}, Y_{2k})\{\mathrm{Cov}(Y_{2k})\}^{-1} = (A_1'\Sigma_{\epsilon\epsilon}A_2)(A_2'\Sigma_{\epsilon\epsilon}A_2)^{-1}$ and the r-dimensional error term ϵ_{1k}^* has zero mean vector and covariance matrix $\Sigma_1^* = (A_1'\Sigma_{\epsilon\epsilon}A_1) - (A_1'\Sigma_{\epsilon\epsilon}A_2)(A_2'\Sigma_{\epsilon\epsilon}A_2)^{-1}(A_2'\Sigma_{\epsilon\epsilon}A_1) \equiv (A'\Sigma_{\epsilon\epsilon}^{-1}A)^{-1}$. The last equality follows from a matrix result presented in Lemma 2b of Rao (1967). The (conditional) model (6.6) is again in the form of the classical regression set up with two sets of predictors. From the discussion in Sections 1.3 and 3.1, it follows that the LS and ML estimators of B in (6.6) are $\widehat{B} = \widehat{\Sigma}_{y_1x.y_2}\widehat{\Sigma}_{xx.y_2}^{-1}$, where $\widehat{\Sigma}_{y_1x.y_2}$ and $\widehat{\Sigma}_{xx.y_2}$ are sample partial covariance matrices adjusted for Y_{2k}. It will be established below that this estimator can be expressed equivalently in the form in (6.2), namely $\widehat{B} = (A'\widetilde{\Sigma}_{\epsilon\epsilon}^{-1}A)^{-1}A'\widetilde{\Sigma}_{\epsilon\epsilon}^{-1}\widetilde{C}$ with $\widetilde{C} = \widehat{\Sigma}_{yx}\widehat{\Sigma}_{xx}^{-1}$. Because the marginal distribution of Y_{2k} does not involve B, it follows that the ML estimator of B from the conditional model (6.6) is also the ML estimator for the unconditional model (6.1), as has already been established.

To verify the equivalence in the two forms of the estimator $\widehat{B}$, we first note that the LS estimator of Θ in (6.6) is obtained as

$$\widehat{\Theta} = \widehat{\Sigma}_{y_1y_2.x}\widehat{\Sigma}_{y_2y_2.x}^{-1} = (A_1'\widehat{\Sigma}_{yy.x}A_2)(A_2'\widehat{\Sigma}_{yy.x}A_2)^{-1}$$

$$\equiv (A_1' \widetilde{\Sigma}_{\epsilon\epsilon} A_2)(A_2' \widetilde{\Sigma}_{\epsilon\epsilon} A_2)^{-1}, \tag{6.7}$$

where $\widehat{\Sigma}_{yy.x} = \widehat{\Sigma}_{yy} - \widehat{\Sigma}_{yx} \widehat{\Sigma}_{xx}^{-1} \widehat{\Sigma}_{xy} \equiv \widetilde{\Sigma}_{\epsilon\epsilon}$. It then follows, from the normal equations for LS estimation, that the LS estimator of B in (6.6), $\widehat{B} = \widehat{\Sigma}_{y_1 x.y_2} \widehat{\Sigma}_{xx.y_2}^{-1}$, can be equivalently expressed as

$$\begin{aligned}
\widehat{B} &= (\widehat{\Sigma}_{y_1 x} - \widehat{\Theta} \widehat{\Sigma}_{y_2 x}) \widehat{\Sigma}_{xx}^{-1} = [A_1' \widehat{\Sigma}_{yx} - (A_1' \widetilde{\Sigma}_{\epsilon\epsilon} A_2)(A_2' \widetilde{\Sigma}_{\epsilon\epsilon} A_2)^{-1} A_2' \widehat{\Sigma}_{yx}] \widehat{\Sigma}_{xx}^{-1} \\
&= [A_1' \widetilde{\Sigma}_{\epsilon\epsilon} - (A_1' \widetilde{\Sigma}_{\epsilon\epsilon} A_2)(A_2' \widetilde{\Sigma}_{\epsilon\epsilon} A_2)^{-1} A_2' \widetilde{\Sigma}_{\epsilon\epsilon}] \widetilde{\Sigma}_{\epsilon\epsilon}^{-1} \widehat{\Sigma}_{yx} \widehat{\Sigma}_{xx}^{-1} \\
&= (A_1' A_1)(A_1' \widetilde{\Sigma}_{\epsilon\epsilon}^{-1} A_1)^{-1} A_1' \widetilde{\Sigma}_{\epsilon\epsilon}^{-1} \widehat{\Sigma}_{yx} \widehat{\Sigma}_{xx}^{-1} \equiv (A' \widetilde{\Sigma}_{\epsilon\epsilon}^{-1} A)^{-1} A' \widetilde{\Sigma}_{\epsilon\epsilon}^{-1} \widetilde{C},
\end{aligned} \tag{6.8}$$

where $\widetilde{C} = \widehat{\Sigma}_{yx} \widehat{\Sigma}_{xx}^{-1}$, and the result in Lemma 2b of Rao (1967) has again been used. Thus the LS estimator of B from the conditional model approach of (6.6) is the ML estimator of B in model (6.1) and can be expressed in the equivalent form as in (6.2).

We note from (6.8) that the ML estimator of B in model (6.1) can be expressed as

$$\begin{aligned}
\widehat{B} &= A_1' \widehat{\Sigma}_{yx} \widehat{\Sigma}_{xx}^{-1} - \widehat{\Theta} A_2' \widehat{\Sigma}_{yx} \widehat{\Sigma}_{xx}^{-1} \\
&= (A'A)^{-1} A' \mathbf{Y}\mathbf{X}' (\mathbf{X}\mathbf{X}')^{-1} - \widehat{\Theta} A_2' \mathbf{Y}\mathbf{X}' (\mathbf{X}\mathbf{X}')^{-1},
\end{aligned} \tag{6.9}$$

with $\widehat{\Theta} = (A_1' \widetilde{\Sigma}_{\epsilon\epsilon} A_2)(A_2' \widetilde{\Sigma}_{\epsilon\epsilon} A_2)^{-1}$. In this form, $\widehat{B}$ is interpretable as a *covariance adjusted* estimator of the simple LS estimator for model (6.1), $\widetilde{B} = (A'A)^{-1} A' \mathbf{Y}\mathbf{X}' (\mathbf{X}\mathbf{X}')^{-1}$, where the adjustment involves use of the (mean-zero) covariates $\mathbf{Y}_2 = A_2' \mathbf{Y}$ through their (estimated) regression relationship with $\mathbf{Y}_1 = A_1' \mathbf{Y} = (A'A)^{-1} A' \mathbf{Y}$. The interpretation and development of properties of the estimator $\widehat{B}$ through the covariance adjustment point of view is examined in more detail by Rao (1967) and Grizzle and Allen (1969). This includes the notion of constructing estimators of B through covariance adjustment as in (6.9), but using only a subset of the covariates in $\mathbf{Y}_2 = A_2' \mathbf{Y}$, whose proper choice would depend on knowledge of special structure for the covariance matrix $\Sigma_{\epsilon\epsilon}$. Specifically, covariates having zero (or very small) covariance with the response variables $\mathbf{Y}_1 = A_1' \mathbf{Y}$ and with the remaining covariates in $\mathbf{Y}_2 = A_2' \mathbf{Y}$ would be omitted in (6.9), the idea being that the use of "irrelevant" covariates in (6.9) would result in an increase in variability of the estimator $\widehat{B}$, and hence, such covariates should be eliminated in the estimation of B.

Notice that an alternate way to view the growth curve model (6.1) is as having the form of the standard multivariate regression model $Y_k = C X_k + \epsilon_k$, with the regression coefficient matrix $C \equiv AB$ constrained or specified to satisfy the linear restriction $A_2' C = 0$ (since $A_2' A = 0$). From the result in (1.18) of Section 1.3 (with $F_1 \equiv A_2'$ and $G_2 \equiv I_n$), it follows that the constrained ML estimator of C under $A_2' C = 0$ can be written as

$$\widehat{C} = \widetilde{C} - SA_2(A_2'SA_2)^{-1}A_2'\widetilde{C},$$

where $\widetilde{C} = \mathbf{YX}'(\mathbf{XX}')^{-1}$ is the usual LS estimator of C. Hence, because in the growth curve model, C is represented in the form $C = AB$ with A known, this implies that the ML estimator of B can be obtained from $\widehat{C}$ ($\equiv A\widehat{B}$) as

$$\widehat{B} = (A'A)^{-1}A'\widehat{C} = (A'A)^{-1}A'\widetilde{C} - (A'A)^{-1}A'SA_2(A_2'SA_2)^{-1}A_2'\widetilde{C},$$

which is identical to the expression previously obtained in (6.9), recalling that $A_1' = (A'A)^{-1}A'$.

Under normality of the ϵ_k, the ML estimator $\widehat{B}$ given by (6.2) or (6.8) can be seen to be unbiased, i.e., $E(\widehat{B}) = B$, using the property that $\widetilde{C}$ and $\widetilde{\Sigma}_{\epsilon\epsilon}$ are independently distributed and $E(\widetilde{C}) = C \equiv AB$. The independence property can also be used to help establish that the covariance matrix of $\widehat{B}$ is given by

$$\text{Cov}[\text{vec}(\widehat{B})] = \frac{T-n-1}{T-n-1-(m-r)}(\mathbf{XX}')^{-1} \otimes (A'\Sigma_{\epsilon\epsilon}^{-1}A)^{-1},$$

and an unbiased estimator of $(A'\Sigma_{\epsilon\epsilon}^{-1}A)^{-1}$ is $\frac{T}{T-n-(m-r)}(A'\widetilde{\Sigma}_{\epsilon\epsilon}^{-1}A)^{-1}$. (For example, see Rao (1967) and Grizzle and Allen (1969) for details.) Although the first two moments of $\widehat{B}$ are known, the complete distribution is not available in simple form. Gleser and Olkin (1966, 1970) and Kabe (1975) obtained expressions for the probability density of the ML estimator $\widehat{B}$.

6.3 Likelihood Ratio Testing of Linear Hypotheses in Growth Curve Model

We now present and derive the LR test statistic for $H_0 : F_1BG_2 = 0$, where F_1 and G_2 are known full-rank matrices of dimensions $r_1 \times r$ and $n \times n_2$, respectively, with $r_1 \leq r$ and $n_2 \leq n$. We also want to give the associated asymptotic distribution theory for the LR statistic. These results can be obtained from testing results for the classical multivariate linear regression model, as given in Section 1.3, applied to the conditional model (6.6). The hypothesis $H_0 : F_1BG_2 = 0$ for model (6.1) can be expressed in terms of the conditional model (6.6) as

$$H_0 : F_1[B, \Theta]\begin{bmatrix} G_2 \\ 0 \end{bmatrix} \equiv F_1DG_2^* = 0,$$

where $D = [B, \Theta]$ and $G_2^* = [G_2', 0']'$. Note that the conditional model (6.6) can be written as $Y_{1k} = [B, \Theta]X_k^* + \epsilon_{1k}^* = DX_k^* + \epsilon_{1k}^*$, where $X_k^* = [X_k', Y_{2k}']'$, which is in the framework of the classical multivariate regression model (1.1). So $H_0 : F_1DG_2^* = 0$ can be tested using the results given in Section 1.3.

The unconstrained ML estimator of D is the LS estimator

$$\widetilde{D} = \mathbf{Y}_1\mathbf{X}^{*\prime}(\mathbf{X}^*\mathbf{X}^{*\prime})^{-1},$$

where $\mathbf{X}^{*\prime} = [\mathbf{X}', \mathbf{Y}_2']$, which yields the estimators $\widehat{B}$ and $\widehat{\Theta}$ discussed in Section 6.2. That is, $\widetilde{D} = [\widehat{B}, \widehat{\Theta}]$ where $\widehat{B}$ and $\widehat{\Theta}$ are the (LS) estimators given in (6.7) and (6.8) of Section 6.2. The constrained ML estimator of D, and hence of B, under $H_0 : F_1BG_2 \equiv F_1DG_2^* = 0$ can be obtained from the expression in (1.18) of Section 1.3 with the appropriate changes in notation. The error sum of square matrix from (unconstrained) LS estimation of the conditional model is $S_1^* = \mathbf{Y}_1[I - \mathbf{X}^{*\prime}(\mathbf{X}^*\mathbf{X}^{*\prime})^{-1}\mathbf{X}^*]\mathbf{Y}_1'$, where $\mathbf{Y}_1 = A_1'\mathbf{Y}$. However, notice that

$$\begin{aligned}
\mathbf{Y}_1[I - \mathbf{X}^{*\prime}(\mathbf{X}^*\mathbf{X}^{*\prime})^{-1}\mathbf{X}^*] &= \mathbf{Y}_1 - \widetilde{D}\mathbf{X}^* = A_1'\mathbf{Y} - \widehat{B}\mathbf{X} - \widehat{\Theta}\mathbf{Y}_2 \\
&= [A_1' - (A_1'SA_2)(A_2'SA_2)^{-1}A_2']\mathbf{Y} - \widehat{B}\mathbf{X} \\
&= (A'S^{-1}A)^{-1}A'S^{-1}\mathbf{Y} \\
&\quad - (A'S^{-1}A)^{-1}A'S^{-1}\mathbf{Y}\mathbf{X}'(\mathbf{X}\mathbf{X}')^{-1}\mathbf{X} \\
&= (A'S^{-1}A)^{-1}A'S^{-1}\mathbf{Y}[I - \mathbf{X}'(\mathbf{X}\mathbf{X}')^{-1}\mathbf{X}],
\end{aligned}$$

again using the result from Lemma 2b of Rao (1967) in the same way as in (6.8). Hence, the error sum of squares matrix can be expressed as

$$\begin{aligned}
S_1^* &= \mathbf{Y}_1[I - \mathbf{X}^{*\prime}(\mathbf{X}^*\mathbf{X}^{*\prime})^{-1}\mathbf{X}^*]\mathbf{Y}_1' \\
&= (A'S^{-1}A)^{-1}A'S^{-1}\mathbf{Y}[I - \mathbf{X}'(\mathbf{X}\mathbf{X}')^{-1}\mathbf{X}]\mathbf{Y}'A_1 \\
&= (A'S^{-1}A)^{-1}A'S^{-1}SA_1 = (A'S^{-1}A)^{-1},
\end{aligned} \tag{6.10}$$

recalling that $A'A_1 = I_r$. The hypothesis sum of squares matrix associated with $H_0 : F_1BG_2 \equiv F_1DG_2^* = 0$ is, from expression (1.19) of Section 1.3, given by

$$\begin{aligned}
H &= S_1^*F_1'(F_1S_1^*F_1')^{-1}(F_1\widetilde{D}G_2^*)[G_2^{*\prime}(X^*X^{*\prime})^{-1}G_2^*]^{-1} \\
&\quad \times (F_1\widetilde{D}G_2^*)'(F_1S_1^*F_1')^{-1}F_1S_1^* \\
&= S_1^*F_1'(F_1S_1^*F_1')^{-1}(F_1\widehat{B}G_2)[G_2'\Omega G_2]^{-1} \\
&\quad \times (F_1\widehat{B}G_2)'(F_1S_1^*F_1')^{-1}F_1S_1^*,
\end{aligned} \tag{6.11}$$

where $\Omega = [I_n, 0](\mathbf{X}^*\mathbf{X}^{*\prime})^{-1}[I_n, 0]'$ is the leading $n \times n$ submatrix of $(\mathbf{X}^*\mathbf{X}^{*\prime})^{-1}$, from which it is known that

$$\Omega = \{\mathbf{X}\mathbf{X}' - \mathbf{X}\mathbf{Y}_2'(\mathbf{Y}_2\mathbf{Y}_2')^{-1}\mathbf{Y}_2\mathbf{X}'\}^{-1} \equiv (1/T)\widehat{\Sigma}_{xx.y_2}^{-1}.$$

In addition, from a standard result on the inverse of a partitioned matrix (e.g., see Rao 1973, p. 33), with $E = \mathbf{Y}_2\mathbf{Y}_2' - \mathbf{Y}_2\mathbf{X}'(\mathbf{X}\mathbf{X}')^{-1}\mathbf{X}\mathbf{Y}_2' \equiv A_2'SA_2$, we also find that

$$\begin{aligned}\Omega &= (\mathbf{X}\mathbf{X}')^{-1} + (\mathbf{X}\mathbf{X}')^{-1}\mathbf{X}\mathbf{Y}_2'E^{-1}\mathbf{Y}_2\mathbf{X}'(\mathbf{X}\mathbf{X}')^{-1} \\ &= (\mathbf{X}\mathbf{X}')^{-1} + (\mathbf{X}\mathbf{X}')^{-1}\mathbf{X}\mathbf{Y}'A_2(A_2'SA_2)^{-1}A_2'\mathbf{Y}\mathbf{X}'(\mathbf{X}\mathbf{X}')^{-1} \\ &= (\mathbf{X}\mathbf{X}')^{-1} + (\mathbf{X}\mathbf{X}')^{-1}\mathbf{X}\mathbf{Y}' \\ &\quad \times [S^{-1} - S^{-1}A(A'S^{-1}A)^{-1}A'S^{-1}]\mathbf{Y}\mathbf{X}'(\mathbf{X}\mathbf{X}')^{-1} \\ &= (\mathbf{X}\mathbf{X}')^{-1} + (\mathbf{X}\mathbf{X}')^{-1}\mathbf{X}\mathbf{Y}'S^{-1}\mathbf{Y}\mathbf{X}'(\mathbf{X}\mathbf{X}')^{-1} \\ &\quad - \widehat{B}'(A'S^{-1}A)\widehat{B},\end{aligned} \tag{6.12}$$

using the matrix result from Lemma 1 of Khatri (1966) that

$$A_2(A_2'SA_2)^{-1}A_2' = S^{-1} - S^{-1}A(A'S^{-1}A)^{-1}A'S^{-1}$$

since $A_2'A = 0$.

As in Section 1.3, the LR test of H_0 is based on the ordered eigenvalues $\widehat{\lambda}_1^2 > \cdots > \widehat{\lambda}_{r_1}^2$ of the $r_1 \times r_1$ matrix $S_*^{-1}H_*$, where

$$H_* = F_1HF_1' = (F_1\widehat{B}G_2)[G_2'\Omega G_2]^{-1}(F_1\widehat{B}G_2)' \tag{6.13}$$

and $S_* = F_1S_1^*F_1' = F_1(A'S^{-1}A)^{-1}F_1'$. The LR test statistic is

$$\begin{aligned}\mathcal{M} &= -[T - n - (m - r) + \frac{1}{2}(n_2 - r_1 - 1)]\log(U) \\ &\equiv [T - n - (m - r) + \frac{1}{2}(n_2 - r_1 - 1)]\sum_{i=1}^{r_1}\log(1 + \widehat{\lambda}_i^2),\end{aligned}$$

which is asymptotically distributed as $\chi^2_{r_1n_2}$, where $U^{-1} = |I + S_*^{-1}H_*| = \prod_{i=1}^{r_1}(1+\widehat{\lambda}_i^2)$. The distribution of S_* is Wishart, $W(F_1\Sigma_1^*F_1', T - n - (m-r))$, and under H_0, the conditional distribution of H_* is also Wishart with matrix $F_1\Sigma_1^*F_1'$ and n_2 degrees of freedom, where $\Sigma_1^* = (A'\Sigma_{\epsilon\epsilon}^{-1}A)^{-1}$. Since this conditional null distribution for H_* does not depend on $\mathbf{X}^*$ (i.e., on $\mathbf{Y}_2$), it is also the unconditional null distribution of H_*. The above form of the LR test of $H_0 : F_1BG_2 = 0$ was first derived by Khatri (1966). Some additional aspects of the testing procedure were further examined by Grizzle and Allen (1969).

Recall that in the growth curve model (6.1), A represents the design matrix of the within-individual model and $\mathbf{X}$ is the usual across-individual design matrix. Various choices of the matrices F_1 and G_2 in $H_0 : F_1BG_2 = 0$ lead to special cases of tests of hypothesis of particular interest. These include testing for the order of polynomial in the within-individual model and testing for homogeneity or similarity in regression coefficients across different treatment groups. For instance, the within-individual model is often represented in terms of a polynomial regression over time points, of degree $r-1$, and suppose we wish to test for lower degree $r-2$ versus $r-1$. That is, we test for the significance of the highest order coefficient of the within-individual model, over all individuals. This can be represented as $H_0 : \beta'_{(r)} = 0$, where $\beta'_{(r)}$ denotes the rth row of B, and this is equivalent to $H_0 : F_1 B G_2 = 0$ with $F_1 = (0, \ldots, 0, 1)$ and $G_2 = I_n$. For this, we see from (6.13) that $H_* = (F_1 \widehat{B})\Omega^{-1}(F_1 \widehat{B})' = \widehat{\beta}'_{(r)}\Omega^{-1}\widehat{\beta}_{(r)}$, where $\widehat{\beta}'_{(r)}$ is the rth row of $\widehat{B}$, and $S_* = F_1(A'S^{-1}A)^{-1}F'_1$ is the rth diagonal element of $(A'S^{-1}A)^{-1}$. Hence, since $r_1 = 1$, the LR test is based on the scalar $H_*/S_* = \widehat{\beta}'_{(r)}\Omega^{-1}\widehat{\beta}_{(r)}/S_*$, which is distributed as $n/(T-n-(m-r))$ times the F-distribution with n and $T-n-(m-r)$ degrees of freedom under H_0.

For another example, suppose that the across-individual design matrix $\mathbf{X}$ corresponds to n distinct treatment groups as in a one-way multivariate ANOVA model, and we wish to test the null hypothesis of homogeneity of the growth curve parameters (or at least of a subset of these parameters) over the n groups. This is equivalent to the condition that the n columns of B be equal (or at least the lower $r_1 \times n$ submatrix of B has equal columns). This hypothesis can be expressed in the form $F_1 B G_2 = 0$ with $F_1 = [0, I_{r_1}]$ and

$$G_2 = \begin{bmatrix} 1 & 0 & \cdot & \cdot & 0 & 0 \\ 0 & 1 & \cdot & \cdot & 0 & 0 \\ \cdot & \cdot & \cdot & \cdot & \cdot & \cdot \\ \cdot & \cdot & \cdot & \cdot & \cdot & \cdot \\ 0 & 0 & \cdot & \cdot & 0 & 1 \\ -1 & -1 & \cdot & \cdot & -1 & -1 \end{bmatrix}$$

of dimension $n \times (n-1)$. For example, in the case with only $n = 2$ groups, we have $G_2 = (1, -1)'$ and $G'_2 \Omega G_2 = (1/T_1) + (1/T_2) + (1/T)\{\bar{Y}_1 - \bar{Y}_2 - A(\widehat{\beta}_1 - \widehat{\beta}_2)\}'\widetilde{\Sigma}_{\epsilon\epsilon}^{-1}\{\bar{Y}_1 - \bar{Y}_2 - A(\widehat{\beta}_1 - \widehat{\beta}_2)\}$, and with $r_1 = r$, the test statistic for $H_0 : BG_2 = 0$ (i.e., $\beta_1 - \beta_2 = 0$, where $B = [\beta_1, \beta_2]$) is based on the single nonzero eigenvalue of $S_*^{-1}H_*$, that is,

$$(\widehat{B}G_2)'(A'S^{-1}A)(\widehat{B}G_2)[G'_2\Omega G_2]^{-1} = \frac{1}{T}(\widehat{\beta}_1 - \widehat{\beta}_2)'A'\widetilde{\Sigma}_{\epsilon\epsilon}^{-1}A(\widehat{\beta}_1 - \widehat{\beta}_2)/[G'_2\Omega G_2].$$

Also notice that a test of goodness-of-fit or adequacy of specification of the within-individual model can be tested in the framework of the general multivariate linear model $Y_k = CX_k + \epsilon_k$, with C specified to be of the form $C = AB$ under the growth curve model (6.1). The LR test can be carried out using the standard methods discussed in Section 1.3 as a test of $H_0 : A'_2 C = 0$ in the model $Y_k = CX_k + \epsilon_k$, since this hypothesis is equivalent to the hypothesis that C has the form AB with A specified (and A_2 is the $m \times (m-r)$ matrix such that $A'_2 A = 0$). Thus, from Section 1.3, we find that the hypothesis sum of squares matrix for H_0 is

$$H_* = A_2'\widetilde{C}(\mathbf{XX}')\widetilde{C}'A_2 = A_2'\mathbf{YX}'(\mathbf{XX}')^{-1}\mathbf{XY}'A_2,$$

while the error sum of squares matrix is $S_* = A_2'SA_2$, with $S_* + H_* = A_2'\mathbf{YY}'A_2$. The LR test procedure uses the statistic $\mathcal{M} = -[T-n+(n-(m-r)-1)/2]\log(U)$, where $U = |S_*|/|S_* + H_*| = 1/|I_{m-r} + S_*^{-1}H_*|$, with $\mathcal{M}$ having an approximate $\chi^2_{n(m-r)}$ distribution under H_0. The LR can also be expressed in a form that avoids the explicit use of the matrix A_2. Specifically, from (6.3), we see that the $m \times m$ error sum of squares matrix under H_0 is

$$\begin{aligned} S_1 &\equiv S + H = (\mathbf{Y} - A\widehat{B}\mathbf{X})(\mathbf{Y} - A\widehat{B}\mathbf{X})' \\ &= S + (I_m - A(A'S^{-1}A)^{-1}A'S^{-1})\mathbf{YX}'(\mathbf{XX}')^{-1}\mathbf{XY}' \\ &\times (I_m - A(A'S^{-1}A)^{-1}A'S^{-1})'. \end{aligned}$$

Therefore, it follows that

$$\begin{aligned} U &= |S|/|S_1| = 1/|I_m + S^{-1}H| \\ &= 1/|I_m + S^{-1}(I_m - A(A'S^{-1}A)^{-1}A')S^{-1}\mathbf{YX}'(\mathbf{XX}')^{-1}\mathbf{XY}'|, \end{aligned}$$

so that U can be expressed in terms of the $(m - r)$ nonzero eigenvalues of the matrix $S^{-1}(I_m - A(A'S^{-1}A)^{-1}A')S^{-1}\mathbf{YX}'(\mathbf{XX}')^{-1}\mathbf{XY}'$. It is also shown by Kshirsagar and Smith (1995, p. 194–195) that U can be expressed equivalently as $U = \{|S||A'S^{-1}A|\}/\{|\mathbf{YY}'||A'(\mathbf{YY}')^{-1}A|\}$.

6.4 An Extended Model for Growth Curve Data

An extension of the basic growth curve (or generalized MANOVA) model (6.1) to allow a mixture of growth curve and MANOVA components has been considered by Chinchilli and Elswick (1985). In this extended model, the mean of the response vector Y_k is represented by the sum of two components, a growth curve portion and a standard MANOVA portion. The model could be viewed as an analysis of covariance model that adjusts the growth curve structure for the influence of additional covariates for the response vector. Specifically, the model considered is

$$Y_k = ABX_k + DZ_k + \epsilon_k, \qquad k = 1, \ldots, T, \tag{6.14}$$

where the $m \times 1$ response vector Y_k depends on the $n_1 \times 1$ vector X_k in the same way as in the basic growth curve model (6.1) and also on the $n_2 \times 1$ vector Z_k of additional covariates. However, the second set of covariates is not specified to enter the model with growth curve structure but simply with the standard multivariate linear model structure. This model may have appeal when it seems appropriate to express the effects of X_k in terms of growth curve structure over time, whereas it may not be desirable to express the effects of the additional covariates Z_k in this manner. Notice that the form of model (6.14) is similar to the reduced-rank model (3.3) of Section 3.1, but in the present growth curve model context, the matrix A is known. As usual, the error terms ϵ_k are assumed to follow a multivariate normal distribution with zero mean vector and positive-definite covariance matrix $\Sigma_{\epsilon\epsilon}$.

The maximum likelihood estimators of the unknown parameters B, D, and $\Sigma_{\epsilon\epsilon}$ of model (6.14) are derived by Chinchilli and Elswick (1985). We can use some of the tools that have been developed in Sections 3.1 and 6.2 to readily present the results. Let $C = AB$ and denote the LS estimates of C and D in the model $Y_k = CX_k + DZ_k + \epsilon_k$ by $\widetilde{C} = \widehat{\Sigma}_{yx.z}\widehat{\Sigma}_{xx.z}^{-1}$ and $\widetilde{D} = \widehat{\Sigma}_{yz.x}\widehat{\Sigma}_{zz.x}^{-1}$, respectively. The ML estimates of model (6.14) are obtained by minimizing $|W|$ or, equivalently, $|\widetilde{\Sigma}_{\epsilon\epsilon}^{-1}W|$, where

$$W = (1/T)(\mathbf{Y} - AB\mathbf{X} - D\mathbf{Z})(\mathbf{Y} - AB\mathbf{X} - D\mathbf{Z})'$$

and $\widetilde{\Sigma}_{\epsilon\epsilon} = (1/T)(\mathbf{Y} - \widetilde{C}\mathbf{X} - \widetilde{D}\mathbf{Z})(\mathbf{Y} - \widetilde{C}\mathbf{X} - \widetilde{D}\mathbf{Z})' \equiv \widehat{\Sigma}_{yy.(x,z)}$ as given in (3.10). As shown in Section 3.1, we have the decomposition

$$\begin{aligned} W = {} & \widetilde{\Sigma}_{\epsilon\epsilon} + (\widetilde{C} - AB)\widehat{\Sigma}_{xx.z}(\widetilde{C} - AB)' \\ & + (\widetilde{D}^* - D - C\widehat{\Sigma}_{xz}\widehat{\Sigma}_{zz}^{-1})\widehat{\Sigma}_{zz}(\widetilde{D}^* - D - C\widehat{\Sigma}_{xz}\widehat{\Sigma}_{zz}^{-1})', \end{aligned}$$

where $\widetilde{D}^* = \widetilde{D} + \widetilde{C}\widehat{\Sigma}_{xz}\widehat{\Sigma}_{zz}^{-1} \equiv \widehat{\Sigma}_{yz}\widehat{\Sigma}_{zz}^{-1}$. For any C, the last component above can be uniformly minimized, as the zero matrix, by the choice of $\widehat{D} = \widetilde{D}^* - C\widehat{\Sigma}_{xz}\widehat{\Sigma}_{zz}^{-1}$. Thus, the ML estimation of minimizing $|\widetilde{\Sigma}_{\epsilon\epsilon}^{-1}W|$ is reduced to minimizing

$$|I_m + \widetilde{\Sigma}_{\epsilon\epsilon}^{-1}(\widetilde{C} - AB)\widehat{\Sigma}_{xx.z}(\widetilde{C} - AB)'|$$

with respect to B. This is the same form of minimization problem as considered at the beginning of Section 6.2 for the basic growth curve model. Hence, from the same arguments as given there, it follows that $\widehat{B} = (A'\widetilde{\Sigma}_{\epsilon\epsilon}^{-1}A)^{-1}A'\widetilde{\Sigma}_{\epsilon\epsilon}^{-1}\widetilde{C}$, with $\widetilde{C} = \widehat{\Sigma}_{yx.z}\widehat{\Sigma}_{xx.z}^{-1}$, is the ML estimator of B in model (6.14). The corresponding ML estimator of D then is

$$\begin{aligned} \widehat{D} = \widetilde{D}^* - A\widehat{B}\widehat{\Sigma}_{xz}\widehat{\Sigma}_{zz}^{-1} & = \widehat{\Sigma}_{yz}\widehat{\Sigma}_{zz}^{-1} - A\widehat{B}\widehat{\Sigma}_{xz}\widehat{\Sigma}_{zz}^{-1} \\ & \equiv \widetilde{D} + (\widetilde{C} - A\widehat{B})\widehat{\Sigma}_{xz}\widehat{\Sigma}_{zz}^{-1}. \end{aligned} \quad (6.15)$$

It follows from the above decomposition that the ML estimate of the error covariance matrix is

$$\begin{aligned} \widehat{\Sigma}_{\epsilon\epsilon} & = \widetilde{\Sigma}_{\epsilon\epsilon} + (\widetilde{C} - A\widehat{B})\widehat{\Sigma}_{xx.z}(\widetilde{C} - A\widehat{B})' \\ & = \widetilde{\Sigma}_{\epsilon\epsilon} + (I_m - A(A'\widetilde{\Sigma}_{\epsilon\epsilon}^{-1}A)^{-1}A'\widetilde{\Sigma}_{\epsilon\epsilon}^{-1})\widetilde{C}\widehat{\Sigma}_{xx.z}\widetilde{C}' \end{aligned}$$

$$\times (I_m - A(A'\widetilde{\Sigma}_{\epsilon\epsilon}^{-1} A)^{-1} A'\widetilde{\Sigma}_{\epsilon\epsilon}^{-1})'.$$

As can be observed from the above results, the main differences between ML estimates of the parameters of the basic growth curve model (6.1) and ML estimates of the parameters of the extended model (6.14) are that the estimate $\widetilde{\Sigma}_{\epsilon\epsilon}$ of the error covariance matrix in (6.14) is obtained from the LS regression of Y_k on X_k and Z_k and the estimate of B involves the LS regression coefficient matrix $\widetilde{C} = \widehat{\Sigma}_{yx.z}\widehat{\Sigma}_{xx.z}^{-1}$ that is obtained after adjusting for the effect of Z_k on Y_k and X_k.

It is also possible to develop the conditional model approach for (6.14) in a similar way to model (6.6) for the basic growth curve model. Hypothesis testing under the extended model (6.14) can be developed along the same lines as in Section 6.3 for the model (6.1). In addition to the adjustments that were mentioned at the end of the last paragraph, the LR tests will involve an adjustment in the error degrees of freedom associated with the inclusion of the additional covariates Z_k in the model. We will not pursue the details for LR testing of linear hypotheses here. However, we do note that there are two separate tests concerning the issue of model adequacy that might be of interest for the extended model (6.14). The first type of test of hypothesis is similar to that mentioned at the end of Section 6.3, namely, a test of the adequacy of the growth curve feature (involving the term ABX_k) in the extended model (6.14) relative to a standard multivariate regression model $Y_k = CX_k + DZ_k + \epsilon_k$. The second type of test of interest would be the adequacy of a "combined" growth curve structure model of the form $Y_k = A(BX_k + B^* Z_k) + \epsilon_k \equiv A[B, B^*][X_k', Z_k']' + \epsilon_k$ relative to the extended model (6.14) that is a "mixture" of growth curve structure and standard multivariate regression.

Verbyla and Venables (1988) examined another extension of growth curve model (6.1) that allows for the sum of two (or more) growth curve or time profile model structures. For the sum of two component structures, the model has the form $Y_k = ABX_k + A^*B^*Z_k + v_k$, where A and A^* are known matrices specifying the growth curve or time profile structures of the coefficients for the two sets of regressors X_k and Z_k. This type of model can accommodate features such as allowance for different forms of growth profile for different sets of regressors (e.g., different forms for treatment design variables compared to measured covariates) and time-varying covariates (i.e., different covariate values at the different response times). An important special case occurs when the columns of A represent a subset of or are "nested" within the column space of A^*, such as when a lower-degree polynomial structure is specified for the coefficients of X_k than for those of Z_k. The extended model (6.14) of Chinchilli and Elswick (1985) can be seen to be a particular case of this situation with $A^* = I_m$. The ML estimation of these sum of profiles models in general requires iterative methods, although explicit closed-form estimation results can be obtained in special cases such as the nested model situation. We remark that the reduced-rank structures could also be considered for the parameter matrices B and B^* in the general sum of profiles setting. This could be viewed as an extension of the model in Chapter 3 and Velu (1991). A special case of the reduced-rank growth curve model setting for two sets of regressors, with $A = A^*$ and B of reduced rank, will be considered later.

6.5 Modification of Basic Growth Curve Model to Reduced-Rank Model

In order to relate the concepts in the growth curve model context to the reduced-rank modeling of Chapter 2, we need to recall the role of the matrix components A and B that appear in the basic growth curve model (6.1). As emphasized previously, the $m \times r$ matrix A is known or specified in the growth curve context and is taken to be the design matrix of the within-individual model. Because the elements of Y_k consist of repeated measurements of the same response variable over time, the columns of A are specified as known functions of time, typically polynomial terms. The $r \times n$ matrix B represents the regression coefficients associated with the known functions of time. The $n \times T$ matrix $\mathbf{X}$ is the usual across-individual design matrix. When the individuals are from n different groups and there are no additional explanatory variables, then $\mathbf{X}$ is simply a matrix of indicator variables as in the one-way ANOVA model.

To focus further on this specific situation, suppose there are n different experimental or treatment groups and T_i vector observations Y_{ij} are available for the ith group, for $j = 1, \ldots, T_i$. With observations in each group assumed to be measured at the same m time points, $t_1 < t_2 < \cdots < t_m$, the basic (polynomial) growth curve model can be written as $\mathbf{Y} = AB\mathbf{X} + \boldsymbol{\epsilon}$, where $\mathbf{Y} = [Y_{11}, \ldots, Y_{1T_1}, \ldots, Y_{n1}, \ldots, Y_{nT_n}]$,

$$A = \begin{bmatrix} 1 & t_1 & \cdot & \cdot & t_1^{r-1} \\ 1 & t_2 & \cdot & \cdot & t_2^{r-1} \\ \cdot & \cdot & \cdot & \cdot & \cdot \\ \cdot & \cdot & \cdot & \cdot & \cdot \\ 1 & t_m & \cdot & \cdot & t_m^{r-1} \end{bmatrix}, \qquad B = \begin{bmatrix} \beta_{11} & \cdot & \cdot & \cdot & \beta_{1n} \\ \beta_{21} & \cdot & \cdot & \cdot & \beta_{2n} \\ \cdot & \cdot & \cdot & \cdot & \cdot \\ \cdot & \cdot & \cdot & \cdot & \cdot \\ \beta_{r1} & \cdot & \cdot & \cdot & \beta_{rn} \end{bmatrix},$$

$$\mathbf{X} = \begin{bmatrix} \mathbf{1}'_{T_1} & 0' & 0' & \cdot & \cdot & 0' \\ 0' & \mathbf{1}'_{T_2} & 0' & \cdot & \cdot & 0' \\ \cdot & \cdot & \cdot & \cdot & \cdot & \cdot \\ \cdot & \cdot & \cdot & \cdot & \cdot & \cdot \\ 0' & 0' & 0' & \cdot & \cdot & \mathbf{1}'_{T_n} \end{bmatrix},$$

and $\mathbf{1}_T$ denotes a $T \times 1$ vector of ones.

In Section 6.3, we indicated how various forms of hypotheses about the matrix B might be of interest for the analysis of the growth curve model. The degree $r - 1$ of the polynomial and the number n of different treatment groups both play roles in determining the number of parameters ($= rn$) to be estimated and compared in the growth curve analysis. If the growth curves could be adequately expressed by a lower-degree polynomial and if the number of distinct groupings could be reduced on the basis of similarity of growth curve features, a more parsimonious model would emerge. In practice, the choice of degree $r - 1$ might be made initially on the basis of plots of mean growth curves of each of the n groups and would subsequently be more formally chosen on the basis of goodness-of-fit LR tests as discussed in Section 6.3. However, in some situations, a relatively low degree polynomial may not be adequate to represent the growth curve patterns over time. Equivalently, we may not want to specify the mathematical form of the growth curve function over time as a specific degree polynomial or any other specified function of time. For such situations, in this section, we explore how the reduced-rank regression model and methods could be applied to build a parsimonious model.

In the reduced-rank regression model framework, $\mathbf{Y} = AB\mathbf{X} + \boldsymbol{\epsilon}$, of Chapter 2, both matrices A and B are unknown, and the rank restrictions on $C = AB$ are meant to imply certain consequences. When considered for use in the growth curve data setting, the roles of the component matrices A and B cannot be viewed in exactly the same way as the matrices A and B in the basic growth curve model (6.1). However, certain similarities and correspondences do exist. For the reduced-rank model, the $m \times r$ matrix A is unknown, but the r columns of A could be viewed as a (low-dimensional) set of basis curves over time (with general unspecified forms) such that each group's growth curve over time can be represented adequately by some linear combination of the basis set of curves. The coefficients in the ith column of the $r \times n$ matrix B then correspond to the coefficients of the linear combination of basis curves for the ith treatment group. The across-individual model could also be parameterized in terms of an overall mean growth curve and $n - 1$ among-group contrasts, with a corresponding design matrix $\mathbf{X}$ whose first row has all elements equal to one. That is, for example, we can take

$$\mathbf{X} = \begin{bmatrix} \mathbf{1}'_{T_1} & \mathbf{1}'_{T_2} & \mathbf{1}'_{T_3} & \cdot & \cdot & \mathbf{1}'_{T_{n-1}} & \mathbf{1}'_{T_n} \\ \mathbf{1}'_{T1} & 0' & 0' & \cdot & \cdot & 0' & -\mathbf{1}'_{T_n} \\ 0' & \mathbf{1}'_{T_2} & 0' & \cdot & \cdot & 0' & -\mathbf{1}'_{T_n} \\ \cdot & \cdot & \cdot & \cdot & \cdot & \cdot & \cdot \\ \cdot & \cdot & \cdot & \cdot & \cdot & \cdot & \cdot \\ 0' & 0' & 0' & \cdot & \cdot & \mathbf{1}'_{T_{n-1}} & -\mathbf{1}'_{T_n} \end{bmatrix}$$

with $B = [\bar{\boldsymbol{\beta}}, \boldsymbol{\beta}_1 - \bar{\boldsymbol{\beta}}, \ldots, \boldsymbol{\beta}_{n-1} - \bar{\beta}]$, where $\boldsymbol{\beta}_i$ denotes the ith column of the original matrix B, that is, the $r \times 1$ vector of growth curve parameters for the ith group, and $\bar{\boldsymbol{\beta}} = n^{-1} \sum_{i=1}^{n} \beta_i$. Then in this parameterization, the first column of AB gives the general mean growth curve over all groups, and the remaining columns provide for various deviations or contrasts of the group means from the overall mean growth curve. Thus, the methodology of the reduced-rank regression modeling of Chapter 2 not only seeks for a basic set of curves over time, and the associated coefficients, it also searches for possible reductions in the number of basic curves required for adequate representation of all the groups' growth curves. However, it does not seem that the unknown matrix A in the reduced-rank model should be viewed as a substitute for the known matrix A of polynomial terms in the basic growth curve model. The use of the reduced-rank modeling methodology might be considered when specification of the exact mathematical form of the growth curves, as polynomials of a certain degree, is not desirable or adequate, as mentioned.

We now recall the fundamental results from Chapter 2 concerning estimation of the reduced-rank regression model $Y_k = ABX_k + \epsilon_k$, $k = 1, \dots, T$, for application to the growth curve data situation. The ML estimators of the component matrices A and B are as given in Section 2.3,

$$\widehat{A} = \widetilde{\Sigma}_{\epsilon\epsilon}^{1/2}\widehat{V}, \qquad \widehat{B} = \widehat{V}'\widetilde{\Sigma}_{\epsilon\epsilon}^{-1/2}\widehat{\Sigma}_{yx}\widehat{\Sigma}_{xx}^{-1}, \tag{6.16}$$

where $\widehat{V} = [\widehat{V}_1, \dots, \widehat{V}_r]$ and $\widehat{V}_j$ is the normalized eigenvector corresponding to the jth largest eigenvalue $\widehat{\lambda}_j^2$ of the matrix $\widetilde{\Sigma}_{\epsilon\epsilon}^{-1/2}\widehat{\Sigma}_{yx}\widehat{\Sigma}_{xx}^{-1}\widehat{\Sigma}_{xy}\widetilde{\Sigma}_{\epsilon\epsilon}^{-1/2}$. The estimates satisfy the normalizations $\widehat{A}'\widetilde{\Sigma}_{\epsilon\epsilon}^{-1}\widehat{A} = I_r$ and $\widehat{B}\widehat{\Sigma}_{xx}\widehat{B}' = \widehat{\Lambda}^2$. The ML estimate of the error covariance matrix $\Sigma_{\epsilon\epsilon}$ can be expressed as

$$\widehat{\Sigma}_{\epsilon\epsilon} = \widetilde{\Sigma}_{\epsilon\epsilon} + (\widetilde{C} - \widehat{C})\widehat{\Sigma}_{xx}(\widetilde{C} - \widehat{C})',$$

where $\widehat{C} = \widehat{A}\widehat{B}$ and $\widetilde{C} = \widehat{\Sigma}_{yx}\widehat{\Sigma}_{xx}^{-1}$. A "goodness-of-fit" test for the adequacy of the model with reduced rank r, that is, of H_0 : rank$(C) = r$ against the general alternative of a full-rank regression matrix C is provided by the LR test statistic given in Section 2.6,

$$\mathcal{M} = -[T - n + (n - m - 1)/2] \sum_{j=r+1}^{m} \log(1 - \widehat{\rho}_j^2),$$

where $\widehat{\rho}_1^2 > \cdots \geq \widehat{\rho}_m^2 \geq 0$ are the ordered eigenvalues of the matrix

$$\widehat{\Sigma}_{yy}^{-1/2}\widehat{\Sigma}_{yx}\widehat{\Sigma}_{xx}^{-1}\widehat{\Sigma}_{xy}\widehat{\Sigma}_{yy}^{-1/2}.$$

The asymptotic distribution of the ML estimators $\widehat{A}$ and $\widehat{B}$ of the reduced-rank component matrices is given in Theorem 2.4 of Section 2.5. For the growth curve setting, one may also want to consider the extended version of the reduced-rank model as discussed in Section 3.1, $Y_k = ABX_k + DZ_k + \epsilon_k$, where Z_k represents a portion or subset of the total set of explanatory variables that are not postulated to enter the model with a growth curve or reduced-rank structure. As noted earlier, this model is in the same spirit as the extended growth curve model discussed in Section 6.4. For example, in the ANOVA situation with n treatment groups, it may be desired to separate the overall mean response curve from the growth curve structure and incorporate growth curve structure only for the remaining $n-1$ among-group contrasts. This corresponds to a case with $Z_k \equiv 1$ for all k, and X_k are $(n-1)$-dimensional vectors of group indicators.

6.6 Reduced-Rank Growth Curve Models

For longitudinal or "growth curve" data Y_k, as we have seen in earlier parts of this chapter, it is often appropriate to specify a growth curve model structure,

$$Y_k = ABX_k + \epsilon_k, \qquad k = 1, \dots, T,$$

where A is a known or specified $m \times r$ matrix and the $r \times n$ matrix B is unknown. The matrix B plays the role of summarizing the mean structure of the response variables over time (i.e., the growth curve structure), typically in the form of regression coefficients of polynomial functions over time. For example, BX_k can be viewed as the mean vector of the r-dimensional summary of the response vector, $A_1' Y_k = (A'A)^{-1} A' Y_k$, and the "complete" response mean vector can be expressed in terms of this through $E(Y_k) = A(BX_k)$. In this way, the mean vectors of the Y_k are specified to have a reduced-rank structure of a specified form. However, the $r \times n$ regression coefficient matrix B, which summarizes the mean vector structure, could itself also have reduced rank. Thus, in this section, we consider the possibility of reduced-rank structure for B in the basic growth curve model and consider the assumption that $\text{rank}(B) = r_1 < \min(r, n)$. As in previous chapters, this assumption of reduced rank yields the matrix decomposition as

$$B = B_1 B_2,$$

where B_1 and B_2 are $r \times r_1$ and $r_1 \times n$ matrices, respectively, of unknown parameters. Therefore, the reduced-rank growth curve model can be written as

$$Y_k = AB_1 B_2 X_k + \epsilon_k, \qquad k = 1, \dots, T. \tag{6.17}$$

The reduced rank of B implies that the vectors of growth curve parameters exist in r_1-dimensional space. The r_1 columns of B_1 provide a basis for this space; the elements in the ith column of B_2 provide the coefficients in the linear combination of basis growth curve parameters for the ith treatment group, in the n-group ANOVA situation, for example. The model also retains the usual reduced-dimension interpretation that the $r_1 \times 1$ vector of linear combinations, $X_k^* = B_2 X_k$, is sufficient to represent the linear relationships between Y_k and X_k. The analysis of model (6.17), including estimation of the unknown parameters B_1, B_2, and $\Sigma_{\epsilon\epsilon}$, and extensions and applications of this model, will be the focus of this section.

To derive the ML estimates of B_1 and B_2, we will again utilize the estimation results of Sections 6.2 and 2.3. The ML estimates are obtained by minimizing $|(\mathbf{Y} - AB_1 B_2 \mathbf{X})(\mathbf{Y} - AB_1 B_2 \mathbf{X})'|$ with respect to B_1 and B_2, where A is a known matrix. Following the derivation given in Section 6.2 for the ML estimate of B in the model (6.1) where B is assumed to be of full rank, we obtain the relation

$$
\begin{aligned}
&|(\mathbf{Y} - AB_1B_2\mathbf{X})(\mathbf{Y} - AB_1B_2\mathbf{X})'| \\
&\quad = |S||I_n + \widehat{\Sigma}_{xx}\boldsymbol{\Psi}|\, |I_n + (\widehat{\Sigma}_{xx}^{-1} + \boldsymbol{\Psi})^{-1}(\widetilde{B} - B)'(A'\widetilde{\Sigma}_{\epsilon\epsilon}^{-1}A)(\widetilde{B} - B)| \\
&\quad = |S|\, |I_n + \widehat{\Sigma}_{xx}\boldsymbol{\Psi}| \\
&\qquad \times |I_r + (A'\widetilde{\Sigma}_{\epsilon\epsilon}^{-1}A)(\widetilde{B} - B_1B_2)(\widehat{\Sigma}_{xx}^{-1} + \boldsymbol{\Psi})^{-1}(\widetilde{B} - B_1B_2)'|, \qquad (6.18)
\end{aligned}
$$

where the "full-rank" ML estimate of B is now denoted as

$$\widetilde{B} = (A'\widetilde{\Sigma}_{\epsilon\epsilon}^{-1}A)^{-1}A'\widetilde{\Sigma}_{\epsilon\epsilon}^{-1}\widetilde{C}$$

with $\widetilde{C} = \mathbf{YX}'(\mathbf{XX}')^{-1}$, and $\boldsymbol{\Psi} = (\widetilde{C} - A\widetilde{B})'\widetilde{\Sigma}_{\epsilon\epsilon}^{-1}(\widetilde{C} - A\widetilde{B})$. Thus, ML estimation for B_1 and B_2 is equivalent to minimizing the last term on the right side of (6.18). Note that this term has the same form as the quantity involved in the expressions following (2.17) in Section 2.3 and that the minimization is achieved by simultaneous minimization, with respect to B_1 and B_2, of all the eigenvalues of the matrix

$$(A'\widetilde{\Sigma}_{\epsilon\epsilon}^{-1}A)^{1/2}(\widetilde{B} - B_1B_2)(\widehat{\Sigma}_{xx}^{-1} + \boldsymbol{\Psi})^{-1}(\widetilde{B} - B_1B_2)'(A'\widetilde{\Sigma}_{\epsilon\epsilon}^{-1}A)^{1/2}, \qquad (6.19)$$

where $\widetilde{\Sigma}_{\epsilon\epsilon} = (1/T)S$. The matrix in (6.19) can be expressed as $(S_{(r)} - P_{(r)})(S_{(r)} - P_{(r)})'$, where

$$
\begin{aligned}
S_{(r)} &= (A'\widetilde{\Sigma}_{\epsilon\epsilon}^{-1}A)^{1/2}\widetilde{B}(\widehat{\Sigma}_{xx}^{-1} + \boldsymbol{\Psi})^{-1/2} \\
&= (A'\widetilde{\Sigma}_{\epsilon\epsilon}^{-1}A)^{-1/2}A'\widetilde{\Sigma}_{\epsilon\epsilon}^{-1}\widetilde{C}(\widehat{\Sigma}_{xx}^{-1} + \boldsymbol{\Psi})^{-1/2}
\end{aligned}
$$

and $P_{(r)} = (A'\widetilde{\Sigma}_{\epsilon\epsilon}^{-1}A)^{1/2}B_1B_2(\widehat{\Sigma}_{xx}^{-1} + \boldsymbol{\Psi})^{-1/2}$. From the same arguments as in Section 2.3, using the result of Lemma 2.2 in particular, the minimizing matrix $P_{(r)}$ is chosen as the rank-r_1 approximation of $S_{(r)}$ obtained through the singular value decomposition of $S_{(r)}$ as $P_{(r)} = \widehat{V}_{(r_1)}\widehat{V}_{(r_1)}'S_{(r)}$, where $\widehat{V}_{(r_1)} = [\widehat{V}_1, \ldots, \widehat{V}_{r_1}]$ and $\widehat{V}_j$ is the normalized eigenvector of the matrix

$$S_{(r)}S_{(r)}' = (A'\widetilde{\Sigma}_{\epsilon\epsilon}^{-1}A)^{-1/2}A'\widetilde{\Sigma}_{\epsilon\epsilon}^{-1}\widetilde{C}(\widehat{\Sigma}_{xx}^{-1}+\boldsymbol{\Psi})^{-1}\widetilde{C}'\widetilde{\Sigma}_{\epsilon\epsilon}^{-1}A(A'\widetilde{\Sigma}_{\epsilon\epsilon}^{-1}A)^{-1/2} \qquad (6.20)$$

corresponding to the jth largest eigenvalue $\widehat{\lambda}_j^2$. The ML estimates of B_1 and B_2 can then be written explicitly, in a similar form as (2.17) of Section 2.3, as

$$\widehat{B}_1 = (A'\widetilde{\Sigma}_{\epsilon\epsilon}^{-1}A)^{-1/2}\widehat{V}_{(r_1)}, \qquad (6.21a)$$

$$\widehat{B}_2 = \widehat{V}_{(r_1)}'(A'\widetilde{\Sigma}_{\epsilon\epsilon}^{-1}A)^{1/2}\widetilde{B} = \widehat{V}_{(r_1)}'(A'\widetilde{\Sigma}_{\epsilon\epsilon}^{-1}A)^{-1/2}A'\widetilde{\Sigma}_{\epsilon\epsilon}^{-1}\widetilde{C}. \qquad (6.21b)$$

The ML estimate of $B = B_1B_2$ under the reduced-rank assumption is therefore given by

$$\widehat{B} = \widehat{B}_1\widehat{B}_2 = (A'\widetilde{\Sigma}_{\epsilon\epsilon}^{-1}A)^{-1/2}\widehat{V}_{(r_1)}\widehat{V}'_{(r_1)}(A'\widetilde{\Sigma}_{\epsilon\epsilon}^{-1}A)^{1/2}\widetilde{B}$$
$$= (A'\widetilde{\Sigma}_{\epsilon\epsilon}^{-1}A)^{-1/2}\widehat{V}_{(r_1)}\widehat{V}'_{(r_1)}(A'\widetilde{\Sigma}_{\epsilon\epsilon}^{-1}A)^{-1/2}A'\widetilde{\Sigma}_{\epsilon\epsilon}^{-1}\widetilde{C}.$$

Since the $\widehat{V}_j$ are normalized eigenvectors, implicit in the construction of the estimators $\widehat{B}_1$ and $\widehat{B}_2$ are the usual normalizations

$$\widehat{B}_2(\widehat{\Sigma}_{xx}^{-1} + \boldsymbol{\Psi})^{-1}\widehat{B}'_2 = \widehat{\Lambda}^2 \equiv \text{diag}\{\widehat{\lambda}_1^2, \ldots, \widehat{\lambda}_{r_1}^2\}, \qquad \widehat{B}'_1(A'\widetilde{\Sigma}_{\epsilon\epsilon}^{-1}A)\widehat{B}_1 = I_{r_1}.$$

The ML estimator of $\Sigma_{\epsilon\epsilon}$ for model (6.17) can be written as

$$\widehat{\Sigma}_{\epsilon\epsilon} = \frac{1}{T}(\mathbf{Y} - A\widehat{B}_1\widehat{B}_2\mathbf{X})(\mathbf{Y} - A\widehat{B}_1\widehat{B}_2\mathbf{X})'$$
$$= \widetilde{\Sigma}_{\epsilon\epsilon} + (\widetilde{C} - A\widehat{B}_1\widehat{B}_2)\widehat{\Sigma}_{xx}(\widetilde{C} - A\widehat{B}_1\widehat{B}_2)'$$
$$= \widetilde{\Sigma}_{\epsilon\epsilon} + (I_m - P)(\widetilde{C}\widehat{\Sigma}_{xx}\widetilde{C}')(I_m - P)',$$

where $P = A(A'\widetilde{\Sigma}_{\epsilon\epsilon}^{-1}A)^{-1/2}\widehat{V}_{(r_1)}\widehat{V}'_{(r_1)}(A'\widetilde{\Sigma}_{\epsilon\epsilon}^{-1}A)^{-1/2}A'\widetilde{\Sigma}_{\epsilon\epsilon}^{-1}$.

It is easy to see that the results derived for the reduced-rank growth curve model (6.17) simplify to the results for the standard reduced-rank regression model discussed in Chapter 2. These are obtained by setting $A = I_m$ with $r = m$, so that $\widetilde{B} \equiv \widetilde{C}$ and $\boldsymbol{\Psi} = 0$. The matrix $S_{(r)}S'_{(r)}$ that is involved in the reduced-rank growth curve model simplifies to $\widehat{R} = \widetilde{\Sigma}_{\epsilon\epsilon}^{-1/2}\widehat{\Sigma}_{yx}\widehat{\Sigma}_{xx}^{-1}\widehat{\Sigma}_{xy}\widetilde{\Sigma}_{\epsilon\epsilon}^{-1/2}$, the matrix through which the standard reduced-rank model calculations were made in Chapter 2. We also readily see that the above expression for $\widehat{\Sigma}_{\epsilon\epsilon}$ simplifies to (2.18) when $A = I_m$ for the standard reduced-rank model.

The same estimation results can be derived using the conditional model approach as presented in Section 6.2. The conditional model (6.6) when B is assumed to be of reduced rank can be written as

$$Y_{1k} = B_1B_2X_k + \Theta Y_{2k} + \epsilon_{1k}^*, \tag{6.22}$$

which is in the form of the reduced-rank model (3.3) of Section 3.1. From (3.11), it follows that the ML estimates of B_1 and B_2 from this model are given by

$$\widehat{B}_1 = (A'\widetilde{\Sigma}_{\epsilon\epsilon}^{-1}A)^{-1/2}\widehat{V}, \qquad \widehat{B}_2 = \widehat{V}'(A'\widetilde{\Sigma}_{\epsilon\epsilon}^{-1}A)^{1/2}\widehat{\Sigma}_{y_1x.y_2}\widehat{\Sigma}_{xx.y_2}^{-1},$$

recalling from (6.10) that $\widetilde{\Sigma}_1^* = (A'\widetilde{\Sigma}_{\epsilon\epsilon}^{-1}A)^{-1}$ is the "unrestricted" ML estimate of the error covariance matrix $\Sigma_1^* = \text{Cov}(\epsilon_{1k}^*)$ in model (6.6), and $\widehat{V} = [\widehat{V}_1, \ldots, \widehat{V}_{r_1}]$, where $\widehat{V}_j$ is the normalized eigenvector of the matrix

$$(A'\widetilde{\Sigma}_{\epsilon\epsilon}^{-1}A)^{1/2}\widehat{\Sigma}_{y_1x.y_2}\widehat{\Sigma}_{xx.y_2}^{-1}\widehat{\Sigma}_{xy_1.y_2}(A'\widetilde{\Sigma}_{\epsilon\epsilon}^{-1}A)^{1/2} \tag{6.23}$$

corresponding to the jth largest eigenvalue. However, from (6.8), we know that $\widehat{\Sigma}_{y_1x.y_2}\widehat{\Sigma}_{xx.y_2}^{-1} \equiv \widetilde{B} \equiv (A'\widetilde{\Sigma}_{\epsilon\epsilon}^{-1}A)^{-1}A'\widetilde{\Sigma}_{\epsilon\epsilon}^{-1}\widetilde{C}$, and it has also been established in Section 6.3 that $\widehat{\Sigma}_{xx.y_2} = (\widehat{\Sigma}_{xx}^{-1} + \boldsymbol{\Psi})^{-1}$ (noting the correspondence $\widehat{\Sigma}_{xx}^{-1} + \boldsymbol{\Psi} \equiv T\Omega = \widehat{\Sigma}_{xx.y_2}^{-1}$ from (6.12)). So we see that the matrix in (6.23) is identical to the matrix $S_{(r)}S_{(r)}'$ in (6.20) that is involved in ML estimation through the previous direct approach. Hence, it immediately follows that the above expressions for $\widehat{B}_1$ and $\widehat{B}_2$, obtained from the conditional model approach, can be expressed in an identical form as in (6.21). In addition, note that the "unrestricted" ML estimate of the error covariance matrix Σ_1^* in model (6.6) can be expressed in the form $\widetilde{\Sigma}_1^* = (A'\widetilde{\Sigma}_{\epsilon\epsilon}^{-1}A)^{-1} \equiv \widehat{\Sigma}_{y_1y_1.y_2} - \widehat{\Sigma}_{y_1x.y_2}\widehat{\Sigma}_{xx.y_2}^{-1}\widehat{\Sigma}_{xy_1.y_2}$ and that

$$\widehat{\Sigma}_{y_1y_1.y_2} = \frac{1}{T}[A_1'\mathbf{YY}'A_1 - A_1'\mathbf{YY}'A_2(A_2'\mathbf{YY}'A_2)^{-1}A_2'\mathbf{YY}'A_1]$$
$$= \frac{1}{T}(A'(\mathbf{YY}')^{-1}A)^{-1} \equiv (A'\widehat{\Sigma}_{yy}^{-1}A)^{-1},$$

with $\widehat{\Sigma}_{yy} = \frac{1}{T}\mathbf{YY}'$, using Lemma 1 of Khatri (1966). Hence, we see that

$$\widehat{\Sigma}_{y_1x.y_2}\widehat{\Sigma}_{xx.y_2}^{-1}\widehat{\Sigma}_{xy_1.y_2} = \widehat{\Sigma}_{y_1y_1.y_2} - (A'\widetilde{\Sigma}_{\epsilon\epsilon}^{-1}A)^{-1}$$
$$= (A'\widehat{\Sigma}_{yy}^{-1}A)^{-1} - (A'\widetilde{\Sigma}_{\epsilon\epsilon}^{-1}A)^{-1},$$

so it immediately follows that another equivalent form to represent the matrix in (6.23) is as $(A'\widetilde{\Sigma}_{\epsilon\epsilon}^{-1}A)^{1/2}[(A'\widehat{\Sigma}_{yy}^{-1}A)^{-1} - (A'\widetilde{\Sigma}_{\epsilon\epsilon}^{-1}A)^{-1}](A'\widetilde{\Sigma}_{\epsilon\epsilon}^{-1}A)^{1/2}$, which is similar to the form given by Albert and Kshirsagar (1993) in related work.

The formulation of the reduced-rank growth curve model (6.17) in the conditional model form (6.22) leads more directly to LR testing for the rank of the matrix B. Because (6.22) is in the framework of the reduced-rank model (3.3) of Section 3.1, a LR test of the rank of B can be obtained through the eigenvalues $\widehat{\lambda}_j^2$ of the matrix in (6.23), which is equivalent to the matrix $S_{(r)}S_{(r)}'$ in (6.20) as noted. These eigenvalues are also directly related, in the usual way, to the sample partial canonical correlations between Y_{1k} and X_k after adjusting for the effects of Y_{2k}. Following the developments in Section 3.1, the LR test for H_0 : rank$(B) = r_1$ is based on the test statistic

$$\mathcal{M} = [T - n - (m - r) + (n - r - 1)/2] \sum_{j=r_1+1}^{r} \log(1 + \widehat{\lambda}_j^2), \tag{6.24}$$

whose null distribution is approximated by the χ^2-distribution with $(r - r_1)(n - r_1)$ degrees of freedom.

Alternatively, the formulation of the reduced-rank growth curve model (6.17) in the conditional model form (6.6) also leads directly to LR testing for the rank of B. Because (6.6) is in the framework of the reduced-rank regression model considered in Chapter 3, a LR test of the rank of B can be obtained through results derived there. Specifically, a LR test is obtained using the eigenvalues $\widehat{\lambda}_j^2$ of the matrix in (6.23), which is equivalent to the matrix $S_{(r)}S'_{(r)}$ as noted. These eigenvalues are also directly related, in the usual way, to the sample partial canonical correlations between Y_{1k} and X_k after adjusting for Y_{2k}. Following the details of Anderson's model discussed in Chapter 3, we are led to the LR test for H_0 : rank$(B) \leq r_1$ that is based on the same test statistic M as in (6.24).

Fujikoshi (1974) derived the form of the LR statistic for testing H_0 : rank$(FBH) \leq r_1$, where F and H are known full-rank matrices, using the conditional model approach. The LR test obtained was of the form (6.14), with the $\widehat{\lambda}_j^2$ being the eigenvalues of the matrix $S_e^{-1/2}S_hS_e^{-1/2} \equiv \{F(A'\widetilde{\Sigma}_{\epsilon\epsilon}^{-1}A)^{-1}F'\}^{-1/2}$ $(F\widetilde{B}H)\{H'\widehat{\Sigma}_{xx.y_2}^{-1}H\}^{-1}(F\widetilde{B}H)'\{F(A'\widetilde{\Sigma}_{\epsilon\epsilon}^{-1}A)^{-1}F'\}^{-1/2}$ and $\widehat{\Sigma}_{xx.y_2}^{-1}$ expressed in the form $[\widehat{\Sigma}_{xx}^{-1} + \Psi]$. The matrix $S_e^{-1/2}S_hS_e^{-1/2}$ reduces to that in (6.23), and the preceding results are obtained as a special case when $F = I_r$ and $H = I_n$.

Note that LR testing for the rank of the matrix $C(= AB)$ in the standard reduced-rank regression model, $Y_k = CX_k + \epsilon_k = ABX_k + \epsilon_k$, ignoring the knowledge that the matrix A is known in the growth curve model, still could serve as a valid initial testing procedure, for convenience. Under the assumed reduced-rank growth curve model (6.17), the additional rank deficiency of $C = AB = AB_1B_2$ is due to the component $B = B_1B_2$, since A is specified to be of full rank r. Hence, LR testing in the standard regression model (as summarized in Section 6.5) would be valid to suggest a possible rank for the matrix B, initially. However, this procedure would not be as efficient as the LR testing procedure discussed in the preceding paragraph, since it does not take into account the known specified value of the component matrix A in model (6.17).

We comment briefly on the relationship between the LR test for specified linear constraints on the coefficient matrix B and the LR test results above for the rank of B. The LR test for $H_0 : FB = 0$, where F is a specified $(r - r_1) \times r$ full-rank matrix, is derived in Section 6.3 to have the same form as the statistic $\mathcal{M}$ in (6.24), with r replaced by $(r - r_1)$, and the statistic is asymptotically distributed as $\chi^2_{(r-r_1)n}$. In this context, the $\widehat{\lambda}_j^2$ are the eigenvalues of the matrix

$$S_*^{-1}H_* \equiv \{F(A'\widetilde{\Sigma}_{\epsilon\epsilon}^{-1}A)^{-1}F'\}^{-1}F(A'\widetilde{\Sigma}_{\epsilon\epsilon}^{-1}A)^{-1/2}S_{(r)}S'_{(r)}(A'\widetilde{\Sigma}_{\epsilon\epsilon}^{-1}A)^{-1/2}F'.$$

When B is of reduced rank $r_1 < r$, this is equivalent to the condition related to H_0 above that B satisfies $FB = 0$ for an unknown $(r - r_1) \times r$ full-rank matrix F. The LR test statistic for the rank of B is given by (6.24) and has an asymptotic $\chi^2_{(r-r_1)(n-r_1)}$ distribution. Thus we see the similarity between the two testing situations, with the matrix $S_{(r)}S'_{(r)}$ playing a main role in both cases. Because the constraints are unknown in the reduced-rank case and need to be estimated, the degrees of freedom $(r - r_1)n$ in the known constraints case are decreased to $(r - r_1)(n - r_1)$ for the reduced-rank case.

6.6.1 Extensions of the Reduced-Rank Growth Curve Model

In the same spirit as the extended model (6.14) considered in Section 6.4, the reduced-rank growth curve model (6.17) can be extended to allow for the effect of some predictor variables on the response without growth curve structure. That is, we consider the model

$$Y_k = AB_1B_2X_k + DZ_k + \epsilon_k, \qquad k = 1, \ldots, T, \tag{6.25}$$

where all terms are similar to those in (6.14) and (6.17). Maximum likelihood estimation of the component matrices B_1 and B_2, as well as D, can be easily carried out along the same lines as the relevant calculations for model (6.17). The main difference is that the basic calculations now will be performed after adjustment for the additional predictor variables Z_k, as discussed in Section 6.4. Hence, the procedures are similar to the preceding ones except that various sample covariance matrices are replaced by sample partial covariance matrices, adjusting for Z_k. In particular, ML estimation of B_1 and B_2, similar to the expressions in (6.21), is now based on $\widetilde{B} = (A'\widetilde{\Sigma}_{\epsilon\epsilon}^{-1}A)^{-1}A'\widetilde{\Sigma}_{\epsilon\epsilon}^{-1}\widetilde{C}$ with $\widetilde{C} = \widehat{\Sigma}_{yx.z}\widehat{\Sigma}_{xx.z}^{-1}$, and the matrix similar to that in (6.19) now has $\widehat{\Sigma}_{xx.z}^{-1}$ in place of $\widehat{\Sigma}_{xx}^{-1}$.

Another extension of the model (6.17) concerns the inclusion of the additional predictor variables Z_k in the growth curve structure, but without any reduced-rank feature. This is equivalent to representing D in model (6.25) in the form $D = AB^*$, so that the resulting model can be represented as

$$\begin{aligned} Y_k &= A\ B_1B_2X_k + AB^*Z_k + \epsilon_k \\ &\equiv A[B_1B_2X_k + B^*Z_k] + \epsilon_k, \qquad k = 1, \ldots, T. \end{aligned} \tag{6.26}$$

This can be viewed as analogous to the reduced-rank model (3.3) of Section 3.1, but where the reduced-rank and "standard" regression structures are in terms of the growth curve parameters of the responses Y_k rather than in terms of the response vectors Y_k themselves. A special case of model (6.26) was the focus of work by Albert and Kshirsagar (1993). Their study was in the context of a multivariate one-way ANOVA situation involving n treatment groups, with $Z_k \equiv 1$ and X_k an $(n-1)$-dimensional vector of indicators. They considered the model of the form of (6.26) for the purpose of combining the features of growth curve with the application of reduced-rank regression to linear discriminant analysis. In Section 3.2, we discussed the application of the reduced-rank regression model for data from a multivariate one-way ANOVA situation and indicated its relation to linear discriminant analysis in the case of m-dimensional unstructured mean vectors. The above model (6.26) can be applied to growth curve structures in a similar way, with emphasis on discriminant analysis through use of the r-dimensional growth curve parameter vectors of each group to summarize their mean structure.

We will first discuss the ML estimation of parameters in model (6.26) and then examine in more detail the special case of the multivariate one-way ANOVA situation. For model (6.26), let $D^* = D + C\widehat{\Sigma}_{xz}\widehat{\Sigma}_{zz}^{-1}$ with $C = AB = AB_1B_2$ and $D = AB^*$, so that $D^* = AB^* + AB_1B_2\widehat{\Sigma}_{xz}\widehat{\Sigma}_{zz}^{-1} = A(B^* + B_1B_2\widehat{\Sigma}_{xz}\widehat{\Sigma}_{zz}^{-1}) \equiv AB^{**}$. Also let $\widetilde{D}^* = \widehat{\Sigma}_{yz}\widehat{\Sigma}_{zz}^{-1}$ and $\widetilde{C} = \widehat{\Sigma}_{yx.z}\widehat{\Sigma}_{xx.z}^{-1}$ denote the unrestricted LS estimators of D^* and C. For ML estimation of (6.26), we need to minimize $|W|$, where $W = (1/T)(\mathbf{Y} - AB_1B_2\mathbf{X} - AB^*\mathbf{Z})(\mathbf{Y} - AB_1B_2\mathbf{X} - AB^*\mathbf{Z})'$. As in previous cases, ML estimation and LR testing results for the extended reduced-rank growth curve model (6.26) can be derived using the conditional model approach. Analogous to (6.22), the conditional model corresponding to (6.26) can be written as

$$\begin{aligned} Y_{1k} &= B_1B_2X_k + B^*Z_k + \Theta Y_{2k} + \epsilon_{1k}^* \\ &= B_1B_2X_k + \Theta^* Y_{2k}^* + \epsilon_{1k}^*, \end{aligned} \tag{6.27}$$

where $\Theta^* = [B^*, \Theta]$ and $Y_{2k}^* = [Z_k', Y_{2k}']'$. This model is in the form of the reduced-rank model (3.3) of Section 3.1. From (3.11) and recalling from (6.10) that $\widetilde{\Sigma}_1^* = (A'\widetilde{\Sigma}_{\epsilon\epsilon}^{-1}A)^{-1}$ is the "unrestricted" ML estimate of the error covariance matrix $\Sigma_1^* = \mathrm{Cov}(\epsilon_{1k}^*)$ in the conditional models (6.6) and (6.27), it follows that the ML estimates of B_1 and B_2 from this model are given by

$$\widehat{B}_1 = (A'\widetilde{\Sigma}_{\epsilon\epsilon}^{-1}A)^{-1/2}\widehat{V}_{(r_1)}^*, \qquad \widehat{B}_2 = \widehat{V}_{(r_1)}^{*\prime}(A'\widetilde{\Sigma}_{\epsilon\epsilon}^{-1}A)^{1/2}\widehat{\Sigma}_{y_1x.y_2^*}\widehat{\Sigma}_{xx.y_2^*}^{-1}, \tag{6.28}$$

where $\widehat{V}_{(r_1)}^* = [\widehat{V}_1^*, \ldots, \widehat{V}_{r_1}^*]$ and $\widehat{V}_j^*$ is the normalized eigenvector of the matrix

$$(A'\widetilde{\Sigma}_{\epsilon\epsilon}^{-1}A)^{1/2}\widehat{\Sigma}_{y_1x.y_2^*}\widehat{\Sigma}_{xx.y_2^*}^{-1}\widehat{\Sigma}_{xy_1.y_2^*}(A'\widetilde{\Sigma}_{\epsilon\epsilon}^{-1}A)^{1/2}$$

corresponding to the jth largest eigenvalue $\widehat{\lambda}_j^{*2}$. However, from previous results associated with the reduced-rank growth curve model (6.17), it follows that $\widehat{\Sigma}_{y_1x^*.y_2}\widehat{\Sigma}_{x^*x^*.y_2}^{-1} = (A'\widetilde{\Sigma}_{\epsilon\epsilon}^{-1}A)^{-1}A'\widetilde{\Sigma}_{\epsilon\epsilon}^{-1}\widetilde{G} \equiv \widetilde{B}_*$, with $\widetilde{G} = [\widetilde{C}, \widetilde{D}]$, and that $\widehat{\Sigma}_{x^*x^*.y_2} = (\widehat{\Sigma}_{x^*x^*}^{-1} + \mathbf{\Psi}_*)^{-1}$, where $\mathbf{X}^* = [\mathbf{X}', \mathbf{Z}']'$ and $\mathbf{\Psi}_* = (\widetilde{G} - A\widetilde{B}_*)'\widetilde{\Sigma}_{\epsilon\epsilon}^{-1}(\widetilde{G} - A\widetilde{B}_*)$. From these results, it can readily be established that $\widehat{\Sigma}_{y_1x.y_2^*}\widehat{\Sigma}_{xx.y_2^*}^{-1} = \widetilde{B} \equiv (A'\widetilde{\Sigma}_{\epsilon\epsilon}^{-1}A)^{-1}A'\widetilde{\Sigma}_{\epsilon\epsilon}^{-1}\widetilde{C}$ and that $\widehat{\Sigma}_{xx.y_2^*} = (\widehat{\Sigma}_{xx.z}^{-1} + \mathbf{\Psi})^{-1}$, where $\mathbf{\Psi} = (\widetilde{C} - A\widetilde{B})'\widetilde{\Sigma}_{\epsilon\epsilon}^{-1}(\widetilde{C} - A\widetilde{B})$. So we find that, equivalently, the eigenvectors $\widehat{V}_j^*$ involved in the ML estimates $\widehat{B}_1$ and $\widehat{B}_2$ in (6.28) are those obtained from the matrix

$$S_{(r)}^* S_{(r)}^{*\prime} = (A'\widetilde{\Sigma}_{\epsilon\epsilon}^{-1}A)^{1/2}\widetilde{B}(\widehat{\Sigma}_{xx.z}^{-1} + \mathbf{\Psi})^{-1}\widetilde{B}'(A'\widetilde{\Sigma}_{\epsilon\epsilon}^{-1}A)^{1/2} \tag{6.29}$$

with $\widetilde{B} \equiv (A'\widetilde{\Sigma}_{\epsilon\epsilon}^{-1}A)^{-1}A'\widetilde{\Sigma}_{\epsilon\epsilon}^{-1}\widetilde{C}$. Similar to the arguments following Eq. (6.23), the matrix in (6.29) can be expressed equivalently in the form $(A'\widetilde{\Sigma}_{\epsilon\epsilon}^{-1}A)^{1/2}[(A'\widehat{\Sigma}_{yy.z}^{-1}A)^{-1} - (A'\widetilde{\Sigma}_{\epsilon\epsilon}^{-1}A)^{-1}](A'\widetilde{\Sigma}_{\epsilon\epsilon}^{-1}A)^{1/2}$.

The ML estimate of $B = B_1B_2$ under the reduced-rank assumption is thus given by

$$\widehat{B} = \widehat{B}_1\widehat{B}_2 = (A'\widetilde{\Sigma}_{\epsilon\epsilon}^{-1}A)^{-1/2}\widehat{V}_{(r_1)}^{*}\widehat{V}_{(r_1)}^{*\prime}(A'\widetilde{\Sigma}_{\epsilon\epsilon}^{-1}A)^{1/2}\widetilde{B}$$
$$= (A'\widetilde{\Sigma}_{\epsilon\epsilon}^{-1}A)^{-1/2}\widehat{V}_{(r_1)}^{*}\widehat{V}_{(r_1)}^{*\prime}(A'\widetilde{\Sigma}_{\epsilon\epsilon}^{-1}A)^{-1/2}A'\widetilde{\Sigma}_{\epsilon\epsilon}^{-1}\widetilde{C}.$$

In addition, from results in Section 3.1 for the model (3.3), we obtain the ML estimate of $\Theta^* = [B^*, \Theta]$ from the conditional model (6.27) as $\widehat{\Theta}^* = \widehat{\Sigma}_{y_1y_2^*}\widehat{\Sigma}_{y_2^*y_2^*}^{-1} - \widehat{B}\widehat{\Sigma}_{xy_2^*}\widehat{\Sigma}_{y_2^*y_2^*}^{-1}$. From this, we can show that the ML estimate of the component B^* is obtained as

$$\widehat{B}^* = \widetilde{B}^{**} - \widehat{B}_1\widehat{B}_2\{\widehat{\Sigma}_{xz}\widehat{\Sigma}_{zz}^{-1} - \widehat{\Sigma}_{xy.z}\widehat{\Sigma}_{yy.z}^{-1}(\widetilde{D}^* - A\widetilde{B}^{**})\}, \tag{6.30}$$

where $\widetilde{D}^* = \widehat{\Sigma}_{yz}\widehat{\Sigma}_{zz}^{-1}$ and $\widetilde{B}^{**} = (A'\widehat{\Sigma}_{yy.z}^{-1}A)^{-1}A'\widehat{\Sigma}_{yy.z}^{-1}\widetilde{D}^*$. Since $\widetilde{C} = \widehat{\Sigma}_{yx.z}\widehat{\Sigma}_{xx.z}^{-1}$, $\widetilde{\Sigma}_{\epsilon\epsilon} = \widehat{\Sigma}_{yy.z} - \widehat{\Sigma}_{yx.z}\widehat{\Sigma}_{xx.z}^{-1}\widehat{\Sigma}_{xy.z}$, and $\widetilde{C}\widehat{\Sigma}_{xz}\widehat{\Sigma}_{zz}^{-1} = \widetilde{D}^* - \widetilde{D}$, where $\widetilde{D} = \widehat{\Sigma}_{yz.x}\widehat{\Sigma}_{zz.x}^{-1}$, it follows that

$$(A'\widetilde{\Sigma}_{\epsilon\epsilon}^{-1}A)^{-1}A'\widetilde{\Sigma}_{\epsilon\epsilon}^{-1}\widetilde{C}\{\widehat{\Sigma}_{xz}\widehat{\Sigma}_{zz}^{-1} - \widehat{\Sigma}_{xy.z}\widehat{\Sigma}_{yy.z}^{-1}(\widetilde{D}^* - A\widetilde{B}^{**})\}$$
$$= (A'\widetilde{\Sigma}_{\epsilon\epsilon}^{-1}A)^{-1}A'\widetilde{\Sigma}_{\epsilon\epsilon}^{-1}\{\widetilde{D}^* - \widetilde{D} + (\widetilde{\Sigma}_{\epsilon\epsilon} - \widehat{\Sigma}_{yy.z})\widehat{\Sigma}_{yy.z}^{-1}(\widetilde{D}^* - A\widetilde{B}^{**})\}$$
$$= \widetilde{B}^{**} - (A'\widetilde{\Sigma}_{\epsilon\epsilon}^{-1}A)^{-1}A'\widetilde{\Sigma}_{\epsilon\epsilon}^{-1}\widetilde{D}.$$

Therefore, the ML estimator $\widehat{B}^*$ in (6.30) can also be expressed in the equivalent form

$$\widehat{B}^* = \widetilde{B}^{**} + (A'\widetilde{\Sigma}_{\epsilon\epsilon}^{-1}A)^{-1/2}\widehat{V}_{(r_1)}^{*}\widehat{V}_{(r_1)}^{*\prime}(A'\widetilde{\Sigma}_{\epsilon\epsilon}^{-1}A)^{1/2}(\widetilde{B}^* - \widetilde{B}^{**}), \tag{6.31}$$

where $\widetilde{B}^* = (A'\widetilde{\Sigma}_{\epsilon\epsilon}^{-1}A)^{-1}A'\widetilde{\Sigma}_{\epsilon\epsilon}^{-1}\widetilde{D}$. Finally, the ML estimate of the error covariance matrix $\Sigma_{\epsilon\epsilon}$ is

$$\widehat{\Sigma}_{\epsilon\epsilon} = \frac{1}{T}(\mathbf{Y} - A\widehat{B}_1\widehat{B}_2\mathbf{X} - A\widehat{B}^*\mathbf{Z})(\mathbf{Y} - A\widehat{B}_1\widehat{B}_2\mathbf{X} - A\widehat{B}^*\mathbf{Z})'$$
$$= \widetilde{\Sigma}_{\epsilon\epsilon} + \frac{1}{T}(\widetilde{C}\mathbf{X} + \widetilde{D}\mathbf{Z} - A\widehat{B}_1\widehat{B}_2\mathbf{X} - A\widehat{B}^*\mathbf{Z})$$
$$\times (\widetilde{C}\mathbf{X} + \widetilde{D}\mathbf{Z} - A\widehat{B}_1\widehat{B}_2\mathbf{X} - A\widehat{B}^*\mathbf{Z})'$$
$$= \widetilde{\Sigma}_{\epsilon\epsilon} + (\widetilde{C} - A\widehat{B}_1\widehat{B}_2)\widehat{\Sigma}_{xx.z}(\widetilde{C} - A\widehat{B}_1\widehat{B}_2)'$$
$$+ (I_m - A\widehat{B}_1\widehat{B}_2\widehat{\Sigma}_{xy.z}\widehat{\Sigma}_{yy.z}^{-1})(\widetilde{D}^* - A\widetilde{B}^{**})\widehat{\Sigma}_{zz}$$
$$\times (\widetilde{D}^* - A\widetilde{B}^{**})'(I_m - A\widehat{B}_1\widehat{B}_2\widehat{\Sigma}_{xy.z}\widehat{\Sigma}_{yy.z}^{-1})'.$$

The formulation of the extended reduced-rank growth curve model (6.26) in the conditional model form (6.27) also leads directly to LR testing for the rank of the matrix B. Because (6.27) is in the framework of the reduced-rank model (3.3) of

Section 3.1, a LR test of the rank of B can be obtained through the eigenvalues $\widehat{\lambda}_j^{*2}$ of the matrix $S^*_{(r)} S^{*\prime}_{(r)}$ in (6.29). These eigenvalues are also directly related, in the usual way, to the sample partial canonical correlations between Y_{1k} and X_k after adjusting for the effects of Y^*_{2k}. Following the developments in Section 3.1, the LR test for H_0 : rank$(B) = r_1$ is based on the test statistic

$$\mathcal{M} = [T - n - (m - r) + (n_1 - r - 1)/2] \sum_{j=r_1+1}^{r} \log(1 + \widehat{\lambda}_j^{*2}), \tag{6.32}$$

whose null distribution is asymptotically the χ^2-distribution with $(r - r_1)(n_1 - r_1)$ degrees of freedom.

Note that the form of the LR statistic for this situation also follows from results of Fujikoshi (1974) mentioned earlier. Because model (6.26) can be expressed as $Y_k = A[B \quad B^*][X_k' \quad Z_k']' + \epsilon_k$, so $B = [B \quad B^*]H$ with $H = [I_{n_1} \quad 0]'$, the LR statistic can be obtained by using $F = I_r$ and $H = [I_{n_1} \quad 0]'$ in the notation for the work of Fujikoshi (1974) discussed earlier.

6.7 Application to One-way ANOVA and Linear Discriminant Analysis

We now elaborate on the multivariate one-way ANOVA case and consider the situation and details similar to those from Section 3.2. Suppose we have independent random samples $Y_{i1}, \ldots, Y_{iT_i}$, $i = 1, \ldots, n$, from n multivariate normal distributions $N(\mu_i, \Sigma)$, and set $T = T_1 + \cdots + T_n$. We suppose the mean vectors have the growth curve structure $\mu_i = A\beta_i$, where A is the known $m \times r$ within-subject design matrix and β_i is the $r \times 1$ vector of unknown growth curve parameters for the ith group. It will be assumed that the β_i lie in some r_1-dimensional linear subspace with $r_1 \le \min(n - 1, r)$. This is equivalent to the condition that the matrix of contrasts $[\beta_1 - \beta_n, \ldots, \beta_{n-1} - \beta_n]$ is of reduced rank r_1. As mentioned above, the model (6.26) accommodates this special situation where the model is expressed as $Y_{ij} = A[BX_i + B^*Z_i] + \epsilon_{ij}$, with $B = B_1B_2$, and the choices of B and B^* as $B = [\beta_1 - \beta_n, \ldots, \beta_{n-1} - \beta_n]$ and $B^* = \beta_n$, and $Z_i \equiv 1$. The LS estimates of $C = AB$ and $D = AB^*$ are as in (3.14), $\widetilde{C} = [\bar{Y}_1 - \bar{Y}_n, \ldots, \bar{Y}_{n-1} - \bar{Y}_n]$ and $\widetilde{D} = \bar{Y}_n$, and the corresponding estimate of $\Sigma_{\epsilon\epsilon}$ is $\widetilde{\Sigma}_{\epsilon\epsilon} = \frac{1}{T}\sum_{i=1}^{n}\sum_{j=1}^{T_i}(Y_{ij} - \bar{Y}_i)(Y_{ij} - \bar{Y}_i)'$. The (full-rank) estimates of B and B^* are obtained directly from the results of Section 6.2 as

$$\widetilde{B} = (A'\widetilde{\Sigma}_{\epsilon\epsilon}^{-1}A)^{-1}A'\widetilde{\Sigma}_{\epsilon\epsilon}^{-1}\widetilde{C}, \qquad \widetilde{B}^* = (A'\widetilde{\Sigma}_{\epsilon\epsilon}^{-1}A)^{-1}A'\widetilde{\Sigma}_{\epsilon\epsilon}^{-1}\widetilde{D}. \tag{6.33}$$

The calculations related to the reduced-rank growth curve aspect, that is, ML estimation of the component matrices B_1 and B_2, follow from the preceding general results. From (6.28) and (6.29), the ML estimates of the component matrices are based on the normalized eigenvectors of

$$(A'\widetilde{\Sigma}_{\epsilon\epsilon}^{-1}A)^{-1/2}A'\widetilde{\Sigma}_{\epsilon\epsilon}^{-1}\widetilde{C}(\widehat{\Sigma}_{xx.z}^{-1}+\boldsymbol{\Psi})^{-1}\widetilde{C}'\widetilde{\Sigma}_{\epsilon\epsilon}^{-1}A(A'\widetilde{\Sigma}_{\epsilon\epsilon}^{-1}A)^{-1/2}. \tag{6.34}$$

Let $\widehat{V}_{(r_1)}^*$ denote the $r \times r_1$ matrix whose r_1 columns are the normalized eigenvectors of the matrix above corresponding to its r_1 largest eigenvalues. Then it follows from (6.28) that the ML estimates of the component matrices are

$$\widehat{B}_1 = (A'\widetilde{\Sigma}_{\epsilon\epsilon}^{-1}A)^{-1/2}\widehat{V}_{(r_1)}^*, \qquad \widehat{B}_2 = \widehat{V}_{(r_1)}^{*\prime}(A'\widetilde{\Sigma}_{\epsilon\epsilon}^{-1}A)^{-1/2}A'\widetilde{\Sigma}_{\epsilon\epsilon}^{-1}\widetilde{C}. \tag{6.35}$$

We set $\widehat{V}_{(r_1)}^{**} = (A'\widetilde{\Sigma}_{\epsilon\epsilon}^{-1}A)^{1/2}\widehat{V}_{(r_1)}^*$ so that it satisfies the normalization $\widehat{V}_{(r_1)}^{**\prime}(A'\widetilde{\Sigma}_{\epsilon\epsilon}^{-1}A)^{-1}\widehat{V}_{(r_1)}^{**} = I_{r_1}$. Then we can express the ML estimates in (6.35) as $\widehat{B}_1 = (A'\widetilde{\Sigma}_{\epsilon\epsilon}^{-1}A)^{-1}\widehat{V}_{(r_1)}^{**}$ and $\widehat{B}_2 = \widehat{V}_{(r_1)}^{**\prime}(A'\widetilde{\Sigma}_{\epsilon\epsilon}^{-1}A)^{-1}A'\widetilde{\Sigma}_{\epsilon\epsilon}^{-1}\widetilde{C}$. The corresponding ML reduced-rank estimate of B is

$$\widehat{B} = \widehat{B}_1\widehat{B}_2 = (A'\widetilde{\Sigma}_{\epsilon\epsilon}^{-1}A)^{-1}\widehat{V}_{(r_1)}^{**}\widehat{V}_{(r_1)}^{**\prime}(A'\widetilde{\Sigma}_{\epsilon\epsilon}^{-1}A)^{-1}A'\widetilde{\Sigma}_{\epsilon\epsilon}^{-1}\widetilde{C}. \tag{6.36}$$

The ML estimate of $B^* = \beta_n$ is, from (6.31), given by

$$\begin{aligned}\widehat{\beta}_n &= \widetilde{B}^{**} + (A'\widetilde{\Sigma}_{\epsilon\epsilon}^{-1}A)^{-1/2}\widehat{V}_{(r_1)}^*\widehat{V}_{(r_1)}^{*\prime}(A'\widetilde{\Sigma}_{\epsilon\epsilon}^{-1}A)^{1/2}(\widetilde{B}^* - \widetilde{B}^{**})\\ &= \widetilde{B}^{**} + (A'\widetilde{\Sigma}_{\epsilon\epsilon}^{-1}A)^{-1}\widehat{V}_{(r_1)}^{**}\widehat{V}_{(r_1)}^{**\prime}(\widetilde{B}^* - \widetilde{B}^{**}),\end{aligned}$$

where $\widetilde{B}^* = (A'\widetilde{\Sigma}_{\epsilon\epsilon}^{-1}A)^{-1}A'\widetilde{\Sigma}_{\epsilon\epsilon}^{-1}\bar{Y}_n$ and $\widetilde{B}^{**} = (A'\widehat{\Sigma}_{yy.z}^{-1}A)^{-1}A'\widehat{\Sigma}_{yy.z}^{-1}\bar{Y}$, since $\widetilde{D}^* = \widehat{\Sigma}_{yz}\widehat{\Sigma}_{zz}^{-1} = \bar{Y}$, with $\widehat{\Sigma}_{yy.z} = (1/T)\sum_{i=1}^{n}\sum_{j=1}^{T_i}(Y_{ij}-\bar{Y})(Y_{ij}-\bar{Y})'$. Because the ith column of $\widetilde{C}$ is $\bar{Y}_i - \bar{Y}_n$ and the ML reduced-rank growth curve estimate is $\widehat{B} \equiv \widehat{B}_1\widehat{B}_2 = [\widehat{\beta}_1 - \widehat{\beta}_n, \ldots, \widehat{\beta}_{n-1} - \widehat{\beta}_n]$, we also have

$$(\widehat{\beta}_i - \widehat{\beta}_n) = (A'\widetilde{\Sigma}_{\epsilon\epsilon}^{-1}A)^{-1}\widehat{V}_{(r_1)}^{**}\widehat{V}_{(r_1)}^{**\prime}(A'\widetilde{\Sigma}_{\epsilon\epsilon}^{-1}A)^{-1}A'\widetilde{\Sigma}_{\epsilon\epsilon}^{-1}(\bar{Y}_i - \bar{Y}_n). \tag{6.37}$$

Combining the last two expressions, we have the ML reduced-rank estimate of β_i as

$$\begin{aligned}\widehat{\beta}_i = {} & (A'\widehat{\Sigma}_{yy.z}^{-1}A)^{-1}A'\widehat{\Sigma}_{yy.z}^{-1}\bar{Y} + (A'\widetilde{\Sigma}_{\epsilon\epsilon}^{-1}A)^{-1}\widehat{V}_{(r_1)}^{**}\widehat{V}_{(r_1)}^{**\prime}\\ & \times \{(A'\widetilde{\Sigma}_{\epsilon\epsilon}^{-1}A)^{-1}A'\widetilde{\Sigma}_{\epsilon\epsilon}^{-1}\bar{Y}_i - (A'\widehat{\Sigma}_{yy.z}^{-1}A)^{-1}A'\widehat{\Sigma}_{yy.z}^{-1}\bar{Y}\}.\end{aligned} \tag{6.38}$$

Thus, the $\beta_i - \beta_n$ are estimated using only the r_1 linear combinations $\widehat{V}_{(r_1)}^{**\prime}(A'\widetilde{\Sigma}_{\epsilon\epsilon}^{-1}A)^{-1}A'\widetilde{\Sigma}_{\epsilon\epsilon}^{-1}(\bar{Y}_i - \bar{Y}_n)$ of the differences in the sample mean vectors. Observe that when there is no growth curve structure, hence $r = m$ and $A = I_m$, the result in (6.38) reduces to the result (3.18) in Section 3.2.

Note that the matrix in (6.34), whose eigenvectors and eigenvalues are needed, involves the matrix $\widetilde{C}(\widehat{\Sigma}_{xx.z}^{-1} + \boldsymbol{\Psi})^{-1}\widetilde{C}'$, where

$$\Psi = (\widetilde{C} - A\widetilde{B})'\widetilde{\Sigma}_{\epsilon\epsilon}^{-1}(\widetilde{C} - A\widetilde{B})$$

and $(\widetilde{C} - A\widetilde{B}) = (I_m - P)\widetilde{C}$ with $P = A(A'\widetilde{\Sigma}_{\epsilon\epsilon}^{-1}A)^{-1}A'\widetilde{\Sigma}_{\epsilon\epsilon}^{-1}$. Now the matrix $\widetilde{C}\widehat{\Sigma}_{xx.z}\widetilde{C}' = \frac{1}{T}S_B$ from (3.15), where $S_B = \sum_{i=1}^{n} T_i(\bar{Y}_i - \bar{Y})(\bar{Y}_i - \bar{Y})'$ is the "between-group" sum of squares matrix. Therefore, using the following standard matrix inversion result (Rao 1973, p. 33),

$$(\widehat{\Sigma}_{xx.z}^{-1} + \Psi)^{-1} = \widehat{\Sigma}_{xx.z} - \widehat{\Sigma}_{xx.z}(\widetilde{C} - A\widetilde{B})' \times \{(\widetilde{C} - A\widetilde{B})\widehat{\Sigma}_{xx.z}(\widetilde{C} - A\widetilde{B})' + \widetilde{\Sigma}_{\epsilon\epsilon}\}^{-1}(\widetilde{C} - A\widetilde{B})\widehat{\Sigma}_{xx.z},$$

it follows that the matrix involved in (6.34) can be expressed in a more explicit form as

$$\widetilde{C}(\widehat{\Sigma}_{xx.z}^{-1} + \Psi)^{-1}\widetilde{C}' = \frac{1}{T}S_B - \frac{1}{T}S_B(I_m - P)' \times \left\{(I_m - P)\left[\frac{1}{T}S_B\right](I_m - P)' + \widetilde{\Sigma}_{\epsilon\epsilon}\right\}^{-1}(I_m - P)\frac{1}{T}S_B.$$

Finally, we briefly discuss aspects of linear discriminant analysis within the reduced-rank growth curve context for the multivariate one-way ANOVA growth curve model, $Y_{ij} = \mu_i + \epsilon_{ij} \equiv A\beta_i + \epsilon_{ij}$, $j = 1, \ldots, T_i$, $i = 1, \ldots, n$. Under the reduced-rank assumptions above for the growth curve parameters β_i, for $i = 1, \ldots, n-1$, we have

$$\beta_i = \beta_n + (\beta_i - \beta_n) = \beta_n + B_1 B_2^{(i)} \quad \text{and} \quad \mu_i = A\beta_i = A\beta_n + AB_1 B_2^{(i)},$$

where $B_2^{(i)}$ is the ith column of the matrix B_2. Let $B_{1*} = [B_1, B_*]$ be an $r \times r$ full-rank augmentation of the matrix B_1, normalized by $B_{1*}'(A'\widetilde{\Sigma}_{\epsilon\epsilon}^{-1}A)B_{1*} = I_r$, and set $L_*' = [\ell_1, \ldots, \ell_r]' = B_{1*}'A'\widetilde{\Sigma}_{\epsilon\epsilon}^{-1}$. Then considering the r linear combinations

$$L_*'\mu_i = B_{1*}'A'\widetilde{\Sigma}_{\epsilon\epsilon}^{-1}(A\beta_n + AB_1 B_2^{(i)}) = L_*'A\beta_n + \begin{bmatrix} I_{r_1} \\ 0 \end{bmatrix} B_2^{(i)}, \quad i = 1, \ldots, n-1,$$

we see that only the first r_1 vectors $\ell_1, \ldots, \ell_{r_1}$ yield different mean vectors among the n groups. So these first r_1 vectors represent the set of potentially useful information for discrimination among the n groups and thus may be regarded as an adequate set of linear discriminant functions. The ML estimate of these r_1 discriminant vectors $L' = [\ell_1, \ldots, \ell_{r_1}]' = B_1'A'\widetilde{\Sigma}_{\epsilon\epsilon}^{-1}$ is

$$\widehat{L}' = \widehat{B}_1'A'\widetilde{\Sigma}_{\epsilon\epsilon}^{-1} = \widehat{V}_{(r_1)}^{**\prime}(A'\widetilde{\Sigma}_{\epsilon\epsilon}^{-1}A)^{-1}A'\widetilde{\Sigma}_{\epsilon\epsilon}^{-1},$$

where $\widehat{B}_1 = (A'\widetilde{\Sigma}_{\epsilon\epsilon}^{-1}A)^{-1}\widehat{V}_{(r_1)}^{**}$ is the ML estimate of B_1. Also notice that the ML estimates of the corresponding linear combinations of mean vectors, $L'\mu_i$, are obtained using (6.38) as

$$\begin{aligned}\widehat{L}'\widehat{\mu}_i &= \widehat{L}'A\widehat{\beta}_i = \widehat{V}_{(r_1)}^{**\prime}\widehat{\beta}_i \\ &= \widehat{V}_{(r_1)}^{**\prime}(A'\widetilde{\Sigma}_{\epsilon\epsilon}^{-1}A)^{-1}A'\widetilde{\Sigma}_{\epsilon\epsilon}^{-1}\bar{Y}_i \equiv \widehat{V}_{(r_1)}^{**\prime}\widetilde{\beta}_i, \quad i = 1, \ldots, n, \qquad (6.39)\end{aligned}$$

where $\widetilde{\beta}_i = (A'\widetilde{\Sigma}_{\epsilon\epsilon}^{-1}A)^{-1}A'\widetilde{\Sigma}_{\epsilon\epsilon}^{-1}\bar{Y}_i$ is the "full-rank" ML estimate of the growth curve parameters for the ith group. So the r_1 column vectors of $\widehat{V}_{(r_1)}^{**}$ may be referred to as the discriminant vectors for the space of growth curve parameters (e.g., Albert and Kshirsagar 1993, p. 351). Thus for an individual with response vector Y_{ij}, $\widetilde{\beta}_{ij} = (A'\widetilde{\Sigma}_{\epsilon\epsilon}^{-1}A)^{-1}A'\widetilde{\Sigma}_{\epsilon\epsilon}^{-1}Y_{ij}$ can be regarded as a weighted LS estimate of its individual growth curve parameters, and the ML estimates of its linear discriminants are $\widehat{L}'Y_{ij} = \widehat{V}_{(r_1)}^{**\prime}(A'\widetilde{\Sigma}_{\epsilon\epsilon}^{-1}A)^{-1}A'\widetilde{\Sigma}_{\epsilon\epsilon}^{-1}Y_{ij} \equiv \widehat{V}_{(r_1)}^{**\prime}\widetilde{\beta}_{ij}$, r_1 linear combinations of its estimated growth curve parameters.

6.8 A Numerical Example

In this section, we present details for a numerical example on multivariate growth curve and reduced-rank growth curve analysis for some bioassay data, to illustrate some of the methods and results described in the previous sections of this chapter.

We consider the basic multivariate growth curve model analysis methods of the previous sections applied to some bioassay data taken from a study by Volund (1980). The data consist of measurements on blood sugar concentration of 36 rabbits at 0, 1, 2, 3, 4, and 5 h after administration of an insulin dose. The data are presented in Table A.4 of the Appendix. The 36 rabbits comprise four groups of nine animals according to a 2×2 factorial with two types of insulin preparation, a "standard" and a "test" preparation, crossed with two dose levels, 0.75 and 1.50 units per rabbit, based on an assumed potency of 26 units per mg insulin. In fact, the "standard" preparation was a preparation of MC porcine insulin, and the "test" was a preparation from another batch of MC porcine insulin, so we will not expect any substantial "type of treatment" effect.

Figure 6.1 shows the average blood sugar concentrations for each of the four groups at each time point.

We consider the vector $Y = (y_1, y_2, \ldots, y_5)'$ of measurements at the times 1, 2, 3, 4, and 5 h as the $m = 5$ dimension response vector and use as predictor variables the indicator variable for insulin type ($x_1 = \pm 1$), the indicator variable for dose level ($x_2 = \pm 1$), the initial blood sugar concentration at time 0 minus 100 ($x_3 \equiv y_0 - 100$), and a term for "interaction" between dose level and initial concentration ($x_4 = x_2 x_3$). Thus, the initial multivariate linear regression model that we consider for the kth vector of responses is of the form

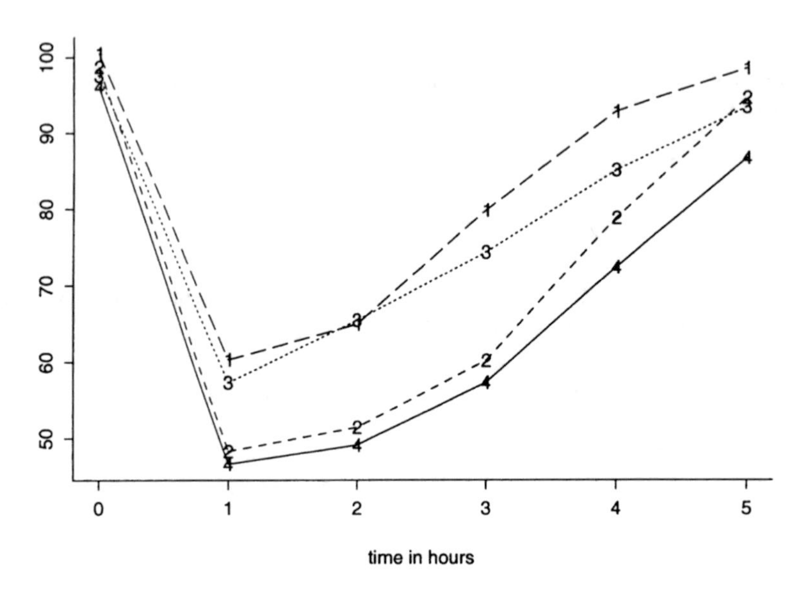

Fig. 6.1 Average blood sugar concentrations over time for each of four treatment groups. Different groups indicated by symbols: 1, standard treatment at low dose; 2, standard treatment at high dose; 3, test treatment at low dose; 4, test treatment at high dose

$$Y_k = c_0 + c_1 x_{1k} + c_2 x_{2k} + c_3 x_{3k} + c_4 x_{4k} + \epsilon_k \equiv C X_k + \epsilon_k, \quad k = 1, \ldots, T,$$

where $C = [c_0, c_1, \ldots, c_4]$ and $X_k = (1, x_{1k}, \ldots, x_{4k})'$, and $T = 36$. With $\mathbf{Y}$ and $\mathbf{X}$ denoting 5×36 data matrices, respectively, the least squares estimate $\widetilde{C} = \mathbf{Y}\mathbf{X}'(\mathbf{X}\mathbf{X}')^{-1}$ of the regression coefficient matrix C is obtained as

$$\widetilde{C} = \begin{bmatrix} 54.5405 & -0.0500 & -5.0428 & 0.7879 & -0.0062 \\ (1.4840) & (1.4530) & (1.4820) & (0.2300) & (0.2242) \\ 59.0460 & 0.6836 & -7.0757 & 0.8073 & -0.1451 \\ (1.5700) & (1.5380) & (1.5680) & (0.2434) & (0.2372) \\ 69.8572 & -0.6324 & -8.1554 & 1.0494 & 0.0948 \\ (1.9110) & (1.8710) & (1.9080) & (0.2962) & (0.2887) \\ 85.1540 & -1.7673 & -4.4398 & 1.3019 & 0.6732 \\ (2.1070) & (2.0630) & (2.1040) & (0.3266) & (0.3183) \\ 95.9062 & -1.6904 & -0.5283 & 1.1409 & 0.6809 \\ (1.8180) & (1.7810) & (1.8160) & (0.2818) & (0.2747) \end{bmatrix},$$

with estimated standard deviations given in parentheses below the corresponding estimate. As was anticipated, it is seen that the insulin type "treatment" effect variable x_1 is not a significant factor for these data, whereas the dose variable (x_2) and initial concentration (x_3) are quite important as expected, and the interaction (x_4) of the dose variable with initial concentration is a relevant factor mainly for the later responses at times of 4 and 5 h. The ML estimate of the 5×5 covariance matrix of the errors ϵ_k (with correlations shown above the diagonal) is given by

$$\widetilde{\Sigma}_{\epsilon\epsilon} = \frac{1}{36}\widehat{\epsilon\epsilon}' = \begin{bmatrix} 62.1769 & 0.8552 & 0.5617 & 0.3149 & 0.1569 \\ 56.2715 & 69.6356 & 0.7907 & 0.5422 & 0.3865 \\ 44.9707 & 66.9940 & 103.0968 & 0.8106 & 0.5045 \\ 27.8012 & 50.6523 & 92.1435 & 125.3396 & 0.7572 \\ 11.9579 & 31.1633 & 49.5009 & 81.9090 & 93.3698 \end{bmatrix},$$

where $\widehat{\boldsymbol{\epsilon}} = \mathbf{Y} - \widetilde{C}\mathbf{X}$, with $\log(|\widetilde{\Sigma}_{\epsilon\epsilon}|) = 17.7096$.

We next allow for a growth curve model structure of the form of (6.1), $Y_k = ABX_k + \epsilon_k$, to represent the pattern of mean blood sugar concentrations over the times 1 through 5 h. With the columns of the within-subject design matrix A specified as orthogonal polynomial terms, it is found that a polynomial model of degree 3 ($r = 4$) is required for an adequate fit. The goodness-of-fit chi-squared test statistic, discussed in Section 6.3, for adequacy of this model gives the value of $\mathcal{M} = 6.22$ with 5 degrees of freedom. For the matrix A as shown, we obtain the ML estimate for the growth curve coefficient matrix B as follows:

$$A' = \begin{bmatrix} 1 & 1 & 1 & 1 & 1 \\ -2 & -1 & 0 & 1 & 2 \\ 2 & -1 & -2 & -1 & 2 \\ -1 & 2 & 0 & -2 & 1 \end{bmatrix},$$

$$\widehat{B} = \begin{bmatrix} 72.9036 & -0.6908 & -5.0451 & 1.0176 & 0.2599 \\ 10.6675 & -0.6092 & 0.9131 & 0.1137 & 0.1932 \\ 1.0211 & -0.1128 & 0.9673 & -0.0307 & 0.0220 \\ -1.0303 & 0.3352 & -0.0117 & -0.0620 & -0.0884 \end{bmatrix},$$

where $\widehat{B} = (A'\widetilde{\Sigma}_{\epsilon\epsilon}^{-1}A)^{-1}A'\widetilde{\Sigma}_{\epsilon\epsilon}^{-1}\widetilde{C}$, with ML error covariance matrix estimate $\widehat{\Sigma}_{\epsilon\epsilon}$ such that $\log(|\widehat{\Sigma}_{\epsilon\epsilon}|) = 17.9010$. It is found that the growth curve model reproduces the mean blood sugar concentration curves over time, in the sense that the estimates of mean concentration obtained from the growth curve model, for each group, are quite close to the mean concentrations estimated from the LS regression results.

Next we explore the possibility of a reduced-rank regression model form, without specification of any specific polynomial growth curve model structure. That is, as suggested in Section 6.5, we consider models of the form of Chapter 2, $Y_k = ABX_k + \epsilon_k$, where the multivariate regression coefficient matrix $C = AB$ is assumed to have reduced rank r and A and B are both unknown (unspecified) matrices of parameters to be estimated (subject to normalization constraints, of course). LR testing procedures for the rank of C, as described in Section 2.6, indicate that a coefficient matrix of rank $r = 2$ will be adequate, with the LR statistic for testing $H_0 : \text{rank}(C) \leq 2$ having the value $\mathcal{M} = 8.62$ with $(5-2)(5-2) = 9$ degrees of freedom. The ML estimates of the component matrices A and B are obtained as in (2.17) of Section 2.3 (also (6.16) in Section 6.5) and are given by

$$\widehat{A}' = \begin{bmatrix} 4.6510 & 5.0493 & 5.9560 & 7.2172 & 8.1490 \\ -4.2176 & -6.3067 & -6.5796 & -2.1946 & 1.3839 \end{bmatrix},$$

$$\widehat{B} = \begin{bmatrix} 11.7598 & -0.2109 & -0.2252 & 0.1454 & 0.0656 \\ 0.0601 & -0.2346 & 0.9531 & -0.0078 & 0.0770 \end{bmatrix},$$

with ML error covariance matrix estimate $\widehat{\Sigma}_{\epsilon\epsilon}$ having $\log(|\widehat{\Sigma}_{\epsilon\epsilon}|) = 17.9922$. Notice, first, that the two column vectors of the estimated matrix $\widehat{A}$ can be viewed as the two "basis" functions over time needed to represent the mean blood sugar concentration curves over time for each of the groups, and these provide essentially as good a representation as the third degree polynomial growth curve model. Also, taking into account the scales of the variables x_{ik}, the first linear combination $x^*_{1k} = \widehat{\beta}'_1 X_k$ of predictor variables in $X^*_k = \widehat{B} X_k$ consists mainly of a constant term and contributions from the initial concentration (x_3) and interaction (x_4) variables, whereas the second linear combination $x^*_{2k} = \widehat{\beta}'_2 X_k$ is dominantly the dose level variable (x_2). We also mention that although a rank 2 model provides an acceptable fit, a rank 3 model produces an excellent fit with $\log(|\widehat{\Sigma}_{\epsilon\epsilon}|) = 17.7404$.

We can also consider a reduced-rank structure within the growth curve model framework, as detailed in Section 6.6. Thus, we consider a model in the form of (6.17), $Y_k = A\, B_1 B_2 X_k + \epsilon_k$, with the within-subject growth curve design matrix A specified to have the third degree orthogonal polynomial terms as in the previous full-rank growth curve model. The LR testing procedure for the rank associated with this model again indicates that a rank 2 model would be adequate, with the LR test statistic from Section 6.6 for a test of H_0 : rank$(B) \leq 2$ giving the value $\mathcal{M} = 7.74$ with $(4-2)(5-2) = 6$ degrees of freedom. The ML estimates of the component growth curve coefficient matrices B_1 and B_2 are obtained from (6.21) and are given as

$$\widehat{B}'_1 = \begin{bmatrix} 6.2299 & 0.8919 & 0.0813 & -0.0785 \\ -3.7281 & 1.3797 & 0.9835 & -0.2211 \end{bmatrix},$$

$$\widehat{B}_2 = \begin{bmatrix} 11.7553 & -0.2105 & -0.2389 & 0.1453 & 0.0644 \\ 0.1224 & -0.2746 & 0.8787 & -0.0094 & 0.0696 \end{bmatrix},$$

with ML error covariance matrix estimate $\widehat{\Sigma}_{\epsilon\epsilon}$ having $\log(|\widehat{\Sigma}_{\epsilon\epsilon}|) = 18.1590$. Notice that the two estimated linear combinations of predictor variables that occur in this model, $x^*_{ik} = \widehat{\beta}'_i X_k$, $i = 1, 2$, are very similar to those estimated for the previous reduced-rank regression model with a rank of 2 (especially the first linear combination x^*_{1k}). In addition, it is found that the factor $A\widehat{B}_1$ from the reduced-rank growth curve model above is fairly similar to the estimated left coefficient matrix $\widehat{A}$ in the previous reduced-rank regression model (especially the first column), with

$$(A\widehat{B}_1)' = \begin{bmatrix} 4.6871 & 5.0997 & 6.0674 & 7.1976 & 8.0977 \\ -4.2995 & -6.5335 & -5.6950 & -2.8896 & 0.7772 \end{bmatrix}.$$

Finally, we can also consider the "extended" versions of the reduced-rank regression and reduced-rank growth curve models, as discussed in Sections 3.1 and 6.6, respectively. The extended reduced-rank regression model we considered is $Y_k = C_1 X_{1,k} + D Z_k + \epsilon_k$, with $C_1 = AB$, where $Z_k \equiv 1$ corresponds to the constant terms in the model and $X_{1,k} = (x_{1k}, x_{2k}, x_{3k}, x_{4k})'$ contains the remaining predictor variables. LR testing procedures for the appropriate rank of C_1, through canonical correlation analysis calculations as described in Section 3.1, indicate that a rank of $r_1 = 2$ would be appropriate for C_1, with a LR test statistic for H_0 : rank$(C_1) \leq 2$ having the value $\mathcal{M} = 1.93$ on $(5-2)(4-2) = 6$ degrees of freedom. The ML estimates of the component matrices A and B, obtained from Eq. (3.11) as described in Section 3.1, are given by

$$\widehat{A}' = \begin{bmatrix} 5.3545 & 5.7591 & 8.0997 & 9.2386 & 6.9661 \\ -3.3313 & -5.4329 & -5.0752 & -0.3182 & 2.5203 \end{bmatrix},$$

$$\widehat{B} = \begin{bmatrix} -0.2229 & -0.4256 & 0.1489 & 0.0643 \\ -0.2865 & 0.8867 & 0.0192 & 0.0917 \end{bmatrix},$$

with $\widehat{D} = (54.4812, 58.9653, 69.9291, 85.1844, 95.8072)'$, and ML error covariance matrix estimate $\widehat{\Sigma}_{\epsilon\epsilon}$ having $\log(|\widehat{\Sigma}_{\epsilon\epsilon}|) = 17.7738$. Note that the ML estimate $\widehat{D}$ in this model is nearly identical to the first column (corresponding to the constant terms) of the LS estimate $\widetilde{C}$ in the multivariate regression model. In addition, notice that the estimated coefficients in $\widehat{B}$, especially those in the first row, that determine the two linear combinations of the predictor variables $X_{1,k}$ are fairly similar to the corresponding values in the previous reduced-rank model, which did not separate out the constant term from the remaining predictor variables in formulating the reduced-rank structure. The current extended model does provide for a slightly better fitting model, however.

The extended version of the reduced-rank growth curve model, $Y_k = ABX_{1,k} + AB^* Z_k + \epsilon_k$, with $B = B_1 B_2$, was also considered. A rank of 2 was selected for the growth curve coefficient matrix B (LR statistic for testing H_0 : rank$(B) \leq 2$ of $\mathcal{M} = 1.15$ with 4 degrees of freedom). The ML estimates were obtained from (6.28) as

$$\widehat{B}_1' = \begin{bmatrix} 7.0023 & 0.7475 & -0.4158 & -0.5541 \\ -2.8220 & 1.4915 & 0.8697 & -0.3841 \end{bmatrix},$$

$$\widehat{B}_2 = \begin{bmatrix} -0.2312 & -0.3852 & 0.1498 & 0.0684 \\ -0.3156 & 0.8303 & 0.0099 & 0.0810 \end{bmatrix},$$

Table 6.1 Summary of fitting results of several multivariate regression and growth curve models estimated for the blood sugar concentrations data

Type of model	Rank of C or B	Number of parameters, n^*	$\log \lvert\widehat{\Sigma}_{\epsilon\epsilon}\rvert$	AIC = $\log \lvert\widehat{\Sigma}_{\epsilon\epsilon}\rvert + 2n^*/T$	BIC = $\log \lvert\widehat{\Sigma}_{\epsilon\epsilon}\rvert + n^* \log T/T$
Regression, full rank	5	25	17.7096	19.0985	20.1981
Regression, reduced rank	2	16	17.9922	18.8811	19.5849
Regression, extended reduced rank	2	19	17.7738	18.8293	19.6651
Growth curve, full rank	4	20	17.9010	19.0121	19.8918
Growth curve, reduced rank	2	14	18.1590	18.9368	19.5526
Growth curve, extended reduced rank	2	16	17.9400	18.8289	19.5327

with $\widehat{B}^* = (72.8995, 10.6707, 0.9950, -1.0610)'$, and ML error covariance matrix estimate $\widehat{\Sigma}_{\epsilon\epsilon}$ having $\log(|\widehat{\Sigma}_{\epsilon\epsilon}|) = 17.9400$. Once again, the implied two linear combinations of predictor variables determined under this model are similar to those estimated in the previous reduced-rank models. In addition, the ML estimate $\widehat{B}^*$ in this model is very similar to the first column of the estimate $\widehat{B}$ in the previous "full-rank" growth curve model.

A summary of the fits of several multivariate regression and growth curve models considered for the blood sugar concentration data is presented in Table 6.1. Given in the table are, values of $\log|\widehat{\Sigma}_{\epsilon\epsilon}|$ and of the AIC and BIC information criteria for each model, where AIC $= \log|\widehat{\Sigma}_{\epsilon\epsilon}| + 2n^*/T$ and BIC $= \log|\widehat{\Sigma}_{\epsilon\epsilon}| + n^* \log(T)/T$, and n^* is the number of estimated regression parameters in the model. From this summary of results, it can be seen that the models with reduced ranks offer desirable improvements (and simplifications) over the full-rank models. Thus, we have illustrated that more parsimonious models can be developed in the growth curve context through the use of reduced-rank regression methods.

6.9 Some Recent Developments

Unbalanced Data The models studied in this chapter all correspond to a single response characteristic over "m" distinct times on "T" subjects. It is possible to collect several response variables, and these data may not be all available over the same number of times; response variables may be measured at distinct frequency.

Hwang and Takane (2004) study a multivariate reduced-rank growth curve model in such a setting for a parsimonious representation.

It is taken that for T independent subjects, $Y_k^{(j)}$ is a m_j-dimensional jth response vector, and $X_k^{(j)}$ is a n_j-dimensional set of time-invariant explanatory variables. This setup is general as data on response variables do not have to be collected at the same time points and not necessarily at the same number of times. Thus, the model (6.1) can be written in a general form,

$$Y_k^{(j)} = A^{(j)} B^{(j)} X_k^{(j)} + \epsilon_k^{(j)}, \quad j = 1, 2, \ldots, J. \tag{6.40}$$

It is important to note that $Y_k^{(j)}$ and $X_k^{(j)}$ are of varying dimensions, varying over the jth response variables. As in model (6.17), a reduced-rank structure on $B^j = B_1^{(j)} B_2^{(j)}$ can be imposed for dimension reduction. It is also possible that "J" predictor sets, $X_k^{(1)} = \cdots = X_k^{(J)} = X_k$, and if $A^{(j)\prime}$s are known, the model is the standard growth curve model. This case can be further simplified when all known basis functions across the "J" response variables, $A^{(1)} = \cdots = A^{(J)}$. The estimation and inference details are all equal and can be found in Hwang and Takane (2004). In some practical applications, it is possible to consider additional constraints on B_1 and B_2 matrices across the response variables.

GAMANOVA-MANOVA von Rosen and von Rosen (2017) consider an extension of the growth curve model with two sets of regressors, one set representing the growth curve part and the other representing the variance part. Imposing column-wise nested reduced-rank structure, the authors demonstrate how it provides interpretation in terms of latent variables.

A Mixture Model The traditional growth curve model specifies a set of r "basis" functions of time, such as (orthogonal) polynomial terms up to degree $r - 1$, to represent the form of mean response over time. The completely prescribed parametric model then takes the form (6.1), $\mathbf{Y}_k = \mathbf{ABX}_k + \epsilon_k$, $k = 1, \ldots, T$, where $\mathbf{A}$ is a $m \times r$ matrix whose columns contain prescribed values of the set of "basis" functions of time, and $\mathbf{B}$ is a $r \times n$ matrix of unknown parameters to be estimated, with $r < m$.

Sometimes a rather large number of basis functions of a particular class might need to be specified to obtain adequate representation for the mean response function, e.g., high degree polynomial curves may be needed. Thus, it might not be desired to specify an entire set of basis functions of time in advance, but to let some functions be left unspecified of a general form to be determined. So, as a generalization, we can consider a model that has a mixture of pre-specified and unspecified basis functions. In particular, to motivate, we could partition $\mathbf{A}$ as $\mathbf{A} = [\mathbf{A}_1, \mathbf{A}_2]$, where $\mathbf{A}_1$ is $m \times r_1$ and is prescribed, with $\mathbf{A}_2$ being $m \times r_2$ but "general" and not prescribed. For instance, we could take $r_1 = 2$ with the columns of $\mathbf{A}_1$ corresponding to a constant term and linear term as the only pre-specified functions. Then partitioning $\mathbf{B}$ compatible with the partition of $\mathbf{A}$, the generalized model is

$$\mathbf{Y}_k = \mathbf{A}_1\mathbf{B}_1\mathbf{X}_k + \mathbf{A}_2\mathbf{B}_2\mathbf{X}_k + \epsilon_k. \tag{6.41}$$

This model can be modified or generalized further to allow for (at least partly) different predictor variable effects for different parts of the model as

$$\mathbf{Y}_k = \mathbf{A}_1\mathbf{B}_1\mathbf{X}_k + \mathbf{A}_2\mathbf{B}_2\mathbf{Z}_k + \epsilon_k, \tag{6.42}$$

where $\mathbf{Z}_k$ could represent a n_2-dimensional vector of additional covariates (some of which could be in common with $\mathbf{X}_k$). Such a modification might be desired if we wish to treat certain of the explanatory variables in a different manner than others, e.g., the effects of treatment group indicator variables might be modeled differently than effects of continuous measurement covariates. We see that the generalized model is a mixture of a *traditional growth curve model* component, $\mathbf{A}_1\mathbf{B}_1\mathbf{X}_k$ with $\mathbf{A}_1$ prescribed, and a *reduced-rank regression* component, $\mathbf{A}_2\mathbf{B}_2\mathbf{Z}_k$ with $r_2 \leq \min(m, n_2)$, and $\mathbf{A}_2$ as well as $\mathbf{B}_2$ unknown. The two extremes of the traditional growth curve model and the reduced-rank regression are included as special cases of the more general model.

Interest will be in model specification/selection, estimation of the unknown regression parameters $\mathbf{B}_1$, $\mathbf{A}_2$, and $\mathbf{B}_2$ (with the latter two parametric matrices subject to some normalization conditions), and estimation of the error covariance matrix $\Sigma_{\epsilon\epsilon}$. The model could also be modified to allow for some forms of random effects covariance structure, particularly for the prescribed component of the model $\mathbf{A}_1\mathbf{B}_1\mathbf{X}_k$, such that $\text{Cov}(\mathbf{Y}_k) = \mathbf{A}_1\boldsymbol{\Gamma}\mathbf{A}_1' + \sigma^2\mathbf{I}$, for example.

The models developed in this chapter have scope for further extension and applications.

(Some of the results in this chapter have appeared in Reinsel and Velu 2003.)

Chapter 7
Seemingly Unrelated Regressions Models With Reduced Ranks

7.1 Introduction and the Seemingly Unrelated Regressions Model

The classical multivariate linear regression model discussed in Chapter 1 can be generalized by allowing the different response variables y_{ik} to have different input or predictor variables $X_{ik} = (x_{i1k}, \ldots, x_{ink})'$ for different i, so that $y_{ik} = X'_{ik}C_{(i)} + \epsilon_{ik}$, $i = 1, \ldots, m$, and the errors ϵ_{ik} are contemporaneously correlated across the different response variables. Multivariate linear regression models of this form were considered by Zellner (1962), who referred to them as *seemingly unrelated regression* (SUR) equation models. In experimental design situations, the model is also referred to as the multiple design multivariate linear model (e.g., Srivastava 1967; Roy et al. 1971). In the notation of Section 1.2, the linear regression model for the $T \times 1$ vector of values for the ith response variable is

$$\mathbf{Y}_{(i)} = \mathbf{X}'_i C_{(i)} + \boldsymbol{\epsilon}_{(i)}, \qquad \text{for} \quad i = 1, \ldots, m, \tag{7.1}$$

where $\mathbf{X}_i = [X_{i1}, \ldots, X_{iT}]$ is $n \times T$, while $\mathbf{Y}_{(i)} = (y_{i1}, \ldots, y_{iT})'$, $\boldsymbol{\epsilon}_{(i)} = (\epsilon_{i1}, \ldots, \epsilon_{iT})'$, and $C_{(i)}$ are the same as in the regression model of Chapter 1. Stacking the m vector equations of (7.1) together, the SUR model can be expressed in vector form as

$$y = \overline{\mathbf{X}}c + e, \tag{7.2}$$

where $\overline{\mathbf{X}} = \text{Diag}(\mathbf{X}'_1, \mathbf{X}'_2, \ldots, \mathbf{X}'_m)$, while the vectors $c = (C'_{(1)}, \ldots, C'_{(m)})'$, $y = (\mathbf{Y}'_{(1)}, \ldots, \mathbf{Y}'_{(m)})'$, and $e = (\boldsymbol{\epsilon}'_{(1)}, \ldots, \boldsymbol{\epsilon}'_{(m)})'$ are the same as in Section 1.3, and hence, $\text{Cov}(e) = \Omega = \Sigma_{\epsilon\epsilon} \otimes I_T = [(\sigma_{ij} I_T)]$.

Before we proceed with discussion of statistical properties associated with analysis of the SUR model, we want to motivate the usefulness of such models.

G. C. Reinsel et al., *Multivariate Reduced-Rank Regression*, Lecture Notes in Statistics 225, https://doi.org/10.1007/978-1-0716-2793-8_7

The dependence of the m equations in (7.1) through the contemporaneous correlations of the error terms $\boldsymbol{\epsilon}_{(i)}$ is a special feature of the SUR model. As will be demonstrated below, use of this additional information across the equations can improve the statistical properties of the estimators of the regression coefficients. The econometrics literature contains many applications of the SUR model (e.g., see Srivastava and Giles 1987, p. 2–3). The first example used by Zellner (1962) is based on data that appear in Boot and de Wit (1960). For two large corporations, General Electric and Westinghouse, both from the same type of industry, data were taken on the investment functions of the firms annually from 1935 through 1954. It seems reasonable to assume that the error terms associated with the investment functions of the two companies may be contemporaneously correlated because of the influence of common market factors. Another example where the SUR specification is used is in the context of explaining certain economic activities in different geographic regions. Giles and Hampton (1984) estimated production functions for several regions in New Zealand assuming the possibility of inter-regional dependency through the correlations in the error terms. These examples suggest that the SUR model is appropriate and useful for a wide range of applications. The least square (LS) estimators of the $C_{(i)}$ in model (7.1) are similar to those in (1.8),

$$\widetilde{C}_{(i)} = (\mathbf{X}_i\mathbf{X}_i')^{-1}\mathbf{X}_i\mathbf{Y}_{(i)}, \qquad i = 1, \ldots, m.$$

The LS estimators have the properties that $E(\widetilde{C}_{(i)}) = C_{(i)}$ and

$$\text{Cov}(\widetilde{C}_{(i)}, \widetilde{C}_{(j)}) = \sigma_{ij}(\mathbf{X}_i\mathbf{X}_i')^{-1}\mathbf{X}_i\mathbf{X}_j'(\mathbf{X}_j\mathbf{X}_j')^{-1}, \qquad i, j = 1, \ldots, m.$$

In particular, we have the usual result that $\text{Cov}(\widetilde{C}_{(i)}) = \sigma_{ii}(\mathbf{X}_i\mathbf{X}_i')^{-1}$. However, unlike the previous classical multivariate linear model, in the SUR model, the LS estimators are not the same as the generalized least squares (GLS) estimator, which is given by

$$\widehat{c} = \{\overline{\mathbf{X}}'(\Sigma_{\epsilon\epsilon}^{-1} \otimes I_T)\overline{\mathbf{X}}\}^{-1}\overline{\mathbf{X}}'(\Sigma_{\epsilon\epsilon}^{-1} \otimes I_T)y. \tag{7.3}$$

Hence, the GLS estimator for this model does depend on the covariance matrix $\Sigma_{\epsilon\epsilon}$ and does not coincide with the individual LS estimators of the m regression models in (7.1), unless $\overline{\mathbf{X}} = I_m \otimes \mathbf{X}'$, that is, $\mathbf{X}_1 = \mathbf{X}_2 = \cdots = \mathbf{X}_m$, so that the m linear regression models each involve the same design matrix $\mathbf{X}$ of input variables. The GLS estimator in (7.3) has mean vector $E(\widehat{c}) = c$ and covariance matrix given by

$$\text{Cov}(\widehat{c}) = \{\overline{\mathbf{X}}'(\Sigma_{\epsilon\epsilon}^{-1} \otimes I_T)\overline{\mathbf{X}}\}^{-1}. \tag{7.4}$$

The difficulty with implementing the GLS estimator in (7.3) is that the covariance matrix $\Sigma_{\epsilon\epsilon}$ will typically be unknown. A two-step estimation procedure was originally proposed (e.g., Zellner 1962) in which an estimate of $\Sigma_{\epsilon\epsilon}$ is obtained from the residuals from the LS estimators $\widetilde{C}_{(i)}$ as $\widetilde{\Sigma}_{\epsilon\epsilon} = T^{-1}\widehat{\boldsymbol{\epsilon}}\widehat{\boldsymbol{\epsilon}}'$, where $\widehat{\boldsymbol{\epsilon}} =$

$[\widehat{\epsilon}_{(1)}, \ldots, \widehat{\epsilon}_{(m)}]'$ with $\widehat{\epsilon}_{(i)} = \mathbf{Y}_{(i)} - \mathbf{X}_i' \widetilde{C}_{(i)}$, $i = 1, \ldots, m$, being the vectors of residuals from the LS estimation of the individual linear regression models in (7.1). Then, the GLS estimator of the form of (7.3) is obtained, but with $\Sigma_{\epsilon\epsilon}$ replaced by the estimate $\widetilde{\Sigma}_{\epsilon\epsilon}$, that is,

$$\widetilde{c} = \{\overline{\mathbf{X}}'(\widetilde{\Sigma}_{\epsilon\epsilon}^{-1} \otimes I_T)\overline{\mathbf{X}}\}^{-1}\overline{\mathbf{X}}'(\widetilde{\Sigma}_{\epsilon\epsilon}^{-1} \otimes I_T)y. \tag{7.5}$$

The covariance matrix of the estimator $\widetilde{c}$ is then approximated by $\widehat{\mathrm{Cov}(\widetilde{c})} = \{\overline{\mathbf{X}}'(\widetilde{\Sigma}_{\epsilon\epsilon}^{-1} \otimes I_T)\overline{\mathbf{X}}\}^{-1}$. The two-step procedure can also be iterated, with the current GLS estimator $\widetilde{c}$ in (7.5) at any given stage used to form the estimate $\widetilde{\Sigma}_{\epsilon\epsilon}$, and this will lead to the maximum likelihood estimators of c and $\Sigma_{\epsilon\epsilon}$ under the normality assumption of the errors $\epsilon_{(i)}$ in (7.1). In principle, the GLS estimator $\widehat{c} = (\widehat{C}'_{(1)}, \ldots, \widehat{C}'_{(m)})'$ in (7.3) offers gains in efficiency, in terms of reduced variance, over the LS estimators. The largest gains in efficiency can be expected for situations when the design matrices satisfy $\mathbf{X}_i \mathbf{X}_j' \approx 0$, and the squared correlations $\rho_{ij}^2 = \sigma_{ij}^2/(\sigma_{ii}\sigma_{jj})$ of the errors are large (e.g., see Zellner 1962, 1963; Revankar 1974). For illustration, consider the case with $m = 2$ equations. Then, from (7.4), the covariance matrix of the GLS estimator can be written as

$$\mathrm{Cov}(\widehat{c}) = \mathrm{Cov}\begin{bmatrix}\widehat{C}_{(1)} \\ \widehat{C}_{(2)}\end{bmatrix} = \begin{bmatrix}\sigma^{11}\mathbf{X}_1\mathbf{X}_1' & \sigma^{12}\mathbf{X}_1\mathbf{X}_2' \\ \sigma^{12}\mathbf{X}_2\mathbf{X}_1' & \sigma^{22}\mathbf{X}_2\mathbf{X}_2'\end{bmatrix}^{-1},$$

where we denote $\Sigma_{\epsilon\epsilon}^{-1} = [(\sigma^{ij})]$, with $\sigma^{11} = \sigma_{22}/(\sigma_{11}\sigma_{22} - \sigma_{12}^2)$, $\sigma^{22} = \sigma_{11}/(\sigma_{11}\sigma_{22} - \sigma_{12}^2)$, and $\sigma^{12} = -\sigma_{12}/(\sigma_{11}\sigma_{22} - \sigma_{12}^2)$. In particular, using a standard matrix inversion result, we have that

$$\begin{aligned}\mathrm{Cov}(\widehat{C}_{(1)}) &= \{\sigma^{11}\mathbf{X}_1\mathbf{X}_1' - [(\sigma^{12})^2/\sigma^{22}]\mathbf{X}_1\mathbf{X}_2'(\mathbf{X}_2\mathbf{X}_2')^{-1}\mathbf{X}_2\mathbf{X}_1'\}^{-1} \\ &= \sigma_{11}(1 - \rho_{12}^2)\{\mathbf{X}_1\mathbf{X}_1' - \rho_{12}^2\mathbf{X}_1\mathbf{X}_2'(\mathbf{X}_2\mathbf{X}_2')^{-1}\mathbf{X}_2\mathbf{X}_1'\}^{-1},\end{aligned}$$

noting that $1/\sigma^{11} = (\sigma_{11}\sigma_{22} - \sigma_{12}^2)/\sigma_{22} = \sigma_{11}(1 - \rho_{12}^2)$ and $(\sigma^{12})^2/(\sigma^{11}\sigma^{22}) = (\sigma_{12})^2/(\sigma_{11}\sigma_{22}) = \rho_{12}^2$. Compared to the covariance matrix of the LS estimator, $\mathrm{Cov}(\widetilde{C}_{(1)}) = \sigma_{11}(\mathbf{X}_1\mathbf{X}_1')^{-1}$, we readily see that the largest reduction in variance for the GLS estimator will occur when $\mathbf{X}_1\mathbf{X}_2' = 0$ and ρ_{12}^2 is large. In most practical situations, the predictor variables (the design matrices) are likely to be similar, and therefore, the condition $X_1X_2' = 0$ is not generally possible except in certain experimental conditions, where the design matrices can be so constructed. In practice, for the GLS estimator in (7.5), the use of the estimated error covariance matrix $\widetilde{\Sigma}_{\epsilon\epsilon}$ will cause some increase in the covariance matrix of the resulting GLS estimator above the ideal result in (7.4).

7.2 Relation of Growth Curve Model to the Seemingly Unrelated Regressions Model

It has been noted by Stanek III and Koch (1985) that the basic growth curve (or GMANOVA) model considered in Chapter 6 can be interpreted as a special case of a SUR model. Recognition of this relationship between the growth curve and SUR models can be useful for better appreciation of estimation procedures under the two models and for methods of extension to more flexible growth curve models. Thus, we briefly discuss the relationship in this section. Consider from Section 6.2 the basic growth curve model in matrix form, $\mathbf{Y} = AB\mathbf{X} + \boldsymbol{\epsilon}$, and let $H = [A_1, A_2]$, where $A_1 = A(A'A)^{-1}$ and A_2 is $m \times (m - r)$ such that $A_2'A = 0$. Then under the transformation of the model as in (6.5), with $\mathbf{Y}^* = H'\mathbf{Y} = [\mathbf{Y}^*_{(1)}, \ldots, \mathbf{Y}^*_{(m)}]'$, we have

$$\mathbf{Y}^* = H'AB\mathbf{X} + H'\boldsymbol{\epsilon} = P'B\mathbf{X} + \boldsymbol{\epsilon}^* = \begin{bmatrix} B\mathbf{X} \\ 0 \end{bmatrix} + \boldsymbol{\epsilon}^*, \tag{7.6}$$

where $P' = H'A = [I_r, 0]'$ and $\boldsymbol{\epsilon}^* = H'\boldsymbol{\epsilon} = [\boldsymbol{\epsilon}^*_{(1)}, \ldots, \boldsymbol{\epsilon}^*_{(m)}]'$. Set $\Sigma^* = \text{Cov}(\epsilon^*_k) = H'\Sigma_{\epsilon\epsilon}H$, and write the $r \times n$ matrix B of unknown growth curve coefficients as $B = [B_{(1)}, \ldots, B_{(r)}]'$. The transformed growth curve model (7.6) can then be expressed as

$$\mathbf{Y}^*_{(i)} = \mathbf{X}'B_{(i)} + \boldsymbol{\epsilon}^*_{(i)}, \qquad i = 1, \ldots, r, \tag{7.7a}$$

and

$$\mathbf{Y}^*_{(i)} = \boldsymbol{\epsilon}^*_{(i)}, \qquad i = r + 1, \ldots, m. \tag{7.7b}$$

The (transformed) growth curve model, expressed in the form (7.7), can be seen to be a particular version of a SUR model (7.1). The model can be written in vector form similar to (7.2) as

$$y^* = \overline{\mathbf{X}}b + e^* = (P' \otimes \mathbf{X}')b + e^*, \tag{7.8}$$

with $\overline{\mathbf{X}} = (P' \otimes \mathbf{X}') = [I_r \otimes \mathbf{X}, 0]'$, $b = \text{vec}(B') = (B'_{(1)}, \ldots, B'_{(r)})'$, and $\text{Cov}(e^*) = \Sigma^* \otimes I_T$. For this model, a two-stage GLS estimator can be constructed by using an estimate of Σ^* based on LS estimation of each of the m regression equations separately, as discussed at the end of Section 7.1, with the common design matrix $\mathbf{X}$. The resulting estimator of Σ^* is $\widetilde{\Sigma}^* = (1/T)\mathbf{Y}^*[I_T - \mathbf{X}'(XX')^{-1}\mathbf{X}]\mathbf{Y}^* = H'\widetilde{\Sigma}_{\epsilon\epsilon}H$. From the last relation and the nonsingularity of H, it follows that $H\widetilde{\Sigma}^{*-1}H' = \widetilde{\Sigma}^{-1}_{\epsilon\epsilon}$. Thus, when GLS estimation is applied to model (7.8), with the estimate $\widetilde{\Sigma}^*$, we obtain

$$\widetilde{b} = [(P \otimes \mathbf{X})(\widetilde{\Sigma}^{*-1} \otimes I_T)(P' \otimes \mathbf{X}')]^{-1}(P \otimes \mathbf{X})(\widetilde{\Sigma}^{*-1} \otimes I_T)y^*$$

$$= [(P\widetilde{\Sigma}^{*-1}P')^{-1}P\widetilde{\Sigma}^{*-1} \otimes (XX')^{-1}\mathbf{X}]y^*.$$

In matrix form, using the relations $P = A'H$ and $H\widetilde{\Sigma}^{*-1}H' = \widetilde{\Sigma}_{\epsilon\epsilon}^{-1}$, we find that this GLS estimator can be written equivalently as

$$\begin{aligned}\widetilde{B} &= (P\widetilde{\Sigma}^{*-1}P')^{-1}P\widetilde{\Sigma}^{*-1}\mathbf{Y}^*\mathbf{X}'(XX')^{-1} \\ &= (A'\widetilde{\Sigma}_{\epsilon\epsilon}^{-1}A)^{-1}A'\widetilde{\Sigma}_{\epsilon\epsilon}^{-1}\mathbf{YX}'(XX')^{-1}.\end{aligned} \tag{7.9}$$

Notice that this last expression in (7.9) is identical to the ML estimator of B for the growth curve model (6.1), as given in (6.2) of Section 6.2. Hence, we see that the basic growth curve model can be transformed and reexpressed as a particular SUR model (7.7) and that a corresponding two-step GLS estimation for the SUR model yields the ML estimate for the growth curve model. Since SUR models can be of much more general form than in (7.7), the connection displayed above between growth curve models and SUR models suggests that, perhaps, more general and flexible growth curve models could be formulated and analyzed through the methods of SUR models.

7.3 Reduced-Rank Coefficient in Seemingly Unrelated Regressions Model

Before we formally introduce the reduced-rank aspects for the SUR model (7.1), we want to discuss the need and usefulness of such a feature particularly in modeling large-scale data sets. From the previous discussion thus far, it can be seen that the main idea is that the m regression equations in (7.1) are indeed related to each other, and this relationship is captured through the correlations between the error terms. The aspect of reduced rank that we will propose here captures additional relationship in a more "structured" way. To expand on this, consider the investment function example used by Zellner (1962). Denote I_t as the current gross investment of the firm, C_{t-1} as the firm's beginning of the year capital stock, and F_{t-1} as the value of the firm's outstanding shares at the beginning of the year. These variables are measured for the two electric companies annually from 1935 through 1954. Observe that each company has its own values of the predictor variables C_{t-1} and F_{t-1}, but they are conceptually the same type of variables. General Electric's gross investment, on the average, is about twice as large as that of Westinghouse. For these data, the GLS estimates obtained from (7.5) are given as

$$\begin{aligned}\text{General Electric}: I_{1t} &= 0.139C_{1,t-1} + 0.038F_{1,t-1} - 27.7 \\ \text{Westinghouse}: I_{2t} &= 0.064C_{2,t-1} + 0.058F_{2,t-1} - 1.3.\end{aligned}$$

The estimated correlation between the estimated residual in the two equations is approximately 0.73. The standard errors of the regression coefficients from the SUR model approach are generally smaller than the standard errors of the ordinary LS coefficient estimates. Thus the SUR model approach provides more efficient estimates. From the above estimated equations, we observe that although the response variable I_t is of different magnitudes for the two companies, the regression coefficients of the predictor variables are roughly of the same magnitude. This raises the possibility that apart from the constant terms, the coefficients of certain of the predictor variables, for example F_{t-1}, could be constrained to be the same, or more generally, the matrix formed from the coefficients of each SUR equation could be of reduced rank. The possible relations or constraints in regression coefficients across equations are somewhat suggested even in the two equation examples, but we might expect such features to be more prominent when the number of SUR equations is larger. Thus the dependence among the SUR equations can be more directly modeled through dependency of the regression coefficient vectors from the different equations because the predictor variables are conceptually similar. While this dependency can be estimated mainly through the available data on the observed response and predictor variables, the dependency through correlations among the error terms as formulated by Zellner (1962) might be thought of as representing the unknown or unobservable common factors, possibly not accounted for by the available data or the regression component of the model (7.1). The approach we will take here has the features of the reduced-rank regression model first discussed in Chapter 2. It provides an opportunity for dimension reduction as well as a way to make interpretations of linear combinations of predictor variables as canonical variables in a certain sense. These features will be emphasized as we develop the model. As a starting point for the development of the reduced-rank model, we restate the SUR model (7.1) as

$$\mathbf{Y}_{(i)} = \mathbf{X}_i' C_{(i)} + \boldsymbol{\epsilon}_{(i)}, \qquad \text{for} \qquad i = 1, \ldots, m.$$

Arranging the regression coefficient vectors $C_{(i)}$ in matrix form, we have $C = [C_{(1)}, C_{(2)}, \ldots, C_{(m)}]'$ as an $m \times n$ matrix. As noted previously, the model (7.1) becomes equivalent to the standard multivariate regression model when the predictor variables are the same for all m equations, that is, $\mathbf{X}_1 = \mathbf{X}_2 = \cdots = \mathbf{X}_m = \mathbf{X}$, with the equations $\mathbf{Y}'_{(i)} = C'_{(i)}\mathbf{X} + \boldsymbol{\epsilon}'_{(i)}$ then being stacked to yield the matrix equation $\mathbf{Y} = C\mathbf{X} + \boldsymbol{\epsilon}$ as in (1.4), where $\mathbf{Y} = [\mathbf{Y}_{(1)}, \ldots, \mathbf{Y}_{(m)}]'$ and $\boldsymbol{\epsilon} = [\boldsymbol{\epsilon}_{(1)}, \ldots, \boldsymbol{\epsilon}_{(m)}]'$. For the SUR equation model, we shall assume that the $m \times n$ matrix C is of reduced rank. More formally, we assume that

$$\text{rank}(C) = r \leq \min(m, n). \tag{7.10}$$

Recall that an implication of this assumption is that C can be written as $C = AB$, where A is a full-rank matrix of dimension $m \times r$ and B is full rank of dimension $r \times n$. Then, because $C_{(i)} = B' a_{(i)}$, where $A' = [a_{(1)}, \ldots, a_{(m)}]$, and each $a_{(i)}$ is of dimension $r \times 1$, we can write the reduced-rank version of the SUR model (7.1) as

$$\mathbf{Y}_{(i)} = \mathbf{X}_i' B' a_{(i)} + \boldsymbol{\epsilon}_{(i)}, \qquad i = 1, \ldots, m. \tag{7.11}$$

From the discussion in Chapter 2, it may be recalled that when the matrices $\mathbf{X}_i$ are the same, the linear combinations of $\mathbf{X}$ that are most useful for modeling the relationships with the response variables $\mathbf{Y}$ are given by $B\mathbf{X}$. Depending on the dimensions involved, substantial reduction in parameterization is possible, and $B\mathbf{X}$ can also be related to canonical variables. We see from (7.11) that such an idea can be extended to the more general multivariate regression system (7.1), and the coefficients (the matrix B) of a type of canonical variables are fixed to be the same for all m seemingly unrelated data sets. The distinctions among these regression equations are reflected in the coefficients $a_{(i)}$ that are allowed to differ among the regression equations in (7.1). Thus the reduced-rank regression approach to analyze the regression models in (7.1) provides a canonical method to link these models through the structure of the coefficients in addition to taking into account the correlations among the error terms. We will focus on the estimation of the component matrices A and B and also on the subsequent estimation of the matrix $C (= AB)$. Similar to the previous reduced-rank model situations, we need to impose normalization conditions as in (2.14) to uniquely identify the parameters A and B. Because of the different matrices $\mathbf{X}_i$ that appear in model (7.1), natural normalization conditions for B are not obvious. Recall that we earlier argued that dimension reduction through reduced rank seemed sensible because of the "similarity" of the predictor variable sets among the m different regression equations. Hence, for the normalization condition, it appears reasonable to use the average of the second moment matrices of the predictor variables as follows:

$$A'\Gamma A = I_r \qquad \text{and} \qquad B\Sigma_{xx}B' = \Lambda_r^2, \tag{7.12}$$

where Γ is a positive-definite matrix that will typically be specified as $\Gamma = \Sigma_{\epsilon\epsilon}^{-1}$ and

$$\Sigma_{xx} = \lim_{T\to\infty} \frac{1}{mT} \sum_{i=1}^{m} \mathbf{X}_i \mathbf{X}_i'.$$

Note that these conditions correspond to (2.14) when $\mathbf{X}_i = \mathbf{X}$ for all i.

7.4 Maximum Likelihood Estimators for Reduced-Rank Model

We now consider derivation of efficient estimators of the coefficient matrices A and B in the reduced-rank SUR model under the normality assumption for the error terms in (7.1). Let $\theta = (\alpha', \beta')'$, where $\alpha = \text{vec}(A')$ and $\beta = \text{vec}(B')$. The criterion to be minimized can be stated as

$$S_T(\theta) = \frac{1}{2T} e'(\Gamma \otimes I_T)e, \tag{7.13}$$

where $e = y - \overline{\mathbf{X}}c$, subject to the normalization conditions as in (7.12) that are based on sample versions of the normalizing matrices. We take $\Gamma = \widetilde{\Sigma}_{\epsilon\epsilon}^{-1}$, the inverse of the error covariance matrix estimate based on LS residuals, and Σ_{xx} is replaced by $\widehat{\Sigma}_{xx} = \frac{1}{mT}\sum_{i=1}^{m} \mathbf{X}_i\mathbf{X}_i'$. It appears that it is not possible to obtain an explicit solution for A and B simultaneously. Therefore, we consider the first order conditions for minimization of (7.13) from which one can solve for either one of the component matrices when the other is known.

Since $C = AB$, we have $c = \text{vec}(C') = (A \otimes I_n)\text{vec}(B') = (I_m \otimes B')\text{vec}(A')$, and so $e = y - \overline{\mathbf{X}}(A \otimes I_n)\beta = y - \overline{\mathbf{X}}(I_m \otimes B')\alpha$. Therefore, the first order conditions from (7.13) are given as

$$\frac{\partial S_T(\theta)}{\partial \alpha} = -\frac{1}{T}(I_m \otimes B)\overline{\mathbf{X}}'(\Gamma \otimes I_T)[y - \overline{\mathbf{X}}(I_m \otimes B')\alpha], \tag{7.14a}$$

$$\frac{\partial S_T(\theta)}{\partial \beta} = -\frac{1}{T}(A' \otimes I_n)\overline{\mathbf{X}}'(\Gamma \otimes I_T)[y - \overline{\mathbf{X}}(A \otimes I_n)\beta]. \tag{7.14b}$$

Introduce the notations $\overline{\mathbf{X}}(B) = \overline{\mathbf{X}}(I_m \otimes B') = \text{Diag}\{\mathbf{X}_1'B', \ldots, \mathbf{X}_m'B'\}$ and $\overline{\mathbf{X}}(A) = \overline{\mathbf{X}}(A \otimes I_n)$. Then the solutions to (7.14) for α and β, each in terms of the other parameter, can be expressed as

$$\widehat{\alpha} = [\overline{\mathbf{X}}(B)'(\Gamma \otimes I_T)\overline{\mathbf{X}}(B)]^{-1}\overline{\mathbf{X}}(B)'(\Gamma \otimes I_T)y, \tag{7.15a}$$

$$\widehat{\beta} = [\overline{\mathbf{X}}(A)'(\Gamma \otimes I_T)\overline{\mathbf{X}}(A)]^{-1}\overline{\mathbf{X}}(A)'(\Gamma \otimes I_T)y. \tag{7.15b}$$

It is easy to see that when $\mathbf{X}_1 = \cdots = \mathbf{X}_m = \mathbf{X}, \overline{\mathbf{X}}' = (I_m \otimes \mathbf{X})$ and the above equations reduce to the partial least squares equation results for the standard multivariate regression model case as given in Section 2.3. That is, we then have $\overline{\mathbf{X}}(B) = (I_m \otimes \mathbf{X}'B')$ and $\overline{\mathbf{X}}(A) = (A \otimes \mathbf{X}')$, so that the solutions in (7.15) reduce to $\widehat{\alpha} = [I_m \otimes (B\boldsymbol{X}\boldsymbol{X}'B')^{-1}B\mathbf{X}]y$ and $\widehat{\beta} = [(A'\Gamma A)^{-1}A'\Gamma \otimes (\boldsymbol{X}\boldsymbol{X}')^{-1}\mathbf{X}]y$, equivalently, $\widehat{A} = \boldsymbol{Y}\boldsymbol{X}'B'(B\boldsymbol{X}\boldsymbol{X}'B')^{-1}$ and $\widehat{B} = (A'\Gamma A)^{-1}A'\Gamma\mathbf{Y}\mathbf{X}'(\boldsymbol{X}\boldsymbol{X}')^{-1}$ with $y = \text{vec}(\mathbf{Y}')$.

In the above expressions, we will take $\Gamma = \widetilde{\Sigma}_{\epsilon\epsilon}^{-1}$, where $\widetilde{\Sigma}_{\epsilon\epsilon}$ is obtained from the LS residuals as defined in Section 7.1. As noted in previous chapters, the first order equations in (7.14) refer only to necessary conditions for a minimum to occur and so do not guarantee the global minimum. We suggest two computational schemes for solving the first order equations to obtain the efficient estimator.

The partial least squares procedure is perhaps computationally the easiest to implement. For a given estimate $\widehat{\beta}$ or $\widehat{B}$, the equation in (7.15a) is used to compute $\widehat{\alpha}$ or $\widehat{A}$, and similarly, given the estimate $\widehat{\alpha}$, Eq. (7.15b) is used to arrive at an estimate

$\widehat{\beta}$. Once these are computed, we use the equations

$$\widehat{A}'\widetilde{\Sigma}_{\epsilon\epsilon}^{-1}\widehat{A} = I_r \qquad \text{and} \qquad \widehat{B}\widehat{\Sigma}_{xx}\widehat{B}' = \Lambda_r^2 \tag{7.16}$$

to normalize the estimates, with $\widehat{\Sigma}_{xx} = \frac{1}{mT}\sum_{i=1}^{m}\mathbf{X}_i\mathbf{X}_i'$. To perform the normalizations based on given unnormalized estimates $\widehat{A}$ and $\widehat{B}$, we first obtain the normalized eigenvectors $V_1, \ldots, V_r$ of $\widehat{A}'\widetilde{\Sigma}_{\epsilon\epsilon}^{-1}\widehat{A}$ with corresponding eigenvalues $d_1, \ldots, d_r$ and define the $r \times r$ matrices $V = [V_1, \ldots, V_r]$ and $D = \operatorname{diag}(d_1, \ldots, d_r)$. Next we obtain the normalized eigenvectors $U_1, \ldots, U_r$ of the matrix $D^{1/2}V'\widehat{B}\widehat{\Sigma}_{xx}\widehat{B}'VD^{1/2}$, with corresponding eigenvalues $\lambda_1^2, \ldots, \lambda_r^2$, and set $U = [U_1, \ldots, U_r]$. Then the normalized versions of the estimates of A and B are given by $\widehat{A}_* = \widehat{A}VD^{-1/2}U$ and $\widehat{B}_* = U'D^{1/2}V'\widehat{B}$, such that $\widehat{C} \equiv \widehat{A}\widehat{B} = \widehat{A}_*\widehat{B}_*$, and $\widehat{A}_*$ and $\widehat{B}_*$ satisfy the normalization conditions (7.16). The partial least squares procedure then continues to iterate between the two solutions $\widehat{\alpha}$ and $\widehat{\beta}$, and at each step, the above normalizations are imposed.

The second computational procedure is essentially a Newton–Raphson method based on the Taylor expansion of the first order equations. Thus, it is based on the same relations as given in (3.29) of Section 3.5.1 and in (4.11) of Section 4.3. Similar to the procedures given in Sections 3.5.1 and 4.4, the ith step of the iterative scheme is given by

$$\begin{bmatrix} \widehat{\theta}_i \\ \widehat{\lambda}_i \end{bmatrix} = \begin{bmatrix} \widehat{\theta}_{i-1} \\ 0 \end{bmatrix} - \begin{bmatrix} W_{i-1} & Q_{i-1} \\ Q_{i-1}' & -I_{r^2} \end{bmatrix} \begin{bmatrix} \partial S_T(\theta)/\partial\theta|_{\widehat{\theta}_{i-1}} \\ h(\theta)|_{\widehat{\theta}_{i-1}} \end{bmatrix}, \tag{7.17}$$

where

$$W = (B_\theta + H_\theta H_\theta')^{-1}B_\theta(B_\theta + H_\theta H_\theta')^{-1} \tag{7.18}$$

and $Q = -(B_\theta + H_\theta H_\theta')^{-1}H_\theta$, with $H_\theta = \{\partial h_j(\theta)/\partial\theta_i\}$, where $h(\theta) = 0$ denotes the $r^2 \times 1$ vector of normalization condition (7.12). In (7.18), the matrix B_θ is $B_\theta = \lim_{T\to\infty}\partial^2 S_T(\theta)/\partial\theta\partial\theta'$ as defined below in Eq. (7.20) in Theorem 7.1. Notice also that the vector of first partial derivatives in (7.17) can be expressed as

$$\frac{\partial S_T(\theta)}{\partial\theta} = -\frac{1}{T}M'\overline{\mathbf{X}}'(\Sigma_{\epsilon\epsilon}^{-1} \otimes I_T)e, \tag{7.19}$$

where $M = [(I_m \otimes B'), (A \otimes I_n)]$. The iterations are carried out until successive changes in $\widehat{\theta}_i$ or $\partial S_T(\widehat{\theta}_i)/\partial\theta$ are reasonably small.

We now state a result on the asymptotic properties of the efficient estimator. For this, we take $\Gamma = \Sigma_{\epsilon\epsilon}^{-1}$ as known in (7.13) and (7.12), for convenience of presentation. Again, the theoretical results follow from the work of Robinson (1973) and Silvey (1959), and the proof is essentially of the same lines as those of Theorems 3.2 and 4.1. Therefore, details of the proof are omitted; we note only that the main asymptotic distributional result needed in the proof is that

$\frac{1}{\sqrt{T}}\overline{\mathbf{X}}'(\Sigma_{\epsilon\epsilon}^{-1} \otimes I_T)e \xrightarrow{d} N(0, V)$ as $T \rightarrow \infty$ where V is defined below in Theorem 7.1 (e.g., see Srivastava and Giles 1987, p. 28).

Theorem 7.1 *Consider the model (7.1) under the stated assumptions, with* $C = AB$*, where* A *and* B *satisfy the normalization conditions (7.12), and let* $h(\theta) = 0$ *denote the* $r^2 \times 1$ *vector of normalization conditions obtained by stacking the condition (7.12) without duplication. Assume the limits* $\lim_{T\rightarrow\infty} \frac{1}{T}\mathbf{X}_i\mathbf{X}_j' = \lim_{T\rightarrow\infty} \frac{1}{T}\sum_{k=1}^{T} X_{ik}X_{jk}' = \Omega_{ij}$*,* $i, j = 1, \ldots, m$*, exist almost surely and* $\Sigma_{\epsilon\epsilon}$ *is a positive-definite matrix. Let* $\widehat{\theta}$ *be the estimator that minimizes the criterion (7.13) subject to conditions (7.12) for the choice* $\Gamma = \Sigma_{\epsilon\epsilon}^{-1}$*. Then as* $T \rightarrow \infty$*,* $\widehat{\theta} \rightarrow \theta$ *almost surely and* $\sqrt{T}(\widehat{\theta} - \theta)$ *has a limiting multivariate normal distribution with zero mean vector and (singular) covariance matrix given by the matrix* $W = (B_\theta + H_\theta H_\theta')^{-1}B_\theta(B_\theta + H_\theta H_\theta')^{-1}$*, as in (7.18), where*

$$B_\theta = \lim_{T\rightarrow\infty} \frac{\partial^2 S_T(\theta)}{\partial\theta\partial\theta'} = M'VM, \qquad M = [(I_m \otimes B'), (A \otimes I_n)], \tag{7.20}$$

and $V = \{(V_{ij})\}$ *is* $mn \times mn$*, where* $V_{ij} = \sigma^{ij}\Omega_{ij}$ *is an* $n \times n$ *matrix,* σ^{ij} *denotes the* (i, j)*th element of* $\Sigma_{\epsilon\epsilon}^{-1}$*, and* $H_\theta = \{\partial h_j(\theta)/\partial\theta_i\}$ *is of dimension* $r(m+n) \times r^2$*.*

The asymptotic covariance matrix W can be used to make inferences on the parameters of the component matrices A and B based on the efficient estimators $\widehat{A}$ and $\widehat{B}$. The inferential aspects associated with the overall reduced-rank regression coefficient matrix estimator $\widehat{C} = \widehat{A}\widehat{B}$ are also of interest. Because $\widehat{C} - C = (\widehat{A} - A)B + A(\widehat{B} - B) + o_p(T^{-1/2})$, it follows as in previous cases that $\sqrt{T}\text{vec}(\widehat{C}' - C') \xrightarrow{d} N(0, MWM')$, and $MWM' = M(M'VM)^{-}M'$. Inferences related to $\widehat{C}$ can be based on this asymptotic distribution result.

7.5 An Alternate Estimator and Its Properties

The computation of the ML estimator in the reduced-rank SUR model requires iterative procedures, as noted in the previous section, and for good convergence of the procedures, a reasonably accurate starting value is needed. Therefore, initial estimates of the component matrices A and B should be chosen with some care. In this section, we seek an alternate to the ML estimator that is computationally simpler and that can be used to specify initial values for the component matrices in the ML procedures. In addition, such an alternate estimator can be used in a more computationally convenient way than the ML estimator to initially specify or identify the appropriate rank of the coefficient matrix C through hypothesis testing procedures. As in the approach discussed for the model considered in Chapter 4, a full-rank estimator can be first computed, and it may be used to help reveal any possible reduced-rank structure in the true coefficient matrix. Hence, we will construct an alternate estimator for the component matrices A and B based on the full-rank GLS estimator of the regression coefficient matrix C for model (7.1).

The motivation for the alternate estimator in the reduced-rank SUR model (7.1) is somewhat similar to the reasoning offered for the alternate estimator considered in Section 4.5 for the reduced-rank regression model with autoregressive errors of Chapter 4. Observe that the criterion (7.13), with the choice $\Gamma = \widetilde{\Sigma}_{\epsilon\epsilon}^{-1}$, can be written as

$$S_T(\theta) = \frac{1}{2T}(y - \overline{\mathbf{X}}\widetilde{c})'(\widetilde{\Sigma}_{\epsilon\epsilon}^{-1} \otimes I_T)(y - \overline{\mathbf{X}}\widetilde{c}) + \frac{1}{2T}(\widetilde{c} - c)'\overline{\mathbf{X}}'(\widetilde{\Sigma}_{\epsilon\epsilon}^{-1} \otimes I_T)\overline{\mathbf{X}}(\widetilde{c} - c), \tag{7.21}$$

where $\widetilde{c} = \{\overline{\mathbf{X}}'(\widetilde{\Sigma}_{\epsilon\epsilon}^{-1} \otimes I_T)\overline{\mathbf{X}}\}^{-1}\overline{\mathbf{X}}'(\widetilde{\Sigma}_{\epsilon\epsilon}^{-1} \otimes I_T)y$ is the full-rank GLS estimator in (7.5). Minimizing $S_T(\theta)$ with respect to A and B (with $C = AB$) is equivalent to minimizing the second term above subject to the normalization conditions in (7.16). As discussed in the previous section, the computational procedures to solve for the efficient estimates of the component matrices are iterative. However, in the special case of the classical multivariate regression setup, we have seen in Chapter 2 that the ML estimates of the component matrices can be calculated simultaneously and explicitly through computation of certain eigenvectors and eigenvalues. In this special case, this is possible because we then have $\overline{\mathbf{X}}' = (I_m \otimes \mathbf{X})$, a specific Kronecker structure, and so the second term in (7.21) reduces to

$$\frac{1}{2T}(\widetilde{c} - c)'(\widetilde{\Sigma}_{\epsilon\epsilon}^{-1} \otimes XX')(\widetilde{c} - c) = \frac{1}{2T}\text{tr}[\widetilde{\Sigma}_{\epsilon\epsilon}^{-1}(\widetilde{C} - AB)XX'(\widetilde{C} - AB)'],$$

where $\widetilde{c} = \text{vec}(\widetilde{C}')$ and $\widetilde{C} = YX'(XX')^{-1}$. The Householder–Young theorem can be used to arrive at explicit solutions for A and B, for example, as given in (2.17) of Section 2.3. These computational simplifications are possible mainly because in the criterion the matrix $\widetilde{\Sigma}_{\epsilon\epsilon}^{-1} \otimes XX'$, which acts as a "weights matrix" for the vector $\widetilde{c} - c$ that indicates the distance between the (full-rank) GLS estimator and the true parameters, is of Kronecker product form, whereas such simplifications do not occur in the SUR model since the matrix $\overline{\mathbf{X}}'(\widetilde{\Sigma}_{\epsilon\epsilon}^{-1} \otimes I_T)\overline{\mathbf{X}}$ is of more complicated (not a Kronecker) form.

For an alternate estimation procedure, we suggest that this latter matrix be replaced in the criterion by a weight matrix of similar form to that which appears in the classical multivariate regression setup. Specifically, in place of $\frac{1}{T}XX'$ in the classical case, we choose the moment matrix of the averages of the predictor variable sets,

$$\widehat{\Sigma}_{\bar{x}\bar{x}} = \frac{1}{T}\bar{\mathbf{X}}\bar{\mathbf{X}}' = \frac{1}{m^2T}\sum_{i=1}^{m}\sum_{j=1}^{m}\mathbf{X}_i\mathbf{X}_j',$$

where $\bar{\mathbf{X}} = \frac{1}{m}\sum_{i=1}^{m}\mathbf{X}_i$, and consider use of the following criterion to be minimized:

$$\frac{1}{2}(\widetilde{c}-c)'(\widetilde{\Sigma}_{\epsilon\epsilon}^{-1}\otimes\widehat{\Sigma}_{\bar{x}\bar{x}})(\widetilde{c}-c)=\frac{1}{2}\mathrm{tr}[\widetilde{\Sigma}_{\widetilde{\epsilon}\widetilde{\epsilon}}^{-1}(\widetilde{C}-AB)\widehat{\Sigma}_{\bar{x}\bar{x}}(\widetilde{C}-AB)']. \tag{7.22}$$

It follows from the Householder–Young theorem and related results in Chapter 2 that the alternate estimates $\widetilde{A}$ and $\widetilde{B}$, which are obtained by minimizing (7.22) and which satisfy similar normalization conditions as in (7.16) for the ML estimates, can be computed in the following way. The columns of $\widehat{\Sigma}_{\bar{x}\bar{x}}^{1/2}\widetilde{B}'$ are chosen to be the eigenvectors corresponding to the r largest eigenvalues of the matrix $\widehat{\Sigma}_{\bar{x}\bar{x}}^{1/2}\widetilde{C}'\widetilde{\Sigma}_{\epsilon\epsilon}^{-1}\widetilde{C}\widehat{\Sigma}_{\bar{x}\bar{x}}^{1/2}$, and $\widetilde{A}=\widetilde{C}\widehat{\Sigma}_{\bar{x}\bar{x}}\widetilde{B}'\widetilde{\Lambda}^{-2}$, where $\widetilde{\Lambda}$ is a diagonal matrix with positive diagonal elements equal to the square roots of the eigenvalues of the matrix. Equivalently,

$$\widetilde{A}=\widetilde{\Sigma}_{\epsilon\epsilon}^{1/2}[\widetilde{V}_1,\ldots,\widetilde{V}_r] \quad \text{and} \quad \widetilde{B}=[\widetilde{V}_1,\ldots,\widetilde{V}_r]'\widetilde{\Sigma}_{\epsilon\epsilon}^{-1/2}\widetilde{C},$$

where the $\widetilde{V}_j$ are normalized eigenvectors of the matrix $\widetilde{\Sigma}_{\epsilon\epsilon}^{-1/2}\widetilde{C}\widehat{\Sigma}_{\bar{x}\bar{x}}\widetilde{C}'\widetilde{\Sigma}_{\epsilon\epsilon}^{-1/2}$ corresponding to its r largest eigenvalues.

It is expected that this alternate procedure might result in estimators of A and B that have fairly high efficiency in cases where the matrices of regressors $\mathbf{X}_i$ have somewhat similar characteristics, in the sense that the terms $\frac{1}{T}\mathbf{X}_i\mathbf{X}_j'$ are moderately similar for all i and j, since then the "weights matrix" $\frac{1}{T}\overline{\mathbf{X}}'(\widetilde{\Sigma}_{\epsilon\epsilon}^{-1}\otimes I_T)\overline{\mathbf{X}}$ might be well approximated by $\widetilde{\Sigma}_{\epsilon\epsilon}^{-1}\otimes\widehat{\Sigma}_{\bar{x}\bar{x}}$. Otherwise, the alternate estimators $\widetilde{A}$ and $\widetilde{B}$ might tend to not have good efficiency, but they still might have reasonable use as initial estimators. We also suggest the following improvement to this initial estimation procedure. Specifically, once the initial estimate $\widetilde{B}$ has been obtained as indicated, the estimate of A is obtained by a (one-step) GLS estimation of the SUR model (7.11) for the $\mathbf{Y}_{(i)}$ on the constructed regressors $(\widetilde{B}\mathbf{X}_i)'$, $i=1,\ldots,m$. That is, we obtain the initial estimate of A using (7.15a), based on the initial estimate $\widetilde{B}$, as

$$\widetilde{\alpha}=[\overline{\mathbf{X}}(\widetilde{B})'(\widetilde{\Sigma}_{\epsilon\epsilon}^{-1}\otimes I_T)\overline{\mathbf{X}}(\widetilde{B})]^{-1}\overline{\mathbf{X}}(\widetilde{B})'(\widetilde{\Sigma}_{\epsilon\epsilon}^{-1}\otimes I_T)y,$$

with $\overline{\mathbf{X}}(\widetilde{B})=\mathrm{Diag}\{\mathbf{X}_1'\widetilde{B}',\ldots,\mathbf{X}_m'\widetilde{B}'\}$ and $\widetilde{\alpha}=\mathrm{vec}(\widetilde{A}')=(\widetilde{a}_{(1)}',\ldots,\widetilde{a}_{(m)}')'$.

Our main interest in developing the alternate estimator is to provide initial values for the component matrices in computing the ML estimates and also for use in preliminary specification of the rank, as will be discussed in Section 7.6. However, the asymptotic distribution of the alternate estimator might be of some interest for comparison with the efficient estimator. Because the details are similar to Theorem 7.1 and to results from Chapter 4, we only briefly mention the final result. We let $\widetilde{\theta}=(\widetilde{\alpha}',\widetilde{\beta}')'$, where $\widetilde{\alpha}=\mathrm{vec}(\widetilde{A}')$ and $\widetilde{\beta}=\mathrm{vec}(\widetilde{B}')$ with the alternate estimators $\widetilde{A}$ and $\widetilde{B}$ as given above. It can then be established that $\sqrt{T}(\widetilde{\theta}-\theta)$ has a limiting multivariate normal distribution with zero mean vector and covariance matrix given by the leading $r(m+n)$-order square submatrix of the $r(m+n+r)$-order matrix

$$\begin{bmatrix} M'(\Sigma_{\epsilon\epsilon}^{-1} \otimes \Sigma_{\bar{x}\bar{x}})M & -H_\theta \\ -H_\theta' & 0 \end{bmatrix}^{-1}$$
$$\times \begin{bmatrix} M'(\sum_{\epsilon\epsilon}^{-1} \otimes \sum_{\bar{x}\bar{x}})V^{-1}(\sum_{\epsilon\epsilon}^{-1} \otimes \sum_{\bar{x}\bar{x}})M & 0 \\ 0 & 0 \end{bmatrix}$$
$$\times \begin{bmatrix} M'(\Sigma_{\epsilon\epsilon}^{-1} \otimes \Sigma_{\bar{x}\bar{x}})M & -H_\theta \\ -H_\theta' & 0 \end{bmatrix}^{-1},$$

where M, V, and H_θ are as defined before in Theorem 7.1, and

$$\Sigma_{\bar{x}\bar{x}} = \lim_{T\to\infty} \frac{1}{m^2 T} \sum_{i=1}^{m} \sum_{j=1}^{m} \mathbf{X}_i \mathbf{X}_j' = \frac{1}{m^2} \sum_{i=1}^{m} \sum_{j=1}^{m} \Omega_{ij}.$$

This covariance matrix is expressible more explicitly as

$$W^* = (B_\theta^* + H_\theta H_\theta')^{-1} M'(\Sigma_{\epsilon\epsilon}^{-1} \otimes \Sigma_{\bar{x}\bar{x}})V^{-1}(\Sigma_{\epsilon\epsilon}^{-1} \otimes \Sigma_{\bar{x}\bar{x}})M(B_\theta^* + H_\theta H_\theta')^{-1},$$

where $B_\theta^* = M'(\Sigma_{\epsilon\epsilon}^{-1} \otimes \Sigma_{\bar{x}\bar{x}})M$. Note that the matrix V is the limiting form of T times the covariance matrix in (7.4). Although the distinction among the different matrices of predictor variables $\mathbf{X}_i$ is not fully accounted for in construction of the component matrix estimates under the alternate procedure, the asymptotic covariance matrix of the alternate estimator $\widetilde{\theta}$ does reflect the distinctness of the $\mathbf{X}_i$ through the matrix V.

7.6 Identification of Rank of the Regression Coefficient Matrix

To carry out the computations related to estimating the component matrices A and B through the maximum likelihood method as outlined in Section 7.4, the rank of the regression coefficient matrix C needs to be determined. In the classical multivariate linear regression model situation, where all $\mathbf{X}_i$ in model (7.1) are equal, the specification of rank could be conveniently carried out because of the correspondence between the number of nonzero canonical correlations and the rank of the regression matrix. For this case, we have used the LR statistics

$$-[(T-n) + \frac{1}{2}(n-m-1)] \sum_{j=r+1}^{m} \log(1 - \widehat{\rho}_j^2),$$

where the $\widehat{\rho}_j^2$ are the squared sample canonical correlations. This form of test does not apply in the SUR model, since each response variable has its own set of predictor variables. We therefore need to develop some large sample procedures to test the hypothesis of the rank of C.

Before we further consider the problem of testing for the rank of the matrix C, we want to briefly comment on procedures for testing usual linear hypotheses for coefficients in the SUR model (7.1). Typically, the model is considered in the form given by (7.2), and the linear hypothesis constraints are stated as $Rc = 0$, where R is a known matrix. A particular application of interest is testing for aggregation bias as discussed by Zellner (1962). Recall that the model (7.1) may often refer to data relating to several different micro-units. Simple aggregation of data from these units into macro-unit data, and subsequent estimation of the aggregate model, will not lead to any aggregation bias if the hypothesis $H_0 : C_{(1)} = C_{(2)} = \cdots = C_{(m)}$ (that is, all regression coefficients are the same) holds, or equivalently, $Rc = 0$, where R is an appropriately defined known $(m-1)n \times mn$ matrix. The usual LR procedures can be applied to test such hypotheses. The linear constraints imposed by a reduced-rank condition are more general than the above equality of coefficients hypothesis, however. If the above hypothesis holds, it implies that the rank of the matrix C is one; the reverse implications are somewhat more general than the coefficient vectors being identical (i.e., they are proportional). Examining the investment function example described in Section 7.3 for aggregation bias, based on the LR test Zellner (1962), concluded that the hypothesis of coefficient vector equality is rejected. Note in our earlier discussion of this example that the coefficients of some predictor variables, not necessarily all, might be constrained to be the same implying that the regression coefficient matrix could be of lower rank. Thus, interest in testing for the rank of the regression coefficient matrix may stem from other than purely statistical modeling considerations. We now discuss procedures for testing the rank of the matrix C of regression coefficients.

We can still consider the LR statistic for testing $H_0 : \text{rank}(C) \leq r$. In general terms, this is given as

$$M = [(T-n) + \frac{1}{2}(n-m-1)] \log(|\widehat{\Sigma}_{\epsilon\epsilon}^{(r)}|/|\widehat{\Sigma}_{\epsilon\epsilon}|), \tag{7.23}$$

where $\widehat{\Sigma}_{\epsilon\epsilon}$ denotes the unrestricted ML estimate of $\Sigma_{\epsilon\epsilon}$ and $\widehat{\Sigma}_{\epsilon\epsilon}^{(r)}$ denotes the restricted estimate based on the reduced-rank ML estimate $\widehat{C} = \widehat{A}\widehat{B}$ of C, subject to the condition that $\text{rank}(C) \leq r$. The statistic M asymptotically will follow a $\chi^2_{(m-r)(n-r)}$ distribution under the null hypothesis. The test statistic could be computed for various values of r and used to test the validity of each rank assumption. Because this procedure would require the somewhat involved calculations for the ML estimates $\widehat{A}$ and $\widehat{B}$ to obtain $\widehat{\Sigma}_{\epsilon\epsilon}^{(r)}$, for a range of possible ranks r, at the initial specification stage, it would not be a convenient procedure for selection of an appropriate rank for C, prior to computing the ML estimate of C. Therefore, for initial specification of possible rank, we also suggest using an alternate testing procedure based on the alternate estimator of C described in Section 7.5. The numerical example presented in the next section indicates that the alternate estimation procedure is somewhat useful for initial testing of rank, as well as for obtaining initial values in iterative computation of the ML estimates of A and B.

The alternate test statistic that we consider is thus $M^* = [(T-n) + \frac{1}{2}(n - m - 1)] \log(|\widetilde{\Sigma}_{\epsilon\epsilon}^{(r)}|/|\widetilde{\Sigma}_{\epsilon\epsilon}|)$, where $\widetilde{\Sigma}_{\epsilon\epsilon}^{(r)}$ is the error covariance matrix estimate constructed, in the usual way, from the residual vectors $\widehat{\boldsymbol{\epsilon}}_{(i)} = \mathbf{Y}_{(i)} - \mathbf{X}_i' \widetilde{C}_{(i)}^{(r)}$, $i = 1, \ldots, m$, where $\widetilde{C}^{(r)} = [\widetilde{C}_{(1)}^{(r)}, \ldots, \widetilde{C}_{(m)}^{(r)}]' = \widetilde{A}^{(r)} \widetilde{B}^{(r)}$ and $\widetilde{A}^{(r)}$ and $\widetilde{B}^{(r)}$ are the alternate estimates of A and B described in Section 7.5. Another possible and convenient alternate test statistic that is motivated by the alternate estimation procedure in Section 7.5 is of the form $M^* = [(T-n) + \frac{1}{2}(n - m - 1)] \sum_{j=r+1}^{m} \log(1 + \widetilde{\lambda}_j^2)$, where the $\widetilde{\lambda}_j^2$ are the eigenvalues of the matrix $\widetilde{\Sigma}_{\epsilon,\epsilon}^{-1/2} \widetilde{C} \widehat{\Sigma}_{\bar{x}\bar{x}} \widetilde{C}' \widetilde{\Sigma}_{\epsilon,\epsilon}^{-1/2}$, $\widehat{\Sigma}_{\bar{x}\bar{x}} = \frac{1}{m^2 T} \sum_{i=1}^{m} \sum_{j=1}^{m} \mathbf{X}_i \mathbf{X}_j'$, and $\widetilde{C}$ is the full-rank GLS estimator obtained from (7.5). Although this test statistic does not involve any quantities that are canonical correlations, because of the analogue to the classical multivariate linear model case we may interpret $\widetilde{\rho}_j^2 = \widetilde{\lambda}_j^2/(1 + \widetilde{\lambda}_j^2)$ as "canonical correlation" quantities. The standardized values $\widetilde{\rho}_j^2$ might be easier to interpret as descriptive tools than the values $\widetilde{\lambda}_j^2$. We thus suggest using the above statistic M^* and the reference $\chi^2_{(m-r)(n-r)}$ distribution to initially test the hypothesis of the rank of C, for various values of r. This alternate "approximate" test procedure will tend to be conservative because the determinant of the estimated error covariance matrix obtained using the alternate estimator will be larger than that obtained using the ML estimator of A and B.

7.7 A Numerical Illustration with Scanner Data

We illustrate the procedures for the seemingly unrelated regressions model developed in this chapter with an application in the area of marketing. Typically, manufacturers and retailers have to make, on a continual basis, important marketing decisions regarding the promotion and pricing of products. They make these decisions based on elasticity estimates related to various promotional activities and the price discounts. Now that scanner data of weekly sales for retail chains are readily available, these elasticities can be easily computed. But these extensive data do not guarantee reliable estimates for individual product brands and generally yield a large set of estimates that may be difficult to fully comprehend. We attempt to address these problems through the reduced-rank SUR models developed in this chapter. The point-of-sale data are collected at the store level for all the brands in a product category for a weekly time interval. In large cities, a retail chain typically owns several stores. For our illustrative example, we aggregate the store level data into data for the chain and compute the elasticity estimates because important marketing negotiations related to trade discounts and promotions are usually made between the manufacturers and the retail chains; the retail chains then decide on marketing plans at the store level.

The product whose data we consider in this analysis is a refrigerated product with several brands; the data are for a large chain located in a highly populated city in the northeastern United States. The weekly data span the period from mid-September 1988 to mid-June 1990, covering 93 weeks. In addition to the number of units sold, for each brand, we consider data on regular price, actual price, whether promotion was present and whether the brand was on display. With the information available, we consider the following model, similar to the model advocated by Blattberg and George (1991) and Blattberg et al. (1994), as appropriate for the store level data analysis:

$$S_t = \beta_0 + \beta_1 S_{t-1} + \beta_2 P_t + \beta_3 PD_t + \beta_4 PR_t + \beta_5 D_t + \beta_6 DD_t + \beta_7 CPD_t + \epsilon_t, \tag{7.24}$$

where

S_t = logarithm of the number of units sold in tth week
P_t = regular price
PD_t = price discount = (regular price—actual price)/regular price
PR_t = indicator for promotions
D_t = indicator for display
DD_t = deal decay code = k if deal is present in $(k+1)$st week
CPD_t = maximum price discount for competing brands in a chain

The "semi-log" model (7.24) has also been used by Blattberg and Wisniewski (1989) and is known to fit better than log–log or linear models. While the regression coefficients of a log–log model can be directly interpreted as elasticities, the coefficients of a semi-log model can be converted into point elasticities.

There are some broad assertions from the marketing literature that can guide us to interpret the coefficients of the above model. Three promotional variables, PR_t, D_t, and DD_t, are included in the model. Although the promotion and display variables had several categories, we use them as binary indicator variables because of a lack of variation in the actual occurrence of these categories. The deal decay index is supposed to reflect the waning effectiveness of promotions. It is postulated that the first weeks of promotions are far more effective than later weeks. Displays are expected to increase sales because the retail space available is limited and expensive; they are usually reserved for high volume items. Two variables are included to measure the price effect, the regular price, and price discount. It is believed that deal elasticities are generally higher than price elasticities. The effect of competition can be measured through their promotional and price cross-elasticities, but here we chose only the competitors' price discount as a variable that may play a significant role. There is some discussion on deciding on which brands compete with each other, but it generally is agreed that brands with similar prices tend to compete with each other. Brand switching based on price discounts occurs mostly toward higher priced brands; if the lower-tier brand promotes, it does not usually attract customers from higher-tier brands. The lagged value of S_t is included in (7.24) to account for serial correlation.

Because we consider the chain level data, the variables in (7.24) are all aggregated over the stores within that chain. The response variable S_t at the chain level is the logarithm of total sales aggregated over the stores, P_t is the average price, PD_t is the average price discount, PR_t is the proportion of stores promoting, D_t is the proportion of stores displaying, DD_t is the deal decay code averaged over stores, and CPD_t is the average price discount offered by the competitors. To get some understanding of the nature of the data, Fig. 7.1 displays time series plots of S_t, log sales for the $m = 8$ brands. For all but brand 6, we notice that the sales generally peak during the summer weeks, especially around July 4th. In Fig. 7.2, we present plots of the predictor variables for brand 1 only, to avoid repetition. The regular price is a non-decreasing step function with relatively few jumps over the duration. The plots show that price discounts and promotions are fairly frequent, and their peaks tend to match peaks in log sales, whereas there was no display activity for brand 1 over this time period. Also, for most weeks, one or more competitors' brands offer price discounts.

All variables are adjusted by subtraction of sample means before we proceed with the analysis. The ordinary LS estimates of the regression model (7.24) were first computed for each of the $m = 8$ brands with $n = 7$ predictor variables, based on $T = 92$ observations, because of the lag term included in the model. The correlations of the residuals across equations are all positive and range from 0.1 to 0.6, indicating similarity of the sales of different brands over time in this product market. This suggests the possibility that SUR equation estimates might be desirable. Both the two-step and the ML estimates of the SUR model were computed. The determinants of the residual covariance matrices are 0.02101, 0.01244, and 0.01141 ($\times 10^{-8}$) for the LS, two-step SUR, and ML SUR estimation, respectively. The (full-rank) ML estimate of the coefficient matrix C is presented in Table 7.1, with expected signs of the coefficients for individual predictor variables indicated in parentheses in the column headings; an asterisk indicates that the coefficient estimate is significant at the 5% level.

For brands 1, 3, and 4, because of no activity in display area, the coefficients for the display indicator D_t cannot be estimated, and they are taken to be "missing." Each type of predictor variable exhibits significance for one or more brands, confirming the usefulness of including them in the model. The brands that have nonsignificant coefficients for lagged sales tend to be influenced by promotions. All significant coefficients have the correct expected signs except for the price discount for brand 3. We are not able to explain this anomaly but observe that having only one "unreliable" estimate among such a large number of estimates might be taken as satisfactory (see Blattberg and George 1991, for related discussion).

To proceed with reduced-rank analysis under the SUR model framework, using the initial alternate estimation procedure, we need to have the full-rank estimate of C with no "missing" elements. Because the display variable is significant for most brands and is potentially important from the marketing area point of view, we want to entertain the possibility of accommodating this predictor variable in the

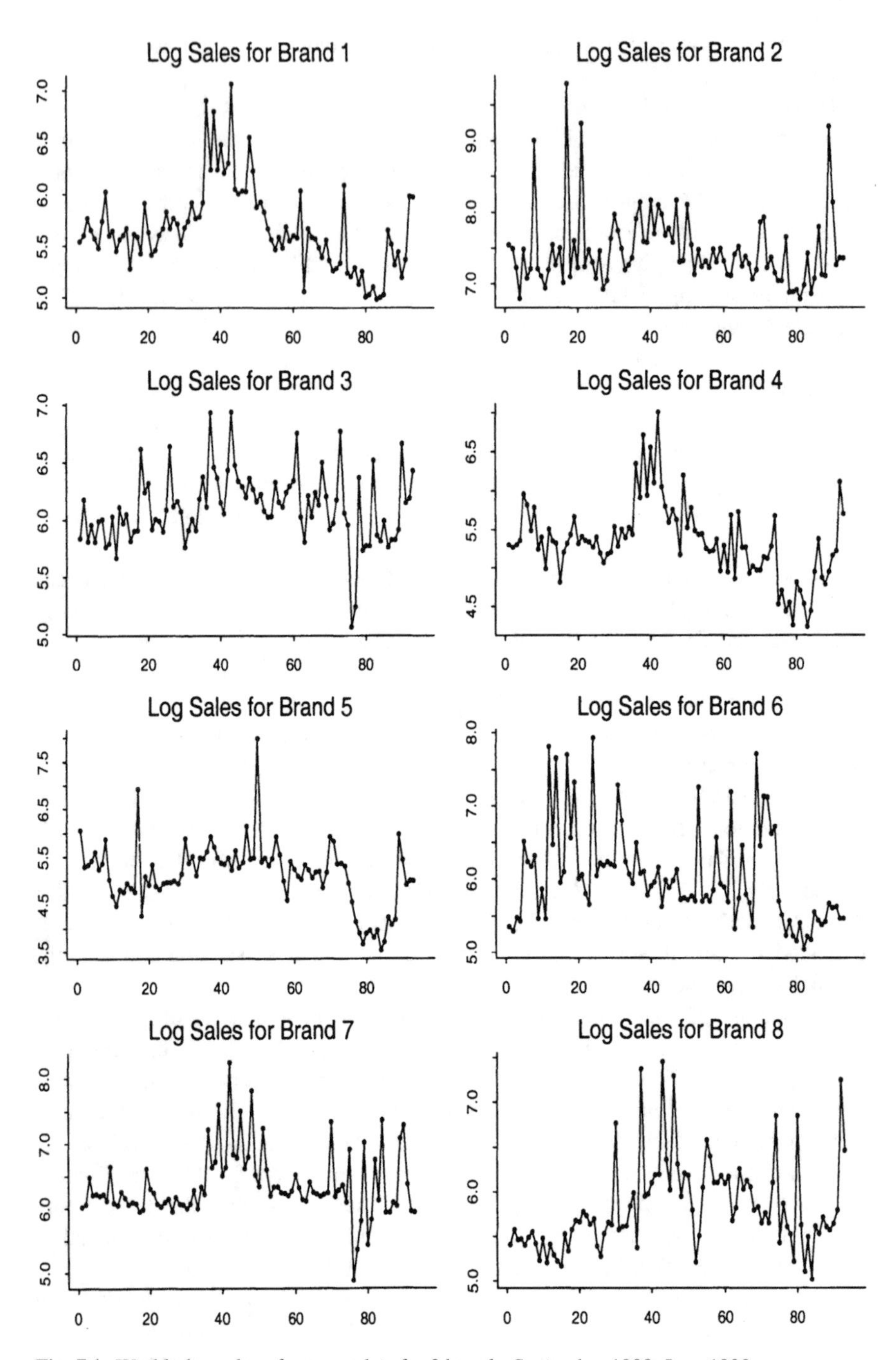

Fig. 7.1 Weekly log sales of scanner data for 8 brands, September 1988–June 1990

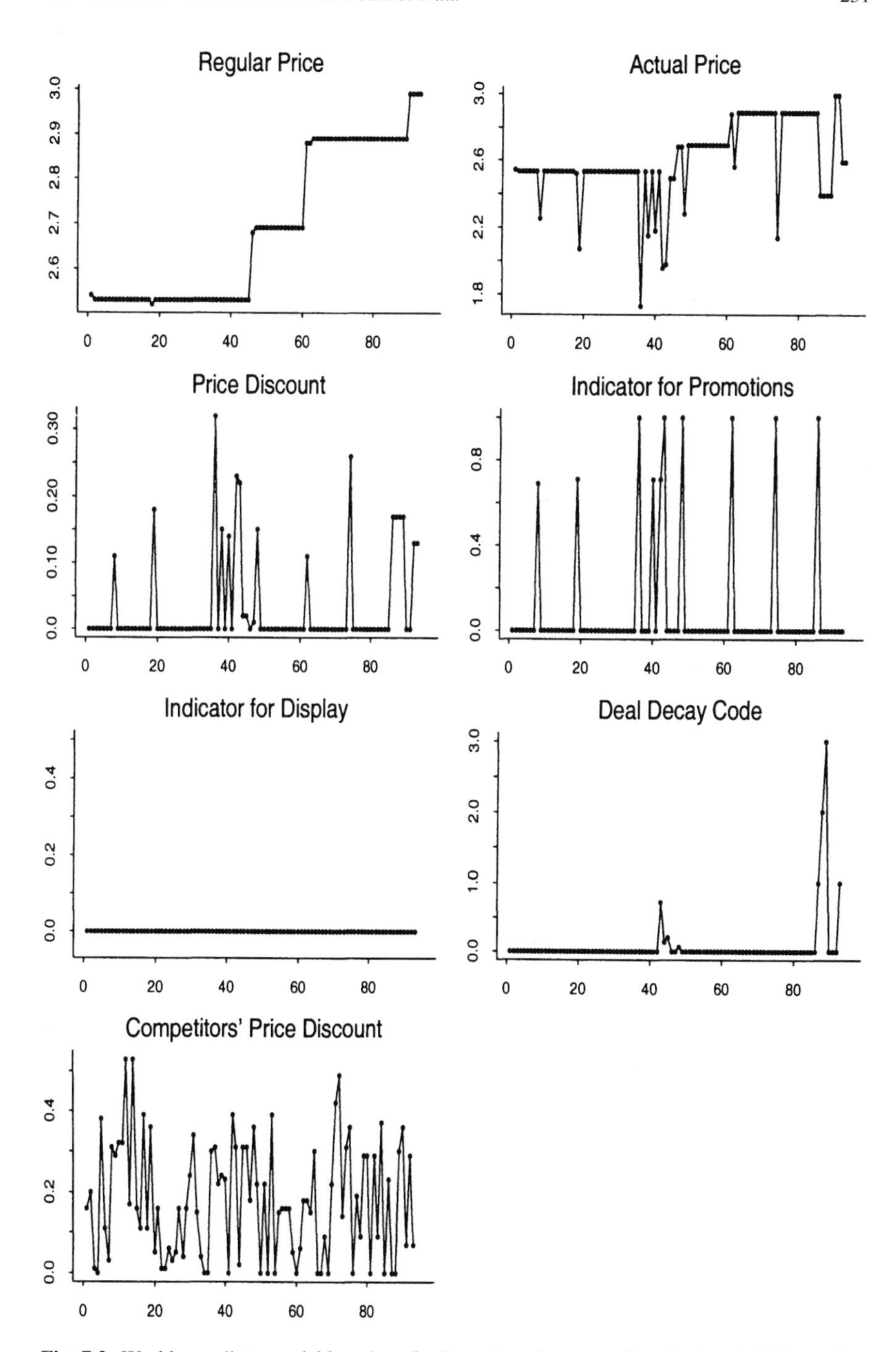

Fig. 7.2 Weekly predictor variable values for log sales of scanner data for brand 1, September 1988–June 1990

Table 7.1 Full-rank ML estimates of coefficient matrix C in SUR model for the scanner data

Brand	Lag Sales (+)	Price (−)	Price Disc (+)	Prom (+)	Disp (+)	DD (−)	CPD (−)
1	0.358*	−0.795*	2.597*	0.213	–	−0.161*	−0.048
2	0.127*	−0.816	0.617	0.587*	2.814*	−0.104	0.304
3	0.041	−0.584*	−1.786*	0.699*	–	0.014	0.166
4	0.311*	−2.736*	2.753*	0.173	–	−0.924	0.150
5	0.244*	−4.696*	1.817*	0.372*	5.646*	−0.047	−0.058
6	0.040	−0.865*	2.367*	0.495*	7.403*	−0.263*	−0.642*
7	0.054	−1.973*	3.136*	0.211	3.111*	−0.116*	−0.164
8	0.292*	−0.621	2.262	0.709	4.633	−2.140	−0.537*

model for all brands. While there are various options to fill in the missing estimates, we need to be sure that the procedure does not unduly affect the estimation of component matrices in the reduced-rank model. We have considered two options: (1) substituting the average value of the available display coefficient estimates for the missing values; (2) because brands tend to compete within a price tier, substituting the available competitor's display variable coefficient estimate for the missing value. Specifically, brands 1 and 2 are in the top price tier, 3, 4, and 5 are in the mid price tier, and brands 6, 7, and 8 are in the low price tier. Since the estimation results were very similar using both options, we will follow option (2) and present the corresponding analysis and results.

As suggested in Section 7.6, we use the results from Section 7.5 that are related to the alternate estimator to obtain information for an initial specification of the rank of C and initial estimates of the coefficient matrices A and B in the reduced-rank SUR model. The nonzero eigenvalues $\widetilde{\lambda}_j^2$ of $\widetilde{\Sigma}_{\epsilon\epsilon}^{-1/2}\widetilde{C}\widehat{\Sigma}_{\bar{x}\bar{x}}\widetilde{C}'\widetilde{\Sigma}_{\epsilon\epsilon}^{-1/2}$ are found to be 2.740, 1.784, 0.691, 0.183, 0.118, 0.067, and 0.005. The corresponding squared "canonical correlation" quantities $\widetilde{\rho}_j^2$ are 0.733, 0.641, 0.409, 0.155, 0.106, 0.062, and 0.005, which may provide a better feel for the possible rank. Formally, using the approximate LR test procedure mentioned in Section 7.6 gives the impression that the rank might need to be taken as large as $r = 4$ or $r = 5$. However, because of the conservative nature of this test procedure and the desire to explore a more parsimonious model, we further investigate the reduced-rank model with specification of lower ranks.

Thus, we examine ML fitting of models of various ranks. The procedure followed in calculation of the ML estimates for the model of each different rank is the partial least squares procedure described in Section 7.4. The BIC criterion is used to select a model of suitable rank, and the results for the BIC as well as the AIC criterion are presented in Table 7.2. It appears from these results that a model of rank 2 might be appropriate based on BIC, although the use of the AIC criterion would again require the use of model with rank as high as 5.

Table 7.2 Summary results on ML estimation of SUR models of various ranks fitted to the scanner data

Rank	$\lvert\widehat{\Sigma}_{\epsilon\epsilon}\rvert \times 10^8$	n^* = number of parameters	AIC	BIC
1	0.081145	14	−2.207	−1.823
2	0.031958	26	−2.878	−2.165
3	0.021850	36	−3.041	−2.054
4	0.014791	44	−3.257	−2.051
5	0.012370	50	−3.305	−1.935
6	0.011406	54	−3.299	−1.819
7	0.011406	56	−3.256	−1.721

The ML estimates of the component matrices B and A in the model of rank $r = 2$ are

$$\widehat{B} = \begin{bmatrix} 1.0226 & -15.0856 & 18.5987 & -0.1566 & 4.7584 & -0.4678 & -0.7757 \\ 0.5840 & 8.3231 & -4.0433 & 5.0278 & 31.8771 & -0.4947 & -2.0161 \end{bmatrix},$$

$$\widehat{A} = \begin{bmatrix} 0.1184 & 0.0945 \\ 0.0770 & 0.1166 \\ 0.0525 & 0.0736 \\ 0.1956 & 0.0008 \\ 0.2095 & 0.0740 \\ 0.1207 & 0.1504 \\ 0.1658 & 0.0510 \\ 0.1273 & 0.1859 \end{bmatrix}.$$

The resulting ML estimate $\widehat{C} = \widehat{A}\widehat{B}$ is displayed in Table 7.3. Comparing elements of the rank 2 ML estimate of C with those of the full-rank estimate given in Table 7.1, we can observe that most of the coefficient estimates that were found to be significant under the full-rank estimation are well recovered within the rank 2 model specification, with the rank 2 model having less than one-half the number of parameters as the full-rank model. Also note that the price discount coefficient estimate for brand 3, which had the "incorrect" sign in the full-rank estimation, now attains the correct sign in the model of rank 2.

From the ML estimate $\widehat{B}$, it is interesting to note that the most important linear combination of the predictor variables (ignoring coefficients with small magnitudes) is $S_{t-1} - 15.1P_t + 18.6PD_t + 4.8D_t$. This linear combination might be interpreted to represent a market factor comprised of price and display variables. The price discount variable PD_t receives a larger weight than the display variable D_t. From the ML estimates in the first column of $\widehat{A}$, we see that this linear combination has the highest impact on sales for brands 4, 5, and 7, medium impact for brands 1, 6, and 8, and smallest impact for brands 2 and 3. The second predictive linear combination includes a large contribution from the display variable D_t, and both the promotion variable PR_t and the competitors' price discount variable CPD_t enter

Table 7.3 ML estimates of coefficient matrix $C = AB$ in reduced-rank SUR model of rank $r = 2$ for the scanner data

Brand	Lag Sales (+)	Price (−)	Price Disc (+)	Prom (+)	Disp (+)	DD (−)	CPD (−)
1	0.176	−0.998	1.819	0.458	3.584	−0.102	−0.283
2	0.146	−0.191	0.960	0.574	4.084	−0.094	−0.295
3	0.096	−0.179	0.679	0.362	2.596	−0.061	−0.189
4	0.205	−2.883	3.605	0.010	1.188	−0.096	−0.168
5	0.257	−2.544	3.597	0.339	3.356	−0.135	−0.312
6	0.211	−0.569	1.637	0.737	5.368	−0.131	−0.397
7	0.199	−2.077	2.878	0.230	2.412	−0.103	−0.231
8	0.238	−0.374	1.616	0.915	6.532	−0.152	−0.474

more substantially than in the first linear combination. This predictive combination seems to have greater influence on brands 8, 6, and 2. To make more serious interpretations of the reduced-rank estimation results for these data, it would be necessary to delve into the extensive marketing literature. A goal here was to illustrate how certain canonical and reduced-rank procedures can be used for the SUR model setup. Further discussion related to interpretation of the estimation results will not be given here.

As a summary of the analysis results, we mention that the values of the ML residual variance estimates of log sales for the 8 brands under the full-rank SUR model fit were 0.0389, 0.1311, 0.0518, 0.0622, 0.2337, 0.1717, 0.0637, and 0.0625, whereas the corresponding values under the reduced-rank ($r = 2$) fit were 0.0520, 0.1318, 0.0620, 0.0674, 0.2670, 0.1720, 0.0593, and 0.0665. It is observed that the magnitudes of the residual variance estimates under the rank 2 model are comparable to those of the full-rank model. Taking into account the considerable reduction in the number of parameters that need to be estimated for the rank 2 model, slight increases in residual variances may be a small price to pay. An additional appeal of reduced-rank modeling of scanner data is that it provides a way to summarize, for managers, the large number of regression coefficient (elasticity) estimates through the "canonical quantities," and these canonical quantities can also be used to keep track of market changes over time. Thus, the interpretation of large amounts of data can be simplified for managerial use. A version of the results in the chapter is published by Velu and Richards (2008). In this chapter, we further demonstrate how the quantities "$\widehat{a}_{(i)}$'s" can be related to store level demographics. In marketing studies, $\widehat{c}_{(i)}$'s are used to group the stores and then common regression coefficients are estimated by pooling. Note that these coefficients are of dimension "n." With the use of reduced-rank idea, first we isolate the common effect via B' in (7.11) and the individual store effect via r-dimensional $a_{(i)}$. Because $r < n$, the clustering is easier to construct and interpret.

7.8 Some Recent Developments

The model developed and studied in this chapter has broader applications as the recent literature would suggest. The basic structure of the model is specified in (7.1) along with (7.10), which results in (7.11) in the formulation of seemingly unrelated reduced-rank regression (SURERR). The basis of the formulation is that regression coefficients from each unit's regression can be pooled together via reduced-rank model inducing some commonality among them. This is more general than simply testing if the coefficients are equal over all units. Recall that the basic data matrix of response variables is still $(m \times T)$ Y-matrix. Because of the structure in (7.1), each row of Y-matrix is associated with $(T \times n)$ X' matrix of predictors. These predictors are also time-varying but take different values for different response units. The reduced-rank constraint in (7.10) essentially pools the information from "m" time series regression models in (7.1). If "m" and "T" are fairly large, it is also possible to consider "T" cross-sectional regressions and pool the information across units via reduced-rank constraint. More specifically, if we denote each column of the data matrix Y as Y_k, then

$$Y_k = Z_k \beta_k + \varepsilon_k, \quad k = 1, 2, \ldots, T. \tag{7.25}$$

Observe that Y_k is a "m"-dimensional vector, Z_k is a "$m \times n$" dimensional matrix of predictors, β_k is a "n"-dimensional coefficient vector, to be estimated at each time point. Although we can add some dependence structure to ε_k', we assume here that the ε_k's are independent and follow, $N(0, \Sigma)$; the variance–covariance matrix captures contemporaneous correlations among the "m" units. Stacking the "β" coefficient as $C = [\beta_1, \ldots, \beta_T]'$, a $T \times n$ matrix, we can impose that

$$\text{rank}(C) = r \leq n. \tag{7.26}$$

This results in the decomposition $\beta_k = B' a_k$, where B' is a $n \times r$ matrix that binds the estimates over time and "a_k" is a r-dimensional vector that drives the model:

$$Y_k = Z_k B' a_k + \varepsilon_k. \tag{7.27}$$

We elaborate on this model form now. While the motivation for the formulation of model (7.1) is given from economics point of view in the seminal paper by Zellner (1962), the motivation for model (7.25) comes from financial economics and can be directly related to the principal component analysis. If Y_k denotes the returns of "m" assets at the kth time unit, then

$$Y_k = A f_k + \varepsilon_k, \tag{7.28}$$

where A is known as the factor loading matrix of order $m \times r$ and f_k is a r-dimensional factor where $r < m$. As shown in Section 2.4.1, the loading matrix and the factors all can be estimated through the eigenvectors of the covariance matrix of Y_k. As noted in Model (2.22),

$$Y_k = ABY_k + \varepsilon_k. \tag{7.29}$$

Thus, the model (7.28) can be written in the form of the reduced-rank regression model. The model (7.29) can be stated as a model with unknown factors; instead if "f_k" can be related to known predictors, X_k, then

$$Y_k = ABX_k + \varepsilon_k \tag{7.30}$$

resulting in the classical reduced-rank model studied in detail in Chapter 2. Interestingly in financial economics, the interest has been on relating the factor loading matrix to asset characteristics that can be both static and dynamic. This approach is outlined in Connor and Linton (2007) and Connor et al. (2012) and further extended by Fan et al. (2016) as "projected principal component analysis." The discussion that follows next is based on the work of Kelly et al. (2019) that introduces the concept of Instrumental Principal Component Analysis (IPCA). The specification of the IPCA model is as follows. Let

$$Y_k = A_k f_k + \varepsilon_k, \quad A_k = Z_k B' + \varepsilon_k^*. \tag{7.31}$$

Here Z_k is $(m \times n)$ time-varying predictors, where each of its rows denotes the asset-specific variables that change over time and B' is $m \times r$ matrix that is fixed; combining the two equations in (7.31), we have the model

$$Y_k = Z_k B' f_k + \varepsilon_k, \tag{7.32}$$

which is exactly in the form of (7.11), except that these are "T" cross-sectional regressions. Thus, the IPCA model can be estimated using the methodology given in this chapter. The criterion to be minimized can be stated as

$$\begin{aligned} S(\theta) &= \frac{1}{2T} e'(I_T \otimes \Gamma)e \\ &= \frac{1}{2T} \sum_{k=1}^{T} (Y_k - Z_k\beta_k)'\Gamma(Y_k - Z_k\beta_k). \end{aligned} \tag{7.33}$$

The above is possible mainly because of the Kronecker structure of the errors and because $\Gamma = \Sigma_{\epsilon\epsilon}^{-1}$ goes with each equation in (7.27), unlike the SURE model in (7.11), where the elements of $\Sigma_{\epsilon\epsilon}$ came from different equations in (7.1). Observe that the GLS estimate of β_k is $\widehat{\beta}_k = (Z_k'\Gamma Z_k)^{-1} Z_k'\Gamma Y_k$ and $\widehat{\beta}_k \sim N(\beta_k, (Z_k'\Gamma Z_k)^{-1})$. The criterion in (7.33) is indeed equivalent to minimizing (apart from the divisor in (7.33))

$$W = \sum_{k=1}^{T} (\widehat{\beta}_k - B'a_k)'(Z_k' \Gamma Z_k)(\widehat{\beta}_k - B'a_k). \tag{7.34}$$

The first order equation is $\frac{\partial W}{\partial a_k} = -B(Z_k' \Gamma Z_k)\widehat{\beta}_k + [BZ_k' \Gamma Z_k B']\widehat{a}_k$, which leads to (for a given "B")

$$\widehat{a}_k = [BZ_k' \Gamma Z_k B']^{-1}(BZ_k' \Gamma Z_k)\widehat{\beta}_k. \tag{7.35}$$

Substituting $\widehat{a}_k$ in (7.34) leads to

$$\min W = \max \operatorname{Tr}\left[\sum_{1}^{T} \left[B(Z_k' \Gamma Z_k)\widehat{\beta}_k \widehat{\beta}_k'(Z_k' \Gamma Z_k)B'\right]\left[B(Z_k' \Gamma Z_k)B'\right]^{-1}\right]. \tag{7.36}$$

Remark The above expression when $\Gamma = I$ is equivalent to Kelly et al. (2019, p 507, equation (8)). As noted in their paper, a closed-form solution is not possible mainly because of the second term in (7.36). But if that term $B(Z_k' \Gamma Z_k)B'$ is replaced by $B(\overline{Z}' \Gamma \overline{Z})B'$, where $\overline{Z}$ is the average of Z_k, then it can be shown that B' can be extracted from the first "r" eigenvectors of $\frac{1}{T}\sum_1^T (Z_k' \Gamma Z_k)\widehat{\beta}_k \widehat{\beta}_k'(Z_k' \Gamma Z_k)$ with respect to $\overline{Z}' \Gamma \overline{Z}$.

To get the first derivative of W with respect to B', we write the criterion in terms of $\text{vec}(B)$. Note that

$$\begin{aligned} \operatorname{Tr} W = &\sum_{1}^{T} \widehat{\beta}_k (Z_k' \Gamma Z_k)\widehat{\beta}_k - 2\sum_{1}^{T} \left[\beta_k'(Z_k' \Gamma Z_k) \otimes a_k'\right] \text{vec}(B) \\ &+ \text{vec'}(B)\left(\sum_{1}^{T} \left(Z_k' \Gamma Z_k \otimes a_k a_k'\right)\right) \text{vec}(B). \end{aligned}$$

We used the matrix result, $\text{tr}(AZ'BZC) = \text{vec}'(Z)(CA \otimes B')\text{vec}(Z)$ in arriving at the above expression. It can be seen for a given "a_k,"

$$\text{vec}(B) = \left[\sum_{1}^{T} \left(Z_k' \Gamma Z_k \otimes a_k a_k'\right)\right]^{-1} \left[\sum_{1}^{T} \left(Z_k' \Gamma Z_k \otimes a_k'\right)\widehat{\beta}_k\right]. \tag{7.37}$$

The solutions to B and a_k can be obtained by alternating (7.35) and (7.37) and imposing normalization conditions. If we set $\widehat{A} = (\widehat{a}_1, \ldots, \widehat{a}_T)$ as a $r \times T$ matrix, then the following conditions are made to hold:

$$\widehat{A}\widehat{A}' = \widehat{\Lambda}_r \quad \text{and} \quad BB' = I_r, \tag{7.38}$$

where Λ_r is a diagonal matrix. Iterative procedure based on the second derivatives can also be derived as given in Theorem 7.1, and additional details can be found in Luca et al. (2022) on the properties of the resulting estimators and for an application on asset pricing models in the finance area. In fact an extended model considered by Kelly et al. (2019) where the non-time varying intercept-like term is included amounts to decomposing the $m \times n$ coefficient matrix, $C = (\beta_1, \ldots, \beta_m)' = AB + \alpha 1_n'$. This can also be addressed using the methodology given in this chapter.

Remark While the research on multivariate regression model we began with in Chapter 1 has been extensively studied and extended, research on seemingly unrelated regression models discussed in this chapter warrants further work. These models have a number of practical applications. They can incorporate time-varying, unit-varying, as well as fixed characteristics. Adding more variables to model (7.1) or to (7.27) that are not part of the dimension reduction aspect requires further study. Note that reduced-rank model of Anderson (Section 3.1) can be identified through partial canonical correlations. Such elegant interpretations are not available for the seemingly unrelated regression models. With the two formulations presented here in this chapter, these models can easily be applied to study the panel data. For example, the model with time-varying coefficients $Y_{ik} = X_{ik}' \beta_k + \mu_k + U_{ik}$ can be accommodated. Here "μ_k" unobservable time effect and β_k can be assumed to be random, time-varying but is the same across the cross-sectional units. For the SURE models, variable selection methods such as LASSO need to be explored and studied. Some ideas toward this direction are covered in later chapters.

Chapter 8
Applications of Reduced-Rank Regression in Financial Economics

8.1 Introduction to Asset Pricing Models

In previous chapters, we have developed reduced-rank regression models of various forms, which have wide applications in a variety of contexts in the physical and social sciences. In this chapter, we focus on the area of financial economics, where several applications of the reduced-rank regression models arise in a fairly natural way. Thus, the topics that will be examined in this chapter involve consideration of the contexts in which financial models arise from economic theories, the form of the models, and the empirical verification of these models through reduced-rank regression methods. In earlier chapters, the application of reduced-rank regression methods was mainly motivated from an empirical dimension reduction aspect, whereas the use of reduced-rank models presented in this chapter results from a priori economic theory. With the high volume of financial data that are now routinely becoming available, these theories can be examined through empirical tests of the models (see Section 7.8). We point out that, for the most part, only the basic reduced-rank methods developed in Chapter 2 and the methods related to cointegrated modeling discussed in Section 5.6 are needed for the presentations considered here.

Of fundamental interest to financial economists is to examine the relationship between the risk of a financial security and its return. While it is obvious that risky assets can generally yield higher returns than risk-free assets, a quantification of the tradeoff between risk and expected return was made through the development of the capital asset pricing model (CAPM), for which the groundwork was laid in the seminal work by Markowitz (1959). A central feature of the CAPM is that the expected return is a linear function of the risk. The risk of an asset typically is measured by the covariability between its return and that of an appropriately defined "market" portfolio. Examples of expected return-risk model relationships include the Sharpe (1964) and Lintner (1965) CAPM, the zero-beta CAPM of

G. C. Reinsel et al., *Multivariate Reduced-Rank Regression*, Lecture Notes in Statistics 225, https://doi.org/10.1007/978-1-0716-2793-8_8

Black (1972), the arbitrage pricing theory (APT) due to Ross (1976), and the intertemporal asset pricing model by Merton (1973). The economy-wide models developed by Sharpe (1964), Lintner (1965), and Black (1972) are based on the work by Markowitz (1959) that assumes that investors would hold a mean–variance efficient portfolio. They showed that if investors have homogeneous expectations and hold efficient portfolios, then the market portfolio will itself be a mean–variance efficient portfolio. The CAPM equation (8.3), given below, is a direct implication of the mean–variance efficiency of the market portfolio. The main difference between the work of Sharpe (1964) and Lintner (1965) and the work of Black (1972) is that the former assume the existence of a risk-free lending and borrowing rate and the rates are the same, whereas the latter derived a more general version of the CAPM in the absence of a risk-free asset. Initially, we focus on the work of Black (1972) as it readily fits into the reduced-rank regression framework.

Gibbons (1982) developed the Black (1972) CAPM hypothesis through the multivariate regression model. Suppose there are m assets whose returns are observed over T time periods, and let r_{ik} denote the return on asset i in period k. The model is specified as follows:

$$r_{ik} = \alpha_i + \beta_i x_k + \epsilon_{ik}, \quad i = 1, \ldots, m, \quad k = 1, \ldots, T, \tag{8.1}$$

where x_k denotes the return on the market portfolio in period k. The coefficient β_i in (8.1) is the "risk" of the ith asset, equal to

$$\beta_i = \mathrm{Cov}(r_{ik}, x_k)/\mathrm{Var}(x_k), \quad i = 1, \ldots, m,$$

and the ϵ_{ik} are assumed to have zero mean and be uncorrelated over time, with the random vector $\epsilon_k = (\epsilon_{1k}, \ldots, \epsilon_{mk})'$ having positive-definite $m \times m$ covariance matrix $\Sigma_{\epsilon\epsilon} = \mathrm{Cov}(\epsilon_k)$. The above Eqs. (8.1) can be written in the form of a multivariate regression model, given in Chapter 1, and represent "a statistical statement rather than one derived from financial theory." With the zero mean assumption on the errors in (8.1), we have

$$E(r_{ik}) = \alpha_i + \beta_i E(x_k), \quad i = 1, \ldots, m. \tag{8.2}$$

The Black (1972) CAPM requires, for the mean–variance efficiency of the market portfolio, that the following relationship holds between the expected returns and the risks:

$$E(r_{ik}) = \gamma_0 + \beta_i [E(x_k) - \gamma_0], \quad i = 1, \ldots, m, \tag{8.3}$$

where γ_0 is the expected return on the "zero-beta" portfolio or any portfolio whose return is not correlated with the return on the market portfolio. The zero-beta portfolio is defined to be the portfolio whose returns have the minimum variance of all portfolios whose returns are uncorrelated with the returns of the market portfolio. Comparing (8.2) with (8.3), we see that (8.3) implies that the intercepts of the model (8.1) must satisfy the constraints

$$\alpha_i = \gamma_0(1 - \beta_i), \quad \text{for all } i = 1, \ldots, m, \tag{8.4}$$

which is used as a basis for empirical testing of the zero-beta CAPM model, that is, testing the efficiency of the market portfolio. A similar formulation of the problem of testing the efficiency of the market (or any specified) portfolio is obtained by Gibbons et al. (1989), but based on the assumptions of the Sharpe (1964) and Lintner (1965) CAPM. In this case, the regression uses excess returns (returns in excess of the risk-free rate), and the null hypothesis is simply that all the alphas are zeros.

To connect the constraints (8.4) with a reduced-rank feature, we first observe that the model (8.1) can be written in a slightly different form as

$$y_{ik} \equiv r_{ik} - x_k = \alpha_i + (\beta_i - 1)x_k + \epsilon_{ik}.$$

We will then express this in the form of the multivariate regression model in (1.1). Define $Y_k = (r_{1k} - x_k, \ldots, r_{mk} - x_k)'$ and $X_k = (1, x_k)'$, where Y_k is the $m \times 1$ vector of "excess" returns on the m assets. Further, write the $m \times 2$ coefficient matrix as

$$C = \begin{bmatrix} \alpha_1 & \alpha_2 & \cdots & \alpha_m \\ \beta_1 - 1 & \beta_2 - 1 & \cdots & \beta_m - 1 \end{bmatrix}' \equiv [\alpha, \beta - \mathbf{1}_m],$$

where $\mathbf{1}_m$ is an $m \times 1$ vector of ones. Now the model can be restated as the multivariate regression model

$$Y_k = CX_k + \epsilon_k, \quad k = 1, \ldots, T, \tag{8.5}$$

where ϵ_k is the $m \times 1$ vector of random errors, with $E(\epsilon_k) = 0$ and $\mathrm{Cov}(\epsilon_k) = \Sigma_{\epsilon\epsilon}$. Notice that under the constraints (8.4), the coefficient matrix C in (8.5) is expressible as

$$\begin{aligned} C = [\alpha, \beta - \mathbf{1}_m] &\equiv [\gamma_0(\mathbf{1}_m - \beta), \beta - \mathbf{1}_m] \\ &= [\beta - \mathbf{1}_m][-\gamma_0, 1] = \beta^* \gamma', \end{aligned} \tag{8.6}$$

where $\gamma' = [-\gamma_0, 1]$ and $\beta^* = \beta - \mathbf{1}_m$. Thus, the matrix C is of reduced- rank one, although there is some slight difference in the reduced-rank setup presented here compared to the approach for the model discussed in Chapter 2. While the form of (8.6) is similar to the form $C = AB$ in (2.4), the right-hand factor γ' in this setup (i.e., $C = \beta^* \gamma'$) is structured in that a normalization is directly incorporated into the second element of the vector γ. Recall, also, that the reduced-rank restriction on C implies that there is one constraint on the matrix C that can be stated as

$$C\gamma^* = 0, \tag{8.7}$$

where $\gamma^* = (1, \gamma_0)'$ is "structured" as well. A main difference between (8.7) and (2.2) is that the constraints in (2.2) operate on the rows of the $m \times n$ matrix C, whereas the above constraints operate on the columns of the $m \times 2$ matrix C. This difference in emphasis is due mainly to the convention adopted in Chapter 2, assumed for convenience of exposition, that $m \leq n$, while $m > n = 2$ in the present situation. For either case, methods developed in Chapters 1 and 2 are useful here.

8.2 Estimation and Testing in the Asset Pricing Model

Gibbons (1982) suggested the following iterative procedure for estimation of the coefficient matrix C in (8.5), subject to the constraints (8.4), which is very much in the spirit of the partial least squares method. Let $\mathbf{Y}$ and $\mathbf{X}$ denote the $m \times T$ and $2 \times T$ data matrices, respectively.

(i) In the first step, obtain the ordinary LS estimator $\widetilde{C} = \mathbf{YX}'(\mathbf{XX}')^{-1}$ as given in (1.7), which is not constrained by (8.4). This yields the LS estimates $\widetilde{\alpha}$ and $\widetilde{\beta}$ of α and β. Then obtain $\widetilde{\Sigma}_{\epsilon\epsilon} = (1/T)(\mathbf{Y} - \widetilde{C}\mathbf{X})(\mathbf{Y} - \widetilde{C}\mathbf{X})'$, the estimate of the contemporaneous error covariance matrix. (To avoid singularity, m must be greater than T, a condition that might not hold; but in the finance setting because of availability of data on a large number m of securities, one might typically have $m > T$.)
(ii) In step two, motivated by (8.4), compute the estimate $\widehat{\gamma}_0 = \{\widetilde{\alpha}'\widetilde{\Sigma}_{\epsilon\epsilon}^{-1}(\mathbf{1}_m - \widetilde{\beta})\}/(\mathbf{1}_m - \widetilde{\beta})'\widetilde{\Sigma}_{\epsilon\epsilon}^{-1}(\mathbf{1}_m - \widetilde{\beta})$, a GLS estimate of γ_0 obtained by regressing the $\widetilde{\alpha}_i$ on the $(1 - \widetilde{\beta}_i)$, $i = 1, \ldots, m$, with the weights matrix $\widetilde{\Sigma}_{\epsilon\epsilon}^{-1}$.
(iii) Using (8.3) and the estimate $\widehat{\gamma}_0$ from (ii), perform m individual "market model" regressions of $r_{ik} - \widehat{\gamma}_0$ on $x_k - \widehat{\gamma}_0$, without a constant term, to reestimate the β_i.
(iv) With the new estimates $\widetilde{\beta}_i$, iterate the steps in (ii) and (iii) that yield $\widehat{\gamma}_0$ and the new $\widetilde{\beta}_i$ until convergence.

In several studies in finance (see Gibbons et al. 1989), following Sharpe (1964) and Lintner (1965), it is assumed that there exists a riskless asset that is usually taken as the U.S. Treasury bill rate. For this choice, the value of γ_0 is known and is set equal to the rate of return on the riskless asset. Then testing for the (known) constraints (8.4) can be done using the results in Chapter 1. For known γ_0, therefore known γ^*, the constraints (8.7) can be tested through the LR statistic in (1.16) and the approximate chi-squared distribution theory, or in this case that involves only a single constraint vector, through Hotelling's T^2-statistic given in (1.22). This follows from the same reasoning as in (1.22), and with $\overline{\Sigma}_{\epsilon\epsilon} = \{T/(T-2)\}\widetilde{\Sigma}_{\epsilon\epsilon}$, the appropriate test statistic is

$$\mathcal{T}^2 = (\gamma^{*\prime}\widetilde{C}'\overline{\Sigma}_{\epsilon\epsilon}^{-1}\widetilde{C}\gamma^*)/\gamma^{*\prime}(\mathbf{XX}')^{-1}\gamma^*, \tag{8.8}$$

which has the Hotelling's T^2-distribution with $T - 2$ degrees of freedom. Thus, $\mathcal{F} = \{(T - m - 1)/m(T - 2)\}\mathcal{T}^2$ has the F-distribution with m and $T - m - 1$

degrees of freedom. This is also known as the Wald test discussed by Gibbons et al. (1989) to test $H_0 : \alpha = 0$. The test is stated as $F = \left(\frac{T-m-1}{m}\right)\left(1+\frac{\overline{x}^2}{s_x^2}\right)\widehat{\alpha}\overline{\Sigma}_{\varepsilon\varepsilon}^{-1}\widehat{\alpha}$, where $\widehat{\alpha}$ is the estimate of α in the regression of excess returns over excess market returns. The interpretation of the noncentrality parameter under the alternative that $C\gamma^* \neq 0$ in terms of the financial context has been of some interest in the finance literature. The constrained ML estimator of C under (8.7) is

$$\widehat{C} = \widetilde{C}\left[I_2 - \frac{\gamma^*\gamma^{*\prime}(\mathbf{XX}')^{-1}}{\gamma^{*\prime}(\mathbf{XX}')^{-1}\gamma^*}\right], \tag{8.9}$$

which follows as a special case of the form given in (1.18) with $F_1 = I_m$ and $G_2 = \gamma^*$. Note that an alternate direct derivation of the test statistic (8.8) can be motivated as follows. With γ_0 known, model (8.1) can be written in the equivalent form as $(r_{ik} - \gamma_0) = \alpha_i^* + \beta_i(x_k - \gamma_0) + \epsilon_{ik}$, where $\alpha_i^* = \alpha_i - \gamma_0(1-\beta_i) \equiv 0$ under conditions (8.4). It follows that if the excess asset returns $r_{ik} - \gamma_0$ are regressed on the excess market return $x_k - \gamma_0$, the resulting intercepts are expected to be zero. The traditional multivariate T^2-test for the m intercepts to be zero yields the statistic in (8.8), noting that $\widetilde{C}\gamma^* = \widetilde{\alpha} - \gamma_0(\mathbf{1}_m - \widetilde{\beta}) \equiv \widetilde{\alpha}^*$ is equal to the vector of LS estimates of the intercepts α_i^*.

When a riskless asset is not assumed to exist, so that γ_0 and hence γ^* are not known, we can obtain the constrained ML estimates and test the CAPM hypothesis of (8.4) using the reduced-rank regression methods developed in Chapter 2. Because of the rank one structure, and the simple nature of both γ in (8.6) and γ^* in (8.7), we can obtain the ML estimator of γ_0 in two ways. From general results in Section 2.3, the ML estimation criterion leads to minimizing (2.16), and the resulting solutions are given in (2.17). It follows, using the notation of the problem discussed here, that the ML estimates of the components β^* and γ in $C = \beta^*\gamma'$ are given by

$$\widehat{\beta}^* = \widetilde{\Sigma}_{\epsilon\epsilon}^{1/2}\widehat{V}_1, \qquad \widehat{\gamma}' = \widehat{V}_1'\widetilde{\Sigma}_{\epsilon\epsilon}^{-1/2}\widetilde{C}, \tag{8.10}$$

where $\widehat{V}_1$ is the (normalized) eigenvector that corresponds to the largest eigenvalue $\widehat{\lambda}_1^2$ of the matrix $\widetilde{\Sigma}_{\epsilon\epsilon}^{-1/2}\widetilde{C}\widehat{\Sigma}_{xx}\widetilde{C}'\widetilde{\Sigma}_{\epsilon\epsilon}^{-1/2}$. These estimates satisfy the normalization $\widehat{\beta}^{*\prime}\widetilde{\Sigma}_{\epsilon\epsilon}^{-1}\widehat{\beta}^* = 1$, while $\widehat{\gamma}'\widehat{\Sigma}_{xx}\widehat{\gamma} = \widehat{\lambda}_1^2$. Because of the disproportionate dimensions of Y_k and X_k involved (m is typically much larger than $n = 2$), we suggest that it would be practical to compute the equivalent ML estimates of β^* and γ based on the eigenvector of the 2×2 matrix $\widehat{\Sigma}_{xx}^{1/2}\widetilde{C}'\widetilde{\Sigma}_{\epsilon\epsilon}^{-1}\widetilde{C}\widehat{\Sigma}_{xx}^{1/2}$. Let $\widehat{V}_1^*$ be the (normalized) eigenvector that corresponds to the largest eigenvalue $\widehat{\lambda}_1^2$ of the latter matrix. Then it follows from discussion given toward the end of Section 2.3 that an equivalent expression for the ML estimates is given by

$$\widehat{\beta}^* = \widetilde{C}\widehat{\Sigma}_{xx}^{1/2}\widehat{V}_1^*, \qquad \widehat{\gamma}' = \widehat{V}_1^{*\prime}\widehat{\Sigma}_{xx}^{-1/2}, \tag{8.11}$$

which satisfy the normalization $\widehat{\gamma}'\widehat{\Sigma}_{xx}\widehat{\gamma} = 1$, while $\widehat{\beta}^{*\prime}\widetilde{\Sigma}_{\epsilon\epsilon}^{-1}\widehat{\beta}^* = \widehat{\lambda}_1^2$. By rescaling the above ML estimate $\widehat{\gamma}$ in (8.10) or (8.11) to have the form $[-\widehat{\gamma}_0,\ 1]'$, we obtain the ML estimate $\widehat{\gamma}_0$ of γ_0. Therefore, the iterative scheme suggested by Gibbons (1982) to obtain estimators of γ_0 and β can be replaced by a one-time eigenvalue and eigenvector calculation to obtain the ML estimates.

If the focus of estimation is on γ_0, this parameter can also be estimated through estimation of the constraints vector γ^* in (8.7). As discussed in Section 2.3, the ML estimate of the constraint corresponds to obtaining the eigenvector $\widehat{V}_2$ (or $\widehat{V}_2^*$) associated with the smallest eigenvalue $\widehat{\lambda}_2^2$ of the matrix $\widetilde{\Sigma}_{\epsilon\epsilon}^{-1/2}\widetilde{C}\widehat{\Sigma}_{xx}\widetilde{C}'\widetilde{\Sigma}_{\epsilon\epsilon}^{-1/2}$ (or of the matrix $\widehat{\Sigma}_{xx}^{1/2}\widetilde{C}'\widetilde{\Sigma}_{\epsilon\epsilon}^{-1}\widetilde{C}\widehat{\Sigma}_{xx}^{1/2}$). This can also be motivated in this context through the T^2-statistic given in (8.8) (e.g., see Anderson, 1951, for a similar argument). For a given vector γ^*, the constrained ML estimate of the error covariance matrix $\Sigma_{\epsilon\epsilon}$ is easily obtained from the constrained ML estimate $\widehat{C}$ given in (8.9) as

$$\widehat{\Sigma}_{\epsilon\epsilon} = \widetilde{\Sigma}_{\epsilon\epsilon} + \frac{1}{\gamma^{*\prime}\widehat{\Sigma}_{xx}^{-1}\gamma^*}\widetilde{C}\gamma^*\gamma^{*\prime}\widetilde{C}'. \tag{8.12}$$

Since

$$\begin{aligned} |\widehat{\Sigma}_{\epsilon\epsilon}| &= |\widetilde{\Sigma}_{\epsilon\epsilon}||I_m + (\gamma^{*\prime}\widehat{\Sigma}_{xx}^{-1}\gamma^*)^{-1}\widetilde{\Sigma}_{\epsilon\epsilon}^{-1/2}\widetilde{C}\gamma^*\gamma^{*\prime}\widetilde{C}'\widetilde{\Sigma}_{\epsilon\epsilon}^{-1/2}| \\ &= |\widetilde{\Sigma}_{\epsilon\epsilon}|(1 + (\gamma^{*\prime}\widehat{\Sigma}_{xx}^{-1}\gamma^*)^{-1}\gamma^{*\prime}\widetilde{C}'\widetilde{\Sigma}_{\epsilon\epsilon}^{-1}\widetilde{C}\gamma^*), \end{aligned}$$

minimizing $|\widehat{\Sigma}_{\epsilon\epsilon}|$ with respect to the unknown constraint vector γ^* to arrive at the ML estimate is equivalent to minimizing the statistic $\mathcal{T}^2$. Therefore, it follows that the estimate $\widehat{\gamma}^*$ can be obtained through the eigenvector that corresponds to the smallest eigenvalue $\widehat{\lambda}_2^2$ of $\widehat{\Sigma}_{xx}^{1/2}\widetilde{C}'\widetilde{\Sigma}_{\epsilon\epsilon}^{-1}\widetilde{C}\widehat{\Sigma}_{xx}^{1/2}$. Again, by appropriately rescaling the components of $\widehat{\gamma}^*$, we obtain the ML estimate $\widehat{\gamma}_0$ of γ_0. In either approach, the likelihood ratio statistic for testing H_0 : rank$(C) = 1$ is obtained from the results of Section 2.6 as $[T - (m + 3)/2]\log(1 + \widehat{\lambda}_2^2)$ and follows an asymptotic χ_{m-1}^2 distribution under the null hypothesis.

Zhou (1991, 1995) formulated the testing of the asset pricing constraints in (8.4) as a problem in reduced-rank regression and investigated small sample distributional properties of the likelihood ratio test (a test essentially based on $\widehat{\lambda}_2^2$) for testing rank$(C) = 1$ versus rank$(C) = 2$. Shanken (1986) as well as Zhou (1991) derived the ML estimate of the parameter γ_0. The focus of Zhou's studies, however, was on testing for the rank of C rather than on estimates of γ_0 and β. Zhou (1991) comments that the accuracy of the ML estimates would be of interest to practitioners in the finance area, but observes that the standard error of the zero-beta rate estimate $\widehat{\gamma}_0$ is not known. Because of the connections with reduced-rank estimation demonstrated here, we mention that the large sample distribution of the ML estimator $\widehat{\gamma}_0$ can be derived using the results in Section 2.5. We state the result.

Result For the model (8.5) where ε_k's satisfy the usual assumptions, let the parameters belong to a compact set defined by the normalization constraints

$\beta^{*\prime}\Sigma_{\varepsilon\varepsilon}^{-1}\beta^* = 1$ and $\gamma'\Sigma_{xx}\gamma = \lambda_1^2$. Then $\sqrt{T}(\widehat{\gamma}_0 - \gamma_0) \sim N(0, \sigma_{\gamma_0}^2)$, where $\sigma_{\gamma_0}^2 = \frac{1}{\lambda_1^2}\left(1 + \dfrac{(\mu_x - \gamma_0)^2}{\sigma_x^2}\right)$.

The proof follows from the discussion in Section 2.5 that the asymptotic variance of $\widehat{\gamma}$ is $\Sigma_{xx}^{-1} + \frac{1}{2}\gamma\gamma'$. Because $\gamma_0 = -\gamma_1/\gamma_2$ is totally differentiable, applying the standard asymptotics (see Rao 1973, p. 387), the result follows. To simplify, we use the condition $\gamma_2^2[\sigma_x^2 + (\mu_x - \gamma_0)^2] = \lambda_1^2$ that results from the normalized condition $\gamma'\Sigma_{xx}\gamma = \lambda_1^2$.

The asymptotic inference on γ_0 can be made using the above result. It should be noted that the asymptotic variance involves the nonzero eigenvalue, λ_1^2. The expression for the asymptotic variance is exactly the same as the one given in Campbell et al. (1997, Eqn 5.3.8.1, p. 202). It follows from the alternative formulation of the ML estimation in (8.11) and the normalization condition $\beta^{*'}\Sigma_{\varepsilon\varepsilon}^{-1}\beta^* = \lambda_1^2$. The quantity $(\mu_x - \gamma_0)/\sigma_x$ for a known risk-free rate γ_0 is the Sharpe ratio.

It is easy to show using (8.5) with the constraint (8.4), a portfolio formed with weight vector, $w = (w_1, \ldots, w_m)'$ as $w'Y_t = w'CX_t + w'\varepsilon_t$, which results in the model for portfolio return as $r_{pt} = \gamma_0 + (w'\beta)(x_t - \gamma_0) - a_t$, where $r_{pt} = \sum_{i=1}^m w_i r_{it}$ and $a_t = w'\varepsilon_t$. Thus the structure under Black-CAPM is preserved at the portfolio level as well. To derive the result, observe that $w'C = w'[\gamma_0(\mathbf{1}_m - \beta), \beta - \mathbf{1}_m] = [\gamma_0(\mathbf{1}_m - w'\beta), (w'\beta - 1)]$ and rearrange the terms.

The methods and results presented for the model (8.1) can easily be generalized to the multi-beta CAPM, which is often considered in the finance literature. The market model for the multi-beta CAPM can be written similar to (8.1) as

$$r_{ik} = \alpha_i + \sum_{j=1}^{n-1} \beta_{ij} x_{jk} + \epsilon_{ik}, \tag{8.13}$$

where $(x_{1k}, \ldots, x_{n-1,k})'$ is a $(n-1) \times 1$ vector of returns on the $(n-1)$ reference portfolios, and $\beta_i = (\beta_{i1}, \ldots, \beta_{i,n-1})'$ is the $(n-1) \times 1$ vector of risk measures. The constraints (8.4) generalize in the multi-beta CAPM to

$$\alpha_i = \gamma_0\left(1 - \sum_{j=1}^{n-1} \beta_{ij}\right) \equiv \gamma_0(1 - \beta_i'\mathbf{1}_{n-1}), \quad i = 1, \ldots, m. \tag{8.14}$$

Similar to the previous case, the model (8.13) can be reexpressed as

$$y_{ik} = r_{ik} - \frac{1}{n-1}\sum_{j=1}^{n-1} x_{jk} = \alpha_i + \sum_{j=1}^{n-1}\left(\beta_{ij} - \frac{1}{n-1}\right) x_{jk} + \epsilon_{ik},$$

and hence, it can be written in the multivariate regression form as $Y_k = CX_k + \epsilon_k$, with $X_k = (1, x_{1k}, \ldots, x_{n-1,k})'$, and the $m \times n$ coefficient matrix C will be of reduced rank $n-1$ ($n < m$ is assumed) under the constraints of (8.14). The reduced-rank methods and results from Chapter 2, in particular, results in Section 2.5, readily extend to this case. Some finite-sample testing procedures related to (8.14) are discussed by Velu and Zhou (1999).

APT Model An alternate financial theory to the CAPM is the arbitrage pricing theory (APT) proposed by Ross (1976). A basic difference between the two theories stems from APT's treatment of the correlations among the asset returns. The APT assumes that returns on assets are generated by a number of industry-wide and market-wide factors; returns on any two securities are likely to be correlated if the returns are affected by the same factors. Although the CAPM allows for correlations among assets, it does not specify the underlying factors that induce the correlations. The APT where the factors are observable economic variables can also be formulated as a multivariate regression model with nonlinear (bilinear) parameter restrictions, that is, with reduced-rank structure. The details of the model as multivariate regression with nonlinear restrictions were first presented by McElroy and Burmeister (1988), and Bekker et al. (1996) further identified the restrictions in the APT model as reduced-rank structure and discussed the estimation of parameters, including the "undersized samples" problem when $m > T$. In the APT, it is assumed that the returns on the m assets at the kth time period are generated by a linear (observable) factor model, with n factors, as

$$r_k = E(r_k) + Bf_k + \epsilon_k, \quad k = 1, \dots, T, \tag{8.15}$$

where r_k is an $m \times 1$ vector of returns, f_k is an $n \times 1$ vector of measured factors representing macroeconomic "surprises," and B is an $m \times n$ matrix with typical element b_{ij} measuring the sensitivity of asset i to factor j. The random errors ϵ_k are assumed to follow the same assumptions as in model (8.5). The APT starts with model (8.15) and explains the cross-sectional variation in the expected returns $E(r_k)$ under the assumption that arbitrage profits are impossible without undertaking some risk and without making some net investment. The fundamental result of APT is that $E(r_k)$ is approximated by

$$E(r_k) = \delta_{0k}\mathbf{1}_m + B\delta,$$

where δ_{0k} is the (possibly time-varying) risk-free rate of return for period k, assumed to be known at the beginning of period k, and δ is an $n \times 1$ vector of premiums earned by an investor for assuming one unit of risk from each of the n factors. Combining the above two models, we have

$$r_k = \delta_{0k}\mathbf{1}_m + B\delta + Bf_k + \epsilon_k. \tag{8.16}$$

Now, we let $Y_k = r_k - \delta_{0k}\mathbf{1}_m$, $X_k = [1, f_k']'$, and $C = [B\delta, B] \equiv B[\delta, I_n]$. Then the model (8.16) is in the same framework as (8.5), $Y_k = CX_k + \epsilon_k$, with the constraint on the coefficient matrix C given by

$$C\begin{bmatrix} 1 \\ -\delta \end{bmatrix} = 0.$$

Thus the $m \times (n+1)$ coefficient matrix C has a rank deficiency of one. The vector of risk premiums δ can be estimated from the eigenvector that corresponds to the smallest eigenvalue of $\widehat{\Sigma}_{xx}^{1/2}\widetilde{C}'\widetilde{\Sigma}_{\epsilon\epsilon}^{-1}\widetilde{C}\widehat{\Sigma}_{xx}^{1/2}$. By normalizing the first element of this eigenvector as equal to -1, the remaining elements give the estimate $\widehat{\delta}$ of δ. The associated estimator of B is then obtained as $\widehat{B} = \mathbf{Y}\widehat{\mathbf{X}}^{*\prime}(\widehat{\mathbf{X}}^{*}\widehat{\mathbf{X}}^{*\prime})^{-1}$, where $\widehat{\mathbf{X}}^{*} = [\widehat{X}_1^{*}, \ldots, \widehat{X}_T^{*}]$ and the $\widehat{X}_k^{*} = \widehat{\delta} + f_k \equiv [\widehat{\delta}, I_n]X_k$ are formed based on the estimator of δ.

McElroy and Burmeister (1988) suggested an iterative procedure to estimate the parameters in (8.16). Geweke and Zhou (1996) provide a Bayesian framework for estimating and making finite sampling inference of the above models. The posterior distributions of the parameters are estimated via Gibbs sampler. Bekker et al. (1996) noted that the problem could be formulated as a reduced-rank regression setup as indicated above. Bekker et al. (1996) also derived the second-order properties of the estimators of B and δ and discussed the case of undersized samples, that is, when there are more assets than time periods or $m > T$. The asymptotic results can be derived directly through the approach of Silvey (1959) as demonstrated in Chapters 3 and 4. The case of undersized samples is important to consider in the finance context because we may often have $m > T$, and then the unconstrained estimate $\widetilde{\Sigma}_{\epsilon\epsilon}$ of $\Sigma_{\epsilon\epsilon}$ will be singular. Two approaches were suggested for this case; one was simply to set $\Sigma_{\epsilon\epsilon} = I_m$, which leads to least squares reduced-rank estimates as discussed in Section 2.3. The second approach is to impose some factor analysis structure on $\Sigma_{\epsilon\epsilon}$, or grouping the assets in terms of industries, thereby reducing the number of unknown parameters and resulting in a nonsingular estimate of $\Sigma_{\epsilon\epsilon}$. The unknown parameters in $\Sigma_{\epsilon\epsilon}$ can be estimated from $\widetilde{\Sigma}_{\epsilon\epsilon}$.

Note that in the model (8.16) above, it is assumed that the risk-free rate δ_{0k}, although time-varying, is known. It is shown that "δ" the risk premium parameters associated with the factors "f_k" are estimated as a constraint on the regression coefficient matrix resulting from regressing $Y_k = r_k - \delta_{0k}\mathbf{1}_m$ on $X_k = [1, f_k']'$. Shanken (1992) and Shanken and Zhou (2007) have studied the estimation of the model $r_k = \delta_0\mathbf{1}_m + B\delta + Bf_k + \varepsilon_k$, where δ_0 is the zero-beta rate that is not known. For a recent discussion of this model, see Gospondinov et al. (2019). We want to observe that this model falls in the framework of Anderson's model discussed in Chapter 3. The (pricing) restriction resulting from the APT model, where "δ_0" is unknown, is

$$E(r_k) = \mu_r = \delta_0\mathbf{1}_m + B\delta,$$

which can also be restated in terms of "α," the intercept term in the regression mode, $Y_k = CX_k + \varepsilon_k$, as

$$\alpha = \mu_r - B\mu_f = \delta_0\mathbf{1}_m + B(\delta - \mu_f).$$

Note the coefficient matrix "C" is associated with $X_k = (1, f_k')'$ that includes the intercept term as $C = (\alpha, B)$. The stated restrictions in this case can also be stated as a constraint on the following expanded matrix:

$$G\gamma = \left(\mathbf{1}_m \ \alpha \ B\right)\begin{pmatrix} -\delta_0 \\ 1 \\ -(\delta - \mu_f) \end{pmatrix} = 0.$$

Using the results of Chapter 3 and Section 8.2 (see Gospondinov et al. 2019, p. 457), with sample estimates substituted, the estimate of γ can be obtained from the eigenvector associated with the smallest eigenvalue of $\widetilde{\Sigma}_{xx}^{*\frac{1}{2}}\widetilde{G}'\widehat{\Sigma}^{-1}\widetilde{G}\widetilde{\Sigma}_{xx}^{*\frac{1}{2}}$, where $\widetilde{\Sigma}_{xx}^{*} = A\widetilde{\Sigma}_{xx}^{-1}A'$ with $A = [\mathbf{0}_n, I_n]'$. From the eigenvector, $\widehat{\gamma}$, with appropriate rescaling, we can recover the estimates for δ_0 and δ.

8.3 Additional Applications of Reduced-Rank Regression in Finance

Additional applications of reduced-rank regression models occur in various areas of finance. In these contexts, the models typically arise as latent variables models, as discussed in Section 2.1, and they are described in Campbell et al. (1997, p. 446–448). Hansen and Hodrick (1983) and Gibbons and Ferson (1985) investigated the time-varying factor risk premiums; Campbell (1987) and Ferson and Harvey (1991) studied the areas of stock returns, forward currency premiums, international equity returns, and capital integration; Stambaugh (1988) tested the theory of the term structure of interest rates. In all these studies, the models involved appear to have the reduced-rank structure. Because of the diverse nature of the various topics above, we refrain from discussing them except for the work of Gibbons and Ferson (1985), which follows somewhat from the models considered earlier.

The Black (1972) CAPM in (8.3) and other related models suffer from some methodological shortcomings; the expected returns $E(r_{ik})$ are assumed to be constant over a sufficient period of time, and the market portfolio of risky assets, on which the return x_k is computed, must be observable. Gibbons and Ferson (1985) developed tests of asset pricing models that do not require the identification of the market portfolio or the assumption of constant risk premiums. These tests can be shown to be related to testing the rank of a regression coefficient matrix. We present only the simpler version of the model.

If γ_0 is assumed to be known in (8.3), then the returns r_{ik} and x_k can be adjusted by subtraction of this riskless rate of return γ_0, and this yields $E(r_{ik}^*) = \beta_i E(x_k^*)$, where $r_{ik}^* = r_{ik} - \gamma_0$ is the excess return on asset i and $x_k^* = x_k - \gamma_0$ is the excess return on the market portfolio. If the expected returns are changing over time and depend on the financial information prescribed by an n-dimensional vector X_k at time $k - 1$, a generalization of the above model can be written as

$$E(r_{ik}^*|X_k) = \beta_i E(x_k^*|X_k), \quad i = 1, \ldots, m, \tag{8.17}$$

where $r_{ik}^* = r_{ik} - \gamma_{0k}$ and $x_k^* = x_k - \gamma_{0k}$ are excess returns. The above holds for all assets, hence for $i = 1$ in particular. Therefore, if $\beta_1 \neq 0$,

$$E(r_{ik}^*|X_k) = (\beta_i/\beta_1) E(r_{1k}^*|X_k), \tag{8.18}$$

which does not involve the expected excess return on the market portfolio. Suppose the observed excess returns are assumed to be a linear function of the financial information X_k,

$$r_{ik}^* = \delta_i' X_k + \epsilon_{ik}, \tag{8.19}$$

where δ_i is an $n \times 1$ vector of regression coefficients, and suppose the assumptions on the ϵ_{ik} hold as stated earlier. Using $i = 1$ as a reference asset and substituting into both sides of (8.18) from (8.19) yield

$$\delta_i = (\beta_i/\beta_1)\delta_1, \quad i \neq 1. \tag{8.20}$$

Observe that the choice of the reference asset is arbitrary, and the coefficients β_i are not known. The conditions (8.20) lead to a reduced-rank regression setup as follows. The multivariate form of model (8.19) is given by

$$Y_k = C X_k + \epsilon_k, \quad k = 1, \ldots, T, \tag{8.21}$$

where $Y_k = (r_{1k}^*, \ldots, r_{mk}^*)'$ is the $m \times 1$ vector of excess returns and $C = [\delta_1, \ldots, \delta_m]'$ is an $m \times n$ regression coefficient matrix. The conditions (8.20) imply that $C' = \delta_1[1, (\beta_2/\beta_1), \ldots, (\beta_m/\beta_1)]$ and hence that $\text{rank}(C) = 1$. It follows that the number of independent parameters to be estimated is $(m + n - 1)$ instead of mn. Gibbons and Ferson (1985) also present a more general version of model (8.17). In this, the excess return on the market portfolio is replaced by excess returns on r hedge portfolios, which are not necessarily observed or by r reference assets. This leads to a multivariate regression model as in (8.21) and to the rank condition on the coefficient matrix that $\text{rank}(C) = r$. Therefore, all the methods that have been developed in Chapter 2 can be readily used in this setup.

8.4 Empirical Studies and Results on Asset Pricing Models

There exists an enormous amount of literature discussing empirical evidence on the CAPM. The early evidence was largely supportive of the CAPM, but in the late 1970s less favorable evidence for the CAPM began to emerge. Because of the availability of voluminous financial data, the statistical tests of hypotheses related to the CAPM and other procedures discussed here can be carried out readily. We will briefly survey some empirical results that have been obtained related to models discussed in this chapter. For additional empirical results, readers may refer to Campbell et al. (1997).

Gibbons (1982) used the monthly stock returns (r_k) provided by the Center for Research in Security Prices (CRSP), and the CRSP equal-weighted index was taken as the return on the market portfolio (x_k). The time period of 1926–1975 was divided into ten equal five-year subperiods. The number of stock returns was chosen to be $m = 40$. Within each subperiod, it is assumed that the parameters of model (8.1) are stable. The general conclusion obtained from the study was that the restrictions (8.4) implied by the CAPM were rejected at "reasonable significance levels in five out of ten subperiods." Some diagnostic procedures were suggested to postulate alternative model specifications. Other studies also do not provide conclusive results. Therefore, it is suggested that in order to analyze financial data more thoroughly, new models need to be specified, or because the testing of restrictions is based on large sample theory, the tests must be evaluated through their small sample properties.

Zhou (1991) used the reduced-rank regression tools for estimation and studied the small sample distributions of the estimators and associated test statistics. In particular, Zhou (1991) obtained results on the small sample distribution of the smallest eigenvalue test statistic $\widehat{\lambda}_2$ described in Section 8.2. He also considered the CRSP data, from 1926–1986, but instead of stock returns that may exhibit more volatility, he used 12 value-weighted industry portfolios. The analysis was again performed on a five-year subperiod basis. Out of 13 subperiods, the hypothesis (8.4) was rejected for all except four. He concluded that "the rejection of efficiency is most likely caused by the fundamentally inefficient behavior of the (market) index."

The widely quoted paper by Gibbons et al. (1989) also considered empirical analyses of similar data. Recall that these authors considered the multivariate Hotelling's T^2-statistic (8.8) for testing for portfolio efficiency in the models discussed by Sharpe (1964) and Lintner (1965), where the risk-free rate of return, γ_0 in (8.4), is known. Although their focus was on studying the sensitivity of the test statistic (8.8), they also considered empirical analyses of monthly returns. The empirical examples that they presented illustrate the effect that different sets of assets can have on the outcome of the test. First they reported their analysis on portfolios that were sorted by their risks. Using monthly returns on $m = 10$ portfolios from January 1931 through December 1965, through the F-statistic indicated in Section 8.2, they confirm that the efficiency of the market portfolio cannot be rejected. For monthly returns data from 1926 through 1982, however, they constructed one set of $m = 12$ industry portfolios and another set of $m = 10$ portfolios based on the relative market value or size of the firms. For industry-based portfolios, the F-statistic rejected the efficiency hypothesis, but for the size-based portfolios, it did not. Thus, the formation of portfolios could influence the outcome of the test procedure. Their overall conclusion is that "the multivariate approach can lead to more appropriate conclusions," because in the case of the industry-based analysis, all 12 univariate t-statistics failed to reject the efficiency hypothesis.

Other empirical results reported by McElroy and Burmeister (1988), Gibbons and Ferson (1985), and Zhou (1995) use the tests for the rank of the coefficient matrix being equal to one in some form or another. McElroy and Burmeister (1988) considered monthly returns from CSRP over January 1972 to December 1982 for a sample of 70 firms. The predictor variables chosen for use in model (8.16) include four macroeconomic "surprises" and the market return. The macro-factors include indices constructed from government and corporate bonds, deflation and treasury bills, and others. These authors focus more on the interpretation of the regression coefficients of the model (8.16); tests for the rank of the resulting coefficient matrix were not reported. Gibbons and Ferson (1985) considered daily returns from 1962 to 1980 for several major companies. While the excess returns on these companies form the response variables, the two predictors used in model (8.19) are taken to be an indicator variable for Monday, when mean stock returns are observed to be lower typically than other days of the week, and a lagged return of a large portfolio on common stocks. The null hypothesis (8.20) is rejected in their analysis. Zhou (1995) applied reduced-rank methodology to examine the number of latent factors in the U.S. equity market. For the 12 constructed industry portfolios, monthly data from October 1941 to September 1986 was examined. Several predictors were included in the model: returns on equal-weighted index, treasury bill rate, yield on Moody's BAA-rated bonds, and so on. The general conclusion reached from the study was that the resulting regression matrix is of rank one consistent with the original CAPM hypothesis.

8.5 Related Topics

Finite-Sample Results When there is a riskless asset, as discussed in Section 8.2, Gibbons et al. (1989) provide an exact finite-sample test of the linear pricing relationship under normality, and Zhou (1993) extends the test to allow for a more general elliptical distribution for the asset returns. Because the zero-beta rate is unobservable and is estimated, it is more difficult to develop a small sample test. As mentioned in the last section, Zhou (1991) provides a finite-sample test for the single beta case. For the multi-beta case, Shanken (1986) provides only bounds on the exact distribution of the likelihood ratio test (LRT). The exact distribution of the LRT for the multi-beta pricing models is given in Velu and Zhou (1999). It is far more complex than the case of risk-free rate and is dependent monotonically on a nuisance parameter. Yet, inferences can be made with a reasonable estimate of the nuisance parameter. Based on the proposed tests, the efficacy of a portfolio of two stock indices, the value-weighted and equal-weighted, is examined using the returns data from January 1926 to December 1994. While most studies reject efficiency of the single index, it is found that the efficacy of a portfolio of the two indices cannot be rejected for all of the ten-year subperiods except for the first one. For more complete details, interested readers can refer to Velu and Zhou (1999). The finite-sample distribution is also studied by Beaulieu et al. (2013) using simulation-based procedures for testing the mean–variance efficiency when the zero-beta rate is unknown and for building confidence intervals for the zero-beta rate. These are shown to be robust and can accommodate a correction for heteroskedasticity and autocorrelations. They are also robust to distributions beyond multivariate normal such as multivariate Student-t, normal mixtures, etc.

Shrinking Factor Dimension In this chapter, the models such as (8.13) or (8.22) are formulated and are used to test pricing restrictions resulting from some a priori theories. These theories commonly identify a select few factors (such as Fama–French factors); in practice, a few hundred factors are considered as potential factors that can affect the stock returns. The key questions then are: how many factors are needed and how much value they add beyond the usual Fama–French factors. This investigation neatly falls into the framework of the model in Chapter 2, where the focus was on dimension reduction, whereas the models considered in this chapter thus far have been on the complementary aspects of the problem, that is, estimating and testing the constraints on the coefficient matrix. He et al. (2022) consider the problem of shrinking the dimension of a large factor space in relation to the returns via a reduced-rank model, but their formulation assumes a more general weight matrix, and they use generalized method of moments (GMM) to estimate the parameters. They follow a two-step approach, where the so-called basis assets are used to estimate the model, to avoid issues related to large unbalanced panel data, and the model then is tested on portfolios of assets. The Instrumented Principal Component Analysis (IPCA) discussed in Section 7.8 is also a dimension reduction technique, but the focus was on changing firm characteristics, but here the focus is on time series factors (refer to Eqs. (7.28) and (7.30)). We begin with the model (7.30) with a constant term:

$$Y_k = \alpha + ABf_k + \varepsilon_k. \tag{8.22}$$

Here B is a $r \times n$ full-rank matrix of factor coefficients. This can also be taken as Anderson's model (3.3) with $D = \alpha$ and $Z_k = 1$. Note the factors "f_k" are observable.

He et al. (2022) consider GMM approach to estimate the model parameters in (8.22). In the GMM framework, the model residuals "ε_k" can have general heteroskedasticity and can represent other deviations from i.i.d. assumption. Using $X_k = (\mathbf{1}, f_k')'$ as instruments, the moment conditions are

$$E[h_k(\alpha, C)] = 0, \qquad h_k(\alpha, C) = \varepsilon_k(\alpha, C) \otimes X_k, \tag{8.23}$$

where

$$h_T(\alpha, C) = \frac{1}{T}\sum_{k=1}^{T} h_k(\alpha, C) = \text{Vec}\left(\frac{X'\varepsilon}{T}\right). \tag{8.24}$$

Here X and ε are the data matrices of predictors and the errors. The estimates of the model (8.22) are obtained by Hansen (1982).

$$\min_{\alpha,C} h_T(\alpha, C)' W_T h_T(\alpha, C), \tag{8.25}$$

where W_T is a $m(n+1) \times m(n+1)$ positive-definite weighting matrix. The main difference in the estimation approach in Chapter 2 and GMM is that the errors "ε_k" can have a covariance structure beyond independent and identical distribution over "k." Because of the complexity of the structure of W_T, there is no elegant solution in terms of eigenvectors, etc., but the estimation can be solved using numerical methods. If the weight matrix, W_T, takes the Kronecker form, $W_T = (W_1 \otimes W_2)$, where W_1 is "$m \times m$" and W_2 is "$(n+1) \times (n+1)$," the solution follows the steps used in Chapter 3. The normalization conditions may differ, but the resulting solutions are equivalent.

We have mentioned elsewhere (see Chapter 4) the importance of having a Kronecker structure for the weight matrix (see Section 4.5) that leads to elegant one-step solution in terms of singular values and the associated eigenvectors of the weighed coefficient matrix. In the case of the reduced-rank regression model, note the weight matrix takes the Kronecker's structure, $\widehat{\Sigma}_{\varepsilon\varepsilon}^{-1} \otimes \widehat{\Sigma}_{xx}$. As noted in Section 4.5, the solution to the estimation of model parameters in (8.23) reduces to finding α, A, and B by minimizing (see 3.4)

$$\text{tr}\{E[\Gamma^{1/2}(Y - ABF - \alpha\mathbf{1}_T')(Y - ABF - \alpha\mathbf{1}_T')'\Gamma^{1/2}]\},$$

where F is the "$n \times T$" matrix of factors in (8.23). With a simple transformation with $F^* = F - \overline{F}\mathbf{1}_T'$ and $\alpha^* = \alpha + C\overline{F}$, the above can be decomposed into two parts as shown in Section 3.1. Apart from a term that does not involve any parameters, the above is equivalent to minimizing,

$$\text{tr}\{E[\Gamma^{1/2}(Y - ABF^*)(Y - ABF^*)\Gamma^{1/2}]\} + \text{tr}\{E[\Gamma^{1/2}(Y - \alpha^*\mathbf{1}_T')(Y - \alpha^*\mathbf{1}_T')'\Gamma^{1/2}]\}.$$

Thus the solution to the reduced-rank part follows from results in Chapter 2. The above can be shown to result in minimizing, $W = (\widetilde{C} - AB)(F^*F^{*\prime})(\widetilde{C} - AB)' + (\overline{Y} - \alpha^*)(\overline{Y} - \alpha^*)'$. The solution is elegantly derived due to the Kronecker structure of the weight matrix, $\widetilde{\Sigma}_{\varepsilon\varepsilon}^{-1} \otimes F^*F^{*\prime}$.

To formulate the solution via GMM, observe that model (8.22) results in $YX' = C^*XX' + \varepsilon X'$, where $C^* = [\alpha\colon AB]$ and $X' = [\mathbf{1}_T, F']$. The covariance matrix of $\text{Vec}(\varepsilon X')$ can be general, accommodating heteroskedasticity, etc., but if it takes a Kronecker form, $W_1 \otimes W_2$, the minimizing criterion is

$$\begin{aligned}&\text{Vec}'(YX' - C^*XX')(W_1 \otimes W_2)\text{Vec}(YX' - C^*XX')\\&= \text{tr}\{W_1^{1/2}(YX' - C^*XX')W_2^{1/2}\, W_2^{1/2}(YX' - C^*XX')'W_1^{1/2}\}.\end{aligned}$$

Write $C^* = [\alpha : C]$ and note only "C" is restricted to reduced-rank condition. The minimizing criterion is then

$$\begin{aligned} W^* &= W_1^{1/2}(Y - CF - \alpha \mathbf{1}_T')(X'W_2X)(Y - CF - \alpha \mathbf{1}_T')W_1^{1/2} \\ &= W_1^{1/2}(Y - CF^* - \alpha^* \mathbf{1}_T')(X'W_2X)(Y - CF^* - \alpha^* \mathbf{1}_T')'W_1^{1/2}. \end{aligned} \tag{8.26}$$

Minimizing tr W^* results in the following: for a given "C,"

$$\widehat{\alpha}^* = \frac{(Y - CF^*)(X'W_2X)\mathbf{1}_T}{\mathbf{1}_T' X' W_2 X \mathbf{1}_T}.$$

Substituting this back in (8.26) and setting $C = AB$, where A and B are of lower rank of "r" matrices, and for a given B, it can be shown that

$$\widehat{A} = YX'W_2^*XF^{*'}B'(BF^*X'W_2^*XF^{*'}B')^{-1}.$$

Observe that

$$W_2^* = (X'W_2X)^{1/2}\left[I_{n+1} - \frac{(X'W_2X)^{1/2}\mathbf{1}_T\mathbf{1}_T'(X'W_2X)^{1/2}}{\mathbf{1}_T'X'W_2X\mathbf{1}_T}\right](X'W_2X)^{1/2}$$

is an idempotent matrix of rank "n." Inserting this back into W^* results in minimizing $(BF^*W_2^*Y'W_1YW_2^*B')(BF^*W_2^*F^{*'}B')^{-1}$, which is in the same framework as in Chapter 2. If we desire the normalization condition as $BF^*W_2^*F^{*'}B' = \Lambda_r$, a diagonal matrix, then the rows of "B" matrix are exactly the "r" eigenvectors that correspond to the "r" largest eigenvalues of $BF^*W_2^*Y'W_1YW_2^*B'$.

We want to conclude this section with some comments. When $W_2 = I_{n+1}$, then the results can be shown to be equivalent to the results for Anderson's model in Chapter 3. Inference on the coefficient matrices A and B and the intercept terms α all can be drawn from the asymptotic results in Chapter 3 (see Theorem 3.2). Because of specific choices of $W_1 (= \widetilde{\Sigma}_{\varepsilon\varepsilon})$ and $W_2 (= \widehat{\Sigma}_{xx.z})$ used in the Anderson model, stronger result such as minimizing simultaneously all the eigenvalues of W^* is possible. It is not clear if such a result will hold in the general $W_1 \otimes W_2$ case. The topic is worth investigating.

8.6 An Application

We use the same data set that was studied in Shanken and Zhou (2007) but extend the duration. The data contains monthly returns of equally weighted 25 portfolios sorted by size and book to market from March 1964 to December 2021. In addition, we have excess market return, SMB and HML factors as well. Thus, we have $m = 25$,

$T = 696$, and the C matrix is of dimensions $m \times 4$, including the intercept terms. The dimensions of C matrix being 25×4, its elements are too numerous to enumerate here, and so we just present the descriptives of the estimated coefficients (see Table 8.1) of the following regression model:

$$r_{ik} = \alpha_i + \beta_{1i}x_{1k} + \beta_{2i}x_{2k} + \beta_{3i}x_{3k} + \epsilon_{ik}, \quad i = 1, 2, \ldots, m, \quad k = 1, 2, \ldots, T. \tag{8.27}$$

Here x_1 is the market return-risk-free rate, x_2 is SMB, small minus big return spread, and x_3 is HML, high minus low return spread.

The asset pricing restriction resulting from model (8.27) is as stated in Shanken and Zhou (2007, Eqn 22),

$$H_0\colon \alpha = \alpha_0 \mathbf{1}_m + \gamma_1\beta_1 + \gamma_2\beta_2 + \gamma_3\beta_3, \tag{8.28}$$

where the γ's are called risk premia. This directly leads to the rank condition as given in (8.14):

$$H_0\colon \alpha = \gamma_0(\mathbf{1}_m - B\mathbf{1}_{n-1}), \tag{8.29}$$

where B is an $m \times (n-1)$ regression coefficient matrix. Taken together with α, $C = [\alpha, B]$ is of reduced rank $(n-1)$. In this example, we can expect $\text{rank}(C) = 3$. Extending the estimates as stated in (8.11) for rank 1 case to rank 3, where $C = \beta^*\gamma'$ with $\widehat{\beta}^* = \widetilde{C}\widehat{\Sigma}_{xx}^{1/2}\widehat{V}^*$, $\widehat{\gamma}' = \widehat{V}^*\widehat{\Sigma}_{xx}^{1/2}$. Here $\widehat{V}^* = [\widehat{V}_1^*, \widehat{V}_2^*, \widehat{V}_3^*]$, the matrix of the three eigenvectors that correspond to the largest three eigenvalues of $\widehat{\Sigma}_{xx}^{1/2}\widetilde{C}'\widetilde{\Sigma}_{\varepsilon\varepsilon}^{-1}\widetilde{C}\widehat{\Sigma}_{xx}^{1/2}$. This approximation of $\widetilde{C}$ by $\widehat{C}$ is fairly close with all the coefficients of the predictors are recovered well. This can be seen from the descriptive statistics of the coefficients of the reduced-rank matrix, $\widehat{C}$, given in Table 8.2 below.

To estimate γ_0, the risk-free rate, we can follow the setup as given in (8.7). With the C matrix of reduced rank, with a rank-deficit by one, we can estimate γ_0 using the relationship

$$C\gamma^* = \alpha\gamma_1^* + B\gamma_2^* = 0, \tag{8.30}$$

Table 8.1 Descriptives of the coefficients in model (8.27)

	$\widehat{\alpha}$	$\widehat{\beta}_1$	$\widehat{\beta}_2$	$\widehat{\beta}_3$
Average	0.352	1.022	0.520	0.305
Std. dev	0.144	0.070	0.502	0.410

Table 8.2 Descriptives of the coefficients of (8.27) with rank constraint

	$\widehat{\alpha}$	$\widehat{\beta}_1$	$\widehat{\beta}_2$	$\widehat{\beta}_3$
Average	0.422	1.018	0.529	0.302
Std. dev	0.027	0.085	0.502	0.411

where γ^* is the eigenvector that corresponds to the smallest eigenvalue of $\Sigma_{xx}^{1/2} C' \Sigma_{\varepsilon\varepsilon}^{-1} C \Sigma_{xx}^{1/2}$. From (8.29), the above equation can be written as

$$\gamma_0 \gamma_1^* (\mathbf{1}_m - B\mathbf{1}_{n-1} + B\gamma_2^*) = 0. \tag{8.31}$$

An alternative method is to simply use the rank 3 based estimator of $\widehat{\alpha}$ and $\widehat{B}$ and regress $\widehat{\alpha}$ on $\widehat{B}\mathbf{1}_{n-1}$ to obtain $\widehat{\gamma}_0$. Such an estimate will also take into account the variation in $\widehat{\alpha}$ as well. For the data set under consideration, $\widehat{\gamma}_0 = 0.419$, and the average risk-free rate of return for the duration is 0.367. Thus the zero-beta return is higher than the average risk-free rate.

8.7 Cointegration and Pairs Trading

While the focus of this Chapter has been on applications of reduced-rank regression in the asset pricing area, there are other areas of finance where the models discussed in this book have natural applications. One such area is pairs trading, an investment strategy, which assumes that if the fundamental values of two stocks are the same, then their prices move together and hence, they are cointegrated. These are sometimes called as "Siamese twins"; examples include, Royal Dutch/Shell and Unilever NV/PLC. In general there is some evidence that asset prices are co-integrated. If the price vector, p_t, is cointegrated with cointegrating rank, r, it implies that there are r linear combinations of p_t that are weakly dependent; thus these assets can be taken as somewhat redundant and any deviation of their price, from the linear combination of the other assets prices are temporary and can be expected to revert. Here we illustrate the application only with two assets.

A model based cointegration approach to pairs trading is elegantly presented in Tsay (2010, p. 313). It is assumed that $p_t = \log(P_t) = \log(\text{Price}_t)$ follows a random walk, but the log prices, p_{1t} and p_{2t} are cointegrated. Assume p_t follows a VAR(1) model, $p_t = \mathbf{\Phi} p_{t-1} + \epsilon_t$ and its error-correction form can be stated as,

$$p_t - p_{t-1} = (\mathbf{\Phi} - I) p_{t-1} + \epsilon_t. \tag{8.32}$$

If rank($\mathbf{\Phi} - I$)=1, then it can be written as

$$\mathbf{\Phi} - I = \alpha\beta'$$

and $w_t = \beta' p_t$ is called the cointegrated series. Writing the model (8.32) as

$$p_t - p_{t-1} = \alpha w_{t-1} + \epsilon_t.$$

Because w_t is stationary, the temporary deviations from its mean can be exploited to make buy and sell decisions between the two assets. The logic is that the price deviations are temporary and the processes will revert back to their means. For an extended discussion of the cointegration applications, see Velu, Hardy and Nehren (2020) and Farago and Hjalmarsson (2019). The discussion presented in Section 5.7 can be applied to more general settings that can use the information from other asset characteristics such as volume, volatility, etc.

Chapter 9
Partially Reduced-Rank Regression with Grouped Responses

An important characteristic of high-dimensional data is the presence of intrinsic group structures. The complex group structures in variables and in units of observations arise in many important scientific fields and they may carry useful information. They reflect correlations among the variables within the groups and ignoring the group structures will lead to inefficient use of available data. For example, in genetics, the DNA markers are grouped into genes which are grouped into biological pathways; these pathways are successfully used to detect rare occurrences of certain health issues in individuals. In macroeconomics, it is postulated that the general economy is driven by two factors or indices, monetary and real; each index has its own set of associated variables. Thus the group structures can be of interest for theoretical reasons as well as for uncovering new rare associations.

In a multivariate response-predictor setup, a widely adopted strategy is to perform predictor variable selection for each response variable. While it may address the predictor group structure, the response group structure is overlooked. This is best dealt with, in the multivariate regression set up that models the simultaneous relationships with appropriate constraints on the regression coefficient matrix, that can represent the block structure. Because in many practical applications, the number of variables can be much greater than the number of observations available, some dimension reductions are required to handle such data. The reduced-rank regression model introduced in Chapter 2 can handle the block structures and the dimension reduction more naturally. While the focus in Chapter 2 was on mostly on the predictor variables selection, in this chapter, the partially reduced-rank model is introduced, where only a subset of response variables may be related to a subset of predictors.

G. C. Reinsel et al., *Multivariate Reduced-Rank Regression*, Lecture Notes in Statistics 225, https://doi.org/10.1007/978-1-0716-2793-8_9

9.1 Introduction: Partially Reduced-Rank Regression Model

We consider the basic multivariate regression model that relates a set of m response variables $Y_k = (y_{1k}, \ldots, y_{mk})'$ to a set of n predictor variables $X_k = (x_{1k}, \ldots, x_{nk})'$,

$$Y_k = CX_k + \epsilon_k, \quad k = 1, \ldots, T, \tag{9.1}$$

where C is a $m \times n$ regression matrix. The $m \times 1$ vectors of errors ϵ_k are assumed to be distributed as i.i.d. multivariate normal with mean vector 0 and $m \times m$ nonsingular covariance matrix $\Sigma = \text{cov}(\epsilon_k)$. We assume that $m \geq n$ and $m + n < T$, but these will be relaxed in later sections.

Recall that the main feature of the reduced-rank linear model is the reduced-rank restriction of the form $\text{rank}(C) = r < \min(m, n)$, which yields the decomposition $C = AB$, where A is $m \times r$ and B is $r \times n$. As noted in Chapter 2, one interpretation provided by this decomposition is that the lower-dimensional set of r predictors $X_k^* = BX_k$ contains all the relevant information in the original set of predictors X_k for representing the variations in the response variables Y_k. Thus the focus was on mainly dimension reduction of the predictor set.

Standard reduced-rank regression methods may not be adequate to identify more specialized structures in the coefficient matrix C, however. For example, suppose the regression structure (as indicated above) of (9.1) were such that the $(m-1)$dimensional subset $Y_{1k} = (y_{1k}, \ldots, y_{m-1,k})'$ has reduced-rank structure of rank 1, $\mathbb{E}(Y_{1k}) = \alpha_1\beta_1' X_k$, and $E(y_{mk}) = \beta_m' X_k$, where β_1 and β_m are linearly independent $n \times 1$ vectors. Then the model for Y_k has reduced-rank structure of rank 2, of the form

$$E(Y_k) = \begin{bmatrix} \boldsymbol{\alpha}_1\boldsymbol{\beta}_1' \\ \boldsymbol{\beta}_m' \end{bmatrix} X_k = \begin{bmatrix} \boldsymbol{\alpha}_1 & 0 \\ 0 & 1 \end{bmatrix} \begin{bmatrix} \boldsymbol{\beta}_1' \\ \boldsymbol{\beta}_m' \end{bmatrix} X_k \equiv A_* B_* X_k.$$

Because $A_* B_* = \left(A_* P^{-1}\right)(P B_*)$ for any nonsingular 2×2 matrix P, standard reduced-rank estimation of a rank 2 model might not reveal the more specialized structure that actually exists. There might be considerable gains in estimation of the model with the more specialized structures specified over estimation of the general form of rank 2 model. Hence, it would be desirable to have a methodology to identify the existence of such specialized structures, and to properly account for it in efficient estimation of regression parameters. Thus, as a variant of the usual reduced-rank model, we consider a somewhat more general situation as above in which the reduced-rank coefficient structure occurs for (or is concentrated on) only a subset of the response variables.

As illustrated, the usual reduced-rank specification may be somewhat restrictive, since it requires that all response variables have regressions that are expressible in terms of a lower-dimensional set of linear combinations of the predictor variables in X_k. In practice, it may be that this feature does not hold for a (small) subset of the response variables, but for the remaining set of response variables it does, possibly with a few relevant linear combinations of the predictor variables. In fact, in such cases the usual reduced-rank procedure may not be very effective or efficient, because the lower-rank feature of the submatrix of C may not be revealed, if the requirement is that the reduced-rank specification be determined on the entire set of response variables. For these cases, the reduced-rank structure should be imposed on a partitioned (row-wise) submatrix of C for more proper modeling.

Initially, for estimation, it will be assumed that the components of $Y_k = (Y'_{1k}, Y'_{2k})'$ have a *known* arrangement so that the first m_1 components Y_{1k} possess a reduced-rank feature separate from the remaining components Y_{2k}. We partition the regression coefficient matrix C in (9.1) as

$$C = \begin{bmatrix} C_1 \\ C_2 \end{bmatrix}, \tag{9.2}$$

where C_1 is $m_1 \times n$ and C_2 is $m_2 \times n$, with $m_1 + m_2 = m$, and assume the reduced-rank feature that

$$\operatorname{rank}(C_1) = r_1 < \min(m_1, n) \tag{9.3}$$

and C_2 is of full rank, and the rows of C_2 are not linearly related to the rows of C_1. Thus, for now it is assumed that the subset of response variables that is of reduced rank is known, but the more practical case where the subset is to be determined has been the subject of much research recently and will be addressed later. Note these assumptions imply $\operatorname{rank}(C) = r_1 + m_2 (\leq n)$ in (9.1). We can write $C_1 = AB$, where A is a $m_1 \times r_1$ matrix and B is a $r_1 \times n$ matrix, both of full ranks, and note that C has the "overall factorization" $C = \operatorname{diag}(A, I_{m_2})[B', C'_2]'$. We consider the estimation of A, B, and hence of C_1 and C_2, and other related inference procedures for this "partially" reduced-rank model.

9.2 Estimation of Parameters

We now obtain the maximum likelihood (ML) estimators of parameters for the model (9.1) with the (partially) reduced-rank restriction (9.3). We represent the $m \times T$ data matrix $\mathbf{Y}$ as $\mathbf{Y} = [\mathbf{Y}'_1, \mathbf{Y}'_2]'$, where $\mathbf{Y}_1$ and $\mathbf{Y}_2$ are $m_1 \times T$ and $m_2 \times T$ matrices of values of the response vectors Y_{1k} and Y_{2k}, respectively. The full-rank or least squares (LS) estimators of C_1 and C_2 are $\widetilde{C}_1 = \mathbf{Y}_1\mathbf{X}'(\mathbf{X}\mathbf{X}')^{-1}$ and $\widetilde{C}_2 = \mathbf{Y}_2\mathbf{X}'(\mathbf{X}\mathbf{X}')^{-1}$, where $\mathbf{X} = [X_1, \ldots, X_n]$. We partition the error vectors as $\epsilon_k = (\epsilon'_{1k}, \epsilon'_{2k})'$ and partition the error covariance matrix $\Sigma = \operatorname{cov}(\epsilon_k)$ such that $\Sigma_{11} = \operatorname{cov}(\epsilon_{1k})$, $\Sigma_{12} = \operatorname{cov}(\epsilon_{1k}, \epsilon_{2k})$, and $\Sigma_{22} = \operatorname{cov}(\epsilon_{2k})$.

We first note that the decomposition $C_1 = AB$ is not unique. Therefore we need to impose some normalization conditions, chosen as follows:

$$B \widehat{\Sigma}_{xx} B' = \Lambda^2 \text{ and } A' \Sigma_{11}^{-1} A = I_{r_1}, \tag{9.4}$$

where $\widehat{\Sigma}_{xx} = (1/T)\mathbf{XX}'$, $\Lambda^2 = \operatorname{diag}(\lambda_1^2, \ldots, \lambda_{r_1}^2)$, and I_{r_1} denotes an $r_1 \times r_1$ identity matrix. We select these because in the basic multivariate reduced-rank regression model, the estimator of A and B can be related to eigenvectors associated with the "r" largest canonical correlations (see Chapter 2). Thus the number of free parameters in the regression coefficient matrix structure of the model is $r_1 (m_1 + n - r_1) + m_2 n$ compared to mn parameters in the full-rank model. Also noted that the number of free parameters in the usual reduced-rank model, which considers only that the rank of the overall matrix C is $r = r_1 + m_2$, is $r(m + n - r) = r_1 (m_1 + n - r_1) + m_2 (m_1 + n - r_1)$. Hence if m_1 is relatively large, there can be a substantial reduction in the number of model parameters.

Assuming the $\boldsymbol{\epsilon}_k$ are i.i.d., following a multivariate normal distribution with mean vector 0 and covariance matrix Σ, apart from irrelevant constants the log-likelihood is

$$L(C_1, C_2, \Sigma) = \left(\frac{T}{2}\right) \left[\log\left|\Sigma^{-1}\right| - \operatorname{tr}\left(\Sigma^{-1} W\right)\right], \tag{9.5}$$

where $W = (1/T)(\mathbf{Y} - C\mathbf{X})(\mathbf{Y} - C\mathbf{X})'$ and $\left|\Sigma^{-1}\right|$ is the determinant of the matrix Σ^{-1}. Maximizing (9.5) with respect to Σ yields $\widehat{\Sigma} = W$. Hence, the concentrated log-likelihood is $L(C_1, C_2, \widehat{\Sigma}) = -(T/2)(\log|W| + m)$. We can proceed to directly derive the ML estimates of $C_1 = AB$ and C_2 as values that minimize $|W|$, but as an alternative we consider an argument based on a conditional distribution approach for additional insight.

Consider maximizing the likelihood expressed in terms of the marginal distribution for $\mathbf{Y}_1$ and the conditional distribution for $\mathbf{Y}_2$ given $\mathbf{Y}_1$. The model for $\mathbf{Y}_1$ is of course the reduced-rank model $Y_{1k} = C_1 X_k + \boldsymbol{\epsilon}_{1k}$, $k = 1, \ldots, T$, with parameters $C_1 = AB$ and $\Sigma_{11} = \operatorname{cov}(\boldsymbol{\epsilon}_{1k})$, and the conditional model for $\mathbf{Y}_2$, given $\mathbf{Y}_1$, is represented as

$$Y_{2k} = C_2 X_k + \Sigma_{21}\Sigma_{11}^{-1}(Y_{1k} - C_1 X_k) + \boldsymbol{\epsilon}_{2k}^* \equiv C_2^* X_k + D^* Y_{1k} + \boldsymbol{\epsilon}_{2k}^* \tag{9.6}$$

with parameters $C_2^* = C_2 - \Sigma_{21}\Sigma_{11}^{-1} C_1$, $D^* = \Sigma_{21}\Sigma_{11}^{-1}$, and $\Sigma_{22}^* = \operatorname{cov}(\boldsymbol{\epsilon}_{2k}) = \Sigma_{22} - \Sigma_{21}\Sigma_{11}^{-1}\Sigma_{12}$, and $\operatorname{cov}(\boldsymbol{\epsilon}_{1k}, \boldsymbol{\epsilon}_{2k}) = 0$. The two sets of parameters $\{C_1, \Sigma_{11}\}$ and $\{C_2^*, D^*, \Sigma_{22}^*\}$ are functionally independent, so ML estimation can be performed separately on them.

ML estimation for C_1 and Σ_{11} is the same as ML estimation of the standard reduced-rank model and is equivalent to simultaneously minimizing the eigenvalues of $\widetilde{\Sigma}_{11}^{-1/2}\left(\widetilde{C}_1 - AB\right)\widehat{\Sigma}_{xx}\left(\widetilde{C}_1 - AB\right)'\widetilde{\Sigma}_{11}^{-1/2}$, where $\widetilde{C} = \left[\widetilde{C}_1', \widetilde{C}_2'\right]' = \mathbf{YX}'\left(\mathbf{XX}'\right)^{-1}$ is the (full-rank) LS estimate of C, and $\widetilde{\Sigma}_{11}$ is the upper-left block of the corresponding estimate $\widetilde{\Sigma} = (1/T)(\mathbf{Y} - \widetilde{C}\mathbf{X})(\mathbf{Y} - \widetilde{C}\mathbf{X})'$. From earlier results in Chapter 2, this yields the ML estimators of A and B as

$$\widehat{A} = \widetilde{\Sigma}_{11}^{1/2}\widehat{V}_{(r_1)}, \quad \widehat{B} = \widehat{V}'_{(r_1)}\widetilde{\Sigma}_{11}^{-1/2}\widetilde{C}_1, \tag{9.7}$$

where $\widehat{V}_{(r_1)} = \left[\widehat{V}_1, \ldots, \widehat{V}_{r_1}\right]$, and $\widehat{V}_j$ is the (normalized) eigenvector that corresponds to the jth largest eigenvalue $\widehat{\lambda}_j^2$ of the matrix $\widehat{R}_1 = \widetilde{\Sigma}_{11}^{-1/2}\widetilde{C}_1\widehat{\Sigma}_{xx}\widetilde{C}_1'\widetilde{\Sigma}_{11}^{-1/2}$. The ML estimator of C_1 is $\widehat{C}_1 = \widehat{A}\widehat{B}$, and the ML estimator of Σ_{11} under the reduced-rank structure is given by $\widehat{\Sigma}_{11} = (1/T)(\mathbf{Y}_1 - \widehat{C}_1\mathbf{X})(\mathbf{Y}_1 - \widehat{C}_1\mathbf{X})'$.

For the conditional model (9.6), notice that the parameters C_2^* and D^* are full rank by assumption, so ML estimation in the conditional model simply yields the usual full-rank LS estimates. These can be expressed in a convenient form (e.g., see Chapter 3) as

$$\widehat{C}_2^* = \widetilde{C}_2 - \widetilde{\Sigma}_{21}\widetilde{\Sigma}_{11}^{-1}\widetilde{C}_1, \quad \widehat{D}^* = \widetilde{\Sigma}_{21}\widetilde{\Sigma}_{11}^{-1}, \tag{9.8}$$

with $\widehat{\Sigma}_{22}^* = \widetilde{\Sigma}_{22} - \widetilde{\Sigma}_{21}\widetilde{\Sigma}_{11}^{-1}\widetilde{\Sigma}_{12}$. The MLE of $C_2 = C_2^* + D^*C_1$ can then be obtained as

$$\widehat{C}_2 = \widehat{C}_2^* + \widehat{D}^*\widehat{C}_1 = \widetilde{C}_2 - \widetilde{\Sigma}_{21}\widetilde{\Sigma}_{11}^{-1}(\widetilde{C}_1 - \widehat{C}_1). \tag{9.9}$$

The MLEs of Σ_{21} and Σ_{22} can be obtained accordingly, from $\widehat{\Sigma}_{21} = \widehat{D}^*\widehat{\Sigma}_{11} = \widetilde{\Sigma}_{21}\widetilde{\Sigma}_{11}^{-1}\widehat{\Sigma}_{11}$ and

$$\widehat{\Sigma}_{22} = \widehat{\Sigma}_{22}^* + \widehat{\Sigma}_{21}\widehat{\Sigma}_{11}^{-1}\widehat{\Sigma}_{12} = \widetilde{\Sigma}_{22} - \widetilde{\Sigma}_{21}\widetilde{\Sigma}_{11}^{-1}\widetilde{\Sigma}_{12} + \widetilde{\Sigma}_{21}\widetilde{\Sigma}_{11}^{-1}\widehat{\Sigma}_{11}\widetilde{\Sigma}_{11}^{-1}\widetilde{\Sigma}_{12}.$$

Thus the MLEs $\widehat{C}_1$ and $\widehat{C}_2$ can be obtained using matrix routines for eigenvector (or singular value) decompositions in a non-iterative fashion. The ML estimation strategy essentially corresponds to the usual reduced-rank estimation to obtain $\widehat{C}_1$ followed by "covariance-adjustment" estimation (9.9) to obtain $\widehat{C}_2$, where the LS estimator $\widetilde{C}_2$ is adjusted by the "covariates" $\mathbf{Y}_1 - \widehat{C}_1\mathbf{X}$.

Note that $\widetilde{C}_1 - \widehat{C}_1 = (I - \widehat{P}_1)\widetilde{C}_1$ and $\widetilde{C}_2 - \widehat{C}_2 = \widetilde{\Sigma}_{21}\widetilde{\Sigma}_{11}^{-1}(\widetilde{C}_1 - \widehat{C}_1) = \widetilde{\Sigma}_{21}\widetilde{\Sigma}_{11}^{-1}(I - \widehat{P}_1)\widetilde{C}_1$, where $\widehat{P}_1 = \widetilde{\Sigma}_{11}^{1/2}\widehat{V}_{(r_1)}\widehat{V}'_{(r_1)}\widetilde{\Sigma}_{11}^{-1/2}$ is an idempotent matrix of rank r_1. Hence, $\widetilde{C} - \widehat{C} = \widetilde{Q}(I - \widehat{P}_1)\widetilde{C}_1$, where $\widetilde{Q}' = [I_{m_1}, \widetilde{\Sigma}_{11}^{-1}\widetilde{\Sigma}_{12}]$, and then from basic results we get the maximum likelihood estimate,

$$\begin{aligned}
\widehat{\Sigma} &= \widetilde{\Sigma} + (\widetilde{C} - \widehat{C})\widehat{\Sigma}_{xx}(\widetilde{C} - \widehat{C})' \\
&= \widetilde{\Sigma} + \widetilde{Q}(I - \widehat{P}_1)\widetilde{C}_1\widehat{\Sigma}_{xx}\widetilde{C}_1'(I - \widehat{P}_1)'\widetilde{Q}' \\
&= \widetilde{\Sigma} + \widetilde{Q}\widetilde{\Sigma}_{11}^{1/2}\left[I - \widehat{V}_{(r_1)}\widehat{V}'_{(r_1)}\right]\widehat{R}_1\left[I - \widehat{V}_{(r_1)}\widehat{V}'_{(r_1)}\right]\widetilde{\Sigma}_{11}^{1/2}\widetilde{Q}'.
\end{aligned} \tag{9.10}$$

Observe the equivalence of (9.10) to (2.18) in the context of the model studied in Chapter 2.

9.3 Test for Rank and Inference Results

We consider the likelihood ratio (LR) test of the hypothesis H_0 : rank$(C_1) \leq r_1$. The LR test statistic for testing rank$(C_1) = r_1$ is $\lambda = U^{T/2}$, where $U = |S|/|S_1| \equiv |\widetilde{\Sigma}|/|\widehat{\Sigma}|$, $S = (\mathbf{Y} - \widetilde{C}\mathbf{X})(\mathbf{Y} - \widetilde{C}\mathbf{X})'$ is the residual sum of squares matrix from fitting the full-rank model, while $S_1 = (\mathbf{Y} - \widehat{C}^{(r)}\mathbf{X})(\mathbf{Y} - \widehat{C}^{(r)}\mathbf{X})'$ is the residual sum of squares matrix from fitting the model under the rank condition on C_1. Here $\widehat{C}^{(r)}$ denotes the estimate of C under rank condition (9.3). It is known that $|\widehat{\Sigma}| = |\widehat{\Sigma}_{11}||\widehat{\Sigma}_{22}^{*}| = |\widehat{\Sigma}_{11}||\widetilde{\Sigma}_{22}^{*}|$ and $|\widetilde{\Sigma}| = |\widetilde{\Sigma}_{11}||\widetilde{\Sigma}_{22}^{*}|$, where $\widehat{\Sigma}_{22}^{*} \equiv \widetilde{\Sigma}_{22}^{*}$ since reduced-rank estimation in the model for $\mathbf{Y}_1$ does not affect the LS estimation for the conditional model (9.6). Therefore, we have $U = |\widetilde{\Sigma}_{11}|/|\widehat{\Sigma}_{11}|$. It follows that the LR testing procedure is the same as in the usual reduced-rank regression model for $\mathbf{Y}_1$ and does not involve the response variables $\mathbf{Y}_2$. Therefore, the criterion $\lambda = U^{T/2}$ is such that (see Section 2.6, eqn (2.40))

$$- 2\log(\lambda) = T \sum_{j=r_1+1}^{m_1} \log\left(1 + \widehat{\lambda}_j^2\right) = -T \sum_{j=r_1+1}^{m_1} \log\left(1 - \widehat{\rho}_j^2\right), \tag{9.11}$$

where $\widehat{\lambda}_j^2$, $j = r_1 + 1, \ldots, m_1$ are the $(m_1 - r_1)$ smallest eigenvalues of $\widehat{R}_1$, and $1 + \widehat{\lambda}_j^2 = 1/(1 - \widehat{\rho}_j^2)$, where the $\widehat{\rho}_j^2$ are the squared sample canonical correlations between Y_{1k} and X_k (adjusting for sample means if constant terms are allowed for). Then (9.11) follows asymptotically the $\chi^2_{(m_1-r_1)(n-r_1)}$ distribution under the null hypothesis (see Anderson (1951), Theorem 3). A simple correction factor for the LR statistic in (9.11), to improve the approximation to the $\chi^2_{(m_1-r_1)(n-r_1)}$ distribution, is given by

$$\begin{aligned} \mathcal{M} &= - 2\left\{[T - n + (n - m_1 - 1)/2]\,/T\right\}\log(\lambda) \\ &= - [T - n + (n - m_1 - 1)/2] \sum_{j=r_1+1}^{m_1} \log\left(1 - \widehat{\rho}_j^2\right). \end{aligned}$$

This approximation is known to work well when T is large (see Anderson 1984b, Chapter 8).

The alternative hypothesis in the above testing procedure is that the matrix C is of full rank. There may be situations where $r_1 + m_2 = r < \min(m, n)$ so that the matrix C would still have reduced rank r to begin with. We might want to test the subset reduced-rank model assumptions of (9.2)–(9.3) against the alternative of the usual reduced-rank model, rank$(C) = r = r_1 + m_2$. The form of the LR statistic for this test can readily be developed, and

$$- 2\log(\lambda) = -T \left[\sum_{j=r_1+1}^{m_1} \log\left(1 - \widehat{\rho}_j^2\right) - \sum_{j=r+1}^{m} \log\left(1 - \widetilde{\rho}_j^2\right) \right], \tag{9.12}$$

where $\widetilde{\rho}_j^2$, $j = r+1, \ldots, m$, are the $(m-r) \equiv (m_1 - r_1)$ smallest squared sample canonical correlations between Y_k and X_k, with $-2\log(\lambda)$ distributed as $\chi^2_{m_2(m_1-r_1)}$ asymptotically.

Concerning distributional properties of the ML estimators $\widehat{C}_1$ and $\widehat{C}_2$, the approximate normal distribution and approximate covariance matrix of the reduced-rank estimator $\widehat{C}_1$ follow directly from results by Anderson (1999a) (see Sections 2.3 and 2.5). In particular, the results from Anderson (1999a, p. 1147–1148) imply that the large sample (as $T \to \infty$) approximate covariance matrix of $\widehat{C}_1$ is given by

$$\begin{aligned}\text{cov}[\text{vec}(\widehat{C}_1)] = \Sigma_{11} \otimes (\mathbf{XX}')^{-1} - \left[\Sigma_{11} - A\left(A'\Sigma_{11}^{-1}A\right)^{-1}A'\right] \\ \otimes \left[(\mathbf{XX}')^{-1} - B'\left(B\left(\mathbf{XX}'\right)B'\right)^{-1}B\right]. \qquad (9.13)\end{aligned}$$

For the general case, arguments similar to those by Ahn and Reinsel (1988) (also see Sec. 3.5) establish that the asymptotic covariance matrix of the joint ML estimator $\widehat{\gamma} = [\text{vec}(\widehat{C}_1'), \text{vec}(\widehat{C}_2')]'$ is given by

$$\text{cov}\left(\widehat{\gamma}\right) = M\left[M'\left(\Sigma^{-1} \otimes \mathbf{XX}'\right)M\right]^{-1}M', \qquad (9.14)$$

where $M = \text{diag}\left(M_1, I_{m_2 n}\right)$ and $M_1 = \partial\gamma_1/\partial\boldsymbol{\theta}'$, with $\gamma_1 = \text{vec}(C_1') \equiv \text{vec}(B'A')$ and $\boldsymbol{\theta} = [\text{vec}(A')', \text{vec}(B')']'$. It can be verified that this approach yields the same result for $\text{cov}[\text{vec}(\widehat{C}_1')]$ as the expression given in (9.13) and, furthermore, that we obtain the asymptotic expression

$$\text{cov}[\text{vec}(\widehat{C}_2')] = \Sigma_{22}^* \otimes (\mathbf{XX}')^{-1} + (\Sigma_{21}\Sigma_{11}^{-1} \otimes I_n)\text{cov}[\text{vec}(\widehat{C}_1')](\Sigma_{11}^{-1}\Sigma_{12} \otimes I_n), \qquad (9.15)$$

where $\text{cov}[\text{vec}(C_1')]$ is given by the expression in (9.13), and $\Sigma_{22}^* = \Sigma_{22} - \Sigma_{21}\Sigma_{11}^{-1}\Sigma_{12}$. Consider, for instance, the extreme case in which $r_1 = \text{rank}(C_1)$ is taken as 0, so that $C_1 = 0$ and estimation of C_1 is not involved. Then, in (9.9), $\widehat{C}_2 = \widehat{C}_2^*$ and (9.15) collapses to $\text{cov}[\text{vec}(\widehat{C}_2')] = \Sigma_{22}^* \otimes (\mathbf{XX}')^{-1}$. This can be recognized as a familiar result in the context of "covariance-adjustment," where in this case the entire set of response variables in Y_{1k} would be used as covariates for Y_{2k} since $C_1 = 0$ implies that the complete vector of responses Y_{1k} is entirely unrelated to X_k.

For the covariance matrix of $\widehat{C}_2$, it may also be instructive to mention a more direct argument. Since $\widetilde{\Sigma}$ is consistent for Σ, we can write the ML estimator $\widehat{C}_2$ as

$$\begin{aligned}\widehat{C}_2 &= \widetilde{C}_2 - \widetilde{\Sigma}_{21}\widetilde{\Sigma}_{11}^{-1}(\widetilde{C}_1 - \widehat{C}_1) = \widetilde{C}_2 - \Sigma_{21}\Sigma_{11}^{-1}(\widetilde{C}_1 - \widehat{C}_1) + o_p\left(T^{-1/2}\right) \\ &= (\widetilde{C}_2 - \Sigma_{21}\Sigma_{11}^{-1}\widetilde{C}_1) + \Sigma_{21}\Sigma_{11}^{-1}\widehat{C}_1 + o_p\left(T^{-1/2}\right) \\ &= H\widetilde{C} + \Sigma_{21}\Sigma_{11}^{-1}\widehat{C}_1 + o_p\left(T^{-1/2}\right),\end{aligned}$$

where $H = \left[-\Sigma_{21}\Sigma_{11}^{-1}, I\right]$ is such that $H\Sigma H' = \Sigma_{22}^{*}$. Because $(\widetilde{C}_2 - \Sigma_{21}\Sigma_{11}^{-1}\widetilde{C}_1)$ and $\widetilde{C}_1$ have zero covariance, it follows that the two terms in (9.3) that comprise $\widehat{C}_2$ are uncorrelated (independent) asymptotically. So we obtain the asymptotic covariance matrix of $\widehat{C}_2$ using (9.3) from $\text{cov}[\text{vec}(\widehat{C}_2')] = \text{cov}[\text{vec}(\widetilde{C}'H')] + \text{cov}[\text{vec}(\widehat{C}_1\Sigma_{11}^{-1}\Sigma_{12})]$, the result in (9.15).

Finally, we can compare the distributional properties of the ML estimators $\widehat{C}_1$ and $\widehat{C}_2$ under the subset reduced-rank model assumptions of (9.2)–(9.3) with those for ML estimators obtained under the usual or standard reduced-rank model assumptions. As noted in 9.1, (9.2)–(9.3) imply that $\text{rank}(C) = r_1 + m_2 = r$ in (9.1) with an overall factorization as $C = \text{diag}(A, I_{m_2})[B', C_2']' \equiv A_* B_*$. When $r = \min(m, n)$ there is no reduced rank overall and one would consider the usual LS estimator $\widetilde{C}$ with $\text{cov}[\text{vec}(\widetilde{C}')] = \Sigma \otimes (\mathbf{XX}')^{-1}$. When $r < \min(m, n)$, however, one may still obtain the ML estimator of C for the usual reduced-rank model, denoted as $\overline{C} = \overline{A}_*\overline{B}_*$ with components $\overline{C}_1$ and $\overline{C}_2$. Using similar methods as previously, it can be shown that the asymptotic covariance matrices of $\text{vec}(\overline{C}_1')$ and $\text{vec}(\overline{C}_2')$ are of the same form as the ML estimators in (9.13) and (9.15) under the subset reduced-rank model, but with $(\mathbf{XX}')^{-1} - B_*'(B_*(\mathbf{XX}')B_*')^{-1}B_*$ in place of $(\mathbf{XX}')^{-1} - B'(B(\mathbf{XX}')B')^{-1}B$. These results can thus be directly compared with the results in (9.13) and (9.15) to indicate the increase in the covariance matrix relative to the subset reduced-rank model.

9.4 Procedures for Identification of Subset Reduced-Rank Structure

In practice, we typically would not know or be able to specify a priori the subset of response variables for which (9.2)–(9.3) hold. The procedure of examining all possible subsets of the response variables for tests of rank, although will result in an optimal solution, involves tedious calculations. For each subset, we need to calculate the canonical correlations or the eigenvalues of $\widehat{R}_1$. Even for modest values of m, this approach leads to excessive calculations because it involves $2^m - m - 1$ subset calculations. For the example we illustrate below, $m = 6$ and we need to consider fifty-seven sets of canonical correlations. As an alternative, similar to the case of multiple regression, we suggest an exploratory procedure, using “Stepwise Backward Elimination” of responsive variables, one in each step. It must be observed that the two concepts are related as they both involve testing for redundancy. Seber (1984, Section 10.1.6) discusses procedures for selecting the best subset of the responsive variables in the case of multivariate regression. We briefly describe them below.

Consider the regression model (9.1) and suppose we want to test the hypothesis $H_0\colon FC = 0$, where F is a known full-rank matrix that provides for restrictions in the coefficients of C across the different response variables. Assume that the dimension of F is $m^* \times m$. The LR test of $H_0\colon FC = 0$ is based on the eigenvalues, $\widehat{\lambda}_i^2$, of $S_*^{-1}H_*$, where $S_* = FSF'$ and $H_* = F\tilde{C}(XX')\tilde{C}'F'$ and $S = T\tilde{\Sigma}$. The test statistic is $M = -(T - n + \frac{1}{2}(n - m^* - 1))\log U$ which has an approximate $\chi^2_{nm^*}$ distribution under H_0, where $U^{-1} = |I + S_*^{-1}H_*| = \prod_{i=1}^{m^*}(1 + \widehat{\lambda}_i^2)$. If we want to eliminate the y-variables, we want to consider those y-variables that make an insignificant contribution to the test of H_0. We have observed elsewhere (see Section 1.3) that the hypothesis that the rank of C is r is equivalent to the above H_0, where F is unknown.

The hypotheses $H_0\colon FC = 0$ can also be tested by considering a set of sequential hypotheses as given in Hawkins (1976). The U statistic can be expressed as $U = \prod_{i=1}^{m} t_{ii}^2$, where the t_{ii}^2's are independent and follow, in small samples, Beta distributions. The quantity t_{ii}^2 measures the strength of the relationship between y_i and X after adjusting for $y_1, \ldots, y_{i-1}$ and generally small values of t_{ii}^2 indicate a significant relationship. The connection between Beta and the F-distribution yields that

$$F_i(m^*, n - m^* - i + 1) = \frac{n - m^* - i + 1}{m^*} \cdot \frac{1 - t_{ii}^2}{t_{ii}^2} \tag{9.16}$$

statistics can be used to test the hypothesis H_0. If the H_0 is true, then each F_i is nonsignificant. As $U_k = \prod_{i=1}^{k} t_{ii}^2$ is the test statistic for H_0 based on k-variables, a criterion for finding the best subset of "k" variables is to find the subset that minimizes U_k. To obtain the overall best subset, the calculation requires $2^m - 1$ values of U_k. This is not feasible if m is large and an alternative is to use a stepwise method. Before we present the backward-elimination method observe that $t_{ii}^2 \leq 1$ and $t_{ii}^2 = \tilde{s}_{ii}/\tilde{s}_{hii}$, where $\tilde{s}_{ii} = s_{ii} - s_{i-1}'S_{i-1}^{-1}s_{i-1}$. The quantities s_{ii} is the ith diagonal element of S and S_{i-1} is the leading $(i-1)\times(i-1)$ submatrix of S. Similarly, s_{hii} is the ith diagonal element of $S+H = S+SF'(FSF')^{-1}F\tilde{C}(XX')\tilde{C}'F'(FSF')^{-1}FS$ and other items can be defined similarly. The decision on inclusion or exclusion of a variable is based on examination of t_{ii}^2. Observe that in the stepwise selection procedure suggested in the literature includes variables for which t_{ii}^2 is small. In Hawkins (1976), it is suggested that we start with a variable "i" for which t_{ii}^2 is minimum. If this quantity is sufficiently small, then the ith variable is included in step 1 and this variable is swept out of S and $S + H$.

We want to describe another procedure similar to above but that assumes that the constraints (F-matrix) are not known. In (9.1) with $m > n$, however, there will exist $m - n$ linear combinations $\ell_j'Y_k$ such that $\ell_j'C = 0$. When (9.2)–(9.3) holds, $m_1 - r_1$ of the vectors ℓ_j can take the form $\ell_j' = [\ell_{1j}', 0']$, with ℓ_{1j} such that $\ell_{1j}'C_1 = 0$. As described in Chapter 2, the ℓ_j can be estimated by $\widehat{\ell}_j = \widetilde{\Sigma}^{-1/2}\widehat{V}_j$, $j = n+1, \ldots, m$, where the $\widehat{V}_j$ are normalized eigenvectors of the matrix $\widehat{R} = \widetilde{\Sigma}^{-1/2}\widetilde{C}\widehat{\Sigma}_{xx}\widetilde{C}'\widetilde{\Sigma}^{-1/2}$ associated with its $m - n$ smallest (zero) eigenvalues. If a certain number $(m_1 - r_1 > m_1 - n)$ of the estimates $\widehat{\ell}_j$ satisfy $\widehat{\ell}_j' \approx [\widehat{\ell}_{1j}', 0']$, for a particular partition of the response variables, then this feature can serve as a preliminary tool to identify the existence and nature of the partially reduced-rank structure for C.

If desired, one can also systematically examine all possible subsets of the response variables for each given dimension $m_1 < m$, and compute the LR statistics for tests of rank and other summary statistics for each subset case. An information criterion such as AIC could then be used to select the "best" subset of response variables that exhibits the most desirable (partially) reduced-rank regression features. The procedure of examining all possible subsets of the response variables for tests of rank and other features may not be practical or desirable when the dimension m is relatively large. Thus we suggest and describe briefly a stepwise "backward-elimination" procedure. In this, we start with all $m(\geq n)$ response variables included in $Y_{1k} \equiv Y_k$ and carry out the LR test procedure for reduced rank. Assume that r is identified as the rank of the overall coefficient matrix C by the LR test at this stage, that is, r is the smallest value such that the hypothesis H_0 : rank$(C) \leq r$ is not rejected by the LR test procedure. Then we consider each of the m distinct subsets of Y_k, of the form $(y_{1k}, \ldots, y_{i-1,k}, y_{i+1,k}, \ldots, y_{mk})'$, obtained by excluding one response variable (the ith variable) at a time. For each $(m - 1)$-dimensional subset we calculate the LR statistic for testing the corresponding hypothesis that rank$(C_1) = r - 1$. If $y_{i'k}$ is the response variable that yields the *smallest* value of the LR statistic for testing rank$(C_1) = r - 1$ and if this value leads to not rejecting this null hypothesis, then the subset Y_{1k} with variable $y_{i'k}$ excluded is chosen. After excluding $y_{i'k}$, we test if it is reasonable to reduce the rank of C_1 without discarding any more y-variables. At the next step we consider each of the $m - 1$ distinct subsets of this Y_{1k} obtained by excluding a remaining response variable one at a time. For each $(m - 2)$-dimensional subset we calculate the LR statistic for testing the corresponding hypothesis rank$(C_1) \leq r - 2$. If $y_{i''k}$ is the response variable that yields the *smallest* value of the LR test statistic and if this leads to not rejecting this null hypothesis, then the new $(m - 2)$-dimensional subset Y_{1k} with variable $y_{i''k}$ excluded (in addition to the previously excluded variable $y_{i'k}$) is chosen. The stepwise procedure continues until no further response variables can be "eliminated." The performance of these procedures needs to be formally evaluated.

9.5 Illustrative Examples

In this section we present two numerical examples to illustrate the (partially) reduced-rank methods. The first example involves macroeconomic time series data for which consideration of subset structure leads to further rank reduction. The second example involves chemometrics data for which ignoring the subset structure leads to no rank reduction.

UK Macroeconomic Data We now consider UK macroeconomic data presented in Chapter 4. Recall that the relationships between endogenous (response) variables and exogenous (prediction) can be taken to reflect the demand side of the macrosystem of the UK economy. The quarterly observations are for the years 1948–1956 and after seasonal adjustment of some series; the plots are presented in Section 4.2.

From preliminary LS regression of each response variable y_{ik} on the predictor variables x_{ik}, some moderate degree of autocorrelation is noted in the residuals, particularly for the second (consumption of food, etc.) and third (total unemployment) response variables. Therefore, based on LS estimation of a model which included a lagged response variable in these two equations, in the analysis described below we use the "adjusted" response variables $y_{2k}^* = y_{2k} + 0.6 y_{2,k-1}$ and $y_{3k}^* = y_{3k} - 0.635 y_{3,k-1}$ to account for autocorrelation. Because of these adjustments, results presented here will differ from the numerical results in Section 4.7.

Now let $Y_k = (y_{ik}, y_{2k}^*, y_{3k}^*, y_{4k}, y_{5k})'$ and $x_k = (x_{1k}, \ldots, x_{5k})'$. For convenience of notation, let $\mathbf{Y}$ and $\mathbf{X}$ denote the 5×36 data matrices of response and predictor variables, respectively, after adjustment by subtraction of overall sample means. The least squares estimate of the 5×5 regression coefficient matrix C is

$$\widetilde{C} = \mathbf{Y}\mathbf{X}'(\mathbf{X}\mathbf{X}')^{-1} = \begin{bmatrix} 5.5010* & -0.8392* & 0.3044 & -0.8910* & 2.0452* \\ 1.9134* & -0.1506 & 0.2745* & -0.5852* & 0.9334* \\ -2.7096 & 0.6824 & -1.2594* & 2.4762* & -2.5368* \\ 4.4956* & -0.5684 & 0.4521 & -0.8408* & 1.5797* \\ 8.9506* & -0.1536 & 0.4797 & -0.4363 & -0.1680 \end{bmatrix},$$

where the entries with asterisk indicate that the usual t-ratios of these estimates are greater than 1.65 in absolute value. The ML estimate $\widetilde{\Sigma} = (1/36)(\mathbf{Y} - \widetilde{C}\mathbf{X})(\mathbf{Y} - \widetilde{C}\mathbf{X})'$ of the 5×5 covariance matrix Σ of the errors $\boldsymbol{\epsilon}_k$ has diagonal elements $\widetilde{\sigma}_{jj}$, $j = 1, \ldots, 5$, given by 10.0280, 2.7598, 69.5145, 25.1619, 34.1522, with $\log(|\widetilde{\Sigma}|) = 13.5484$.

From the values and significance of the elements in $\widetilde{C}$, the regression coefficients corresponding to certain of the response variables y_{ik} share strong similarities. For example, possible similarities among y_{1k}, y_{2k}^*, and y_{4k} indicate that reduced rank may be possible for the set of response variables and/or for some subset of these variables. We perform LR tests of $\text{rank}(C) \le r$ for $r = 4, 3, 2, 1$. For reference, the squared sample canonical correlations between Y_k and X_k (adjusting for sample means) are $\widehat{\rho}_1^2 = 0.96921$, $\widehat{\rho}_2^2 = 0.40003$, $\widehat{\rho}_3^2 = 0.31975$, $\widehat{\rho}_4^2 = 0.11723$, and $\widehat{\rho}_5^2 = 0.01675$. For $r = 3$ the LR test gives a chi-squared test statistic value of $\mathcal{M} = 4.176$ with 4 degrees of freedom, so a hypothesis of reduced rank of three or less for C is clearly not rejected. For $r = 2$, the value $\mathcal{M} = 15.543$ with 9 degrees of freedom is obtained, which hints that a reduced rank of two could even be possible. For the present, however, we adopt the more conservative conclusion that $\text{rank}(C) \le 3$ with $r = 3$.

The ML estimates of the matrix factors $\widehat{A}$ and $\widehat{B}$ in $\widehat{C} = \widehat{A}\widehat{B}$ under the rank 3 model are obtained in the standard way, similar to (9.7). These estimates are

$$\widehat{A}' = \begin{bmatrix} 2.144* & 1.292* & -0.574* & 1.966* & 2.230* \\ 0.907 & -0.001 & -3.829* & 1.044 & 4.964* \\ 0.631 & 0.785* & -7.272* & -0.073 & -2.049 \end{bmatrix},$$

$$\widehat{B} = \begin{bmatrix} 1.845 & -0.187 & 0.125 & -0.303 & 0.669 \\ 0.875 & 0.015 & 0.090 & -0.069 & -0.171 \\ -0.274 & -0.060 & 0.117 & -0.273 & 0.344 \end{bmatrix},$$

and the corresponding ML estimate $\widehat{\Sigma}$ of the error covariance matrix Σ has diagonal elements $\widehat{\sigma}_{jj}$, $j = 1, \ldots, 5$, given by 10.5321, 2.8166, 69.6403, 25.6354, 34.1685, quite similar to the full-rank LS results, with $\log(|\widehat{\Sigma}|) = 13.6900$. For reference, ML estimates under the rank 2 model ($r = 2$) would merely consist of the first two columns and first two rows of $\widehat{A}$ and $\widehat{B}$, respectively. Notice that, in particular, from the estimate $\widehat{A}$ there may still be some strong similarities among its rows of coefficients, except that the third row of estimates, corresponding to response variable y_3^*, does seem to show substantial differences from the remaining rows.

We examine the estimated vectors $\widehat{\ell}'_j = \widehat{V}'_j \widetilde{\Sigma}^{-1/2}$, $j = 4, 5$, associated with the "near" zero eigenvalues $\widehat{\lambda}_j^2 = \widehat{\rho}_j^2/(1 - \widehat{\rho}_j^2)$ of $\widehat{R} = \widetilde{\Sigma}^{-1/2}\widetilde{C}\widehat{\Sigma}_{xx}\widetilde{C}'\widetilde{\Sigma}^{-1/2}$ (such that $\widehat{\ell}'_j\widetilde{C} \approx 0$), which are $\widehat{\ell}'_4 = [-0.3362,\ 0.4722,\ 0.0029,\ -0.0199,\ 0.0678]$ and $\widehat{\ell}'_5 = [0.1334,\ 0.0432,\ 0.0109,\ -0.2004,\ 0.0262]$. These vectors have the feature that coefficients corresponding to the response variable y_3^* are close to zero, there are only moderate values corresponding to variable y_5, and some relatively large nonzero coefficients corresponding to variables y_1, y_2^*, and y_4. Based on the discussion on procedures to identify subset reduced-rank structure, the above features are highly suggestive of a further reduced-rank structure for the coefficient matrix associated with variables y_1, y_2^*, y_4, and y_5, with variable y_3^* excluded. The stepwise backward-elimination procedure discussed earlier also performs quite well in revealing this feature in this example. In particular, at the first stage the LR statistic $\mathcal{M}$ for testing $\text{rank}(C_1) \leq 2$ gives a relatively small and nonsignificant value of 4.363 with 6 degrees of freedom, when the single response variable y_{3k}^* is excluded from Y_k, and this LR test statistic value is smaller than when any of the other single response variables y_{ik}, $i \neq 3$ is excluded.

We rearrange the response variables with $Y_{1k} = (y_{1k}, y_{2k}^*, y_{4k}, y_{5k})'$ and $Y_{2k} = (y_{3k}^*)$, and let C_1 denote the upper 4×5 submatrix of C corresponding to the (rearranged) response variables in Y_{1k}. As indicated, the LR test of $H_0 : \text{rank}(C_1) \leq 2$ gives the test statistic value of $\mathcal{M} = 4.363$ with 6 degrees of freedom, whereas the LR test of $H_0 : \text{rank}(C_1) \leq 1$ gives $\mathcal{M} = 18.542$ with 12 degrees of freedom. The hypothesis of reduced rank of two for C_1 is clearly acceptable, while reduced rank of one might also be plausible but again the more conservative value $r_1 = 2$ is taken. ML estimates of the factors $\widehat{A}$ and $\widehat{B}$ in $\widehat{C}_1 = \widehat{A}\widehat{B}$ for this subset reduced-rank model under $r_1 = 2$ are obtained from (9.7) and, in particular, the estimate $\widehat{A}$ is

$$\widehat{A}' = \begin{bmatrix} 2.372* & 1.429* & 2.172* & 2.475* \\ 0.391 & -0.439 & 0.837 & 5.260* \end{bmatrix}.$$

From these results we see that the coefficient estimates for the second predictive factor (second column of $\widehat{A}$) are generally small (and nonsignificant) for response variables y_{1k}, y^*_{2k}, and y_{4k}, but much more substantial for the last response variable y_{5k}. This suggests an even further reduced-rank feature for the coefficient matrix associated with y_1, y^*_2, and y_4, with variable y_5 excluded (in addition to the previously exclude y^*_3). Thus we entertain continuation of the backward-elimination procedure for reduced rank, leading to eliminating y_{5k} and retaining only y_{1k}, y^*_{2k}, and y_{4k} in Y_{1k}, and identification of coefficient structure of rank one for this subset.

Finally, we rearrange the response variables with $Y_{1k} = (y_{1k}, y^*_{2k}, y_{4k})'$, $Y_{2k} = (y^*_{3k}, y_{5k})'$, and let C_1 denote the upper 3×5 submatrix of C corresponding to the (rearranged) response variables in Y_{1k}. The squared sample canonical correlations between Y_{1k} and X_k (adjusting for sample means) are $\widehat{\rho}_1^2 = 0.96244$, $\widehat{\rho}_2^2 = 0.16460$, and $\widehat{\rho}_3^2 = 0.02684$. The LR test of H_0 : $\text{rank}(C_1) \leq 1$ gives $\mathcal{M} = 6.315$ with 8 degrees of freedom, so the hypothesis of reduced rank of one for C_1 is clearly acceptable. We obtain the ML estimate $\widehat{C}_1 = \widehat{A}\widehat{B}$ under this rank one model, with $\widehat{A} = [2.3808^*,\ 1.4350^*,\ 2.1792^*]'$ and $\widehat{B} = [1.7153,\ -0.1984,\ 0.1738,\ -0.3951,\ 0.7226]$. The associated reduced rank one ML estimate $\widehat{C}_1$ rather accurately represents the corresponding LS estimates of C, displayed previously as the first, third, and fourth rows of $\widetilde{C}$ in the original ordering of response variables. The associated ML estimate of C_2 (last two rows of C under rearrangement) is obtained from (9.9) as

$$\widehat{C}_2 = \begin{bmatrix} -2.4660 & 0.6212 & -1.2741 & 2.4845 & -2.4863 \\ 7.7977 & 0.1343 & 0.5454 & -0.4755 & -0.4024 \end{bmatrix}.$$

The corresponding ML estimate $\widehat{\Sigma}$ of the error covariance matrix Σ has diagonal elements, corresponding to the original ordering of the response variables, of 10.6964, 2.8908, 69.5369, 26.3046, 34.6696, with $\log(|\widehat{\Sigma}|) = 13.7554$, which are close to values from $\widetilde{\Sigma}$ under (full-rank) LS estimation and from the earlier "general" reduced rank 3 model for C. Taking into account the magnitudes of variation of the five predictor variables and (approximate) standard errors of elements of $\widehat{B}$, the single predictive index variable $x^*_{1k} = \widehat{B}X_k$ in the above rank one subset model is composed most prominently of strong contributions from x_{1k}, x_{4k}, and x_{5k}, with much lesser relative weight given to x_{2k} and x_{3k}. Graphical illustration in support of the reduced-rank one model feature can be provided by scatter plots of each of the response variables against the single predictive index variable $x^*_{1k} = \widehat{B}X_k$ determined from the rank one modeling of the coefficient matrix C_1 for the variables y_1, y^*_2, and y_4. Such illustration further confirms that each of y_1, y^*_2, and y_4 has a quite strong linear relationship with the single index x^*_1, whereas response variables y^*_3 and y_5 (especially y^*_3) do not show as strong of relationship with x_1^* because they are also influenced by other factors within the set of predictor variables. This sort of exploration needs further investigation.

Chemometrics Data We consider again the multivariate chemometrics data from Skagerberg et al. (1992) that was analyzed in Chapter 3. The partial least squares (PLS) multivariate regression modeling of the data was used both for predicting properties of the produced polymer and for multivariate process control. The data were also considered by Breiman and Friedman (1997) and Reinsel (1999) to illustrate the relative performance of different multivariate prediction methods. In Chapter 3, we used the data to illustrate the use of partial canonical correlation methods in reduced-rank modeling.

Recall that the data set consists of $T = 56$ multivariate observations, with $m = 6$ response variables and 22 "original" predictor (or process) variables. The response variables are the following output properties of the polymer produced: y_1, number-average molecular weight; y_2, weight-average molecular weight; y_3, frequency of long chain branching; y_4, frequency of short chain branching; y_5, content of vinyl groups in the polymer chain; y_6, content of vinylidene groups in the polymer chain. The process variable measurements employed consist of the wall temperature of the reactor (x_1) and the feed rate of the solvent (x_2) that also acts as a chain transfer agent, complemented with 20 different temperatures measured at equal distances along the reactor. For interpretational convenience, the response variables y_3, y_5, and y_6 were rescaled by the multiplicative factors 10^2, 10^3, and 10^2, respectively, for the analysis presented here, so that all six response variables would have variability of the same order of magnitude. The predictor variable x_1 was also rescaled by the factor 10^{-1}. The temperature measurements in the temperature profile along the reactor are expected to be highly correlated and therefore methods to reduce the dimensionality and complexity of the input or predictor set of data may be especially useful. In Chapter 3, we used partial canonical correlation analysis between the response variables and the 20 temperature measurements, given x_1 and x_2, to exhibit that only the first two (partial) canonical variates of the temperature measurements were necessary for adequate representation of the six response variables y_1 through y_6. We denote these two (partial) canonical variates of the temperature measurements as x_3 and x_4, and suppose that these two (partial canonical) variables together with the two original variables x_1 and x_2 represent the set of predictor variables available for modeling of the six response variables, for purposes of illustration of the partially reduced-rank modeling methodology.

We consider a multivariate linear regression model for the kth vector of responses of the form

$$Y_k = D + CX_k + \epsilon_k, \quad k = 1, \ldots, T, \tag{9.17}$$

with $T = 56$, where $Y_k = (y_{1k}, \ldots, y_{6k})'$ is a 6×1 vector, $X_k = (x_{1k}, \ldots, x_{4k})'$ is a 4×1 vector (hence $n = 4$), and D allows for the 6×1 vector of constant terms in the regression model. For convenience of notation, let $\widetilde{\mathbf{Y}}$ and $\widetilde{\mathbf{X}}$ denote the 6×56 and 4×56 data matrices of response and predictor variables, respectively, after adjustment by subtraction of overall sample means. Then the least squares estimate of the 6×4 regression coefficient matrix C is

$$\widetilde{C} = \widetilde{\mathbf{Y}}\widetilde{\mathbf{X}}'(\widetilde{\mathbf{X}}\widetilde{\mathbf{X}}')^{-1} = \begin{bmatrix} -0.3686 & 5.4768 & 0.0396 & 0.0678 \\ (0.0415) & (0.0779) & (0.0030) & (0.0089) \\ & & & \\ -0.2129 & 27.6520 & 0.1080 & 0.0378 \\ (0.1396) & (0.2621) & (0.0101) & (0.0301) \\ & & & \\ -1.2063 & -0.8007 & -0.0471 & 0.4525 \\ (0.1469) & (0.2757) & (0.0106) & (0.0316) \\ & & & \\ -1.0284 & -1.4021 & 0.1964 & 0.2293 \\ (0.0978) & (0.1835) & (0.0071) & (0.0211) \\ & & & \\ -0.3756 & -0.4906 & 0.0763 & 0.0821 \\ (0.0566) & (0.1062) & (0.0041) & (0.0122) \\ & & & \\ -0.3943 & -0.5444 & 0.0742 & 0.0913 \\ (0.0413) & (0.0776) & (0.0030) & (0.0089) \end{bmatrix},$$

with estimated standard errors of the individual elements of $\widetilde{C}$ displayed in parentheses below the estimates, and

$$\widetilde{D} = [24.525, 18.930, 109.556, 97.631, 41.281, 40.722]'.$$

The ML estimate $\widetilde{\Sigma} = (1/56)(\widetilde{\mathbf{Y}} - \widetilde{C}\widetilde{\mathbf{X}})(\widetilde{\mathbf{Y}} - \widetilde{C}\widetilde{\mathbf{X}})'$ of the 6×6 covariance matrix Σ of the errors $\boldsymbol{\epsilon}_k$ has diagonal elements $\widetilde{\sigma}_{jj}$, $j = 1, \ldots, 6$, given by 0.0262, 0.2964, 0.3280, 0.1453, 0.0487, 0.0260, with $\log(|\widetilde{\Sigma}|) = -16.6529$. Moderate sample correlations, of the order of 0.5, are found between most of the pairs of residual variables $\widehat{\epsilon}_{jk}$, $j = 1, \ldots, 6$, except for residuals $\widehat{\epsilon}_{2k}$ for the second response variable y_{2k} which exhibit little correlation with residuals from response variables y_{3k} through y_{6k}. By comparison with these (partial) correlations among the y_{jk} after adjustment for X_k, the original response variables y_{1k} and y_{2k} show a correlation of about 0.985, variables y_{4k}, y_{5k}, and y_{6k} form another strongly correlated group with correlations of about 0.975, while y_{3k} has moderate negative correlations with y_{4k}, y_{5k}, and y_{6k} of the order of -0.5.

From the values and significance of the elements in $\widetilde{C}$, the regression coefficients corresponding to variables y_4 through y_6 share strong similarities, indicating that rank one may be possible for this (sub)set of response variables. But with the inclusion of response variables y_1 through y_3, the rank of C might be four, that is, full rank. In fact, a LR test of $\text{rank}(C) \leq 3$ versus $\text{rank}(C) = 4$ gives a chi-squared test statistic value of $\mathcal{M} = 10.172$ with 3 degrees of freedom, so a hypothesis of reduced rank of three or less for C is rejected. Moreover, the estimated vectors $\widehat{\ell}'_j = \widehat{V}'_j \widetilde{\Sigma}^{-1/2}$, $j = 5, 6$, associated with the zero eigenvalues of $\widehat{R} = \widetilde{\Sigma}^{-1/2} \widetilde{C} \widehat{\Sigma}_{xx} \widetilde{C}' \widetilde{\Sigma}^{-1/2}$ are found to be

$$\widehat{\ell}'_5 = [0.2398,\ -0.0542,\ 0.0143,\ -2.2035,\ 3.7356,\ 1.9501] \text{ and}$$
$$\widehat{\ell}'_6 = [-0.2420,\ 0.0432,\ 0.1300,\ 0.6732,\ 4.1355,\ -5.8899].$$

These vectors have the feature that coefficients corresponding to the first three response variables y_1, y_2, and y_3 are close to zero, with relatively large nonzero coefficients corresponding to variables y_4, y_5, and y_6. Based on the related discussion in Chapter 3 and earlier in this chapter, this feature is highly suggestive of a reduced-rank structure for the coefficient matrix associated with variables y_4 through y_6. The stepwise backward-elimination procedure discussed in Section 9.4 also performs quite well in this example. In particular, at the first stage the LR statistics $\mathcal{M}$ for testing $\text{rank}(C_1) \leq 3$ give extremely small and nonsignificant values of 0.025, 0.024, and 0.243, each with 2 degrees of freedom, when the single response variable y_{1k}, y_{2k}, or y_{3k}, respectively, is excluded from Y_k. As the procedure continues, it leads to clearly retaining only the variables y_{4k}, y_{5k}, and y_{6k} in Y_{1k} and identification of coefficient structure of rank one for this subset.

We now rearrange the response variables with $Y_{1k} = (y_{4k}, y_{5k}, y_{6k})'$ and $Y_{2k} = (y_{1k}, y_{2k}, y_{3k})'$, and let C_1 denote the upper 3×4 submatrix of C corresponding to the (rearranged) response variables in Y_{1k}. The LR test of $H_0 : \text{rank}(C_1) \leq 2$ gives the test statistic value of $\mathcal{M} = 0.032$ with 2 degrees of freedom, and the LR test of $H_0 : \text{rank}(C_1) \leq 1$ gives the test statistic value of $\mathcal{M} = 2.285$ with 6 degrees of freedom, so the hypothesis of reduced rank of one for C_1 is quite acceptable. The hypothesis that $C_1 = 0$ has a LR statistic value of $\mathcal{M} = 201.310$ with 12 degrees of freedom, so we should clearly retain the rank one hypothesis. We mention that the squared sample canonical correlations between Y_{1k} and X_k (adjusting for sample means) are $\widehat{\rho}_1^2 = 0.97981$, $\widehat{\rho}_2^2 = 0.04322$, and $\widehat{\rho}_3^2 = 0.00063$.

To obtain the ML estimate $\widehat{C}_1 = \widehat{A}\widehat{B}$ under the rank one hypothesis, we find the normalized eigenvector of the matrix $\widehat{R}_1 = \widetilde{\Sigma}_{11}^{-1/2} \widetilde{C}_1 \widehat{\Sigma}_{xx} \widetilde{C}'_1 \widetilde{\Sigma}_{11}^{-1/2}$ associated with the largest eigenvalue $\widehat{\lambda}_1^2 \equiv \widehat{\rho}_1^2/(1 - \widehat{\rho}_1^2) = 48.5232$. This normalized eigenvector is $\widehat{V}_1 = [0.7747,\ 0.2362,\ 0.5866]'$. So as in (9.7) we compute $\widehat{A} = \widetilde{\Sigma}_{11}^{1/2} \widehat{V}_1$ and $\widehat{B} = \widehat{V}'_1 \widetilde{\Sigma}_{11}^{-1/2} \widetilde{C}_1$ to obtain

$$\widehat{A} = [0.3218,\ 0.1266,\ 0.1216]',$$
$$\widehat{B} = [-3.2250,\ -4.4277,\ 0.6104,\ 0.7320],$$

so that the reduced-rank ML estimate of C_1 is

$$\widehat{C}_1 = \widehat{A}\,\widehat{B} = \begin{bmatrix} -1.0376 & -1.4245 & 0.1964 & 0.2355 \\ -0.4082 & -0.5604 & 0.0773 & 0.0926 \\ -0.3920 & -0.5382 & 0.0742 & 0.0890 \end{bmatrix}.$$

This rank one estimate quite accurately recovers the corresponding LS estimates, displayed previously as the last three rows of $\widetilde{C}$ in the original ordering of response variables. The associated ML estimate of C_2 (last three rows of C under rearrangement) is

$$\widehat{C}_2 = \begin{bmatrix} -0.3769 & 5.4591 & 0.0399 & 0.0697 \\ -0.2131 & 27.6516 & 0.1081 & 0.0379 \\ -1.2616 & -0.9190 & -0.0455 & 0.4700 \end{bmatrix}.$$

The corresponding ML estimate $\widehat{\Sigma}$ of the error covariance matrix Σ has diagonal elements, corresponding to the original ordering of the response variables, of $0.0263, 0.2964, 0.3315, 0.1456, 0.0499, 0.0260$, with $\log(|\widehat{\Sigma}|) = -16.6081$, which are very close to values from $\widetilde{\Sigma}$ under (full-rank) LS estimation. For graphical illustration in support of the reduced-rank feature among the subset of response variables y_4, y_5, and y_6, Fig. 9.1 displays scatter plots of each of the response variables against the single predictive index variable $x_{1k}^* = \widehat{B}X_k$ determined from the rank one modeling of the coefficient matrix C_1 for the variables y_4, y_5, and y_6. (The linear fits of each of y_4, y_5, and y_6 with x_1^* obtained from the reduced-rank model are also indicated in the graphs.) This illustration confirms that each of y_4, y_5, and y_6 has a quite strong linear relationship with the single index x_1^*, whereas response variables y_1, y_2, and y_3 do not show any particular relationship with x_1^*, consistent with the modeling results.

9.6 Discussion and Extensions

The model considered in this chapter might be viewed as complementary to the original reduced-rank model of Anderson (1951) (see Chapter 3) wherein for the regression of Y_k the set of *predictor* variables X_k was separated into a subset X_{1k} whose coefficient matrix was taken to be of reduced rank and another subset X_{2k} with full-rank coefficient matrix. This division into separate sets could be based on knowledge of the subject matter. In the current model the role is reversed in that the set of response variables is divided into one subset Y_{1k} having reduced-rank coefficient matrix in its regression on X_k and so being influenced by only a small number of predictive variables constructed as linear combinations of X_k, and another subset Y_{2k} having full-rank coefficient matrix with separate predictors that are linearly independent of those influencing Y_{1k}.

An extended version of the original reduced-rank model was also considered in Chapter 3 and in Velu (1991) where each subset X_{1k} and X_{2k} of the predictors has

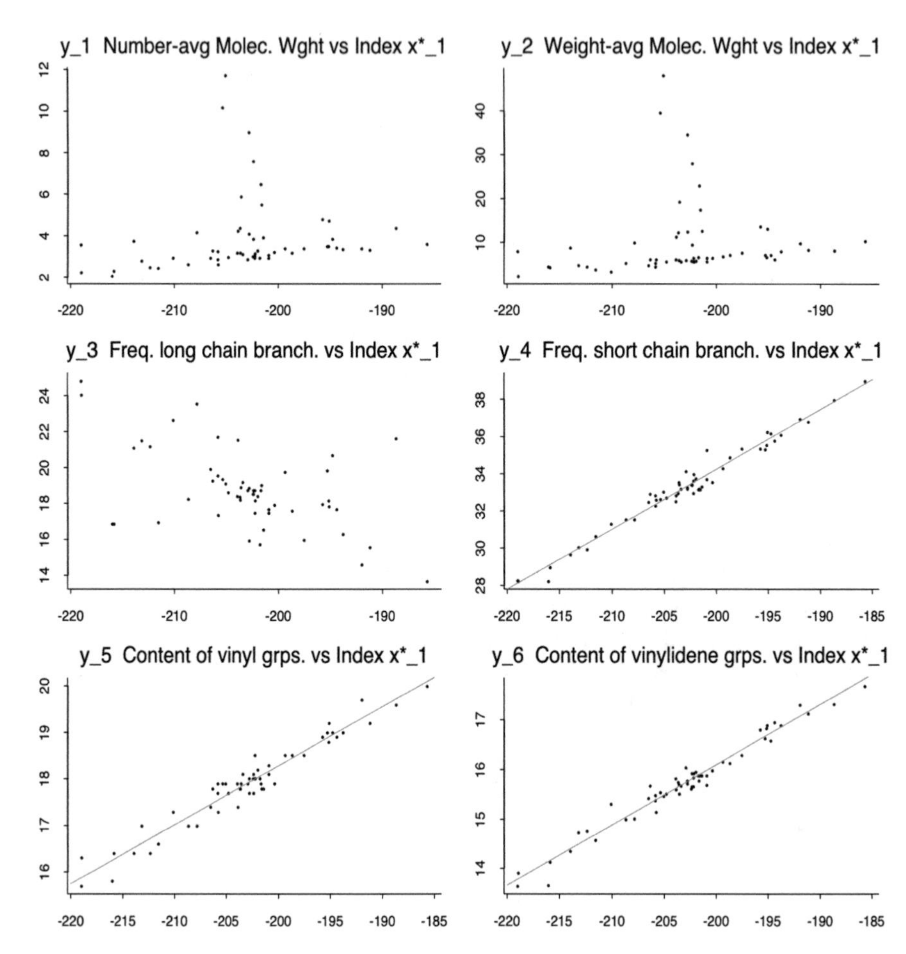

Fig. 9.1 Response variables $y_1, \ldots, y_6$ for the chemometrics data versus the single index predictor variable $x_1^* = \widehat{\mathbf{B}}\mathbf{X}$ determined from the partially reduced-rank regression model

coefficient matrix of reduced rank and hence each subset can be represented by a smaller number of linear combinations to describe the relationship with Y_k. As an analogous extension of the current "partially" reduced-rank model introduced in this section, one could also easily envision situations where the two subsets Y_{1k} and Y_{2k} of response variables each have separate reduced-rank structures, instead of only Y_{1k} having reduced rank. As a motivating example, as stated earlier, in an economic system it might be postulated that a certain subset of the endogenous (response) economic variables are influenced by only a few indices of the exogenous (predictor) economic variables in the system, while another subset of the endogenous variables is influenced only by a few different indices of the exogenous variables. Hence, each subset is influenced by a small number of different (linearly independent) linear combinations of X_k. Such a model adds flexibility to the current model and may be useful in various applications. It would be interesting to explore the statistical procedures of ML estimation and LR testing for ranks associated with models of this form. In particular, we anticipate that simultaneous ML estimation of the two separate components of reduced-rank structure will involve a feature, which we might refer to as "seemingly unrelated reduced-rank regression," analogous to the aspect that occurs in simultaneous estimation of "seemingly unrelated (full-rank) regression" systems. Some of these ideas are currently studied under the topic of sparse group Lasso (see Li et al. 2015) and an identification procedure of possible groups is investigated in Luo and Chen (2020). Because this topic requires introduction of sparseness concepts, we will defer the discussion on this to a later chapter in the book.

Chapter 10
High-Dimensional Reduced-Rank Regression

10.1 Introduction

In various fields of science and engineering, massive amount of data are produced and collected at unprecedented speed. High-dimensional statistical problems, with the number of variables exceeding the sample size, are routinely encountered. Such problems are non-trivial even in the simple case of multiple linear regression; with high-dimensional data, conventional methods, including least squares method and maximum likelihood estimation, may cease to work. Many challenges, such as noise accumulation, spurious correlation, data quality issues, among others, are known collectively as the "curse of dimensionality" (Donoho 2000). Fortunately, such high-dimensional problems are not hopeless; a universal key is to summon the Occam's razor, the fundamental law of parsimony, by assuming that the underlying truth is "simple." That is, the true model lies in a low-dimensional "corner" of the high-dimensional model space, in a proper sense. With this belief, various regularized estimation techniques have been developed, making it possible to recover low-dimensional signal from high-dimensional data.

The *regularized learning paradigm* has a long history. The well-known ridge regression, which was introduced to overcome the deficiencies of multicollinearity among predictors by shrinking the regression coefficients toward zero, dates back to as early as Tikhonov (1943) and was independently developed in several related disciplines (in the field of statistics, see, e.g., Hoerl and Kennard (1970), McDonald and Galarneau (1975), and Brown and Zidek (1980)). With the rise of big data and the prevalence of large-scale learning tasks, this paradigm has undergone an exciting development in recent decades. In particular, various sparse models have been developed and investigated (Tibshirani 1996; Fan and Lv 2010), and shrinkage- or sparsity-inducing regularization has become a standard technique in training of high-dimensional or over-parameterized statistical and machine learning models.

G. C. Reinsel et al., *Multivariate Reduced-Rank Regression*, Lecture Notes in Statistics 225, https://doi.org/10.1007/978-1-0716-2793-8_10

The unprecedented data explosion opens the door for tackling many complicated real-world problems through an integrative learning of the underlying data generation mechanism, which, by its nature, is often multivariate. Indeed, multivariate regression and related techniques emerge as essential tools for uncovering entangled relations between multiple outcomes and multiple features with ever growing dimensionality and complexity. In these problems, the central task usually amounts to recovering one or more large-scale coefficient matrices with some desired low-dimensional structures. In the context of multivariate regression, for example, various sparse estimation methods concern variable selection by pursuing entrywise or blockwise sparse patterns in the coefficient matrix (Turlach et al. 2005; Kim et al. 2009; Peng et al. 2010; Rothman et al. 2010; Obozinski et al. 2011).

What truly distinguishes multivariate problems from their univariate counterparts, and also one of the most fascinating facets of the field of modern multivariate learning, is that matrix-type objects have much more delicate structures than vector objects because of the possible "interactions" between rows and columns. In particular, the notion of "sparsity" for matrix-type objects can take many different forms: while the complexity of a vector can be measured by its sparsity, the complexity of a matrix can also be measured by its rank, which is the main subject of this book. Intriguingly, although the low-rank structure in general does not translate to sparsity on the entries of the matrix, it translates to sparsity on the singular values. As such, reduced-rank regression is recognized as a regularized learning technique and has been revived for high-dimensional multivariate learning (Fazel 2002; Yuan et al. 2007; Bunea et al. 2011).

In this chapter, we start with an overview of the regularized learning paradigm and a few fundamental sparse regression methods in 10.2. In Section 10.3, we cast reduced-rank regression into the regularized learning paradigm and introduce its path of revival to high-dimensional settings under the framework of singular value penalization. Section 10.4 presents adaptive nuclear-norm penalized regression (Chen et al. 2013), a prototype singular value penalization approach for simultaneous rank reduction and model estimation. In Section 10.5, we consider an integrative multi-view regression model (Li et al. 2019), which is capable of estimating view-specific low-rank structures from different views/groups of the predictors. The framework nicely bridges sparse and reduced-rank models and reveals that they can be analyzed in a unified framework. Two applications are discussed in Section 10.6.

10.2 Overview of High-Dimensional Regularized Regression

Throughout this chapter, for any vector $X \in \mathbb{R}^n$, we denote by $\|X\|_0$, $\|X\|_1$, $\|X\|_2$, and $\|X\|_\infty$ the ℓ_0 norm, ℓ_1 norm, ℓ_2 norm, and ℓ_∞ norm, defined, respectively, as $\|X\|_0 = \sum_{j=1}^n I(X_j \neq 0)$, where $I(\cdot)$ is the indicator function, $\|X\|_1 = \sum_{j=1}^n |X_j|$, $\|X\|_2 = (\sum_{j=1}^n X_j^2)^{1/2}$, and $\|X\|_\infty = \max_{j=1}^n |X_j|$.

To illustrate the main ideas and results of regularized learning, consider the fundamental multiple linear regression setup,

$$Y = \mathbf{X}'\beta + \boldsymbol{\epsilon}, \tag{10.1}$$

where $Y = (y_1, y_2, \ldots, y_T)'$ is the $T \times 1$ vector of responses, $\mathbf{X} = [X_1, \ldots, X_T]$ the $n \times T$ design matrix, $\beta = (\beta_1, \ldots, \beta_n)'$ a vector of regression coefficients, and $\boldsymbol{\epsilon} = (\epsilon_1, \epsilon_2, \ldots, \epsilon_T)'$ a vector of random errors. We assume the errors are independently and identically distributed (i.i.d.) standard normal random variables with mean zero and variance σ^2.

In some practical applications, the number of predictors can be large, comparable to or even exceeding the sample size this gives rise to the *high-dimensional regression* problems. A parsimonious model can be achieved by imposing some low-dimensional structure on β and employing regularization methods for its estimation and inference. The general form of a regularized estimation criterion can be expressed as

$$\min_{\beta \in \mathbb{R}^n} \mathcal{J}(\beta; Y, \mathbf{X}) + \lambda \rho(\beta), \tag{10.2}$$

where $\mathcal{J}(C)$ is a loss function measuring how well the model fits the observed data, $\rho(\cdot)$ is a penalty function or regularizer measuring the complexity of the model as a function of the coefficient vector, and λ is a non-negative tuning parameter for balancing model fitting and model complexity. For linear regression, it is conventional and fundamental to consider the sum of squared error loss, i.e., $\mathcal{J}(\beta; Y, \mathbf{X}) = \|Y - \mathbf{X}'\beta\|_2^2/2$.

The well-known ridge regression can be expressed as

$$\min_{\beta \in \mathbb{R}^n} \frac{1}{2}\|Y - \mathbf{X}'\beta\|_2^2 + \lambda\|\beta\|_2^2. \tag{10.3}$$

This corresponds to using a squared ℓ_2 norm regularizer $\rho(\beta) = \|\beta\|_2^2 = \sum_{j=1}^n \beta_j^2$, which overcomes the deficiencies resulting from the multicollinearity among predictors by shrinking the regression coefficients toward zero. The problem admits a unique and explicit solution,

$$\widehat{\beta}_{Ridge} = (\mathbf{X}\mathbf{X}' + \lambda I_n)^{-1}\mathbf{X}Y,$$

where I_n denotes the $n \times n$ identity matrix. As mentioned earlier, this idea dates back to as early as 1943 in Tikhonov (1943). Besides inducing shrinkage estimation, ridge regression, or ℓ_2 regularization in general, has a grouping effect such that highly correlated variables tend to exhibit similar estimated coefficients.

In many applications, there is a need of performing variable selection, that is, to identify a set of features/predictors that are the most relevant to the response. This is based on the belief that the underlying true model is simple and only involves a small set of predictors. In the above linear regression model, it amounts to assuming that the true coefficient vector β is sparse. We use

$$\mathcal{S}^* = \{j : \beta_j \neq 0, j = 1, \ldots, n\}$$

to denote the active set with $s^* = |\mathcal{S}^*|$ being its cardinality, the number of active variables. In high-dimensional settings, although n can grow with T or even exceed T, the number of active variables s^* shall remain to be relatively small. The ridge regression, while can greatly improve prediction, is not suited for the task of variable selection because it does not produce exactly sparse solution. Various sparsity-inducing regularization methods have been developed to enable variable selection, that is, to recover the active set $\mathcal{S}^*$. Below we provide an overview of the ℓ_0 and ℓ_1 norm penalized regression methods, which are the most fundamental.

The ℓ_0 norm regularized regression takes the form

$$\min_{\beta \in \mathbb{R}^n} \frac{1}{2}\|Y - \mathbf{X}'\beta\|_2^2 + \lambda\|\beta\|_0, \tag{10.4}$$

where $\|\beta\|_0 = \sum_{j=1}^n I(\beta_j \neq 0)$. As such, the method directly uses the model dimension as the penalty; this can be immediately seen to be closely related to the best subset selection approach. Foster and George (1994) and Barron et al. (1999) studied the behavior of ℓ_0 regularization, and their main results, in the context of linear regression, give the following prediction error bound: when choosing $\lambda = O(\sigma^2 \log(n))$, the ℓ_0 norm penalized estimator $\widehat{\beta}_{\ell_0}$ satisfies

$$E\left\{\frac{1}{T}\|\mathbf{X}'\widehat{\beta}_{\ell_0} - \mathbf{X}'\beta\|_2^2\right\} \preceq \inf_{\mathcal{S}\subset\{1,\ldots,n\}} \left\{\frac{1}{T}\|(I_T - \mathcal{P}_{\mathcal{S}})\mathbf{X}'\beta\|_2^2 + \frac{\sigma^2(1+\log n)|\mathcal{S}|}{T}\right\},$$

where $\preceq$ means the inequality holds up to a multiplicative constant, I_T is the $T \times T$ identity matrix, and $\mathcal{P}_{\mathcal{S}}$ is the projection matrix onto the row space of the submatrix of $\mathbf{X}$ with row indices in $\mathcal{S}$. The first term quantifies the approximation error or bias when using the predictors in $\mathcal{S}$ to build the regression model; the term vanishes when $\mathcal{S} = \mathcal{S}^*$, leading to a simplified bound $E\{\|\mathbf{X}'\widehat{\beta}_{\ell_0} - \mathbf{X}'\beta\|_2^2/T\} \preceq (1+\log n)s^*/T$. This shows the rate of the prediction error and the price we have to pay for searching over many subset models: it is not much because the prediction error rate is only inflated by a logarithm factor, $\log n$, compared to the oracle rate, the rate that is achieved by the least squares estimation when we knew the active set $\mathcal{S}^*$ a priori. Unfortunately, despite the appealing risk behavior of ℓ_0 regularization, its computation with a naive exhaustive search approach is infeasible even for a moderate number of predictors, say, the value of n is around 50. The computational burden of (10.4) stems from the combinatory nature of the subset selection problem and the non-convexity and discontinuity of the ℓ_0 norm regularizer. Not until recently, many optimization algorithm and strategies are been developed to more efficiently solve the problem, typically in an approximate way (Bertsimas et al. 2016; Zhu et al. 2020).

The infeasibility of ℓ_0 norm regularization naturally leads to the consideration of its convex relaxation. The ℓ_1 norm regularized regression, known as the Lasso (least absolute shrinkage and selection operator) (Tibshirani 1996), takes the form

$$\min_{\beta \in \mathbb{R}^n} \frac{1}{2} \|Y - \mathbf{X}'\beta\|_2^2 + \lambda \|\beta\|_1. \tag{10.5}$$

Due to the properties of the ℓ_1 penalty, Lasso not only is able to perform shrinkage estimation of β toward zero, but also is capable of producing exactly sparse solutions. To understand how Lasso works, it is intriguing to examine its geometry and compare with that of the ridge regression in (10.3). First of all, these two methods can both be equivalently expressed in a constrained form. Lasso can be expressed as

$$\min_{\beta \in \mathbb{R}^n} \|Y - \mathbf{X}'\beta\|_2^2 \qquad \text{s.t. } \|\beta\|_1 \leq \delta,$$

while ridge regression minimizes the same loss subject to a different constraint, $\|\beta\|_2 \leq \delta$, i.e.,

$$\min_{\beta \in \mathbb{R}^n} \|Y - \mathbf{X}'\beta\|_2^2 \qquad \text{s.t. } \|\beta\|_2 \leq \delta.$$

Figure 10.1 shows the geometry of Lasso and ridge regression when $n = 2$. The key difference of the two methods is reflected on the shapes of their constrained regions: the ℓ_1 ball $\|\beta\|_1 \leq \delta$ forms a diamond centered at the origin, while the ℓ_2 ball $\|\beta\|_2 \leq \delta$ forms a disk centered at the origin. Provided that $\mathbf{X}$ is of full row rank, the contours of the loss $\|Y - \mathbf{X}'\beta\|_2^2$ form ellipsoids centered at the least squares solution $\widehat{\beta}_{LS}$. For either method, the solution is obtained when the elliptical contour first hits the constrained region. Both methods perform shrinkage estimation, as these estimators are necessarily closer to the origin than the least squares estimator $\widehat{\beta}_{LS}$, unless $\widehat{\beta}_{LS}$ is already in the constrained region. The Lasso estimator $\widehat{\beta}_{Lasso}$ can be obtained at one of the corners of the diamond with non-trivial probability, and thus it is capable of producing exactly sparse solution. The ridge estimator $\widehat{\beta}_{Ridge}$, on the other hand, has zero probability to be exactly sparse since its constrained region is smooth without any corners.

In terms of computation, the solution paths of Lasso can be efficiently obtained by several methods, including coordinatewise descent algorithms (Tseng 2001; Fu 1998), a modified version of least angle regression (Efron et al. 2004), and several versions of path-following algorithms (Friedman et al. 2007). Here we sketch the main idea of the coordinatewise descent algorithm for an illustration. For any given $\lambda > 0$ and some initial value $\beta^{(0)}$, the algorithm solves (10.5) by minimizing the objective function with respect to one coefficient at a time, with all the other coefficients held fixed at their current values. With some algebra, it can be shown that the one-dimensional problem takes the following general form:

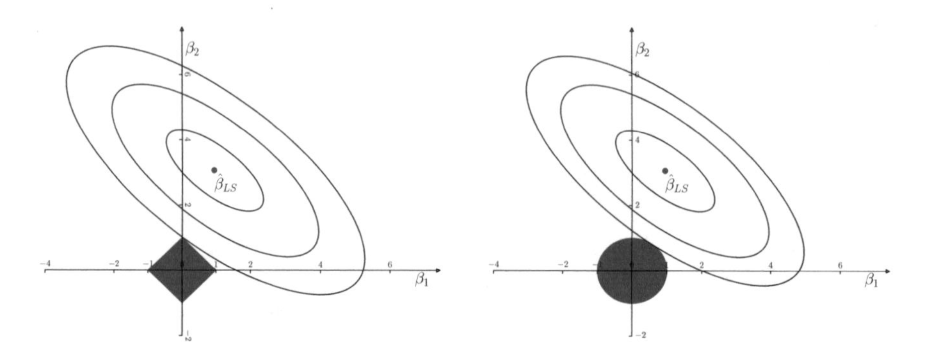

Fig. 10.1 Geometry of the Lasso and the ridge regression methods

$$\min_{\theta \in \mathbb{R}} \frac{1}{2}(y-\theta)^2 + \lambda|\theta|,$$

which admits an explicit solution given by the soft-thresholding rule: $\widehat{\theta} = \text{sgn}(y)(|y| - \lambda)_+$, where $\text{sgn}(\cdot)$ is the sign function and a_+ takes the positive part of a. The algorithm then updates one coefficient at a time as above and cycles through all coefficients until convergence.

The theoretical properties of the Lasso have been thoroughly studied; a comprehensive treatment can be found in Bühlmann and van de Geer (2011). The main theoretical concerns are on the prediction of the response, the estimation of β, and the recovery of $\mathcal{S}^*$, three closely related yet different tasks. Under certain compatibility condition on the design matrix $\mathbf{X}$, such as the restricted eigenvalue (RE) condition (Bickel et al. 2009), the prediction error of Lasso is bounded as $E\{\|\mathbf{X}'\widehat{\beta}_{Lasso} - \mathbf{X}'\beta\|_2^2/T\} \preceq \sigma^2(\log n)s^*/T$ for a suitable choice of λ at the order $\sigma\sqrt{\log(n)T}$. What is remarkable about this result is that it shows that the computationally efficient Lasso method can achieve the same prediction error rate as the computationally infeasible ℓ_0 regularization method, albeit under additional assumptions on the design matrix. Again, this implies an oracle property of the Lasso: up to a $\log(n)$ factor, the mean squared prediction error is of the same order as if one knew a priori that s^* predictors are relevant and only used them to conduct least squares regression. Moreover, the rate is essentially optimal up to the $\log(n)$ factor, in a minimax sense.

As for the estimation accuracy of the Lasso, under the same restrictive eigenvalue condition and choice of λ, we have that $\|\widehat{\beta}_{Lasso} - \beta\|_2^2 \preceq (\log n)s^*/T$ with probability tending to 1. Therefore, the estimation consistency can be achieved even when n grows polynomially in T, i.e., $n = O(T^a)$ for any $a > 0$, as long as s^* stays relatively small, i.e., $s^* = o(T/\log(n))$.

To ensure the success in variable selection or support recovery with the Lasso, much stronger conditions are needed. Intuitively, the so-called β-min condition requires that the non-zero coefficients are sufficiently large, so they can be hopefully distinguished from zero, and the "irrepresentable" condition states that the relevant predictors cannot be too highly correlated and thus be too well represented by the irrelevant predictors (Zhao and Yu 2006). The irrepresentable condition is rather strong and implies the compatibility condition (Bühlmann and van de Geer 2011), and the former is essentially a necessary condition for Lasso to achieve selection consistency. Therefore, a general conclusion is that Lasso for variable selection only works in a rather narrow range of problems, excluding many cases where the predictors exhibit strong correlations.

Beyond the convex Lasso method, there also have been extensive work on non-convex penalization and more structured regularization approaches with auxiliary information on the design; we refer to, e.g., Bühlmann and van de Geer (2011), Fan and Lv (2010), and Huang et al. (2012) for comprehensive reviews on these topics.

10.3 Framework of Singular Value Regularization

In multivariate regression problems, the most critical learning task usually amounts to the recovery of one or more large-scale coefficient matrices with some desired low-dimensional structures. To elaborate, recall the multivariate linear regression framework presented in Section 1.2. Let $Y \in \mathbb{R}^m$ be a multivariate response variable and $X \in \mathbb{R}^n$ a multivariate predictor variable. Observing T independent data pairs (X_k, Y_k), $k = 1, \ldots, T$, the multivariate linear regression (MLR) model is again stated as

$$\mathbf{Y} = C\mathbf{X} + \boldsymbol{\epsilon}, \tag{10.6}$$

where $\mathbf{Y} = [Y_1, \ldots, Y_T] = [Y_{(1)}, \ldots, Y_{(m)}]' \in \mathbb{R}^{m\times T}$ is the response matrix, $\mathbf{X} = [X_1, \ldots, X_T] = [X_{(1)}, \ldots, X_{(n)}]' \in \mathbb{R}^{n\times T}$ the predictor matrix, $C = [C_1, \ldots, C_n] = [C_{(1)}, \ldots, C_{(m)}]' \in \mathbb{R}^{m\times n}$ the unknown coefficient matrix, and $\boldsymbol{\epsilon} = [\boldsymbol{\epsilon}_1, \ldots, \boldsymbol{\epsilon}_T] = [\boldsymbol{\epsilon}_{(1)}, \ldots, \boldsymbol{\epsilon}_{(m)}]' \in \mathbb{R}^{m\times T}$ the random error matrix. Unless otherwise noted, we assume $\boldsymbol{\epsilon}_k$s are i.i.d. and each has mean zero and a positive-definite covariance matrix $\Sigma_{\epsilon\epsilon}$. An intercept term is avoided by either including a vector of ones in $\mathbf{X}$ or by centering the responses and predictors. Without any additional structural assumptions, the least squares estimation of C under Model (10.6) is equivalent to perform m separate least squares for each response on the set of predictors, which does not utilize any multivariate flavor of the problem and may fail miserably in high-dimensional scenarios.

The *reduced-rank regression* (RRR), the main subject of this book, assumes that the coefficient matrix C is rank-deficient, i.e.,

$$r(C) = r,$$

where $r(\cdot)$ denotes the rank of the enclosed matrix, and $0 \leq r \leq \min(m, n)$. This is equivalent to assuming that there exist two matrices $A \in \mathbb{R}^{m\times r}$ and $B \in \mathbb{R}^{r\times n}$, of full column rank and row rank, respectively, such that $C = AB$. The reduced-rank structure mitigates the curses of dimensionality by dramatically reducing the effective number of parameters when r is small. Further, this structure conveniently captures dependencies between responses and induces an appealing latent factor interpretation: the responses are linked to the predictors through r common latent factors/variables as given by BX. Conceptually, a reduced-rank model searches for the most relevant subspace of the predictors when it is believed that they contribute to the prediction of the outcomes collectively rather than sparsely.

A natural question is then whether reduced-rank regression can be cast in the regularized learning paradigm. In the context of multivariate regression in which the main objective is to estimate the coefficient matrix C, the regularized learning framework can be expressed as minimizing a penalized loss function of the following form:

$$\mathcal{J}(C; \mathbf{Y}, \mathbf{X}) + \rho(C; \lambda) \tag{10.7}$$

with respect to $C \in \mathbb{R}^{m \times n}$, where $\mathcal{J}(C)$ is a loss function measuring the goodness-of-fit, and $\rho(\cdot; \lambda)$ is a penalty function or regularizer measuring the complexity of the enclosed matrix, in which λ is a non-negative tuning parameter for balancing between model fitting and model complexity.

Before diving into regularized learning, we make a few remarks on choice of the loss function. Thus far in the formulation of the reduced-rank regression, we have been using a weighted ℓ_2 loss, $\mathcal{J}(C; \mathbf{Y}, \mathbf{X}) = \text{tr}\{(\mathbf{Y} - C\mathbf{X})'\Gamma(\mathbf{Y} - C\mathbf{X})\}$; see Section 2.3. It is common to regard the weighting matrix $\Gamma \in \mathbb{R}^{m \times m}$ as known (Yuan et al. 2007; Izenman 2008). The choice of Γ is flexible and is usually based on a pilot covariance estimate $\widehat{\Sigma}_{\epsilon\epsilon}$. For example, it can be $\widehat{\Sigma}_{\epsilon\epsilon}^{-1}$ when $\widehat{\Sigma}_{\epsilon\epsilon}$ is nonsingular, or a regularized version $(\widehat{\Sigma}_{\epsilon\epsilon} + cI_m)^{-1}$ for some $c > 0$ and I_m being the $m \times m$ identity matrix. When a reliable estimate of $\Sigma_{\epsilon\epsilon}$ is unavailable, a standard practice in finance and econometric forecasting is to simplify Γ to a diagonal matrix, or equivalently, to an identity matrix after robustly scaling the response variables. Although it sounds intriguing to consider jointly estimating the reduced-rank mean structure and the error covariance matrix, the problem is notoriously difficult with high-dimensional data. Potential further improvement in regression coefficient estimation is not warranted without strong technical or structural conditions. Therefore, for ease of presentation, in this chapter, we take Γ as the $m \times m$ identity matrix and mainly focus on the regularized least squares problem of the form

$$\min_{C} \left\{ \|\mathbf{Y} - C\mathbf{X}\|_F^2 + \rho(C; \lambda) \right\},$$

where $\|\cdot\|_F$ denotes the Frobenius norm, i.e., $\|\mathbf{Y} - C\mathbf{X}\|_F^2 = \text{tr}\{(\mathbf{Y} - C\mathbf{X})'(\mathbf{Y} - C\mathbf{X})\}$.

In general, reduced-rank structure of a matrix does not necessarily result in sparsity of its entries. For example, a matrix of ones, which is apparently of unit rank, has no zero entries at all. Rather, the reduced-rank structure directly translates to sparsity of the singular values of the matrix because the rank of a matrix is the same as the number of its nonzero singular values. To be precise, denote $d_j(C)$ as the jth largest singular value of the matrix C and $d(C) = (d_1(C), \dots, d_q(C))'$ the vector of singular values of C with $q = \min(m, n)$. Then, the rank of C can be expressed as the ℓ_0 norm of the singular values

$$r(C) = \|d(C)\|_0 = \sum_{j=1}^{q} I(d_j(C) \neq 0), \qquad q = \min(m, n).$$

Similarly, the nuclear norm of a matrix is defined as the ℓ_1 norm of its singular values,

$$\|C\|_\star = \|d(C)\|_1 = \sum_{j=1}^{q} d_j(C).$$

Both the matrix rank and the nuclear norm are complexity measures of a matrix. Their connections and differences can be understood through considering the fundamental low-rank matrix approximation problem. Write the singular value decomposition of $\mathbf{Y} \in \mathbb{R}^{m \times T}$ as

$$\mathbf{Y} = VDU', \qquad D = \text{diag}\{d(\mathbf{Y})\}, \tag{10.8}$$

where U and V are, respectively, $T \times q$ and $m \times q$ orthonormal matrices and diag$(\cdot)$ denotes a diagonal matrix with the enclosed vector or elements on its diagonal. The following results characterize the solutions of the rank or nuclear-norm penalized matrix approximation problems.

Result 10.1 For any $\lambda \geq 0$ and $\mathbf{Y} \in \mathbb{R}^{m \times T}$, let

$$\mathcal{H}_\lambda(\mathbf{Y}) = \arg\min_C \{\|\mathbf{Y} - C\|_F^2 + \lambda^2 r(C)\},$$

$$\mathcal{S}_\lambda(\mathbf{Y}) = \arg\min_C \{\|\mathbf{Y} - C\|_F^2 + 2\lambda \|C\|_\star\}.$$

The solutions are $\mathcal{H}_\lambda(\mathbf{Y}) = V\mathcal{H}_\lambda(D)U'$ with

$$\mathcal{H}_\lambda(D) = \text{diag}\left\{d_j(\mathbf{Y}) I(d_j(\mathbf{Y}) > \lambda), j = 1, \dots, q\right\}, \tag{10.9}$$

and $\mathcal{S}_\lambda(\mathbf{Y}) = V\mathcal{S}_\lambda(D)U'$ with

$$\mathcal{S}_\lambda(D) = \text{diag}\left\{(d_j(\mathbf{Y}) - \lambda)_+, j = 1, \dots, q\right\}. \tag{10.10}$$

The singular value hard-thresholding estimator, $\mathcal{H}_\lambda(\mathbf{Y})$, eliminates any singular value of $\mathbf{Y}$ that is below the threshold λ, while the singular value soft-thresholding estimator $\mathcal{S}_\lambda(\mathbf{Y})$ eliminates any singular value that is below λ and also shrinks all the other larger singular values of $\mathbf{Y}$ by λ toward zero. These estimators can be considered as natural extensions of the hard/soft-thresholding estimators for scalars and vectors (Donoho and Johnstone 1995; Cai et al. 2010).

With the above connection between rank deficiency and sparsity of singular values, the reduced-rank regression can be recognized as a sparse regression model under a broader concept of sparsity. Consequently, under the regularized learning framework in (10.7), reduced-rank estimation can be achieved through penalizing the singular values of either the coefficient matrix C or the regression component $C\mathbf{X}$; the latter may lead to desirable computational advantages (to be elaborated later in Section 10.4). That is, we can explicitly set $\rho(C;\lambda) = \rho(d(C);\lambda)$ or $\rho(C;\lambda) = \rho(d(C\mathbf{X});\lambda)$, where $\rho(\cdot;\lambda)$ is a sparsity-inducing penalty function. To demonstrate, below we discuss the cases when ρ is the ℓ_0 or the ℓ_1 function.

Bunea et al. (2011) studied the rank-penalized regression criterion, in which the penalty is set to be proportional to the rank, or equivalently, the ℓ_0 norm of the singular values of the coefficient matrix,

$$\|\mathbf{Y} - C\mathbf{X}\|_F^2 + \lambda^2\|d(C)\|_0. \tag{10.11}$$

This work was among the first to cast reduced-rank regression as a penalized regression and generalize it to high-dimensional settings. It is recognized that minimizing (10.11) is equivalent to minimizing

$$\|\mathbf{Y} - C\mathbf{X}\|_F^2 + \lambda^2\|d(C\mathbf{X})\|_0, \tag{10.12}$$

owing to the fact that $r(C\mathbf{X}) = r(C)$ whenever the rank reduction is effective (Chen et al. 2013). The rank-penalized regression problem admits an explicit solution. Specifically, by Pythagoras theorem, the problem is the same as minimizing $\|\widehat{C}_L\mathbf{X} - C\mathbf{X}\|_F^2 + \lambda^2 r(C\mathbf{X})$, where $\widehat{C}_L\mathbf{X}$ denotes the least squares fitted value from minimizing $\|\mathbf{Y} - C\mathbf{X}\|_F^2$ with respect to C. It then follows from Result 10.1 that the solution can be obtained through hard-thresholding the singular value decomposition of $\widehat{C}_L\mathbf{X}$ using (10.9), i.e.,

$$\widehat{C}_H^{(\lambda)} = \widehat{V}\widehat{D}^{-}\mathcal{H}_\lambda(\widehat{D})\widehat{V}'\widehat{C}_L, \tag{10.13}$$

where $\widehat{V}\widehat{D}^2\widehat{V}'$ is the eigenvalue decomposition of $(\widehat{C}_L\mathbf{X})(\widehat{C}_L\mathbf{X})' = \mathbf{Y}\mathcal{P}_X\mathbf{Y}'$ with $\mathcal{P}_X$ being the projection matrix onto the column space of $\mathbf{X}'$, and A^- denotes the Moore–Penrose inverse (Moore 1920; Penrose 1955) of matrix A. Not surprisingly, the set of rank-penalized estimators spans exactly the same solution path as the corresponding rank-constrained formulation, which is $\min_C \|\mathbf{Y} - C\mathbf{X}\|_F^2$ subject to $r(C) \le r$, for $0 \le r \le q$ (see also Theorem 2.2).

Another popular approach under the framework of (10.7) is the nuclear-norm penalized regression (NNP) (Yuan et al. 2007),

$$\frac{1}{2}\|\mathbf{Y} - C\mathbf{X}\|_F^2 + \lambda\|d(C)\|_1. \tag{10.14}$$

It is attractive as a convex relaxation of the rank-penalized criterion in (10.11). This problem, however, generally does not have an explicit solution. Extensive research has been devoted to developing efficient optimization methods (Yuan et al. 2007; Toh and Yun 2010; Udell et al. 2016), one of which is the iterative singular value soft-thresholding algorithm (Cai et al. 2010).

Alternatively, motivated by the equivalency between (10.11) and (10.12), one could consider

$$\frac{1}{2}\|\mathbf{Y} - C\mathbf{X}\|_F^2 + \lambda\|d(C\mathbf{X})\|_1, \tag{10.15}$$

which, based on Pythagoras theorem and Result 10.1, leads to an explicit solution through singular value soft-thresholding,

$$\widehat{C}_S^{(\lambda)} = \widehat{V}\widehat{D}^{-}\mathcal{S}_\lambda(\widehat{D})\widehat{V}'\widehat{C}_L. \tag{10.16}$$

In analogous to the ℓ_1 norm penalization on vectors, nuclear-norm penalization encourages sparsity among the singular values of the coefficient matrix. This results in a continuous solution path and achieves simultaneous rank reduction and shrinkage estimation.

10.4 Reduced-Rank Regression via Adaptive Nuclear-Norm Penalization

It is intriguing that the rank and nuclear-norm penalization approaches can be viewed as ℓ_0 and ℓ_1 singular value penalization methods, respectively. Motivated by these connections, we consider in detail the adaptive nuclear norm regularization method, which is designed with a goal to close the gap between the ℓ_0 and ℓ_1 penalization schemes.

The adaptive nuclear norm of a matrix $C \in \mathbb{R}^{m\times n}$ is defined as a weighted sum of its singular values:

$$\|C\|_{\star w} = \sum_{j=1}^{q} w_j d_j(C), \tag{10.17}$$

where w_js are non-negative weights. We first show that the adaptive nuclear norm is non-convex when the weight of the singular value decreases with the singular value, a condition needed for a meaningful regularization. Despite the non-convexity, the adaptive nuclear-norm penalized estimator has a closed-form solution in matrix approximation problems. We then present the adaptive nuclear-norm penalized regression (ANN) method for conducting simultaneous dimension reduction and coefficient estimation in high-dimensional multivariate regression.

10.4.1 Adaptive Nuclear Norm

The adaptive nuclear norm $\|C\|_{\star w}$ defined in (10.17) forms a rich class of penalty functions indexed by the weights. Clearly, it includes the nuclear norm as a special case with unit weights. Because the nuclear norm is convex and is a matrix norm, an immediate question arises as to whether or not its extension with weights preserves convexity, which is the case for the analogous Lasso and adaptive Lasso penalties (Zou 2006). Convexity of the penalty function has important consequences in both numerical optimization and statistical analysis. There are some advantages and disadvantages: comparing to the non-convex counterparts, convex regularization in general leads to more efficient computation but may need much stronger conditions to achieve desired parameter estimation and model selection performance (Fan and Lv 2010; Zhang and Zhang 2012).

Let us consider an example to demonstrate the convexity issue. If there exists an index k such that $w_k < w_{k+1}$, then $f(C) = \|C\|_{\star w}$ as defined in (10.17) is necessarily a non-convex function. Let C and Z be two $m \times m$ diagonal matrices such that $c_{jj} = j$, for $j = 1, \ldots, m$, while Z equals C but with entries switched at positions $m-k+1$ and $m-k$ for some $1 \leq k \leq m-1$ on the diagonal. It is then easy to verify that

$$f(C) = f(Z) = \sum_{j=1}^{m} w_j(m-j+1),$$

$$f\left(\frac{C+Z}{2}\right) - \frac{f(C)}{2} - \frac{f(Z)}{2} = \frac{1}{2}(w_k + w_{k+1}) - w_k > 0.$$

Therefore, $f(\cdot)$ is non-convex.

More generally, the following result states that the convexity of (10.17) depends on the ordering of the non-negative weights.

Result 10.2 Let $f(C) = \|C\|_{\star w}$ be defined in (10.17) for any matrix $C \in \mathbb{R}^{m \times n}$. Then $f(\cdot)$ is convex in C if and only if $w_1 \geq \cdots \geq w_q \geq 0$.

Hence, for the adaptive nuclear norm (10.17) to be a convex function, the weights must be non-decreasing with the singular values. However, for penalized estimation, the opposite is desirable. We would and shall henceforth impose the order constraint

$$0 \leq w_1 \leq \cdots \leq w_q, \qquad q = \min(m, n), \tag{10.18}$$

which ensures that a larger singular value receives a lighter penalty to help reduce the bias and a smaller singular value receives a heavier penalty to help promote sparsity. The non-convexity of $f(\cdot)$ arises from the constraint (10.18), under which $f(\cdot)$ is no longer a matrix norm. In fact, $f(\cdot)$ is neither convex nor concave under constraint (10.18). Here is an example. Consider $m = n = 2$, and

$$C_1 = \begin{pmatrix} 2 & 0 \\ 0 & 1 \end{pmatrix}, \qquad C_2 = \begin{pmatrix} 1 & 0 \\ 0 & 2 \end{pmatrix}.$$

Let $w_1 = 1$ and $w_2 = 2$. It can be verified that $f(C_1) = f(C_2) = f(-C_2) = 4$, while $f\{(C_1+C_2)/2\} = 4.5 > \{f(C_1)+f(C_2)\}/2$; also, $f\{(C_1-C_2)/2\} = 1.5 < \{f(C_1) + f(-C_2)\}/2$.

The preceding discussion in Section 10.3 on the connections between different thresholding rules and penalty terms motivates us to consider the use of the adaptive nuclear norm to bridge the gap between the ℓ_0 and ℓ_1 singular value penalties and fine-tuning the bias–variance tradeoff. We begin with a result in the context of low-rank matrix approximation; we are able to obtain, with complete characterization, an explicit global minimizer of the least squares criterion penalized by a non-convex adaptive nuclear norm.

Theorem 10.1 *For any $\lambda \geq 0$, $0 \leq w_1 \leq \cdots \leq w_q$ with $q = \min(m, n)$, and $\mathbf{Y} \in \mathbb{R}^{m\times n}$ with its singular value decomposition given by $\mathbf{Y} = VDU'$, a global optimal solution to the optimization problem*

$$\min_C \left\{ \frac{1}{2}\|\mathbf{Y} - C\|_F^2 + \lambda\|C\|_{\star w} \right\} \tag{10.19}$$

is $\mathcal{S}_{\lambda w}(\mathbf{Y})$, where

$$\begin{aligned} \mathcal{S}_{\lambda w}(\mathbf{Y}) &= V\mathcal{S}_{\lambda w}(D)U', \\ \mathcal{S}_{\lambda w}(D) &= \operatorname{diag}\{(d_j(\mathbf{Y}) - \lambda w_j)_+, j = 1, \ldots, q\}. \end{aligned} \tag{10.20}$$

Further, if all the nonzero singular values of $\mathbf{Y}$ are distinct, then $\mathcal{S}_{\lambda w}(\mathbf{Y})$ is the unique optimal solution.

Proof We first prove that $\mathcal{S}_{\lambda w}(\mathbf{Y})$ is indeed a global optimal solution to (10.19). Write $d = (d_1, \ldots, d_q)' = d(C)$, which implies the entries of d are in non-increasing order. Since the penalty term only depends on the singular values of C, (10.19) can be equivalently written as

$$\min_{d:d_1\geq\cdots\geq d_q\geq 0} \left\{ \min_{C\in\mathbb{R}^{m\times n}, d(C)=d} \left(\frac{1}{2}\|\mathbf{Y} - C\|_F^2 + \lambda \sum_{j=1}^{q} w_j d_j \right) \right\}.$$

For the inner minimization problem, we have the inequality

$$\begin{aligned} \|\mathbf{Y} - C\|_F^2 &= \operatorname{tr}(\mathbf{Y}\mathbf{Y}') - 2\operatorname{tr}(\mathbf{Y}C') + \operatorname{tr}(CC') \\ &= \sum_{j=1}^{q} d_j^2(\mathbf{Y}) - 2\operatorname{tr}(\mathbf{Y}C') + \sum_{j=1}^{q} d_j^2 \end{aligned}$$

$$\geq \sum_{j=1}^{q} d_j^2(\mathbf{Y}) - 2d(\mathbf{Y})'d + \sum_{j=1}^{q} d_j^2.$$

The last inequality is due to von Neumann's trace inequality (von Neumann 1937; Mirsky 1975). Equality holds when C admits the singular value decomposition $C = V\text{diag}(d)U'$, where V and U are defined in (10.8) as the left and right singular matrices of $\mathbf{Y}$. Then the optimization reduces to

$$\min_{d:d_1\geq\cdots\geq d_q\geq 0}\left[\sum_{j=1}^{q}\left\{\frac{1}{2}d_j^2 - (d_j(\mathbf{Y}) - \lambda w_j)d_j + \frac{1}{2}d_j^2(\mathbf{Y})\right\}\right]. \quad (10.21)$$

This objective function is completely separable in the d_j's and is minimized only when $d_j = (d_j(\mathbf{Y}) - \lambda w_j)_+$. This is a feasible solution because $\{d_j(\mathbf{Y})\}$ is in non-increasing order, while $\{w_j\}$ is in non-decreasing order. Therefore, $\mathcal{S}_{\lambda w}(\mathbf{Y}) = V\text{diag}\{(d(\mathbf{Y}) - \lambda w)_+\}U'$ is a global optimal solution to (10.19). The uniqueness follows by the equality condition for the von Neumann's trace inequality when $\mathbf{Y}$ has distinct nonzero singular values and the uniqueness of the strictly convex optimization in (10.21). This concludes the proof.

The fact that a closed-form global minimizer can be found for the non-convex matrix approximation problem (10.19) is rather surprising. As mentioned, the result mainly stems from von Neumann's trace inequality (Mirsky 1975). Following Zou (2006), the weights can be set to be $w_j = d_j^{-\gamma}(\mathbf{Y})$, for $j = 1, \ldots, q$, where $\gamma \geq 0$ is a pre-specified constant. In this way, the order constraint (10.18) is automatically satisfied.

10.4.2 Adaptive Nuclear-Norm Penalized Regression

Predictive accuracy and computational efficiency are both pivotal in high-dimensional regression problems. Motivated by criteria (10.11), (10.12), (10.14), and (10.15), we consider the estimation of C by minimizing the following adaptive nuclear-norm penalized regression (ANN) criterion:

$$\frac{1}{2}\|\mathbf{Y} - C\mathbf{X}\|_F^2 + \lambda\|C\mathbf{X}\|_{\star w}, \quad (10.22)$$

where the weights $\{w_j\}$ are required to be non-negative and non-decreasing. In practice, a foremost task of using (10.22) is setting proper adaptive weights. Following Zou (2006), this can be based on the least squares solution,

$$w_j = d_j^{-\gamma}(\mathbf{Y}\mathcal{P}_X) = d_j^{-\gamma}(\widehat{C}_L\mathbf{X}), \quad (10.23)$$

where γ is a non-negative constant, and $\mathcal{P}_X$ denotes the projection matrix onto the row space of $\mathbf{X}$, so that $\mathbf{Y}\mathcal{P}_X = \widehat{C}_L\mathbf{X}$ with $\widehat{C}_L$ being a least squares estimator of the coefficient matrix C.

The ANN method in (10.22) is built on two main ideas. First, the criterion focuses on the fitted values $C\mathbf{X}$ and encourages sparsity among the singular values of $C\mathbf{X}$ rather than those of C. This may yield a low-rank estimator for $C\mathbf{X}$ and hence for C (Koltchinskii et al. 2011). A prominent advantage of this setup is that the problem can then be solved explicitly and efficiently by utilizing the results established in Theorem 10.1. Second, the adaptive penalization of the singular values allows a flexible bias–variance tradeoff: a large singular value receives a small penalty to control for possible bias, and a small singular value receives a large penalty to help induce sparsity and hence reduce the rank. As such, the method builds a bridge between the ℓ_0 and ℓ_1 singular value penalization methods and can be viewed as analogous to the adaptive Lasso method (Tibshirani 1996; Zou 2006; Huang et al. 2008) developed for the case of univariate sparse regression.

The following result shows that the ANN criterion admits an explicit minimizer.

Result 10.3 A minimizer of (10.22), denoted by $\widehat{C}_S^{(\lambda w)}$, is given by

$$\widehat{C}_S^{(\lambda w)} = \widehat{V}\widehat{D}^{-}\mathcal{S}_{\lambda w}(\widehat{D})\widehat{V}'\widehat{C}_L, \qquad \widehat{C}_S^{(\lambda w)}\mathbf{X} = \mathcal{S}_{\lambda w}(\widehat{C}_L\mathbf{X}) = \widehat{V}\mathcal{S}_{\lambda w}(\widehat{D})\widehat{U}', \tag{10.24}$$

where $\widehat{C}_L$ is the least squares estimator of C, $\widehat{V}\widehat{D}\widehat{U}'$ is the singular value decomposition of $\widehat{C}_L\mathbf{X}$, and $\mathcal{S}_{\lambda w}(\cdot)$ is defined in (10.20).

By Pythagoras theorem, minimizing criterion (10.22) is equivalent to minimizing $\|\widehat{C}_L\mathbf{X} - C\mathbf{X}\|_F^2/2 + \lambda\|C\mathbf{X}\|_{\star w}$ with respect to C. The above result then follows directly from Theorem 10.1. The proposed method first projects $\mathbf{Y}$ onto the row space of X, i.e., $\mathbf{Y}\mathcal{P}_X = \widehat{C}_L\mathbf{X}$; the estimator is then obtained as a low-rank approximation of $\mathbf{Y}\mathcal{P}_X$ by adaptively soft-thresholding its singular values. The thresholding level is data-driven: the smaller an initially estimated singular value, the larger its thresholding level.

The estimated rank from ANN corresponds to the smallest nonzero singular value of $\mathbf{Y}\mathcal{P}_X$ that exceeds its thresholding level, i.e., $\widehat{r} = \max\{r : d_r(\mathbf{Y}\mathcal{P}_X) > \lambda w_r\}$. For the choice of the weights in (10.23), the estimated rank is

$$\widehat{r} = \max\{r : d_r(\mathbf{Y}\mathcal{P}_X) > \lambda^{1/(\gamma+1)}\}, \tag{10.25}$$

for $0 \le \lambda < d_1^{\gamma+1}(\mathbf{Y}\mathcal{P}_X)$ and $\widehat{r} = 0$ for $\lambda \ge d_1^{\gamma+1}(\mathbf{Y}\mathcal{P}_X)$. Therefore, the plausible range of the tuning parameters is $\lambda \in [0, d_1^{\gamma+1}(\mathbf{Y}\mathcal{P}_X)]$, with $\lambda = 0$ corresponding to the least squares solution and $\lambda = d_1^{\gamma+1}(\mathbf{Y}\mathcal{P}_X)$ to the null solution. To select an optimal λ and hence an optimal solution, K-fold cross-validation method can be used, based on predictive performance of the models (Stone 1974).

The ANN estimator $\widehat{C}_S^{(\lambda w)}$ in (10.24) and the ℓ_0 estimator $\widehat{C}_H^{(\lambda)}$ in (10.13) differ only in their estimated singular values for $C\mathbf{X}$, but the difference can be consequential. While the solution path of the ℓ_0 method is discontinuous and the number of possible solutions equals the maximum rank, the proposed criterion offers a more flexible bias–variance tradeoff in that the resulting solution path is continuous and is guided by the data-driven weights. The two methods can both be efficiently computed, in contrast to the computationally intensive ℓ_1-based NNP method in (10.14). Moreover, compared to NNP that tends to overestimate the rank, ANN may improve rank determination with the use of the adaptive weights.

10.4.3 Theoretical Analysis

We perform non-asymptotic analysis to evaluate the finite-sample performance of the reduced-rank estimators in high-dimensional settings. In particular, we focus on the rank estimation and prediction properties of the ANN method, which covers a family of reduced-rank estimators based on singular value thresholding. The finite-sample analysis would be particularly insightful in high-dimensional settings, i.e., when the number of observations is much smaller than the number of variables.

10.4.3.1 Setup and Assumptions

To understand the behaviors of ANN on rank estimation and prediction, we need proper measures of the signal strength and the noise level. Recall that $\mathcal{P}_X$ denotes the projection matrix unto the column space of $\mathbf{X}'$ (or the row space of $\mathbf{X}$). From the multivariate linear model in (10.6), it holds that

$$\mathbf{Y}\mathcal{P}_X = C\mathbf{X} + \boldsymbol{\epsilon}\mathcal{P}_X.$$

This suggests that we can use the rth largest singular value of $C\mathbf{X}$, i.e., $d_r(C\mathbf{X})$, to measure the signal strength, and use the largest singular value of the projected error matrix $\boldsymbol{\epsilon}\mathcal{P}_X$, i.e., $d_1(\boldsymbol{\epsilon}\mathcal{P}_X)$, to measure the noise level. To control for the noise level, we make the following conventional yet fundamental assumption on the error distribution.

Assumption 10.1 The entries of the error matrix $\boldsymbol{\epsilon}$ are independently and identically distributed $N(0, \sigma^2)$ random variables.

Result 10.4 Let r_x denote the rank of $\mathbf{X}$ and suppose Assumption 10.1 holds. Then for any $t > 0$, $E(d_1(\boldsymbol{\epsilon}\mathcal{P}_X)) \leq \sigma(\sqrt{r}_x + \sqrt{m})$, and $P(d_1(\boldsymbol{\epsilon}\mathcal{P}_X) \geq E\{d_1(\boldsymbol{\epsilon}\mathcal{P}_X)\} + \sigma t) \leq \exp(-t^2/2)$.

Assumption 10.1 ensures that the noise level $d_1(\boldsymbol{\epsilon}\mathcal{P}_X)$ is of order $\sqrt{r_x}+\sqrt{m}$, and this result remains valid for independent sub-Gaussian errors (Bunea et al. 2011).

Now let us consider the signal strength relative to the noise level. Intuitively, if $d_1(\boldsymbol{\epsilon}\mathcal{P}_X)$ is much larger than the size of the signal, it is possible that part of the low-rank regression components could be masked by the noise and may be lost during the thresholding procedure; as such, the components that can be recovered or distinguished from the noise may have a smaller rank than the true rank r. The results below characterize the true target of the rank estimation in ANN, i.e., the *effective rank*, and its relationship to the signal strength, the noise level, and the adaptive weights.

Result 10.5 Suppose that there exists an index $s \leq r$ such that $d_s(C\mathbf{X}) > (1+\delta)\lambda^{1/(\gamma+1)}$ and $d_{s+1}(C\mathbf{X}) \leq (1-\delta)\lambda^{1/(\gamma+1)}$ for some $\delta \in (0, 1]$. Then s is called the *effective rank*. Moreover, the estimated rank by ANN satisfies that

$$P(\widehat{r} = s) \geq 1 - P(d_1(\boldsymbol{\epsilon}\mathcal{P}_X) \geq \delta\lambda^{1/(\gamma+1)}).$$

Here r is the rank of true coefficient matrix C, $\widehat{r}$ the estimated rank given in (10.25), $\mathcal{P}_X$ the projection matrix onto the row space of $\mathbf{X}$, $\boldsymbol{\epsilon}$ the error matrix in model (10.6), and γ the power parameter in the adaptive weights (10.23).

Proof By (10.25), $\widehat{r} > s$ holds if and only if $d_{s+1}(\mathbf{Y}\mathcal{P}_X) > \lambda^{1/(\gamma+1)}$, and $\widehat{r} < s$ holds if and only if $d_s(\mathbf{Y}\mathcal{P}_X) \leq \lambda^{1/(\gamma+1)}$. Then

$$P(\widehat{r} \neq s) = P\left\{d_{s+1}(\mathbf{Y}\mathcal{P}_X) > \lambda^{1/(\gamma+1)} \text{ or } d_s(\mathbf{Y}\mathcal{P}_X) \leq \lambda^{1/(\gamma+1)}\right\}.$$

Using Weyl's inequalities on singular values (Franklin 2000) and observing that $\mathbf{Y}\mathcal{P}_X = C\mathbf{X} + \boldsymbol{\epsilon}\mathcal{P}_X$, we have $d_1(\boldsymbol{\epsilon}\mathcal{P}_X) \geq d_{s+1}(\mathbf{Y}\mathcal{P}_X) - d_{s+1}(C\mathbf{X})$ and $d_1(\boldsymbol{\epsilon}\mathcal{P}_X) \geq d_s(C\mathbf{X}) - d_s(\mathbf{Y}\mathcal{P}_X)$. Hence, $d_{s+1}(\mathbf{Y}\mathcal{P}_X) > \lambda^{1/(\gamma+1)}$, which implies $d_1(\boldsymbol{\epsilon}\mathcal{P}_X) \geq \lambda^{1/(\gamma+1)} - d_{s+1}(C\mathbf{X})$, and $d_s(\mathbf{Y}\mathcal{P}_X) \leq \lambda^{1/(\gamma+1)}$. This in turn implies that $d_1(\boldsymbol{\epsilon}\mathcal{P}_X) \geq d_s(C\mathbf{X}) - \lambda^{1/(\gamma+1)}$. It then follows that

$$P(\widehat{r} \neq s) \leq P\left[d_1(\boldsymbol{\epsilon}\mathcal{P}_X) \geq \min\{\lambda^{1/(\gamma+1)} - d_{s+1}(C\mathbf{X}), d_s(C\mathbf{X}) - \lambda^{1/(\gamma+1)}\}\right].$$

Note that $\min\{\lambda^{1/(\gamma+1)} - d_{s+1}(C\mathbf{X}), d_s(C\mathbf{X}) - \lambda^{1/(\gamma+1)}\} \geq \delta\lambda^{1/(\gamma+1)}$. This completes the proof.

With the above results, it is clear that we need to have a strong enough signal to avoid the gap between the effective rank and the true rank, and we need to choose an appropriate rate of the tuning parameter to ensure that the estimated rank equals to the true rank with high probability. These considerations lead to the following assumption on the signal strength.

Assumption 10.2 For any $\theta > 0$, let $\lambda = \{(1+\theta)\sigma(\sqrt{r_x} + \sqrt{m})/\delta\}^{\gamma+1}$ with δ defined in Result 10.5, and $d_r(C\mathbf{X}) > 2\lambda^{1/(\gamma+1)}$.

10.4.3.2 Rank Consistency and Prediction Error Bound

Rank determination is an important task in reduced-rank estimation. The rank of the true coefficient matrix C, denoted as r, can be viewed as the number of effective linear combinations of the predictors linked to the responses. Anderson (2002) studied the effect of rank mis-specification on the properties of the estimators in classical settings. Generally speaking, when the specified rank is less than the true rank, biases occur; when the specified rank is greater, variances of the estimators can be unnecessarily large. Hence, it is important to correctly estimate the rank. Here, under the high-dimensional regime, we study the rank determination problem and examine the effects of rank mis-specification through establishing the non-asymptotic prediction error bounds of the reduced-rank estimators.

Theorem 10.2 *Suppose Assumptions 10.1 and 10.2 hold. Let $\widehat{r}$ be the estimated rank given in (10.25), and let $r_x = r(\mathbf{X})$ be the rank of $\mathbf{X}$. Then $P(\widehat{r} = r) \to 1$ as $r_x + m \to \infty$.*

Proof When $d_r(C\mathbf{X}) > 2\lambda^{1/(\gamma+1)}$, we have $d_r(C\mathbf{X}) > 2\lambda^{1/(\gamma+1)} \geq (1+\delta)\lambda^{1/(\gamma+1)}$ and $d_{r+1}(C\mathbf{X}) = 0 \leq (1-\delta)\lambda^{1/(\gamma+1)}$, for some $0 < \delta \leq 1$. The effective rank s defined in Result 10.5 equals the true rank, i.e., $s = r$, and $\min\{\lambda^{1/(\gamma+1)} - d_{r+1}(C\mathbf{X}), d_r(C\mathbf{X}) - \lambda^{1/(\gamma+1)}\} \geq \delta\lambda^{1/(\gamma+1)}$. It then follows from Result 10.4 that

$$\begin{aligned} P(\widehat{r} = r) &\geq 1 - P\{d_1(\boldsymbol{\epsilon}\mathcal{P}_X) \geq \delta\lambda^{1/(\gamma+1)}\} \\ &= 1 - P\{d_1(\boldsymbol{\epsilon}\mathcal{P}_X) \geq (1+\theta)\sigma(\sqrt{r}_x + \sqrt{m})\} \\ &\geq 1 - \exp\{-\theta^2(r_x+m)/2\} \\ &\to 1 \end{aligned}$$

as $r_x + m \to \infty$. This completes the proof.

Theorem 10.2 shows that ANN is able to identify the correct rank with probability tending to 1 as $r_x + m$ goes to infinity. The consistency results can be extended to the case of sub-Gaussian errors and can also be adapted to the case when $r_x + m$ is bounded and the sample size T goes to infinity. The rank consistency of the ANN estimator is thus valid for both classical and high-dimensional asymptotic regimes.

The prediction performance of the ANN estimator is characterized in Theorem 10.3 below. For simplicity, we write $\widehat{C}_S$ for $\widehat{C}_S^{(\lambda w)}$.

Theorem 10.3 *Suppose Assumptions 10.1 and 10.2 hold. Then with probability at least* $1 - \exp\{-\theta^2(r_x+m)/2\}$,

$$\|\widehat{C}_S\mathbf{X} - C\mathbf{X}\|_F^2 \preceq (1+\theta)^2\sigma^2(r_x+m)r, \tag{10.26}$$

where $\preceq$ means the inequality holds up to a multiplicative constant.

The proof starts with the so-called basic inequality, i.e., by the definition of $\widehat{C}_S$ in (10.24),

$$\|\mathbf{Y} - \widehat{C}_S\mathbf{X}\|_F^2 + 2\lambda \sum_{j=1}^{q} w_j d_j(\widehat{C}_S\mathbf{X}) \leq \|\mathbf{Y} - C\mathbf{X}\|_F^2 + 2\lambda \sum_{j=1}^{q} w_j d_j(C\mathbf{X}).$$

That is, the objective function evaluated at the solution is no greater than that evaluated at any other $m \times n$ matrix including the true coefficient matrix C. Noting that

$$\|\mathbf{Y} - \widehat{C}_S\mathbf{X}\|_F^2 = \|\mathbf{Y} - C\mathbf{X}\|_F^2 + \|\widehat{C}_S\mathbf{X} - C\mathbf{X}\|_F^2 + \mathrm{tr}\{\boldsymbol{\epsilon}(C\mathbf{X} - \widehat{C}_S\mathbf{X})'\},$$

we get

$$\begin{aligned}\|\widehat{C}_S\mathbf{X} - C\mathbf{X}\|_F^2 \leq\, & 2\mathrm{tr}\{\boldsymbol{\epsilon}\mathcal{P}_X(\widehat{C}_S\mathbf{X} - C\mathbf{X})'\} \\ & + 2\lambda\{\sum_{j=1}^{q} w_j d_j(C\mathbf{X}) - \sum_{j=1}^{q} w_j d_j(\widehat{C}_S\mathbf{X})\}.\end{aligned}$$

The results can then be established by bounding the stochastic error part, i.e., the first term, with concentration inequalities in Result 10.4, and bounding the difference in penalty part, i.e., the second term, using matrix inequalities such as Weyl's inequality (Franklin 2000). The details can be found in Bunea et al. (2011) and Chen et al. (2013).

The non-asymptotic bound in (10.26) shows that the prediction error is bounded by $d_1^2(\boldsymbol{\epsilon}\mathcal{P}_X)r$ up to some multiplicative constant, with probability $1-\exp\{-\theta^2(r_x + m)/2\}$. The smaller the noise level or the true rank, the smaller the prediction error. The bound is valid for any $\mathbf{X}$ and C. The estimation error bound of $\widehat{C}_S$ can also be readily derived from Theorem 10.3 in low-dimensional scenarios. For example, if $d_{r_x}(\mathbf{X}) \geq c > 0$ for some constant c, then under Assumptions 10.1 and 10.2, $\|\widehat{C}_S - C\|_F^2 \preceq c^{-2}\lambda^{2/(\gamma+1)}r$.

To better understand the effects of rank mis-specification in reduced-rank estimation, we can apply similar techniques as above to examine the performance of the rank restricted estimators obtained as

$$\widehat{C}_k = \arg\min_{C} \|\mathbf{Y} - C\mathbf{X}\|_F^2, \qquad \text{s.t. } r(C) \leq k,$$

for $k = 1 \ldots, \min(r_x, m)$. Bunea et al. (2011) showed that under Assumption 10.1,

$$\|\widehat{C}_k\mathbf{X} - C\mathbf{X}\|_F^2 \preceq \sum_{j=k+1}^{\min(r_x,m)} d_j^2(C\mathbf{X}) + \sigma^2(r_x + m)k,$$

with high probability (tending to one as $r_x + m \rightarrow \infty$). The first term, $\sum_{j=k+1}^{\min(r_x,m)} d_j^2(C\mathbf{X})$, is the approximation error or the bias term, which is decreasing in k and vanishes for $k > r(C\mathbf{X})$. The second term, $\sigma^2(r_x + m)k$, results from the stochastic error and can be regarded as the variance term, which increases in k and is tied to the dimension of the parameter space. Hence, we can interpret the error bound as the squared bias plus the variance in reduced-rank estimation and understand the effects of rank specification from the perspectives of bias–variance tradeoff. This is consistent with earlier work by, e.g., Anderson (2002). Similar results regarding the ANN estimator can be found in Theorem 4 of Chen et al. (2013). With proper choice of the tuning parameter, either the rank-penalized regression or the adaptive nuclear-norm penalized regression can achieve the best bias–variance tradeoff among their corresponding class of estimators.

The error bounds of ANN are comparable to those of the ℓ_0 rank-penalized estimator from (10.11) and the ℓ_1 nuclear-norm penalized estimator from (10.14) (Bunea et al. 2011; Rohde and Tsybakov 2011). The rate of convergence is $(r_x + m)r$, which is the optimal minimax rate for rank sparsity under suitable regularity conditions (Rohde and Tsybakov 2011; Bunea et al. 2012). However, the bounds for the nuclear-norm penalized estimator from (10.14) were obtained with some additional assumptions on the design matrix $\mathbf{X}$, and its tuning sequence for achieving the smallest mean squared error usually does not lead to correct rank recovery (Bunea et al. 2011). While both the rank-penalized estimator and the ANN method are able to achieve correct rank recovery and minimal mean squared error simultaneously, the latter possesses a continuous solution path produced by data-driven adaptive penalization that may lead to improved empirical performance.

Other theoretical aspects of high-dimensional low-rank estimation problems can be found in Negahban and Wainwright (2011), Rohde and Tsybakov (2011), Koltchinskii et al. (2011), and Lu et al. (2012).

10.5 Integrative Reduced-Rank Regression: Bridging Sparse and Low-Rank Models

Multi-view data, or measurements of several distinct yet interrelated sets of characteristics pertaining to the same set of subjects, have become increasingly common in various fields. In a human lung study, for example, segmental airway tree measurements from CT-scanned images, patient behavioral data from questionnaires, gene expressions data, together with multiple pulmonary function test results from spirometry, were all collected. Unveiling lung disease mechanisms then amounts to linking the microscopic lung airway structures, the genetic information, and the patient behaviors to the global measurements of lung functions. In an Internet network analysis, the popularity and influence of a web page are related to its layouts, images, texts, and hyperlinks as well as to the content of other web pages that link back to it. In the Longitudinal Studies of Aging (LSOA)

(Stanziano et al. 2010), the interest is to predict current health conditions of patients using historical information of their living conditions, household structures, habits, activities, medical conditions, among others. The availability of such multi-view data has made tackling many fundamental problems possible through an *integrative statistical learning* paradigm, whose success owes to the utilization of information from various lenses and angles simultaneously.

The aforementioned problems can all be cast under a multivariate regression framework, in which both the responses and the predictors can be high-dimensional, and in addition, the predictors admit some natural grouping structure. We use this simple yet general framework for achieving integrative learning. To formulate, suppose we observe $\mathbf{X}_k \in \mathbb{R}^{n_k \times T}$ for $k = 1, \ldots, K$, each consisting of T copies of independent observations from a set of predictor/feature variables of dimension n_k, and also we observe data on m response variables $\mathbf{Y} \in \mathbb{R}^{m \times T}$. Let $\mathbf{X} = [\mathbf{X}_1', \ldots, \mathbf{X}_K']' \in \mathbb{R}^{n \times T}$ be the design matrix collecting all the predictor sets/groups, with $n = \sum_{k=1}^{K} n_k$. Both n and m can be much larger than the sample size T. Consider the multivariate linear regression model,

$$\mathbf{Y} = C\mathbf{X} + \boldsymbol{\epsilon} = \sum_{k=1}^{K} C_k \mathbf{X}_k + \boldsymbol{\epsilon}, \tag{10.27}$$

where $C = [C_1, \ldots, C_K] \in \mathbb{R}^{m \times n}$ is the unknown regression coefficient matrix partitioned corresponding to the predictor groups, and $\boldsymbol{\epsilon}$ contains independent random errors with zero mean. We denote $q = \min(n, m)$ and $q_k = \min(n_k, m)$. For simplicity, we assume both the responses and the predictors are centered so there is no intercept term. Again, the naive least squares estimation fails miserably in high dimensions as it leverages neither the associations among the responses nor the grouping of the predictors.

When the predictors exhibit a group structure as in model (10.27), a group regularization approach, for example, the convex group Lasso method (Yuan and Lin 2006), can be readily applied to promote groupwise predictor selection; a comprehensive review of these sparse methods is provided by Huang et al. (2012). On the other hand, for problems with multivariate response, reduced-rank regression has been attractive, where a low-rank constraint on the parameter matrix induces an interpretable latent factor formulation and conveniently enables information sharing among the regression tasks.

In essence, to best predict the multivariate response, sparse methods search for the most relevant subset or groups of predictors, while reduced-rank methods search for the most relevant subspace of the predictors. However, neither class of methods can fulfill the needs in the aforementioned multi-view problems. The predictors within each group/view may be strongly correlated, each individual variable may only have weak predictive power, and it is likely that only a few of the views are useful for prediction. Indeed, in many studies, it is largely the collective effort of some groups of predictors that drives the outcomes.

We consider an *integrative reduced-rank regression* (iRRR) model, where the integration is in terms of the multi-view predictors. To be specific, under model (10.27), the iRRR approach assumes that each set of predictors has its own low-rank coefficient matrix, i.e., each C_k is possibly of low rank or even a zero matrix, i.e., $0 \leq r_k \leq q_k$ where $r_k = r(C_k)$, for $k = 1, \ldots, K$. As such, latent features or relevant subspaces are extracted from each predictor set $\mathbf{X}_k$ under the supervision of the multivariate response $\mathbf{Y}$, and the sets of latent variables/subspaces in turn jointly predict $\mathbf{Y}$.

The iRRR model strikes a balance between flexibility and parsimony, as it nicely bridges two seemingly quite different model classes: reduced-rank models and sparse models. On the one hand, iRRR generalizes the two-set regressor model discussed in Chapter 3 by allowing multiple sets of predictors, each of which can correspond to a low-rank coefficient matrix. On the other hand, iRRR subsumes sparse model setup by allowing the rank of C_k be zero, for any $k = 1, \ldots, K$, i.e., the coefficient matrix of a predictor group could be entirely zero.

The groupwise low-rank structure in iRRR is distinctive from a globally low-rank structure for C in classical reduced-rank regression models. The low rankness of C_k's does not necessarily imply that C is of low rank. Conversely, if C is of low rank, i.e., $r = r(C) \ll \min(n, m)$, all we know for sure is that the rank of each C_k is upper bounded by r. Nevertheless, we can first attempt an intuitive understanding of the potential parsimony of iRRR in multi-view settings. The numbers of free parameters in C (the naive degrees of freedom) for an iRRR model, a globally reduced-rank model, and a group-sparse model are $\mathrm{df}_1 = \sum_{k=1}^{K}(n_k + m - r_k)r_k$, $\mathrm{df}_2 = (n + m - r)r$, and $\mathrm{df}_3 = \sum_{k=1}^{K} n_k m I(r_k \neq 0)$, respectively, where $I(\cdot)$ is an indicator function. With high-dimensional multi-view data, consider the scenario that only a few views/predictor groups impact the prediction in a collective way, i.e., r_k's are mostly zero, and each nonzero r_k could be much smaller than $\min(n_k, m)$. Then df_1 could be substantially smaller than both df_2 and df_3. For example, if $r_1 > 0$, while $r_k = 0$ for any $k > 1$ (i.e., $r = r_1$), we have $\mathrm{df}_1 = (n_1 + m - r_1)r_1$, $\mathrm{df}_2 = (n + m - r_1)r_1$, and $\mathrm{df}_3 = n_1 m$, respectively. Another example is when $r = \sum_{k=1}^{K} r_k$, e.g., C_k's in model (10.27) have distinct column spaces. Since $\sum_{k=1}^{K}(n_k + m - r_k)r_k \leq \{m + \sum_{k=1}^{K}(n_k - r_k)\}\{\sum_{k=1}^{K} r_k\} = (n + m - r)r$, iRRR is more parsimonious than the globally reduced-rank model.

10.5.1 Composite Nuclear-Norm Penalization

To recover the view-specific low-rank structure in the iRRR model, consider a convex optimization approach via *composite nuclear-norm penalization*,

$$\widehat{C} \in \arg \min_{C \in \mathbb{R}^{m \times n}} \frac{1}{2}\|\mathbf{Y} - C\mathbf{X}\|_F^2 + \lambda \sum_{k=1}^{K} w_k \|C_k\|_{\star}, \tag{10.28}$$

where $\|C_k\|_\star = \sum_{j=1}^{q_k} d_j(C_k)$ is the nuclear norm of C_k, w_k's are some pre-specified weights, and λ is a tuning parameter controlling the amount of regularization. The use of the weights is to adjust for the dimension and scale differences of $\mathbf{X}_k$s. We choose

$$w_k = d_1(\mathbf{X}_k)\{\sqrt{m} + \sqrt{r(\mathbf{X}_k)}\}, \tag{10.29}$$

based on a concentration inequality of the largest singular value of a Gaussian matrix. This choice balances the penalization of different views and allows the use of only a single tuning parameter to achieve desired statistical performance; see Section 10.5.2 for additional details.

Through the composite nuclear-norm penalization, the iRRR approach can achieve view selection and view-specific subspace selection simultaneously. This shares the same spirit as the bi-level selection methods for univariate sparse regression (Breheny and Huang 2009; Chen et al. 2016). Moreover, iRRR seamlessly bridges group-sparse and low-rank methods as its special cases:

- Case 1: *Nuclear-norm penalized regression (NNP).* When $n_1 = n$ and $K = 1$, (10.28) reduces to the NNP method as in (10.14), which learns a global low-rank association structure.
- Case 2: *Multi-task learning (MTL).* When $n_k = 1$ and $n = K$, (10.28) becomes a special case of MTL (Caruana 1997), in which all the tasks are with the same set of features and the same set of samples. MTL achieves integrative learning by exploiting potential information sharing across the tasks, i.e., all the task models share the same sparsity pattern of the features.
- Case 3: *Lasso and group Lasso.* When $m = 1$, (10.28) becomes a group Lasso method, as $\|C_k\|_\star = \|C_k\|_2$ when $C_k \in \mathbb{R}^{n_k}$. Further, when $n_k = 1$ and $n = K$, (10.28) reduces to a Lasso regression.

The convex optimization of (10.28) has no closed-form solution in general, but one can solve it with an alternating direction method of multipliers (ADMM) algorithm (Boyd et al. 2011). The details can be found in Li et al. (2019), with extensions on handling incomplete data and non-Gaussian response; these topics are also discussed later in Chapter 11.

10.5.2 Theoretical Analysis

We investigate the theoretical properties of the iRRR estimator from solving the composite nuclear-norm penalization problem. In particular, we derive its non-asymptotic performance bounds for estimation and prediction. The results recover performance bounds of several related methods, including Lasso, group Lasso, and NNP. We further show that iRRR is capable of substantially outperforming those methods under realistic settings of multi-view learning.

We mainly consider the multi-view regression model in (10.27), $\mathbf{Y} = \sum_{k=1}^{K} C_k\mathbf{X} + \boldsymbol{\epsilon}$, and the iRRR estimator in (10.28) with the weights defined in (10.29),

$$\widehat{C} \in \arg \min_{C\in\mathbb{R}^{m\times n}} \frac{1}{2}\|\mathbf{Y} - C\mathbf{X}\|_F^2 + \lambda \sum_{k=1}^{K} d_1(\mathbf{X}_k) \left\{\sqrt{m} + \sqrt{r(\mathbf{X}_k)}\right\} \|C_k\|_\star.$$

Define $\mathbf{Z} = \mathbf{X}\mathbf{X}'/T$, and $\mathbf{Z}_k = \mathbf{X}_k\mathbf{X}_k'/T$, for $k = 1, \ldots, K$. We scale the rows of $\mathbf{X}$ such that the diagonal elements of $\mathbf{Z}$ all equal to 1. Denote $\Lambda_l(\mathbf{Z})$ as the lth largest eigenvalue of $\mathbf{Z}$, so that $\Lambda_l(\mathbf{Z}) = d_l(\mathbf{X})^2/T$.

Theorem 10.4 *Assume $\boldsymbol{\epsilon}$ has i.i.d. $N(0, \sigma^2)$ entries. Let $\lambda = (1 + \theta)\sigma$, with $\theta > 0$ arbitrary. Then with probability at least $1-\sum_{k=1}^{K} \exp[-\theta^2\{m+r(\mathbf{X}_k)\}/2]$, we have*

$$\|\widehat{C}\mathbf{X} - C\mathbf{X}\|_F^2 \leq \|B\mathbf{X} - C\mathbf{X}\|_F^2 + 4\lambda \sum_{k=1}^{K} d_1(\mathbf{X}_k) \left\{\sqrt{m} + \sqrt{r(\mathbf{X}_k)}\right\} \|B_k\|_\star,$$

for any $B = [B_1, \ldots, B_K] \in \mathbb{R}^{m\times n}$ with $B_k \in \mathbb{R}^{m\times n_k}$, $k = 1, \ldots, K$.

Theorem 10.4 shows that $\widehat{C}$ balances the bias term $\|B\mathbf{X}-C\mathbf{X}\|_F^2$ and the variance term $4\lambda \sum_{k=1}^{K} d_1(\mathbf{X}_k)\{\sqrt{m} + \sqrt{r(\mathbf{X}_k)}\}\|B_k\|_\star$ in prediction.

Now consider the estimation performance of $\widehat{C}$. For the low-dimensional scenario that $d_n(\mathbf{X}) > 0$, an oracle inequality for $\widehat{C}$ can be readily obtained from Theorem 10.4. We therefore mainly focus on the general high-dimensional scenario. Motivated by Lounici et al. (2011), Negahban and Wainwright (2011), Koltchinskii et al. (2011), among others, we impose a general restricted eigenvalue (RE) condition for the multi-view regression model in (10.27). We say that $\mathbf{X}$ satisfies RE condition over a restricted set $\mathcal{C}(r_1, \ldots, r_K; \delta) \subset \mathbb{R}^{m\times n}$ if there exists some constant $\kappa(\mathbf{X}) > 0$ such that

$$\frac{1}{2T}\|\Delta\mathbf{X}\|_F^2 \geq \kappa(\mathbf{X})\|\Delta\|_F^2, \qquad \text{for all } \Delta \in \mathcal{C}(r_1, \ldots, r_K; \delta).$$

Here each r_k is an integer satisfying $1 \leq r_k \leq \min(n_k, m)$ and δ is a tolerance parameter.

To specify the restricted set $\mathcal{C}$, we need some additional constructions. For each $C_k \in \mathbb{R}^{m\times n_k}$ $(k = 1, \ldots, K)$, let $C_k = V_k D_k U_k'$ be its full singular value decomposition, where $U_k \in \mathbb{R}^{n_k\times n_k}$, $V_k \in \mathbb{R}^{m\times m}$ satisfy $U_k'U_k = I_{n_k}$ and $V_k'V_k = I_m$. For each $r \in \{1, 2, \cdots, q_k\}$, where $q_k = \min(n_k, m)$, let U_k^r, V_k^r be the submatrices of singular vectors associated with the top r singular values of C_k. Define the following subspaces of $\mathbb{R}^{m\times n_k}$:

$$\mathcal{A}(U_k^r, V_k^r) = \{\Delta_k \in \mathbb{R}^{m\times n_k}; \text{col}(\Delta_k) \subset V_k^r, \text{row}(\Delta_k) \subset U_k^r\},$$

$$\mathcal{B}(U_k^r, V_k^r) = \{\Delta_k \in \mathbb{R}^{m \times n_k}; \text{col}(\Delta_k) \perp V_k^r, \text{row}(\Delta_k) \perp U_k^r\},$$

where $\text{row}(\Delta_k)$ and $\text{col}(\Delta_k)$ denote the row space and column space of Δ_k, respectively. We may adopt the shorthand notation $\mathcal{A}_k^r$ and $\mathcal{B}_k^r$ when no confusion arises. Let $\mathcal{P}_{\mathcal{B}_k^{r_k}}$ denote the projection operator onto the subspace $\mathcal{B}_k^{r_k}$, and define $\Delta_k'' = \mathcal{P}_{\mathcal{B}_k^{r_k}}(\Delta_k)$ and $\Delta_k' = \Delta_k - \Delta_k''$. We now define the restricted set

$$\mathcal{C}(r_1, \ldots, r_K; \delta) = \Big\{\Delta \in \mathbb{R}^{m \times n}; \|\Delta\|_F \geq \delta,$$
$$\sum_{k=1}^K w_k \|\Delta_k''\|_\star \leq \sum_{k=1}^K \left(3w_k \|\Delta_k'\|_\star + 4w_k \sum_{j=r_k+1}^{q_k} d_j(C_k)\right)\Big\}, \tag{10.30}$$

where δ is a tolerance parameter and $w_k = d_1(X_k)(\sqrt{m} + \sqrt{r(X_k)})/T$, as defined in (10.29).

We are now ready to present the main results on the estimation error of the integrative reduced-rank estimator $\widehat{C}$.

Theorem 10.5 *Assume that $\boldsymbol{\epsilon}$ has i.i.d. $N(0, \sigma^2)$ entries. Suppose $\mathbf{X}$ satisfies the RE condition with parameter $\kappa(\mathbf{X}) > 0$ over the set $\mathcal{C}(r_1, \ldots, r_K; \delta)$. Let $\lambda = 2(1+\theta)\sigma$ with $\theta > 0$ arbitrary. Then with probability at least $1 - \sum_{k=1}^K \exp\{-\theta^2(m + r(\mathbf{X}_k))/2\}$,*

$$\|\widehat{C} - C\|_F^2 \preceq \max\Bigg\{\delta^2, \sigma^2(1+\theta)^2 \sum_{k=1}^K \frac{\Lambda_1(\mathbf{Z}_k)}{\kappa(\mathbf{X})^2} \frac{(\sqrt{m} + \sqrt{r(\mathbf{X}_k)})^2 r_k}{T},$$
$$\sigma(1+\theta) \sum_{k=1}^K \frac{\sqrt{\Lambda_1(\mathbf{Z}_k)}}{\kappa(\mathbf{X})} \frac{(\sqrt{m} + \sqrt{r(\mathbf{X}_k)})(\sum_{j=r_k+1}^{q_k} d_j(C_k))}{\sqrt{T}}\Bigg\}.$$

On the right-hand side of the above upper bound, the first term is from the tolerance parameter in the RE condition, which ensures that the condition can possibly hold when the true model is not exactly low-rank (Negahban and Wainwright 2011), i.e., when $\sum_{j=r_k+1}^{q_k} d_j(C_k) \neq 0$. The second term gives the *estimation error* of recovering the desired view-specific reduced-rank structure, and the third term gives the *approximation error* incurred due to approximating the true model with the view-specific reduced-rank model. When the true model has exactly the desired structure, i.e., $r(C_k) = r_k$, it suffices to take $\delta = 0$ and the upper bound then yields the estimation error

$$\|\widehat{C} - C\|_F^2 \preceq \sigma^2 \sum_{k=1}^K (m + r(\mathbf{X}_k)) r_k / T,$$

which holds with high probability in the high-dimensional setting that $m + r(\mathbf{X}_k) \to \infty$. In the classical setting of $T \to \infty$ with fixed m and $r(\mathbf{X}_k)$, by choosing $\theta \propto \sqrt{\log T}$, the rate becomes $\sigma^2 \log(T) \sum_{k=1}^{K} r_k / T$ with probability approaching 1.

Intriguingly, the results in Theorem 10.5 can specialize into oracle inequalities of several regularized estimation methods, including NNP, MTL, Lasso, and group Lasso. This is because these models can all be viewed as special cases of the iRRR model. As such, iRRR seamlessly bridges sparse models and low-rank methods and provides a unified theory of the two types of regularization.

First, consider the NNP method defined in (10.14), which corresponds to the special case of $K = 1$ and $w_k = 1$ in iRRR. The restricted set in (10.30) becomes

$$\mathcal{C}(r) = \{\Delta \in \mathbb{R}^{m \times n}; \|\Delta''\|_\star \leq 3\|\Delta'\|_\star\},$$

where $\Delta'' = \mathcal{P}_{\mathcal{B}^r}(\Delta)$ and $\Delta' = \Delta - \Delta''$. Theorem 10.5 then implies that under the RE condition with $\kappa(\mathbf{X}) > 0$ over $\mathcal{C}(r)$, if we choose $\lambda = 2\sigma(1+\theta)d_1(\mathbf{X})\{\sqrt{m} + \sqrt{r(\mathbf{X})}\}$, then with probability at least $1 - \exp\{-\theta^2(m + r(\mathbf{X}))/2\}$, it holds that

$$\|\widehat{C} - C\|_F^2 \preceq \frac{\sigma^2}{\kappa(\mathbf{X})^2} \frac{(\sqrt{m} + \sqrt{r(\mathbf{X})})^2 r}{T}.$$

This bound recovers the known results on NNP, see, e.g., Negahban and Wainwright (2011).

Next, consider the MTL setting, which corresponds to $n_k = 1$ and $n = K$ in iRRR. Write $C = [C_1, \ldots, C_n] \in \mathbb{R}^{m \times n}$, and $\mathcal{S} = \{j; \|C_j\|_2 \neq 0\}$. The restricted set becomes

$$\mathcal{C}(\mathcal{S}) = \left\{\Delta = [\Delta_1, \ldots, \Delta_n] \in \mathbb{R}^{m \times n}; \sum_{k \in \mathcal{S}^c} \|\Delta_k\|_2 \leq 3 \sum_{k \in \mathcal{S}} \|\Delta_k\|_2 \right\}.$$

By choosing $\lambda \propto \sigma\sqrt{\log n/m}$, Theorem 10.5 yields the high-probability bound

$$\|\widehat{C} - C\|_F^2 \preceq \frac{\sigma^2}{\kappa(\mathbf{X})^2} \frac{(\log n + m)|\mathcal{S}|}{T},$$

where $|\mathcal{S}|$ is the cardinality of $\mathcal{S}$. The same bound can be obtained from results in Lounici et al. (2011) in a more general setting of MTL, or from results in Negahban et al. (2012) on group Lasso by vectorizing the MTL problem here into a univariate response regression.

Another example is Lasso, which corresponds to $m = 1$ and $n = K$ in iRRR. It is seen that the integrative model now reduces to a conventional linear regression model with a univariate response, and correspondingly the composite nuclear norm degenerates to the ℓ_1 norm of the coefficient vector $C' \in \mathbb{R}^n$. Let $\mathcal{S} = \{j; c_j \neq 0\}$; then the restricted set becomes

$$\mathcal{C}(\mathcal{S}) = \left\{ \Delta = (\Delta_1, \ldots, \Delta_n)' \in \mathbb{R}^n; \sum_{k \in \mathcal{S}^c} |\delta_k| \leq 3 \sum_{k \in \mathcal{S}} |\delta_k| \right\}.$$

Theorem 10.5 implies that by choosing $\lambda \propto \sigma \sqrt{\log n}$,

$$\|\widehat{C} - C\|_2^2 \preceq \frac{\sigma^2}{\kappa(\mathbf{X})^2} \frac{\log n \cdot |\mathcal{S}|}{T}$$

holds with high probability. Again, this is a well-known result and has been discussed in Section 10.2.

To see the potential advantage of iRRR over NNP or MTL, we make some comparisons of their error rates based on Theorem 10.5. To convey the main message, consider the case where $n_k = n_1$, $r(\mathbf{X}_k) = r(\mathbf{X}_1)$ for $k = 1, \ldots, K$, $r_k = r_1$ for $k = 1, \ldots, s$, and $r_k = 0$ for $k = s + 1, \ldots, K$. The error rate is $\sigma^2 s r_1 (m + r(\mathbf{X}_1))/T, \sigma^2 r(m + r(\mathbf{X})))/T$, for iRRR and NNP, respectively. As long as $s r_1 = O(r)$, iRRR achieves a faster rate because that $r(\mathbf{X}_1) \leq r(\mathbf{X})$ always holds. For comparing iRRR and MTL, we get that with probability $1 - n^{-1}$, iRRR achieves an error rate $\sigma^2 (\log n + m + r(\mathbf{X}_1)) s r_1 / T$ (by choosing $\theta = \sqrt{4 \log n / (m + r(\mathbf{X}_1))}$), while MTL achieves $\sigma^2 (\log n + m + 1) s n_1 / T$. The two rates agree with each other in the MTL setting when $r(X_1) = r_1 = n_1 = 1$, and the former rate can be much faster in the iRRR setting when, for example, $r_1 \ll n_1$ and $r(\mathbf{X}_1) = O(\log(n) + m)$. These theoretical results justify the superiority of iRRR over NNP and MTL for analyzing multi-view data.

10.6 Applications

10.6.1 Breast Cancer Data

We consider a breast cancer data set consisting of gene expression measurements and comparative genomic hybridization measurements for $n = 89$ subjects. The data set is available in the R package PMA (Witten et al. 2009), and a detailed description can be found in Chin et al. (2006).

Prior studies have demonstrated that certain types of cancer are characterized by abnormal DNA copy-number changes (Pollack et al. 2002; Peng et al. 2010). Thus it is of interest to examine the relationship between DNA copy-number variations and gene expression profiles, for which multivariate regression methods can be useful. Biologically, it makes sense to regress gene expression profiles on copy-number variations because the amplification or deletion of the portion of DNA corresponding to a given gene may result in a corresponding increase or decrease in expression of that gene. The reverse approach is also meaningful, in that the resulting predictive model may identify functionally relevant copy-number variations. This approach has been shown to be promising in enhancing the limited comparative genomic hybridization data analysis with the wealth of gene expression data (Geng et al. 2011; Zhou et al. 2012). We have tried both approaches. Setting 1: designating the copy-number variations of a chromosome as predictors and the gene expression profiles of the same chromosome as responses, and Setting 2: reversing the roles of the predictors and the responses. We find that in setting 1, none of the methods provides an adequate fit to the data, and reduced-rank regression may even fail to pick up any signals. The reduced-rank models give much better results under setting 2. We thus report only the results for setting 2.

We focus the analysis on chromosome 21, for which $n = 227$ and $m = 44$. Both the responses and the predictors are standardized. We compare the various reduced-rank methods by the following cross-validation procedure. The data were randomly split into a training set of size $T_{tr} = 79$ and a test set of size $T_{te} = 10$. All model estimation was carried out using the training data, with the tuning parameters selected by ten-fold cross validation. We used the test data to calibrate the predictive performance of each estimator $\widehat{C}$, specifically, by its mean squared prediction error $\|\mathbf{Y}_{te} - \widehat{C}\mathbf{X}_{te}\|_F^2/(mT_{te})$, where $(\mathbf{Y}_{te}, \mathbf{X}_{te})$ denotes the test set. The random-splitting process was repeated 100 times to yield the average mean squared prediction error and the average rank estimate for each method; see the upper panel of Table 10.1.

As the number of predictors is much greater than the sample size, it is reasonable to assume that only a subset of predictors is important. Therefore, a perhaps better modeling strategy is the subset multivariate regression with a selected subset of predictors. Several variable selection methods have been proposed in the context of reduced-rank regression (Chen et al. 2012; Chen and Huang 2012; Bunea et al. 2012). We modified the preceding cross-validation procedure for comparing the reduced-rank subset regression methods. The only modification was that for each random split, we first applied the method of Chen et al. (2012) using the training set to select a set of predictors, with which the reduced-rank methods were subsequently carried out using the training set and calibrated using the test set. Since our main goal here is to compare the various reduced-rank methods, we defer the description of the predictor selection procedures (Chen et al. 2012) to Chapter 13.

The results are summarized in the lower panel of Table 10.1. We consider the nuclear-norm penalized estimator $\widehat{C}_N$, the rank-penalized estimator $\widehat{C}_H$, the adaptive nuclear-norm penalized estimator $\widehat{C}_S$ (with weight parameter $\gamma = 2$), the rank-penalized estimator with an additional ridge penalty $\widehat{C}_{HR}$, and the adaptive nuclear-norm penalized estimator with an additional ridge penalty $\widehat{C}_{SR}$. For each method, we report the average of mean squared prediction errors (MSPE) based on 100 replications of 79/10 splits of the data and the corresponding average of the rank estimates (Rank). The standard errors are reported in parentheses.

Table 10.1 shows that the ANN estimator $\widehat{C}_S$ enjoys slightly better predictive performance than both $\widehat{C}_H$ and $\widehat{C}_N$. The number of selected predictors in the 100 splits ranges from 71 to 102; hence, incorporating variable selection greatly reduces the number of predictors and may potentially improve model interpretation. However, in this example, reduced-rank estimation using a subset of predictors

Table 10.1 Comparison of the model fits to the chromosome 21 data for various reduced-rank methods

	$\widehat{C}_N$	$\widehat{C}_H$	$\widehat{C}_{HR}$	$\widehat{C}_S$	$\widehat{C}_{SR}$
	Full data				
MSPE	0.71 (0.2)	0.69 (0.1)	0.68 (0.1)	0.68 (0.1)	0.68 (0.1)
Rank	6.2 (0.8)	1.0 (0.0)	1.0 (0.0)	1.0 (0.2)	1.8 (0.4)
	Selected predictors				
MSPE	0.85 (0.2)	0.90 (0.2)	0.82 (0.1)	0.85 (0.1)	0.84 (0.1)
Rank	4.6 (0.7)	1.3 (0.5)	1.9 (0.3)	2.1 (0.2)	2.4 (0.5)

results in higher mean squared prediction error than using all predictors, uniformly for all methods, but more so for $\widehat{C}_H$ than for other methods. The nuclear-norm penalized estimator $\widehat{C}_N$ generally yields a higher rank estimate than the other methods. Incorporating ridge penalization improves the predictive performance of the reduced-rank methods. Particularly, $\widehat{C}_{HR}$ may substantially outperform its non-robust counterpart $\widehat{C}_H$. Both $\widehat{C}_N$ and $\widehat{C}_{HR}$ can be computationally intensive for large data sets, while other methods are much faster to compute.

10.6.2 Longitudinal Studies of Aging

The Longitudinal Studies of Aging (LSOA) (Stanziano et al. 2010) had been a collaborative effort of the National Center for Health Statistics and the National Institute on Aging. The study interviewed a large cohort of senior people (70 years of age and over) in 1997–1998 (WAVE II) and 1999–2000 (WAVE III), respectively, and measured their health conditions, living conditions, family situations, health service utilizations, among others. Here our objective is to examine the predictive relationship between health-related events in earlier years and health outcomes in later years, which can be formulated as a multivariate regression problem.

There are $T = 3988$ subjects who participated in both WAVE II and WAVE III interviews. After data pre-processing (Luo et al. 2018), $n = 294$ health risk and behavior measurements in WAVE II are treated as predictors, and $m = 41$ health outcomes in WAVE III are treated as responses. The response variables are binary indicators, characterizing various cognitive, sensational, social, and life quality outcomes, among others. Over 20% of the response data entries are missing. The predictors are multi-view, including housing condition ($\mathbf{X}_1$ with $n_1 = 38$), family structure/status ($\mathbf{X}_2$ with $n_2 = 60$), daily activity ($\mathbf{X}_3$ with $n_3 = 40$), prior medical condition ($\mathbf{X}_4$ with $n_4 = 114$), and medical procedure since last interview ($\mathbf{X}_5$ with $n_5 = 40$). We thus apply the iRRR method to perform the regression analysis. As a comparison, we also implement generalized reduced-rank regression (gRRR) with binary responses (Luo et al. 2018) (to be elaborated in Section 12.4), and both classical and sparse logistic regression methods using the R package `glmnet`, denoted as glm and glmnet, respectively.

We use a random-splitting procedure to evaluate the performance of different methods. More specifically, each time we randomly select $T_{tr} = 3000$ subjects as training samples and the remaining $T_{te} = 988$ subjects as testing samples. For each method, we use fivefold CV on the training samples to select tuning parameters and apply the method to all the training data with the selected tuning parameters to yield its coefficient estimate. The performance of each method is measured by the average deviance between the observed true response values and the estimated probabilities, defined as

$$\text{Average Deviance} = \frac{-2\sum_{k=1}^{T_{te}}\sum_{j=1}^{m}\{y_{jk}\log\widehat{p}_{jk} + (1-y_{jk})\log(1-\widehat{p}_{jk})\}\delta_{jk}}{\sum_{k=1}^{T_{te}}\sum_{j=1}^{m}\delta_{jk}},$$

where δ_{jk} is an indicator of whether y_{jk} is observed. To measure the predictive performance, we also calculate the Area Under the Curve (AUC) value of the Receiver Operating Characteristic (ROC) curve for each outcome variable. This procedure is repeated 100 times and the results are averaged.

In terms of the average deviance, iRRR and glmnet yield very similar results (with mean 0.77 and standard deviation 0.01), and both substantially outperform gRRR (with mean 0.83 and standard deviation 0.01) and glm (fails due to a few singular outcomes). The out-of-sample AUCs for different response variables are shown in Fig. 10.2. The response variables are sorted based on their missing rates from large (over 70%) to small (about 13%). Again, the performance of iRRR is comparable to that of glmnet. The iRRR tends to have a slight advantage over glmnet for responses with high missing rates. This could be due to the fact that iRRR can borrow information from other responses, while the univariate glmnet cannot.

To understand the impact of different views on prediction, we produce heatmaps of the estimated coefficient matrices in Fig. 10.3 (glm is omitted due to its poor performance). The predictors fall into 5 groups, namely, housing condition, family status, daily activity, prior medical condition, and change in medical procedure since last interview, from top to bottom separated by horizontal black lines. For visualization purpose, we also sort the responses based on their grouping structure (e.g., cognition, sensation, social behavior, and life quality). The estimates from iRRR and glmnet show quite similar patterns: it appears that the family structure/status group and the daily activity group have the most predictive power, and the variables within these two groups contribute to the prediction in a collective way. As for the other three views, iRRR yields heavily shrunk coefficient estimates, while glmnet yields very sparse estimates. These agreements partly explain the similarity of the two methods in their prediction performance. In contrast, the gRRR method tries to learn a globally low-rank structure rather than a view-specific structure; consequently, it yields a less parsimonious solution with less competitive prediction performance.

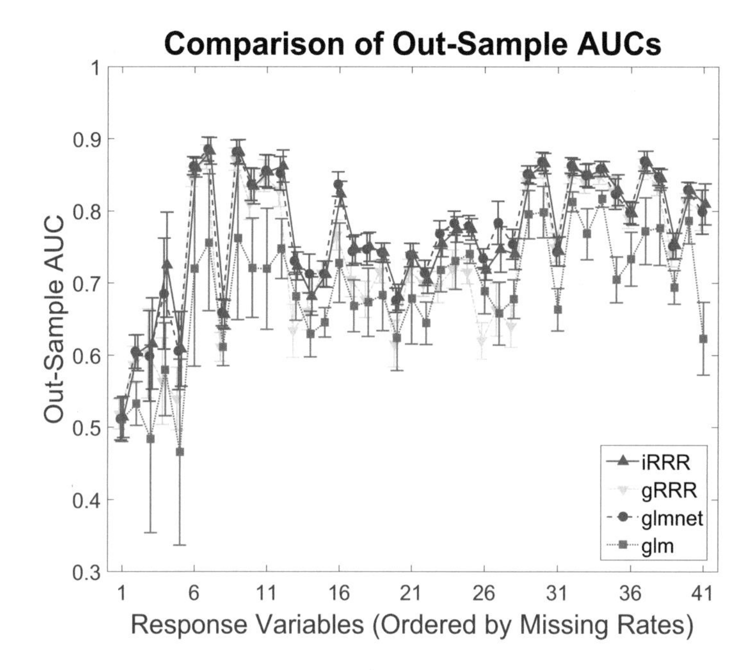

Fig. 10.2 LSOA data analysis. The mean and standard deviation (error bar) of AUC for each response variable over 100 random-splitting procedures. The responses, from left to right, are ordered by missing rates from large to small (Li et al. 2019)

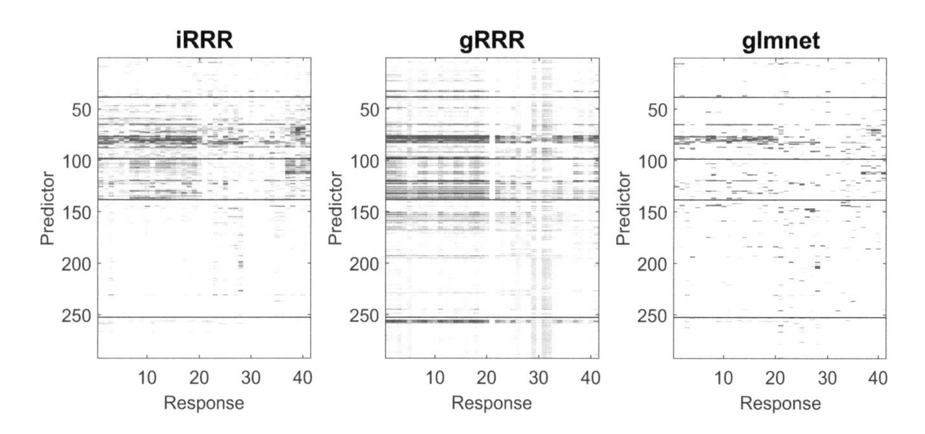

Fig. 10.3 LSOA data analysis. The heat maps of the coefficient matrices estimated from different methods (Li et al. 2019)

Chapter 11
Unbiased Risk Estimation in Reduced-Rank Regression

11.1 Introduction

Reduced-rank regression, similar to any regularized estimation technique, produces a spectrum of low-dimensional models with varying complexities. Its successful application in practice relies on accurate model assessment and selection, for which the quality of a model is objectively measured by its true prediction error, that is, its error for predicting a new observation. Apparently, the true prediction error depends on the underlying true model and has to be estimated from the data. The celebrated Stein's unbiased risk estimation (SURE) framework (Stein 1981) provides a general way to construct an unbiased estimator of the true prediction error of any "nearly arbitrary" estimator (Donoho and Johnstone 1995). It turns out that the unbiased estimator usually is a sum of two parts. The first part is the training error based on the observed data, and if used alone, it inevitably underestimates the true prediction error because the same data is being used to fit the model and assess its error. The second term is the bias correction term or the term of optimism (Hastie et al. 2009), and it is closely related to model complexity: the "harder" we fit the data (using a more complex model), the stronger each observation affects its own prediction, thereby increasing the optimism. This shows that a foremost task in estimating the true prediction error for model selection and assessment is to obtain an accurate estimate of the model complexity, or broadly known as the model degrees of freedom.

Degrees of freedom is a widely used concept in statistics, and in a narrow sense, it simply counts the number of free parameters in a model. As a general measure of the complexity of a model fitting procedure (Hastie and Tibshirani 1990), the concept was generalized and formalized under the aforementioned SURE framework (Stein 1981). An accurate estimation of the degrees of freedom is key to evaluating the true prediction risk for model assessment (Mallows 1973; Efron 2004). In practice, with the estimated degrees of freedom, one can construct information criterion such as

G. C. Reinsel et al., *Multivariate Reduced-Rank Regression*, Lecture Notes in Statistics 225, https://doi.org/10.1007/978-1-0716-2793-8_11

Akaike Information Criterion (AIC) (Akaike 1974), Generalized Cross-Validation (GCV) criterion (Golub et al. 1979), among others, for model selection.

In conventional multiple linear regression model (10.1) with the least squares estimation, the degrees of freedom is the number of regression coefficients, n. However, for general nonlinear estimation procedures, the exact correspondence between the degrees of freedom and the number of free parameters may not hold (Ye 1998). For example, with the best subset selection method (Hocking and Leslie 1967), we search for the best regression model with $n_0 \in \{1, 2, \ldots, n\}$ predictors that minimizes the residual sum of squares. The resulting model has n_0 coefficients, but the degrees of freedom could be much higher than n_0 because intuitively the search over all the subsets of size n_0 increases estimation complexity (Hastie et al. 2009). From an alternative perspective, with the best subset selection, the optimal n_0-dimensional subspace that minimizes the residual sum of squares clearly depends on the response; as such the estimation is highly nonlinear in the response, which results in the loss of correspondence between the degrees of freedom and the number of free parameters.

Similar arguments apply to reduced-rank regression. Instead of searching for the best subset of the predictors, reduced-rank regression searches for the best r-dimensional subspace of the predictors, which leads to increased model complexity. Because the optimal rank-r subspace depends also on the response matrix $\mathbf{Y}$, the correspondence between the number of free parameters and the degrees of freedom need not hold. Indeed, the number of free parameters in a $m \times n$ matrix of rank r, $(n+m-r)r$, has been suggested (see Chapter 2) as a naive estimate of the degrees of freedom of reduced-rank regression with the rank constraint $r \leq \min(m, n)$. More precisely, with an arbitrary and possibly rank-deficit predictor matrix $\mathbf{X}$, the number of free parameters equals to $(r_x + m - r)r$, where $r_x = r(\mathbf{X})$ denotes the rank of $\mathbf{X}$. Henceforth, we refer to $(r_x + m - r)r$ as the naive estimator of the degrees of freedom of a rank-r reduced-rank regression.

In this chapter, we present a finite-sample unbiased estimator of the degrees of freedom for a general class of reduced-rank estimators (Mukherjee et al. 2015). The finding that the unbiased estimator admits an explicit form as a function of the estimated singular values is quite elegant. We demonstrate that using the unbiased estimator leads to more accurate model evaluation and selection in reduced-rank regression.

11.2 Degrees of Freedom

Stein (1981), in his theory of unbiased risk estimation, first introduced a formal definition of the degrees of freedom of a statistical estimation procedure. Later Efron (2004) showed that Stein's treatment can be considered as a special case of a more general notion under the Gaussian assumption. Assume that we have data of the form $Y \in \mathbb{R}^T, \mathbf{X} \in \mathbb{R}^{n \times T}$. Given $\mathbf{X}$, the response originates from the model $Y \sim (\mu, \sigma^2 I)$, where $\mu \in \mathbb{R}^T$ is the true mean that can be a function of $\mathbf{X}$, and σ^2 is the common variance. Then for any estimation procedure $m(\cdot)$ with fitted values $\widehat{\mu} = m(\mathbf{X}, Y) \in \mathbb{R}^T$, the degrees of freedom of $m(\cdot)$ is defined as

$$\mathrm{df}(m) = \sum_{i=1}^{T} \mathrm{cov}(\widehat{\mu}_i, Y_i)/\sigma^2. \tag{11.1}$$

The rationale is that more complex models would provide better fit for the data, and hence the covariance between observed and fitted pairs would be higher. This expression is not directly observable except for certain simple cases, for example, when $m(\mathbf{X}, Y) = SY$ is a linear smoother, with $S \in \mathbb{R}^{T \times T}$ being a function of $\mathbf{X}$ only. In that case, it is not difficult to see that $\mathrm{df}(m) = \mathrm{tr}(S)$, which agrees with the classical definition of the degrees of freedom (Hastie and Tibshirani 1990). More progress can be made for the case when $Y \sim N(\mu, \sigma^2 I)$. It was proven that as long as the partial derivative $\partial\widehat{\mu}_i/\partial Y_i$ exists almost everywhere for all $i \in \{1, \ldots, T\}$, it holds true that

$$\mathrm{cov}(\widehat{\mu}_i, Y_i) = \sigma^2 E\left(\frac{\partial \widehat{\mu}_i}{\partial Y_i}\right).$$

This leads to the following unbiased estimator of the degrees of freedom for the fitting procedure $m(\cdot)$,

$$\widehat{\mathrm{df}}(m) = \sum_{i=1}^{T} \frac{\partial \widehat{\mu}_i}{\partial Y_i}. \tag{11.2}$$

Using the degrees of freedom definition in (11.1), Efron (2004) employed a covariance penalty approach to prove that the $\mathrm{C_P}$-type statistics (Mallows 1973), namely

$$\mathrm{C_P}(\widehat{\mu}) = \frac{1}{T}\|Y - \widehat{\mu}\|_2^2 + \frac{2\mathrm{df}(m)}{T}\sigma^2, \tag{11.3}$$

provide an unbiased estimator of the true prediction error. This demonstrates the important role played by the degrees of freedom in model assessment and gives a principled way to select the optimal model without using more computationally expensive methods such as cross validation (Efron 2004).

Many important works followed Stein (1981). Donoho and Johnstone (1995) used the unbiased risk estimation framework to derive the degrees of freedom for the soft-thresholding operator in wavelet shrinkage. Meyer and Woodroofe (2000) employed this framework to derive the same for shape-restricted regression. Li and Zhu (2008) used this approach to compute an unbiased estimator of the degrees of freedom of penalized quantile regression. Zou et al. (2007) showed that the number of nonzero coefficients provides an unbiased estimator of the degrees of freedom for the Lasso. These results are of great practical importance, allowing one to use model selection criteria such as C_P for selecting the corresponding regularized estimators without incurring extra computational cost.

However, beyond a few special cases, the unbiased estimator in (11.2) in general may not admit an explicit analytical form. To overcome the analytical difficulty in estimating the degrees of freedom for general statistical procedures, Ye (1998) and Shen and Ye (2002) proposed the generalized degrees of freedom approach, where they evaluated partial derivatives in (11.2) numerically, using data perturbation techniques to compute an approximately unbiased estimator. Efron (2004) used a parametric bootstrap to also arrive at an approximately unbiased estimator of (11.1). In this method the objective is to directly estimate $\text{cov}(\widehat{\mu}_i, Y_i)$ by drawing repeated samples from the underlying distribution and fitting the model. In the absence of such samples, this can be achieved by using a parametric bootstrap to simulate data from a larger unbiased model and computing the covariance between the fitted and observed values. Although these simulation approaches allow the estimation of the exact degrees of freedom in many highly nonlinear modeling procedures, they are computationally expensive, and the lack of any closed-form expression makes investigation of the theoretical properties a difficult task.

11.3 Degrees of Freedom of Reduced-Rank Estimation

Estimating the degrees of freedom for reduced-rank regression is a challenging problem because of the nonlinear nature of the estimator. Even though the reduced-rank solution admits a closed-form, it is highly nonlinear and involves complex matrix operations such as the singular value decomposition. Here we study the degrees of freedom of a general class of reduced-rank estimators in the framework of unbiased risk estimation. As it turns out, the unbiased estimator indeed admits an explicit form.

Recall the multivariate linear regression model in (10.6). Let $\widehat{\mathbf{Y}} \in \mathbb{R}^{m\times T}$ be the least squares fitted value matrix, with a singular value decomposition of the form

$$\widehat{\mathbf{Y}} = \mathbf{Y}\mathbf{X}'(\mathbf{X}\mathbf{X}')^{-}\mathbf{X} = \underset{m\times\bar{r}}{V}\ \underset{\bar{r}\times\bar{r}}{D}\ \underset{\bar{r}\times n}{W'}, \tag{11.4}$$

where A^{-} denotes the Moore-Penrose inverse (Moore 1920; Penrose 1955) of matrix A. The Moore-Penrose inverse is well defined for an arbitrary set of dimensions (m, n, T) as well as for a potentially rank-deficient predictor matrix $\mathbf{X}$. In (11.4), V and W are orthogonal matrices that represent the left and right singular vectors of $\widehat{\mathbf{Y}}$ and $D = \text{diag}\{d_i, i = 1, \ldots, \bar{r}\}$, with $d_1 \geq \cdots \geq d_{\bar{r}} > 0$ being the nonzero singular values of $\widehat{\mathbf{Y}}$. Without loss of generality we assume that $r(\widehat{\mathbf{Y}}) = \bar{r} = \min(r_x, m)$, where r_x denotes the rank of $\mathbf{X}$. We denote the kth column of W and V by w_k and v_k, respectively.

Using the Eckart–Young Theorem (Eckart and Young 1936) (see Theorem 2.1), the fitted value matrix from the rank-r reduced-rank regression is given by the best rank-r approximation of the least square fitted value matrix, which can be expressed as

$$\widehat{\mathbf{Y}}(r) = \left(\sum_{k=1}^{r} v_k v_k'\right) \widehat{\mathbf{Y}} = V^{(r)} D^{(r)} W^{(r)'}, \quad r = 1, \ldots, \bar{r}, \tag{11.5}$$

where $A^{(r)}$ denotes the first r columns of A. More generally, we consider a broad class of reduced-rank estimators with fitted value matrix given by

$$\widetilde{\mathbf{Y}}(\lambda) = \widetilde{C}(\lambda)\mathbf{X} = \sum_{k=1}^{\bar{r}} s_k(d_k, \lambda) d_k v_k w_k' = \left(\sum_{k=1}^{\bar{r}} s_k(d_k, \lambda) v_k v_k'\right) \widehat{\mathbf{Y}}, \tag{11.6}$$

where each $s_k(d_k, \lambda) \in [0, 1]$ is a function of d_k and λ, and they satisfy $s_1(d_1, \lambda) \geq \cdots \geq s_{\bar{r}}(d_{\bar{r}}, \lambda) \geq 0$. For simplicity, we write $s_k(d_k, \lambda) = s_k(\lambda) = s_k$. It is immediately seen that both the reduced-rank regression estimator in (11.5) and the adaptive nuclear-norm penalized estimator in (10.24) are special cases. This class of estimators has the same set of singular vectors as the reduced-rank regression estimator in (11.5), but may have different singular value estimates given by shrunk or thresholded versions of the singular values of the least squares estimator. As discussed in Section 10.4, they can be obtained from a non-convex singular value penalization or thresholding operations and possess many desirable theoretical properties.

We now apply the definition in (11.2) to derive an unbiased estimator of the degrees of freedom for the above class of reduced-rank estimators. Hansen (2018) showed that weak differentiability of reduced-rank estimator is sufficient to guarantee that Stein's Lemma applies here. We start by rewriting the multivariate linear regression model (10.6) as a multiple linear regression model

$$\underset{mT\times 1}{\mathrm{vec}(\mathbf{Y}')} = \underset{mT\times mn}{\left(I_m \otimes \mathbf{X}'\right)} \underset{mn\times 1}{\mathrm{vec}(C')} + \underset{mT\times 1}{\mathrm{vec}(\boldsymbol{\epsilon}')},$$

where $\otimes$ denotes the Kronecker product between matrices, and vec$(\cdot)$ stands for the column-wise vectorization operator on a matrix. This allows us to directly apply results derived under the vector case such as those in (11.1) and (11.2).

To facilitate the calculation of the degrees of freedom, let $\mathbf{X}\mathbf{X}' = QS^2Q'$ be the eigen decomposition of $\mathbf{X}\mathbf{X}'$, that is, $Q \in \mathbb{R}^{n\times r_x}$, $Q'Q = I$, and $S \in \mathbb{R}^{r_x\times r_x}$ is a diagonal matrix with positive diagonal elements. Then, the Moore-Penrose inverse of $\mathbf{X}\mathbf{X}'$ can be written as $(\mathbf{X}\mathbf{X}')^- = QS^{-2}Q'$. Define

$$H = S^{-1}Q'\mathbf{X}\mathbf{Y}'.$$

It follows that $H \in \mathbb{R}^{r_x\times m}$ admits a singular value decomposition of the form

$$H = UDV', \tag{11.7}$$

where $U \in \mathbb{R}^{r_x \times \bar{r}}$, $U'U = I$, and V, D are defined in (11.4). The matrix H shares the same set of singular values and right singular vectors with $\widehat{\mathbf{Y}}$ in (11.4), as $H'H = \widehat{\mathbf{Y}}\widehat{\mathbf{Y}}' = \mathbf{Y}\mathbf{X}'(\mathbf{X}\mathbf{X}')^{-}\mathbf{X}\mathbf{Y}'$. Moreover, H is full rank because $\widehat{\mathbf{Y}}$ is of rank $\bar{r} = \min(r_x, m)$. The matrix H plays a key role in deriving a simple form for the degrees of freedom. In particular, this construction allows us to avoid singularities arising from $r_x < m$ in the high-dimensional scenario.

Now, upon using the above construction, we can write $\widehat{\mathbf{Y}} = H'S^{-1}Q'\mathbf{X}$ and

$$\widehat{\mathbf{Y}}(r) = H(r)'S^{-1}Q'\mathbf{X}, \qquad r = 1, \ldots, \bar{r},$$

where $H(r) = U^{(r)}D^{(r)}V^{(r)\prime}$. Then, applying definition (11.2) and using the identities $\mathrm{tr}(AB) = \mathrm{tr}(BA)$, $\mathrm{vec}(ABC) = (C' \otimes A)\mathrm{vec}(B)$ and the chain rule of differentiation, we obtain the following expression of the exact estimator of the degrees of freedom,

$$\begin{aligned}
\widehat{\mathrm{df}}(r) &= \mathrm{tr}\left[\frac{\partial \mathrm{vec}\{\widehat{\mathbf{Y}}(r)'\}}{\partial \mathrm{vec}(\mathbf{Y}')}\right] \\
&= \mathrm{tr}\left(\left(I_m \otimes \mathbf{X}'QS^{-1}\right)\left[\frac{\partial \mathrm{vec}\{H(r)\}}{\partial \mathrm{vec}(H)}\right]\left\{\frac{\partial \mathrm{vec}(H)}{\partial \mathrm{vec}(\mathbf{Y}')}\right\}\right) \\
&= \mathrm{tr}\left(\left(I_m \otimes \mathbf{X}'QS^{-1}\right)\left[\frac{\partial \mathrm{vec}\{H(r)\}}{\partial \mathrm{vec}(H)}\right]\left(I_m \otimes S^{-1}Q'\mathbf{X}\right)\right) \\
&= \mathrm{tr}\left[\frac{\partial \mathrm{vec}\{H(r)\}}{\partial \mathrm{vec}(H)}\right].
\end{aligned} \tag{11.8}$$

Similarly, for the general class of reduced-rank estimators in (11.6), we have

$$\widetilde{\mathbf{Y}}(\lambda) = \left(\sum_{k=1}^{\bar{r}} s_k(d_k, \lambda)v_k v_k'\right) H'S^{-1}Q'\mathbf{X} = V\widetilde{D}(\lambda)U'S^{-1}Q'\mathbf{X},$$

where $\widetilde{D}(\lambda) = \mathrm{diag}\{s_k(d_k, \lambda)d_k, k = 1, \ldots, \bar{r}\}$. Once again using familiar results in matrix algebra we arrive at a simpler expression for the exact estimator of the degrees of freedom

$$\widetilde{\mathrm{df}}(\lambda) = \mathrm{tr}\left[\frac{\partial \mathrm{vec}\{\widetilde{H}(\lambda)\}}{\partial \mathrm{vec}(H)}\right], \tag{11.9}$$

where $\widetilde{H}(\lambda) = U\widetilde{D}(\lambda)V'$.

It is now clear that the problem of estimating the degrees of freedom reduces to the computation of the divergence measure of a low-rank approximation of the matrix H with respect to H itself. This necessarily involves the derivatives of the singular values and singular vectors with respect to the entries of the matrix, which are not only highly nonlinear functions of the matrix, but are also discontinuous on certain subsets of matrices (O'Neil 2005). Fortunately, there has been a considerable amount of work on the smoothness and differentiability of the singular value decomposition of a real matrix, see Magnus and Neudecker (1998), O'Neil (2005). In particular, we refer to Theorem 1 in Mukherjee et al. (2015) on the derivatives of the singular values and singular vectors of a matrix with respect to any entry of the matrix.

Even with such powerful matrix algebra results, it is still not at all clear whether the degrees of freedom estimators in (11.8) and (11.9) may admit explicit forms. Examining the structure of the singular value decomposition sheds light on this. The pairs of singular vectors (u_k, v_k) are orthogonal to each other, representing distinct directions in $\mathbb{R}^{r_x \times m}$ without any redundancy. Intuitively, these directions themselves are indistinguishable, and their relative importance in constituting the matrix H are entirely revealed by the singular values. This suggests that the complexity of reduced-rank estimation, as exhibited by the relative complexity of a low-rank approximation $H(r)$ or $\widetilde{H}(\lambda)$ of H, may only depend on the singular values of the matrix H and the mechanism of singular value shrinkage or thresholding. This is the main intuition that motivated the findings for the explicit forms of (11.8) and (11.9), which are summarized in the following theorems.

Theorem 11.1 *Let* $\widehat{\mathbf{Y}}$ *be the least squares fitted value matrix as given in* (11.4). *Let* $r_x = r(\mathbf{X})$ *and* $\bar{r} = r(\widehat{\mathbf{Y}}) = \min(r_x, m)$. *Suppose the singular values of* $\widehat{\mathbf{Y}}$ *satisfy* $d_1 > \cdots > d_{\bar{r}} > 0$. *Consider the reduced-rank estimator* $\widehat{\mathbf{Y}}(r)$ *in* (11.5). *An unbiased estimator of the effective degrees of freedom is*

$$\widehat{\mathrm{df}}(r) = \begin{cases} \max(r_x, m)r + \sum_{k=1}^{r} \sum_{l=r+1}^{\bar{r}} \dfrac{d_k^2 + d_l^2}{d_k^2 - d_l^2}, & r < \bar{r}; \\ r_x m, & r = \bar{r}. \end{cases}$$

The above results are further generalized to the class of reduced-rank estimators in (11.6). The weights $s_k(d_k, \lambda)$ are treated as random quantities because they are usually functions of the singular values.

Theorem 11.2 *Let* $\widehat{\mathbf{Y}}$ *be the least squares fitted value matrix as given in* (11.4). *Let* $\bar{r} = \mathrm{rank}(\widehat{\mathbf{Y}}) = \min(r_x, m)$ *and suppose the singular values of* $\widehat{\mathbf{Y}}$ *satisfy* $d_1 > \cdots > d_{\bar{r}} > 0$. *Consider the reduced-rank estimator* $\widetilde{\mathbf{Y}}(\lambda)$ *in* (11.6), *and let* $\widetilde{r} = \widetilde{r}(\lambda) = \max\{k : s_k(d_k, \lambda) > 0.\}$. *An unbiased estimator of the effective degrees of freedom is*

$$\widetilde{\mathrm{df}}(\lambda) = \begin{cases} \max(r_x, m) \sum_{k=1}^{\widetilde{r}} s_k + \sum_{k=1}^{\widetilde{r}} \sum_{l=\widetilde{r}+1}^{\bar{r}} \dfrac{s_k(d_k^2 + d_l^2)}{d_k^2 - d_l^2} \\ \quad + \sum_{k=1}^{\widetilde{r}} \sum_{l \neq k}^{\widetilde{r}} \frac{d_k^2(s_k - s_l)}{d_k^2 - d_l^2} + \sum_{k=1}^{\widetilde{r}} d_k s_k', & \widetilde{r} < \bar{r}; \\ \max(r_x, m) \sum_{k=1}^{\widetilde{r}} s_k + \sum_{k=1}^{\widetilde{r}} \sum_{l \neq k}^{\widetilde{r}} \dfrac{d_k^2(s_k - s_l)}{d_k^2 - d_l^2} + \sum_{k=1}^{\widetilde{r}} d_k s_k', & \widetilde{r} = \bar{r}, \end{cases}$$

where for simplicity we write $s_k = s_k(d_k, \lambda)$ *and* $s'_k = \partial s_k(d_k, \lambda)/\partial d_k$, *the partial derivative of* $s_k(d_k, \lambda)$ *with respect to* d_k.

The explicit formulae presented in the above theorems facilitate further exploration of the behaviors and properties of the degrees of freedom. For example, consider the unbiased estimator for reduced-rank regression in Theorem 11.1. It is always true that

$$\widehat{\mathrm{df}}(r) \geq \max(r_x, m)r + \sum_{k=1}^{r} \sum_{l=r+1}^{\bar{r}} \frac{d_k^2 + 0}{d_k^2 - 0} = (r_x + m - r)r, \qquad r = 1, \ldots, \bar{r}. \tag{11.10}$$

This suggests that the exact estimator is always greater than the naive estimator, the number of free parameters $(r_x + m - r)r$. Similar to the Lasso in multiple linear regression problem (Tibshirani 1996; Zou et al. 2007), reduced-rank estimation can be viewed as a latent factor selection procedure, in which we both construct and search over as many as $\bar{r}$ latent linear factors. Therefore, the increments in the degrees of freedom as shown in (11.10) can be interpreted as the price we have to pay for performing this latent factor selection. For the general methods considered in Theorem 11.2, this inequality no longer holds, due to the shrinkage effects induced by the weights, $0 \leq s_k \leq 1$. The reduction in the degrees of freedom due to singular value shrinkage can offset the price paid for searching over the set of latent variables. Therefore, similar to Lasso, adaptive singular value penalization can provide effective control over the model complexity (Tibshirani and Taylor 2012; Mukherjee et al. 2015).

Although the unbiased estimator and naive estimators are quite different, some interesting connections can be made. They are close to each other when they are evaluated at the true underlying rank, especially when the signal is strong relative to the noise level. This phenomenon is also noted in the empirical studies. Suppose the true model rank is $r(C) = r$. Intuitively, the $\bar{r} - r$ smallest singular values from least squares may be close to zero and are not comparable to the r largest ones; using the approximation $d_k \approx 0$, $k = r + 1, \ldots, \bar{r}$, we obtain $\widehat{\mathrm{df}}(r) \approx (r_x + m - r)r$. A more rigorous argument can be made under classical large sample setting where n and m are fixed and $T \to \infty$; under standard assumptions, consistency of the least squares estimation can readily be established. Using techniques such as perturbation expansion of matrices (Izenman 1975), the consistency of $\widehat{\mathbf{Y}}$ implies the consistency of the estimated singular values. The first r estimated singular values converge to their nonzero true counterparts while the rest converge to zero in probability. It follows that

$$\widehat{\mathrm{df}}(r) \to (r_x + m - r)r,$$

as $T \to \infty$ in probability. An immediate implication is that for each $r = 1, \ldots, \bar{r}$, if we assume that the true model is of rank r, then in a classical asymptotic sense, the number of free parameters, $(r_x + m - r)r$, is the correct degrees of freedom.

This clearly relates to the error degrees of freedom of the classical asymptotic χ^2 statistic from the likelihood ratio test of H_0: $r(C) = k$ (Izenman 1975), for each $k = 1, \ldots, \bar{r}$. If n and m are allowed to diverge with sample size T, these classical asymptotic results would break down as the convergence of nonzero singular values fails (Bai and Silverstein 2009). In contrast, the unbiasedness property of the exact estimators is non-asymptotic, and hence it is valid for any given (n, m, T). Furthermore, non-asymptotic prediction error bounds have been developed for reduced-rank estimation methods as discussed in Section 10.4.3, and the minimax convergence rate coincides with the number of free parameters (Rohde and Tsybakov 2011; Bunea et al. 2011). These results do reveal the limitations, the underlying assumptions and the asymptotic nature of the naive estimator and provide further justification of the exact estimators.

The derived formulae also exhibit some interesting behaviors of rank reduction. In essence, reduced-rank methods distinguish the signal from the noise by examining the estimated singular values from least squares estimation: the large singular values more likely represent the signals while the small singular values mostly correspond to the noise. By rank reduction, we aim to recover the signals exceeding the noise level. Suppose that d_k and d_{k+1} are close for some $k = 1, \ldots, \bar{r} - 1$. It can be argued that the true model rank is unlikely to be k, because the $(k+1)$th and the kth layers are hardly distinguishable. Indeed, this is reflected from the degrees of freedom: for $r = k$, the formula includes a term $(d_k + d_{k+1})/(d_k - d_{k+1})$, which can be excessively large. On the other hand, there is no such term for $r = k + 1$. Consequently, the unbiased estimator of the degrees of freedom may not monotonically increase as the rank r increases, in contrast to the naive estimator. In the above scenario, the estimates for $r = k$ can even be larger than that of $r = k + 1$. This automatically reduces the chance of k being selected as the final rank.

We remark that several works, including Candès et al. (2013), Mukherjee et al. (2015), and Yuan (2016), independently studied the degrees of freedom in low-rank estimation under the SURE framework. A rigorous verification on the applicability of Stein's Lemma, however, came later in Hansen (2018). For a most recent investigation of this problem, we refer to Mazumder and Weng (2020), where the authors emphasized the verification of the regularity conditions for applying Stein's lemma on a group of low-rank matrix estimators. Moreover, for those estimators for which Stein's Lemma does not apply readily, the estimator of the degrees of freedom is derived through a Gaussian convolution method.

11.4 Comparing Empirical and Exact Estimators

11.4.1 Simulation Setup

We investigate the properties of the degrees of freedom estimator using numerical simulation. Two scenarios of model dimensions are considered. In Model I, we set

$T = 50$, $n = 20$, $m = 15$, $r_x = 20$, $r = 5$. The design matrix $\mathbf{X}$ is constructed by generating its T columns as i.i.d. samples from $N(0, \Sigma_x)$, where $\Sigma_x = (\sigma_{ij})_{n\times n}$ and $\sigma_{ij} = \rho^{|i-j|}$ with some $0 < \rho < 1$. In Model II, we set $T = 40$, $n = 100$, $m = 50$, $r_x = 15$, $r = 5$. The design matrix $\mathbf{X}$ is generated as $\mathbf{X} = \Sigma_x^{1/2}\mathbf{X}_0$, where Σ_x is defined as above, $\mathbf{X}_0 = \mathbf{X}_1\mathbf{X}_2$, $\mathbf{X}_1 \in \mathbb{R}^{n\times r_x}$, $\mathbf{X}_2 \in \mathbb{R}^{r_x\times T}$, and all entries of $\mathbf{X}_1, \mathbf{X}_2$ are i.i.d. $N(0, 1)$. To control the singular structure of $C\mathbf{X}$, the right singular vectors of C are taken as the eigenvectors of Σ_x. The left singular vectors of C are generated by first generating a matrix of standard normal random samples and then orthogonalizing the matrix by QR decomposition. The first r singular values of C are set to be nonzero values, and the difference between successive singular values is fixed to be two. The data matrix $\mathbf{Y}$ is then generated by $\mathbf{Y} = C\mathbf{X} + \boldsymbol{\epsilon}$, where the elements of $\boldsymbol{\epsilon}$ are i.i.d. samples from $N(0, \sigma^2)$. We define the signal to noise ratio (SNR) as $d_r(C\mathbf{X})/\|\boldsymbol{\epsilon}\|_F$ and set σ^2 accordingly.

Each simulated model is then characterized by the following parameters: T (sample size), n (number of predictors), m (number of responses), r_x (rank of the design matrix), r (rank of the coefficient matrix), ρ (design correlation), and SNR (signal to noise ratio). The experiment was replicated 500 times for each $\rho \in \{0.3, 0.5, 0.9\}$, corresponding to low, moderate, and high correlations among the predictors, and SNR $\in \{0.1, 0.2, 0.3\}$, corresponding to low, moderate, and high signal strength approximately. Three estimation methods are considered, namely the ordinary least squares method (OLS), the classical reduced-rank regression (RRR), and the adaptive nuclear-norm penalized regression (ANN) with $\gamma = 2$.

11.4.2 Comparing Estimators of the Degrees of Freedom

We compare the unbiased estimator of the degrees of freedom with the naive one given by the number of free parameters. Figures 11.1 and 11.2 show the results for the simulation models I and II, respectively, for $\rho = 0.5$ and SNR $= 0.3$. (The results are similar in other settings.) The dashed line with triangles in each figure shows the naive degrees of freedom. The solid line connects a sequence of boxplots covering a range of increasing rank values (for RRR) or decreasing penalty levels (for ANN), each of which shows the empirical distribution of the exact estimator of the degrees of freedom. For better illustration, very few outliers declared by the boxplots are not shown. For the ANN method, the averaged estimated ranks are also shown as a function of the penalty level (the dot-dashed line).

The unbiased estimator is indeed quite different from the naive estimator. For the RRR method, the curve for the exact estimator is always above the one for the naive estimator, and the increments represent the price paid for latent variable selection. When the rank is smaller than the true rank, the two estimates are quite close to each other. When the rank exceeds the true rank, their discrepancy is much larger, so is the variability of the unbiased estimator. This is related to the fact that the estimated singular values exceeding the true rank could be small and close to each

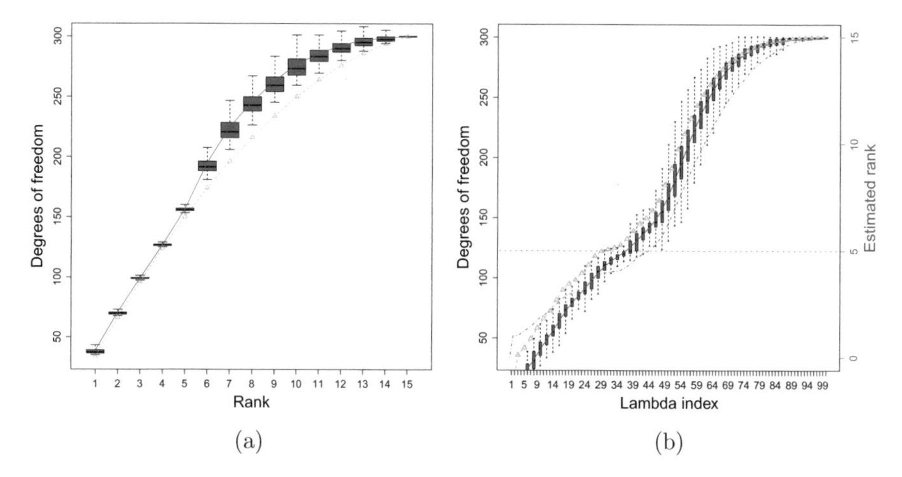

Fig. 11.1 Comparing naive and exact degrees of freedom estimators using Model I. (**a**) Estimated degrees of freedom in the reduced-rank regression method (RRR). (**b**) Estimated degrees of freedom in the adaptive nuclear-norm penalization method (ANN)

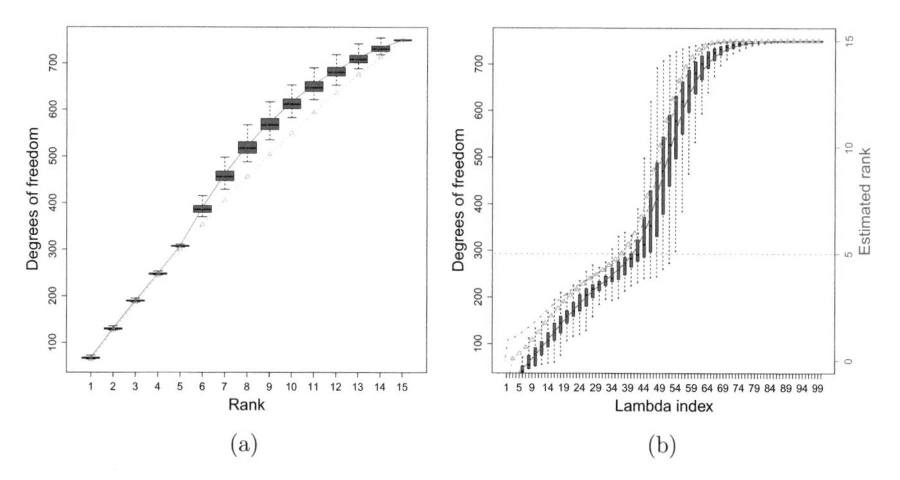

Fig. 11.2 Comparing naive and exact degrees of freedom estimators using Model II. (**a**) Estimated degrees of freedom in the reduced-rank regression method (RRR). (**b**) Estimated degrees of freedom in the adaptive nuclear-norm penalization method (ANN). The settings are the same as in Fig. 11.1

other. For the ANN method, the curve with the exact estimator is mostly lower than the one with the naive estimator, demonstrating that shrinkage estimation provides effective control over the model complexity. Again, as expected, the variability of the unbiased estimator becomes larger when the rank exceeds the true rank.

11.4.3 Performance on Estimating the Prediction Error

We examine the performance of the exact and naive estimators of the degrees of freedom on the estimation of true prediction error. Based on the SURE theory, we use the C_P criterion in (11.3) to estimate the prediction error. For each replication i ($i = 1, \dots, g$, $g = 500$), denote the simulated data as $\mathbf{Y}_{(i)}$, the estimated regression component by RRR as $\widehat{\mathbf{Y}}_{(i)}(r)$ and the estimated degrees of freedom as $\widehat{\mathrm{df}}_{(i)}(r)$, for $r = 1, \dots, \min(r_x, m)$. The average C_P prediction error is then computed as,

$$\widehat{R}(r) = \frac{1}{Tmg}\sum_{i=1}^{g}\left\{\|\mathbf{Y}_{(i)} - \widehat{\mathbf{Y}}_{(i)}(r)\|_F^2 + 2\sigma^2\widehat{\mathrm{df}}_{(i)}(r)\right\}.$$

In practice, the error variance σ^2 is usually unknown and needs to be estimated. However, since our main purpose is to examine the behavior of the estimators of the degrees of freedom, we simply use the true σ^2 value in the above expression.

To assess how well the C_P formula resembles the true prediction error, we also compute another estimate based on independently generated test data. Specifically, in each replication i, we first estimate the model by RRR based on the training data, and then evaluate its prediction performance on an independently generated test set $\mathbf{Y}_{[i]}$ of the same size. We compute

$$R(r) = \frac{1}{Tmg}\sum_{i=1}^{g}\left\{\|\mathbf{Y}_{[i]} - \widehat{\mathbf{Y}}_{(i)}(r)\|_F^2\right\},$$

and refer to it as the average true prediction error.

The average C_P prediction error curves are computed for both the RRR and ANN methods and for both the proposed and naive estimators of the degrees of freedom. Figures 11.3 and 11.4 show the simulation results for $\rho = 0.5$ and SNR $= 0.2$. (The results are similar in other settings.) In each panel, the solid line is the true prediction error curve, the dotted line is the C_P prediction error curve based on the unbiased exact estimator of the degrees of freedom, and the dashed line is the C_P prediction error curve based on the naive estimator. For the ANN method, the dot-dashed line shows the averaged estimated rank as a function of the penalty levels.

The average C_P curve based on the proposed estimator mostly overlaps with the true prediction error curve, demonstrating that the proposed estimator is indeed unbiased. On the other hand, there is an apparent discrepancy between the true prediction error curve and the one computed based on the naive estimator, especially when the estimated rank is close to or larger than the true rank. Moreover, the curve based on the naive degrees of freedom is minimized at a slightly different location than those of the other two curves. These results suggest that the proposed degrees of freedom estimator leads to an accurate assessment of the actual prediction risk, and using the naive estimator may lead to suboptimal performance on both model assessment and model selection.

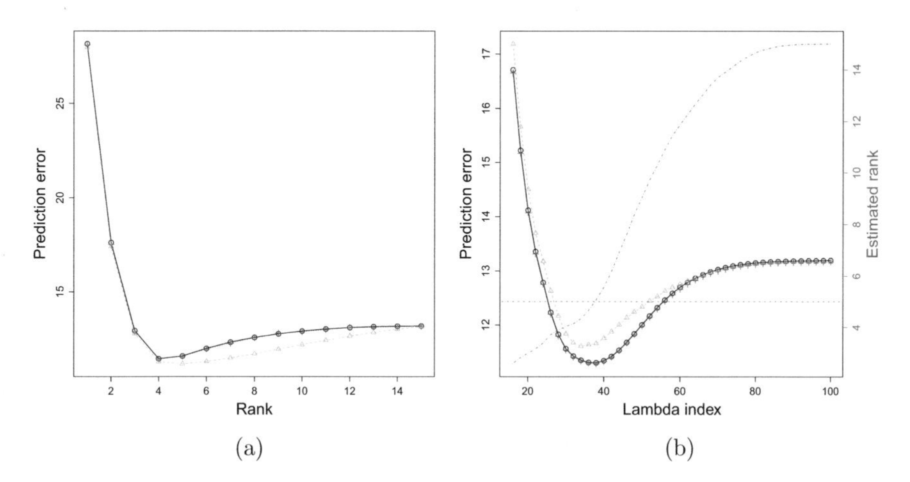

Fig. 11.3 Comparing prediction error estimates using Model I. (**a**) Estimated prediction error curves in the reduced-rank regression method (RRR). (**b**) Estimated prediction error curves in the adaptive nuclear-norm penalization method (ANN)

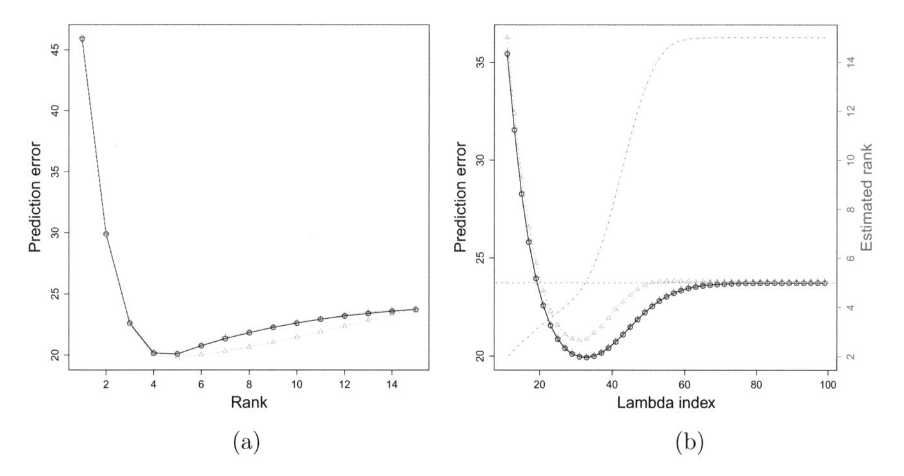

Fig. 11.4 Comparing prediction error estimates using Model II. (**a**) Estimated prediction error curves in the reduced-rank regression method (RRR). (**b**) Estimated prediction error curves in the adaptive nuclear-norm penalization method (ANN). All the settings are the same as in Fig. 11.3

11.4.4 Performance on Model Selection

With the estimated degrees of freedom, information criterion can be constructed to efficiently perform model selection. We can thus examine the performance of various information criteria based on either the unbiased or naive estimators of degrees of freedom. For an illustration, we choose the generalized cross-validation criterion (GCV) (Golub et al. 1979), based on the predictive performance of models.

The results of using AIC, BIC, and C_P are similar. For the RRR method, the GCV criterion based on $\widehat{df}(r)$ is defined as

$$\frac{Tm\|\mathbf{Y} - \widehat{\mathbf{Y}}(r)\|_F^2}{\{Tm - \widehat{\mathrm{df}}(r)\}^2}. \tag{11.11}$$

Note that GCV does not involve the unknown error variance so that its estimation is not required. The model accuracy is measured by the average of the scaled mean squared prediction errors (MSPE) from all replications, i.e., $100\|C\mathbf{X} - \widehat{C}\mathbf{X}\|_F^2/(Tm)$. To evaluate the rank determination performance, we report the averaged rank estimates from all replications (Rank). All the above quantities are similarly defined for the ANN method. The GCV criteria constructed from the unbiased and naive degrees of freedom are denoted as GCV(e) and GCV(n), respectively. The label "(e)" denotes the use of the exact degrees of freedom estimator in computing the model selection criteria while "(n)" denotes the use of the naive degrees of freedom estimator. The results for various SNR and design correlation levels are summarized in Tables 11.1 and 11.2, with the standard errors reported in the parenthesis.

Table 11.1 Model selection performance of GCV(e) and GCV(n) using Model I

				RRR		ANN	
ρ	SNR		OLS	GCV(e)	GCV(n)	GCV(e)	GCV(n)
0.3	0.1	MSPE	1362.5 (299.1)	723.76 (179.1)	739.35 (184.8)	582.64 (131.3)	605.19 (145.9)
		Rank	15 (0.0)	3.28 (0.7)	3.79 (0.6)	3.94 (0.7)	3.79 (0.9)
	0.2	MSPE	348.2 (76.9)	193.09 (48.2)	200.86 (50.8)	171.55 (41.3)	175.47 (42.2)
		Rank	15 (0.0)	4.37 (0.6)	4.82 (0.6)	4.89 (0.7)	4.56 (0.7)
	0.3	MSPE	153.42 (34)	83.62 (20.4)	85.47 (22.6)	77.38 (18.5)	78.24 (19.1)
		Rank	15 (0.0)	4.98 (0.3)	5.14 (0.4)	5.22 (0.5)	5.08 (0.4)
0.5	0.1	MSPE	1476.18 (338.5)	761.8 (196.9)	786.52 (217.6)	634.21 (149.6)	657.39 (172.2)
		Rank	15 (0.0)	3.47 (0.6)	3.94 (0.6)	4.07 (0.7)	3.78 (0.8)
	0.2	MSPE	376.4 (83.1)	206.88 (50.5)	215.47 (54.9)	186.04 (43.6)	189.95 (44.7)
		Rank	15 (0.0)	4.32 (0.5)	4.81 (0.6)	4.88 (0.7)	4.54 (0.7)
	0.3	MSPE	167.62 (36.5)	90.54 (21.9)	92.7 (23.1)	84.38 (19.7)	85.4 (20.7)
		Rank	15 (0.0)	4.98 (0.3)	5.14 (0.4)	5.17 (0.4)	5.06 (0.4)
0.9	0.1	MSPE	491.84 (112)	242.91 (60.3)	251.62 (66.7)	214.49 (50.5)	220.9 (54.6)
		Rank	15 (0.0)	3.67 (0.5)	4.1 (0.5)	4.19 (0.7)	3.82 (0.6)
	0.2	MSPE	124.55 (27.3)	67.69 (17.1)	70.6 (18.2)	61.63 (14.6)	63.09 (15.2)
		Rank	15 (0.0)	4.34 (0.5)	4.82 (0.6)	4.87 (0.7)	4.47 (0.6)
	0.3	MSPE	55.69 (12.1)	29.77 (7.3)	30.64 (8)	27.93 (6.7)	28.4 (7.2)
		Rank	15 (0.0)	4.96 (0.3)	5.15 (0.4)	5.22 (0.5)	5.06 (0.5)

Table 11.2 Model selection performance of GCV(e) and GCV(n) using Model II

				RRR		ANN	
ρ	SNR		OLS	GCV(e)	GCV(n)	GCV(e)	GCV(n)
0.3	0.1	MSPE	2631.71 (1320.4)	1015.06 (518.9)	1046.8 (552.2)	843.65 (392.5)	877.11 (423.6)
		Rank	15 (0.0)	3.51 (0.6)	3.83 (0.5)	4.24 (0.7)	3.58 (0.7)
	0.2	MSPE	687.88 (340.1)	305.14 (158.1)	303.31 (159.4)	261.08 (129.2)	272.84 (135.8)
		Rank	15 (0.0)	4.57 (0.5)	4.92 (0.3)	5.18 (0.5)	4.57 (0.5)
	0.3	MSPE	294.98 (142.9)	122.05 (60)	122.77 (60.3)	114.58 (55.1)	114.44 (55)
		Rank	15 (0.0)	5 (0.0)	5.02 (0.1)	5.15 (0.4)	5 (0.0)
0.5	0.1	MSPE	2646.61 (1345.1)	1026.89 (532.9)	1048.34 (559)	849.66 (400.1)	886.8 (436.8)
		Rank	15 (0.0)	3.49 (0.6)	3.8 (0.5)	4.23 (0.7)	3.55 (0.7)
	0.2	MSPE	645.17 (318.2)	284.59 (143.3)	283.43 (143.8)	243.59 (117.2)	253.44 (120.8)
		Rank	15 (0.0)	4.58 (0.5)	4.93 (0.3)	5.15 (0.4)	4.57 (0.5)
	0.3	MSPE	292.16 (140.5)	121.35 (59.6)	121.56 (59.5)	113.78 (55.1)	113.47 (55)
		Rank	15 (0.0)	5 (0.0)	5.01 (0.1)	5.18 (0.5)	5 (0.0)
0.9	0.1	MSPE	1843.64 (1106.5)	698.59 (417.6)	715.15 (434.5)	589.27 (325)	619.54 (359.3)
		Rank	15 (0.0)	3.5 (0.6)	3.8 (0.5)	4.25 (0.8)	3.51 (0.7)
	0.2	MSPE	449.34 (258.1)	198.09 (118.9)	195.7 (115.3)	169.6 (97.1)	177.1 (103.5)
		Rank	15 (0.0)	4.6 (0.5)	4.91 (0.3)	5.18 (0.5)	4.58 (0.5)
	0.3	MSPE	199.9 (107.6)	82.88 (45.4)	83.2 (45.5)	77.89 (41.6)	77.73 (41.5)

Overall, the GCV(e) outperforms the GCV(n) on model selection in both the RRR and ANN methods. In most of the cases, the models selected by GCV(e) have slightly better predictive performance than those by GCV(n), especially when the SNR level is low to moderate and the correlation between predictors is moderate to high. We note that as expected the ANN method performs slightly better than RRR due to its adaptive shrinkage estimation.

11.5 Applications

11.5.1 Norwegian Paper Quality Data

We consider a data set from a paper quality experiment conducted at a paper-making factory in Norway. The goal was to examine how the $m = 13$ response variables of different characteristics of paper were influenced by $n = 9$ predictor

Table 11.3 Model selection performance of GCV(e) and GCV(n) in the Norwegian paper quality data analysis

			RRR		ANN	
l%		OLS	GCV(e)	GCV(n)	GCV(e)	GCV(n)
50%	MSPE	2.3 (1.2)	1.76 (0.9)	1.81 (1.0)	1.51 (0.8)	1.60 (0.9)
	Rank	9 (0.0)	2.50 (1.4)	2.87 (1.0)	3.01 (1.2)	2.98 (1.4)
70%	MSPE	1.29 (0.6)	1.11 (0.5)	1.12 (0.6)	1.01 (0.4)	1.05 (0.5)
	Rank	9 (0.0)	2.67 (1.1)	2.97 (0.7)	3.22 (0.9)	3.07 (1.1)
90%	MSPE	0.89 (0.3)	0.78 (0.2)	0.78 (0.3)	0.75 (0.2)	0.76 (0.2)
	Rank	9 (0.0)	2.97 (0.2)	3.01 (0.1)	3.05 (0.3)	3.07 (0.4)

variables constructed from three experimental factors and their second-order terms for capturing interaction and nonlinear effects. The sample size is $T = 29$. The data were used in Aldrin (2000), Izenman (2008) and Bunea et al. (2012) to illustrate the efficacy of various reduced-rank approaches. A detailed description of the data set can be found in either Aldrin (2000) or Izenman (2008).

We randomly select a proportion, say, $l\%$ of the data as the training set to fit the regression model and obtain $\widehat{C}$, and then use the rest of the data as the testing set to evaluate the prediction error defined as $\|\mathbf{Y}_{te} - \widehat{C}\mathbf{X}_{te}\|_F^2/(Tm)$. The rank in the RRR method and the tuning parameter in the ANN method are selected by GCV based on either the unbiased or the naive estimator of the degrees of freedom. This process is repeated 100 times for each $l\% \in \{50\%, 70\%, 90\%\}$. Table 11.3 reports the averaged prediction error and the averaged rank estimates of all runs, along with their standard errors. It can be seen that the GCV based on the unbiased estimator of the degrees of freedom performs slightly better in terms of model selection than that based on the naive estimator. For both RRR and ANN, the rank is often estimated to be three, which agrees with previous studies (Izenman 2008; Bunea et al. 2012). The effectiveness and applicability of the unbiased exact estimator of the degrees of freedom are well demonstrated by this data set.

11.5.2 Arabidopsis Thaliana Data

In this application, we use several different information criteria, including Mallow's C_P, GCV, and BIC, each with either the unbiased or the naive estimator of the degrees of freedom, for selecting a reduced-rank model in a genetic association study (Wille et al. 2004).

This is a microarray experiment aimed at understanding the regulatory control mechanisms between the isoprenoid gene network in the Arabidopsis thaliana plant, more commonly known as thale cress or mouse-ear cress. Isoprenoids serve many important biochemical functions in plants. To monitor the gene-expression levels, 118 GeneChip microarray experiments were carried out. The predictors consist of 39 genes from two isoprenoid bio-synthesis pathways, namely MVA and MEP, and the responses consist of gene-expression of 795 genes from 56 metabolic pathways,

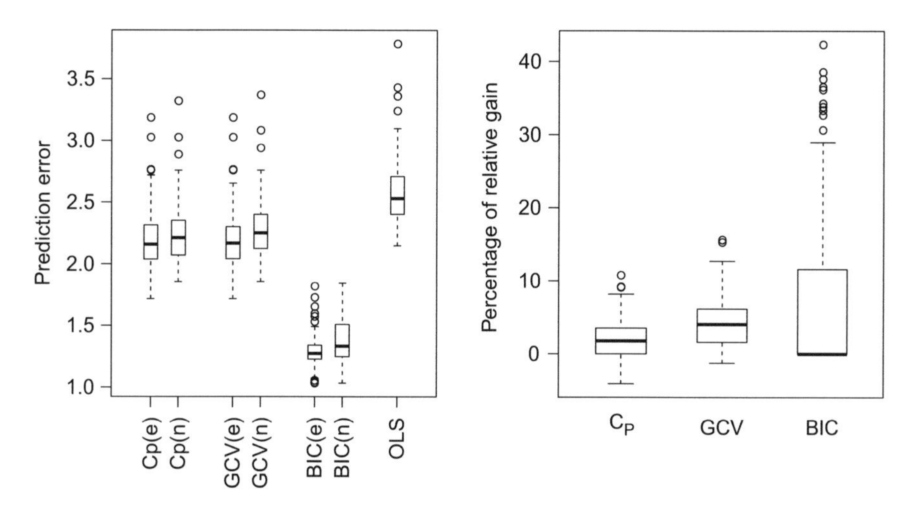

Fig. 11.5 Prediction performance in Arabidopsis Thaliana data (Mukherjee et al. 2015)

many of which are downstream of the two pathways considered as predictors. Thus some of the responses are expected to show significant associations to the predictor genes. We select two downstream pathways, Carotenoid and Phytosterol, as our responses. It has already been proven experimentally that the Carotenoid pathway is strongly attached to the MEP pathway, whereas the Phytosterol pathway is significantly related to the MVA pathway. See Wille et al. (2004) and the references therein for a more detailed discussion. We have $T = 118$ observations on $n = 39$ predictors and $m = 36$ responses, all log-transformed to reduce the skewness. The responses are standardized in order to make them comparable.

For the purpose of illustration, we split the data set randomly into training and test sets of equal size. The model is built using the training samples and the test set is used to evaluate the predictive performance. The performance measure is the mean squared prediction error (MSPE) on the testing data $\|\mathbf{Y}_{te} - \widehat{\mathbf{Y}}_{te}\|_F^2/(mT)$. The entire process is repeated 100 times based on random splits.

The mean squared prediction errors for all methods are summarized in Fig. 11.5. The left panel displays the boxplot of mean square prediction errors of each method over 100 random splits for three model selection criteria. Again, the label "(e)" denotes the use of the exact degrees of freedom estimator in computing the model selection criteria while "(n)" denotes the use of the naive degrees of freedom estimator. The right panel displays the boxplot of relative gain in prediction error for each model selection method by using the exact degrees of freedom estimator over the naive degrees of freedom estimator. For all three model selection criteria considered, using the exact unbiased estimator tends to outperform using the naive estimator in terms of prediction accuracy. The relative gain is almost always positive, shown by the right panel of Fig. 11.5. Also, among the three model selection criteria, the BIC appears to outperform the other criteria and tends to select a very parsimonious model that may be easier to interpret; see Table 11.4.

Table 11.4 Prediction accuracy and rank selection performance for the competing methods on the Arabidopsis thaliana data

	$C_P(e)$	$C_P(n)$	GCV(e)	GCV(n)	BIC(e)	BIC(n)	OLS
Avg(MSPE)	2.20	2.24	2.19	2.28	1.30	1.39	2.59
Std(MSPE)	0.3	0.3	0.3	0.3	0.1	0.2	0.3
Mean(Rank)	8.76	9.71	8.68	10.50	1.09	1.48	–
Std(Rank)	1.2	0.8	1.3	1.0	0.4	0.8	–

Chapter 12
Generalized Reduced-Rank Regression

12.1 Introduction

Our discussion on reduced-rank models thus far is mainly focused on classical multivariate linear regression setup. This chapter further generalizes reduced-rank models to deal with more complicated scenarios that are commonly encountered in real-world applications. In particular, we want to discuss several related aspects that will enhance the use of these models. These include, robust estimation and outlier detection in reduced-rank models with possibly contaminated data, low-rank matrix completion with incomplete data, and generalized reduced-rank models for a mix of continuous and discrete responses. All these topics are discussed under the general framework of high-dimensional multivariate regression.

Outliers and anomalies are bound to occur in many applications, especially with large-scale data. For example, in finance, stock returns of specific sectors or companies sometimes experience dramatic short-term disturbances that may or may not be easily explained or captured by the underlying market factors. These are often short-lived and isolated fluctuations and need to be adequately accounted for in studying the overall market movement that is estimated by reduced-rank models (see Chapter 8). In some applications, besides conducting robust estimation, the interest could also lie in the detection of anomalies hidden behind some low-dimensional "common" signal. For example, in motion detection via video surveillance, a reduced-rank model component is designed to extract the common background of the images/frames, and whatever remains is taken to capture the motion of object. Motivated by these applications, we address a fundamental problem: how and to what extent can outliers affect reduced-rank estimation (She and Chen 2017)? We show that indeed, reduced-rank estimation is non-robust and can breakdown with just one extreme (abnormal) observation. Several questions follow: How to identify potential outliers? What tools can be used for model diagnostics in reduced-rank estimation? How to robustify reduced-rank estimation?

G. C. Reinsel et al., *Multivariate Reduced-Rank Regression*, Lecture Notes in Statistics 225, https://doi.org/10.1007/978-1-0716-2793-8_12

Multivariate data are also generally prone to missing values. There is often a need to either impute the missing values or conduct modeling with incomplete data. A famous example is the Netflix problem. Movie ratings collected from Netflix customers are organized as a matrix, in which each entry represents the rating of a movie by a customer; the value is observed if the customer has watched the movie and otherwise the entry is missing. Because each customer watches only a limited number of movies, this matrix is largely incomplete. The problem is to build a recommender system that predicts the missing entries, that is, to recommend movies based on predicting the movie ratings a customer would give, had the customer watched the movies. Another example is from the longitudinal studies of aging (Stanziano et al. 2010) discussed in Section 10.6.2. Recall that a national representative sample of several thousands of subjects who were seventy years of age or over were interviewed and followed. Their health, functional status, living arrangements, and health services utilization were measured as they aged. It is of interest to examine the changes and the association between their past and current health status. However, there were more than 20% missing values in the outcomes across a range of assessments on health conditions. In these applications, the fundamental problem is termed as "*matrix completion*" (Candès and Recht 2009; Recht 2011). The question then is, how can we recover the missing entries of a partially observed and possibly contaminated low-rank matrix? We present the main theoretical results and computational algorithms developed for matrix completion and make connections to reduced-rank regression modeling with incomplete response.

In many real-world problems, the collected outcomes are often of mixed types including continuous measurements, binary indicators, and counts, rendering the classical multivariate linear regression framework not directly applicable. In the aforementioned study on aging, for example, continuous measurements of health, memory and sensation scores, dichotomous measurements of various medical conditions are to be predicted by subject demographics, their medical history, lifestyle, and social behavior. The mixed outcomes, regardless of their types, are often likely to be interrelated, reflecting diverse yet related aspects of the underlying multivariate data generating mechanism. As such, reduced-rank modeling is expected to be beneficial. We introduce methods that utilize mixed outcomes in reduced-rank estimation and a simple yet effective framework that simultaneously accounts for outliers and missing values (Luo et al. 2018).

12.2 Robust Reduced-Rank Regression

12.2.1 Non-Robustness of Reduced-Rank Regression

We have seen that reduced-rank regression can be very effective in modeling multivariate data. It substantially reduces the number of free parameters. However, it is yet unclear how sensitive reduced-rank estimation is to the presence of outliers, which is likely to occur commonly in practice. How robust is reduced-rank regression? In other words, could the low-rank structure be masked or be distorted by outliers? Here we attempt to provide some answers from both empirical and theoretical perspectives. The contents in this chapter are mainly drawn from She and Chen (2017).

We use a real data example in finance to show the non-robustness of reduced-rank regression as well as the potential power of robustification. Consider the weekly stock log-return (the difference in log price) data for nine of the ten largest (ranked by Fortune magazine) American corporations in 2004 (Yuan et al. 2007), with $Y_k \in \mathbb{R}^9$ $(k = 1, \ldots, T)$ and $T = 52$. Chevron was excluded due to a big drop in its stock price in the 38th week of the year. The nine time series are shown in Fig. 12.1. For the purpose of constructing market factors that drive general stock movements, a reduced-rank vector autoregressive (VAR) model of order 1 can be used ($Y_k = CY_{k-1} + \epsilon_k$, with C of reduced rank). By conditioning on the initial state and assuming the normality of ϵ_k, the conditional likelihood leads to a least squares criterion, so the estimation of C can be formulated as a reduced-rank regression problem. However, as shown in Fig. 12.1, several stock returns have experienced abrupt changes. The autoregressive structure makes any outlier in the time series also a leverage point in the covariates, which in this case are the lagged values.

Using the weekly log-returns in the first 26 weeks for training and those in the last 26 weeks for validation, we can analyze the data with both the classical reduced-rank regression and a robust reduced-rank regression approach (to be introduced later). While both methods result in unit-rank models, the robust reduced-rank regression automatically detects three outliers. These are the log-returns of Ford at weeks 5 and 17 and the log-return of General Motors at week 5. These two weeks correspond to major market disturbances attributed to the auto industry. By taking the outlying samples into account, the robust approach has led to a more reliable model. Table 12.1 displays the factor coefficients indicating how the stock returns are related to the estimated factors, and the p-values for testing the associations between the estimated factors and the individual stock return series using the data in the last 26 weeks. The stock factor estimated robustly has positive influence over all nine companies, and overall, it correlates with the series much better as judged from the reported correlation coefficients and p-values. The out-of-sample mean squared errors for the least squares, reduced-rank regression, and robust reduced-rank regression are 9.97, 8.85, and 6.72, respectively, and the 40% trimmed mean square errors are 5.44, 4.52, and 3.58, respectively. As such, the robustification of reduced-rank estimation has resulted in about 20% improvement in prediction.

Theoretically, the non-robustness of reduced-rank regression can be conceptualized using the notion of *breakdown point* from the robust statistics literature. Denote the reduced-rank regression estimator of the coefficient matrix as $\widehat{C}(r; \mathbf{X}, \mathbf{Y})$,

$$\widehat{C}(r; \mathbf{X}, \mathbf{Y}) \in \arg\min_{C} \|\mathbf{Y} - C\mathbf{X}\|_F^2 \quad \text{s.t. } r(C) \leq r, \tag{12.1}$$

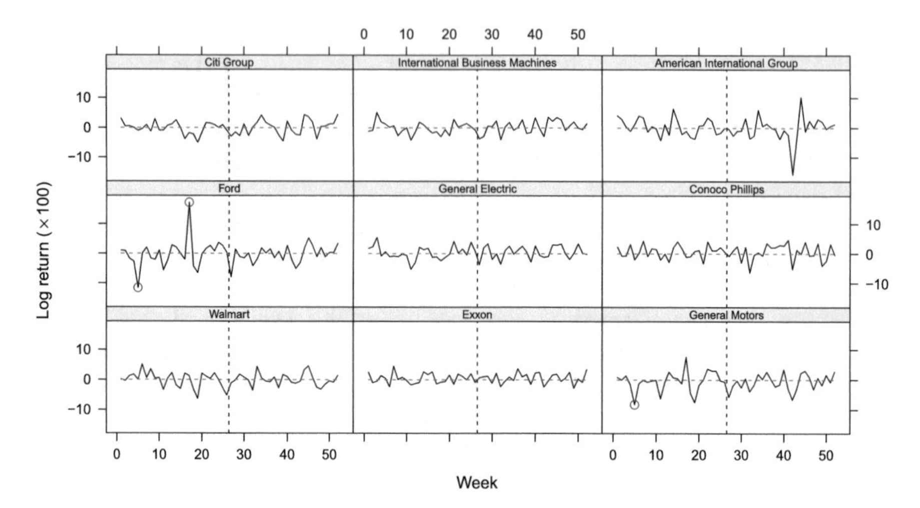

Fig. 12.1 Stock returns example: scaled weekly log-returns of stocks in 2004. The dashed line in each panel separates the series to two parts, i.e., the first 26 weeks for training and the last 26 weeks for testing. The log-returns of Ford at weeks 5 and 17 and the log-return of General Motors at week 5 are captured as outliers by fitting robust reduced-rank regression, which are indicated by the circles (She and Chen 2017)

Table 12.1 Stock returns example: factor coefficients showing how the stock returns load on the estimated factors. Reported p-values are for testing the associations between the estimated factors and the stock returns using the data in the last 26 weeks

	Reduced-rank regression		Robust reduced-rank regression	
	Coefficient	p-value	Coefficient	p-value
Walmart	0.46	0.44	0.36	0.23
Exxon	−0.15	0.32	0.14	0.84
General Motors	0.96	0.42	0.90	0.02
Ford	1.20	0.64	0.59	0.18
General Electric	0.24	0.67	0.32	0.06
Conoco Phillips	−0.04	0.19	0.36	0.08
Citi Group	0.27	0.93	0.45	0.00
IBM	0.36	0.42	0.57	0.13
American International Group	0.19	0.01	0.58	0.00

to emphasize its dependence on the specified rank and the data. We define the finite-sample breakdown point for reduced-rank regression, denoted as $\epsilon^*(\widehat{C})$, in the spirit of Donoho and Huber (1983) as follows:

$$\epsilon^*(\widehat{C}) = \frac{1}{T} \min \left\{ k \in \mathbb{N} \cup \{0\} : \sup_{\widetilde{\mathbf{Y}} \in \mathbb{R}^{m \times T} : \|\widetilde{\mathbf{Y}} - \mathbf{Y}\|_0 \leq k} \|\widehat{C}(r; \mathbf{X}, \widetilde{\mathbf{Y}})\mathbf{X}\|_F = +\infty \right\},$$

where $\mathbb{N}$ is the set of natural numbers, and $\|\cdot\|_0$ is the ℓ_0 norm of the enclosed matrix, i.e., $\|A\|_0 = \|\text{vec}(A)\|_0 = |\{(i, j) : a_{ij} \neq 0\}|$ with $|\cdot|$ denoting the cardinality of the enclosed set. To describe this quantity intuitively, $\epsilon^*(\widehat{C})$ is the minimum proportion of arbitrary outliers needed among the observations such that the model prediction can be completely distorted (with infinite bias). For example, the sample mean has a breakdown point of $1/T$, because as long as one of the observations goes to infinity, the sample mean goes to infinity as well. On the other hand, the sample median, as a robust estimator for the population center, has a breakdown point of 1/2.

Result 12.1 Given any $\mathbf{X} \in \mathbb{R}^{n \times T}$, $\mathbf{Y} \in \mathbb{R}^{m \times T}$ with $\mathbf{X} \neq 0$ and $r \geq 1$, let $\widehat{C}(r; \mathbf{X}, \mathbf{Y})$ be a reduced-rank regression estimator which results from (12.1). Then its finite-sample breakdown point is exactly $1/T$.

This breakdown point analysis indicates that a single outlier is capable of severely distorting reduced-rank estimation. The result still holds for a broad class of the singular value penalized estimators (She and Chen 2017). In other words, the popular means of reduced-rank regularization, although quite effective for dimension reduction, is extremely sensitive to outliers. This conclusion suggests that the conventional rank reduction methods are to be used with caution in real data applications.

Although the above result confirms the non-robustness of reduced-rank estimation, it does not provide any insight into the relationship between the magnitude of the outliers and the amount of distortion in the parameter estimation or prediction. Indeed, classical robust analysis falls in either deterministic worst-case studies or large-T asymptotics with the model dimensions held fixed, which may not meet the needs of modern large-scale reduced-rank modeling. Therefore, robust reduced-rank estimation and a non-asymptotic robust analysis under a general high-dimensional regression framework are very much needed.

12.2.2 Robustification with Sparse Mean Shift

As we have observed in the previous section, the robustification of reduced-rank matrix estimation is non-trivial. A straightforward idea might be to use a robust loss function ℓ in place of the squared error loss in (12.1), leading to

$$\min_{C} \sum_{k=1}^{T} \ell(\|(Y_k - CX_k)\|_2) \qquad \text{s.t. } r(C) \leq r. \tag{12.2}$$

However, such an estimator may result in non-trivial computational and theoretical challenges. Even when ℓ is the well-known Huber's loss (Huber 1981), there may not be a straightforward algorithm for solving (12.2), though non-convex robust loss functions are known to be preferable in dealing with gross outliers with possible high leverage values.

To begin with, we consider a sparse mean-shift regression framework to robustify reduced-rank regression. The framework was used for linear regression (She and Owen 2011; Lee et al. 2012) and was later utilized for high-dimensional reduced-rank estimation in She and Chen (2017). The multivariate mean-shift regression model can be stated as follows,

$$\mathbf{Y} = C\mathbf{X} + S + \boldsymbol{\epsilon}, \tag{12.3}$$

where $C \in \mathbb{R}^{m \times n}$ is the coefficient matrix, $S \in \mathbb{R}^{m \times T}$ describes the outlier effects on $\mathbf{Y}$, and $\boldsymbol{\epsilon} \in \mathbb{R}^{m \times T}$ is the random error matrix with i.i.d. columns following $N(0, \Sigma)$. Obviously, this leads to an over-parameterized model, so we must make some structural assumptions on the unknown matrices (C, S) appropriately. Hence we assume that C is of reduced rank and S is sparse with only a few nonzero entries or nonzero columns of entries because outliers are expected to be rare, not consistent with most of the data.

The robust reduced-rank regression problem is then formulated as

$$\min_{C,S} \left\{ \frac{1}{2} \|\mathbf{Y} - C\mathbf{X} - S\|_F^2 + \rho(S; \lambda) \right\} \quad \text{s.t. } r(C) \leq r. \tag{12.4}$$

Here, $\rho(\cdot; \lambda)$ is a sparsity-inducing penalty function with λ controlling the amount of penalization. The following form of ρ can handle element-wise outliers

$$\rho(S; \lambda) = \sum_{j=1}^{m} \sum_{k=1}^{T} \rho(|s_{jk}|; \lambda). \tag{12.5}$$

This was used to illustrate the stock return analysis in Section 12.2.1. It is more common in studies on robust statistics to assume outlying samples, or outlying columns in $(\mathbf{Y}, \mathbf{X})$, which corresponds to imposing the penalty function

$$\rho(S; \lambda) = \sum_{k=1}^{T} \rho(\|s_k\|_2; \lambda), \tag{12.6}$$

where s_k is the kth column vector of S. Unless otherwise specified, we mainly focus on dealing with column-wise outliers. This covers distortions that occur at specific samples or time points. The analyses to be presented here can handle element-wise outliers as well after some simple modification.

Interestingly, there is a universal connection between the mean-shift robustification and the conventional M-estimation with robust loss functions. As such, the sparse mean-shift estimation approach indeed comes with a guarantee of robustness and generalizes M-estimation to the multivariate reduced-rank setting.

To proceed, the following notations are needed. For $A = [a_1, \ldots, a_T] \in \mathbb{R}^{m \times T}$, let $\|A\|_{2,1} = \sum_{k=1}^{T} \|a_k\|_2$, which gives the sum of the column-wise ℓ_2 norms of A, and let $\|A\|_{2,0} = \sum_{k=1}^{T} I(\|a_k\|_2 \neq 0)$, which gives the number of nonzero columns of A. Given an index set $\mathcal{S} \subset \{1, \ldots, T\}$, we denote $\sum_{k \in \mathcal{S}} \|a_k\|_2$ by $\|A_{\mathcal{S}}\|_{2,1}$.

It is well known that sparsity-inducing penalization is always associated with certain thresholding operation. For example, the convex ℓ_1 penalty yields an estimator that is obtained from a soft-thresholding rule. In fact, this coupling framework is more general. Let us first formally define the threshold functions. A thresholding function is a real-valued (odd, monotonic, and bounded) function $\Theta(t; \lambda)$ defined for $-\infty < t < \infty$ and $0 \leq \lambda < \infty$ such that

$$
\begin{aligned}
&(i)\, \Theta(-t; \lambda) = -\Theta(t; \lambda); \\
&(ii)\, \Theta(t; \lambda) \leq \Theta(t'; \lambda) \text{ for } t \leq t'; \\
&(iii)\, \lim_{t \to \infty} \Theta(t; \lambda) = \infty; \\
&(iv)\, 0 \leq \Theta(t; \lambda) \leq t \text{ for } 0 \leq t < \infty.
\end{aligned}
\tag{12.7}
$$

Given any Θ, $\widetilde{\Theta}$ is defined for any matrix $A = [a_1, \ldots, a_T] \in \mathbb{R}^{m \times T}$ in a column-wise fashion as $\widetilde{\Theta}(A; \lambda) = [\widetilde{\Theta}(a_1; \lambda), \ldots, \widetilde{\Theta}(a_T; \lambda)]$, where

$$
\widetilde{\Theta}(a; \lambda) = \begin{cases} \{\Theta(\|a\|_2; \lambda)/\|a\|_2\}a, & a \neq 0; \\ 0, & a = 0. \end{cases} \tag{12.8}
$$

A penalty ρ can be constructed from an arbitrary thresholding function $\Theta(\cdot; \lambda)$ by

$$
\rho(t; \lambda) - \rho(0; \lambda) = \rho_{\Theta}(t; \lambda) + \delta(t; \lambda), \tag{12.9}
$$

where $\rho_{\Theta}(t; \lambda) = \int_0^{|t|} [\sup\{s : \Theta(s; \lambda) \leq u\} - u]\, \mathrm{d}u$, and $\delta(\cdot; \lambda)$ is a non-negative function satisfying $\delta\{\Theta(s; \lambda); \lambda\} = 0$, for all $s \in \mathbb{R}$ (She 2009). Many popularly used convex and non-convex penalties are all covered by (12.9). For example, the convex group-ℓ_1 penalty $\lambda \sum_{k=1}^{T} \|a_k\|_2$ is associated with the soft-thresholding $\Theta_S(t; \lambda) = \mathrm{sgn}(t)(|t| - \lambda)_+$. The group-$\ell_0$ penalty $(\lambda^2/2) \sum_{k=1}^{T} I(\|a_k\|_2 \neq 0)$ can be obtained from (12.9) with the hard-thresholding $\Theta_H(t; \lambda) = tI(|t| > \lambda)$, and $\delta(t; \lambda) = (\lambda - |t|)^2 I(0 < |t| < \lambda)/2$. This Θ-ρ coupling framework also covers the bridge penalty ℓ_p for $0 < p < 1$, the smoothly clipped absolute deviation (SCAD) penalty (Fan and Li 2001), the minimax concave penalty (Zhang 2010), and the capped ℓ_1 (Zhang 2010) as particular cases.

Theorem 12.1

(a) *Suppose $\Theta(\cdot; \lambda)$ is an arbitrary thresholding rule satisfying* (12.7), *and let ρ be any penalty associated with Θ through* (12.9). *Consider the group penalized form of the robust reduced-rank regression criterion*

$$\min_{C,S}\left\{\frac{1}{2}\|\mathbf{Y}-C\mathbf{X}-S\|_F^2+\sum_{k=1}^{T}\rho(\|s_k\|_2;\lambda)\right\}\quad s.t.\ r(C)\leq r. \tag{12.10}$$

For any fixed C, a globally optimal solution for S is $S(C)=\widetilde{\Theta}(\mathbf{Y}-C\mathbf{X};\lambda)$. By profiling out S with $S(C)$, (12.10) *can be expressed as an optimization problem with respect to C only, and it is equivalent to the robust M-estimation problem in* (12.2)*, where the robust loss function is given by*

$$\ell(t;\lambda)=\int_0^{|t|}\psi(u;\lambda)\,du,\qquad \psi(t;\lambda)=t-\Theta(t;\lambda).$$

(b) Given $\varrho\in\{0,1,\ldots,T\}$, consider the group-$\ell_0$ constrained form of the robust reduced-rank regression criterion

$$\min_{C,S}\|\mathbf{Y}-C\mathbf{X}-S\|_F^2\ \ s.t.\ r(C)\leq r,\ \|S\|_{2,0}\leq\varrho. \tag{12.11}$$

Similarly, after profiling out S, (12.11) *can be expressed as an optimization problem with respect to C only and is equivalent to the rank-constrained trimmed least squares problem*

$$\min_{C}\sum_{k=1}^{T-\varrho}r_{(k)}\quad s.t.\ \ r(C)\leq r, \tag{12.12}$$

where $r_{(1)},\ldots,r_{(T)}$ are the order statistics of the ℓ_2 norms of the residual vectors $r_k=\|Y_k-CX_k\|_2$, $k=1,\ldots,T$, satisfying $r_{(1)}\leq\cdots\leq r_{(T)}$.

Theorem 12.1 connects sparse mean-shift estimation with a penalty ρ to robust M-estimation with a loss function ℓ. The universal link between (12.2) and (12.10) provides insight into the choice of the penalty form. It is easy to verify that the ℓ_1-norm penalty leads to Huber's loss, which is known to be prone to masking and swamping and may fail in the presence of even moderately leveraged outliers. To handle gross outliers, redescending ψ function is often advocated. This amounts to using non-convex penalties in (12.10). For example, Hampel's three-part ψ (Hampel et al. 2005) corresponds to the SCAD penalty (Fan and Li 2001), the skipped mean ψ corresponds to the exact ℓ_0 penalty, and the rank-constrained least trimmed squares can be rephrased as the ℓ_0-constrained form as in (12.11). The connection is also valid in the case of element-wise outliers, with ρ and ℓ applied in an element-wise manner. Thus the framework provided here covers other widely discussed methods in the literature.

An important feature of the mean-shift robustification is that it leaves the original loss function intact. This approach not only provides a unified way to robustify the estimation of low-rank matrix but also facilitates theoretical analysis and computation of reduced-rank M-estimators in high dimensions.

Algorithm 1 Robust reduced-rank regression algorithm

Input $\mathbf{X}$, $\mathbf{Y}$, $C^{(0)}$, $S^{(0)}$; $t = 0$.
Repeat
(a) $S^{(t+1)} \leftarrow \widetilde{\Theta}(\mathbf{Y} - C^{(t)}\mathbf{X}; \lambda)$, according to (12.8);
(b) $\mathbf{C}^{(t+1)} \leftarrow \widehat{C}(r; \mathbf{X}, \mathbf{Y} - S^{(t+1)})$, according to (12.1);
$t \leftarrow t + 1$;
Until convergence, e.g., $\|C^{(t+1)} - C^{(t)}\|_F / \|C^{(t)}\|_F + \|S^{(t+1)} - S^{(t)}\|_F / \|S^{(t)}\|_F < \epsilon$.

Now it is in order to consider the computational aspect of the problem. Compared with the M-estimation approach, the mean-shift formulation (12.3) simplifies computation and parameter tuning. Consider the penalized form of the robust reduced-rank regression problem in (12.10), where ρ is constructed by (12.9). A simple iterative algorithm is described in Algorithm 1, where the two matrices C and S are alternatingly updated while the other held fixed until convergence.

Step (a) performs simple multivariate thresholding operations and Step (b) conducts reduced-rank regression on the adjusted response matrix $\mathbf{Y} - S^{(t+1)}$. It is important to observe that there is no need to explicitly compute $C^{(t)}$ for updating S in the iterative process. In fact, only $C^{(t)}\mathbf{X}$ is needed, which depends on $\mathbf{X}$ through $\mathcal{P}_X$, or I when $n \gg T$. The eigenvalue decomposition used in (12.1) has low computational complexity because the rank values of practical interest are often small. Algorithm 1 is simple to implement and is cost-effective. The model tuning can be based on cross validation or some information criterion as discussed in Chapter 10. More details on computation can be found in She and Chen (2017).

It is worth noting that the robust reduced-rank regression subsumes a special but important case, $\mathbf{Y} = C + S + \boldsymbol{\epsilon}$, which aims to decompose an observed matrix to be the sum of a low-rank matrix C, a sparse matrix S, and a stochastic error matrix $\boldsymbol{\epsilon}$. This problem is perhaps less challenging than its supervised counterpart, but has wide applications in computer vision and machine learning (Wright et al. 2009; Candès et al. 2011). Furthermore, the approach can be extended to reduced-rank vector generalized linear models; see, e.g., Yee and Hastie (2003), She (2013), and Luo et al. (2018). In these scenarios, directly robustifying the loss function becomes even more complicated, while a sparse mean-shift term can always be introduced without altering the form of the given loss function, so that many algorithms designed for fitting generalized linear models can still be seamlessly utilized.

12.2.3 Theoretical Analysis

Theorem 12.1 provides some helpful intuition for the robust reduced-rank regression, but it might be inadequate from a theoretical perspective. For example, how can one justify the need for robustification in estimating a matrix of low rank? Is using redescending ψ functions still preferable in rank-deficient settings? We cannot assume, as in traditional robust analysis, an infinite sample size and a fixed number of predictors or response variables, because n and/or m can be much larger than T in modern applications. The key is then to conduct a non-asymptotic analysis to "quantify" the robustness.

Assume that the mean-shift model is given by $\mathbf{Y} = C\mathbf{X} + S + \boldsymbol{\epsilon}$, where $\boldsymbol{\epsilon}$ has i.i.d. $N(0, \sigma^2)$ entries, and consider the robust reduced-rank regression problem in (12.4). We use $r = r(C)$ to denote the rank of the true coefficient matrix, $r_x = r(\mathbf{X})$ to denote the rank of $\mathbf{X}$, and $\|S\|_{2,0}$ to denote the number of nonzero columns in S, i.e., the number of outliers.

Denote $\Delta(C, S) = \|C\mathbf{X} + S\|_F^2$. Given an estimator $(\widehat{C}, \widehat{S})$, we focus on its prediction accuracy measured by

$$\Delta(\widehat{C} - C, \widehat{S} - S) = \|(\widehat{C} - C)\mathbf{X} + (\widehat{S} - S)\|_F^2. \tag{12.13}$$

This predictive learning perspective is useful for evaluating the performance of an estimator, and it requires no signal strength or parameter uniqueness assumptions. The ℓ_2 prediction bound of $\Delta(\widehat{C} - C, \widehat{S} - S)$ is fundamental, in that such a bound, together with additional compatibility conditions on the design, can lead to estimation error bounds for C and S in different norms as well as selection consistency properties (Ye and Zhang 2010; Lounici et al. 2011).

We present finite-sample oracle inequalities that apply to the problem of robust reduced-rank regression in arbitrary dimensions. The following theorem concerns a general penalty form $\rho(S; \lambda) = \sum_{k=1}^{T} \rho(\|s_k\|_2; \lambda)$. We assume that $\rho(\cdot; \lambda)$, with λ as the threshold parameter, satisfies

$$\rho(0; \lambda) = 0, \quad \rho(t; \lambda) \geq \rho_H(t; \lambda), \tag{12.14}$$

where $\rho_H(t; \lambda) = (-t^2/2 + \lambda|t|)1_{|t|<\lambda} + (\lambda^2/2)1_{|t|\geq\lambda}$. The latter inequality is natural in view of (12.9), because a shrinkage estimator with λ as the threshold is always bounded above by the hard-thresholding function $\Theta_H(\cdot, \lambda)$.

Theorem 12.2 *Let* $\lambda = a\sigma(r_x + \log T)^{1/2}$ *with* $a > 0$ *being a constant and let* $(\widehat{C}, \widehat{S})$ *be a global minimizer of* (12.4). *Then, for sufficiently large a, the following inequality holds for any* $\widetilde{C} \in \mathbb{R}^{m\times n}$, $\widetilde{S} \in \mathbb{R}^{m\times T}$ *satisfying* $r(\widetilde{C}) \leq \widetilde{r}$:

$$E\{\Delta(\widehat{C} - C, \widehat{S} - S)\} \preceq \Delta(\widetilde{C} - C, \widetilde{S} - S) + \sigma^2(r_x + m)\widetilde{r} + \rho(\widetilde{S}; \lambda) + \sigma^2, \tag{12.15}$$

where $\preceq$ *means the inequality holds up to a multiplicative constant.*

The bound in (12.15) involves a bias term $\Delta(\widetilde{C} - C, \widetilde{S} - S)$, which ensures applicability of robust reduced-rank regression to approximately sparse S. Similarly, $\widetilde{r}$ may also deviate from the true rank r to some extent, as a benefit from the bias-variance tradeoff. The second term $\sigma^2(r_x + m)\widetilde{r}$ is from the estimation of the reduced-rank component, and the rest is due to estimation of the sparse component. In particular, by setting $\widetilde{r} = r$, $\widetilde{C} = C$, and $\widetilde{S} = S$ in (12.15) and assuming ρ is a bounded non-convex penalty satisfying $\rho(t; \lambda) \preceq \lambda^2$ for any $t \in \mathbb{R}$, we obtain a prediction error bound

$$E\{\Delta(\widehat{C} - C, \widehat{S} - S)\} \preceq \sigma^2(r_x + m)r + \sigma^2(r_x + \log T)\|S\|_{2,0}. \tag{12.16}$$

Observe that the bias term vanishes. The first term coincides with the prediction error rate for reduced-rank regression as obtained in Theorem 10.3, while the second term quantifies the additional error from accommodating the outliers.

We remark that Theorem 12.2 does not place any requirement on the design matrix $\mathbf{X}$. Similar conclusions can be shown for the constrained form in (12.11) and for any doubly penalized form with sparse penalization on S and singular value penalization on C.

The potential benefit of applying a redescending ψ is revealed by Theorem 12.2. As an example, for Huber's ψ, which corresponds to the popular convex ℓ_1 penalty due to Theorem 12.1, $\rho(S; \lambda)$ on the right-hand side of (12.15) is unbounded, while Hampel's three-part ψ gives a finite rate as shown in (12.16). However, convex methods are not hopeless. In some less challenging problems, where some incoherence condition is satisfied by the augmented design matrix, Huber's ψ can achieve the same low error rate. Furthermore, it can be shown that in a minimax sense, the error rate obtained in (12.16) is essentially optimal.

The non-asymptotic robustness analysis can also reveal the necessity of robust estimation when outliers are present. Similar to Theorem 12.2, it can be shown that the conventional reduced-rank regression, which sets $\widehat{S} = 0$, satisfies

$$E\{\Delta(\widehat{C} - C, \widehat{S} - S)\} \preceq \inf_{r(\widetilde{C}) \leq \widetilde{r}} \|\widetilde{C}\mathbf{X} - (C\mathbf{X} + S)\|_F^2 + \sigma^2(r_x + m)\widetilde{r} + \sigma^2. \tag{12.17}$$

Taking $\widetilde{r} = r$ and evaluating at the optimal $\widetilde{C}$ satisfying $\widetilde{C}\mathbf{X} = C\mathbf{X} + S\mathcal{P}_{C\mathbf{X}}$ and $r(\widetilde{C}) \leq r$, the error bound of reduced-rank regression is of order

$$\sigma^2(r_x + m)r + \|S(I_T - \mathcal{P}_{C\mathbf{X}})\|_F^2. \tag{12.18}$$

Because $C\mathbf{X}$ has low rank, $I_T - \mathcal{P}_{C\mathbf{X}}$ is not null in general. Since notable outliers that affect the projection subspace in performing rank reduction tend to occur in the orthogonal complement of the range of $C\mathbf{X}$, (12.18) can be arbitrarily large. This result thus echoes the same non-robustness conclusion of reduced-rank regression reached in Theorem 12.1 through the analysis of breakdown point.

One may argue that in the presence of outliers, applying the conventional reduced-rank regression with a larger rank value may better control the size of the bias term. To examine this, we can set $\widetilde{C} = C + S\mathbf{X}'(\mathbf{X}\mathbf{X}')^-$ in (12.17), which yields

$$\sigma^2 r_x \|S\|_{2,0} + \sigma^2 m \|S\|_{2,0} + \sigma^2(r_x + m)r + \|S(I - \mathcal{P}_X)\|_F^2. \tag{12.19}$$

We used $r(\widetilde{C}) \leq r + \|S\|_{2,0}$ to derive the about result. When $n > T$, we have that $\mathcal{P}_X = I$, and so (12.19) offers an improvement over (12.18) by giving a finite error rate of

$$\sigma^2(r_x + m)r + \sigma^2(m + r_x)\|S\|_{2,0}.$$

However, the robust reduced-rank regression guarantees a consistently lower rate at $\sigma^2(r_x + m)r + \sigma^2(m + \log T)\|S\|_{2,0}$, because $\sigma^2 r_x \|S\|_{2,0} \gg \sigma^2(\log T)\|S\|_{2,0}$. The performance gain can be large in big data applications, where the predictor matrix can be of high dimensions and typically, multiple outliers are bound to be present.

We remark that although the above results are for the global minimizers of (12.4), they remain valid under mild conditions for local solutions that can be achieved by Algorithm 1 presented in Section 12.2.2. More recently, Tan et al. (2022) considered robust sparse reduced-rank regression with heavy-tailed random noise via a convex relaxation approach. Their approach minimizes the Huber's loss function subject to both the entrywise ℓ_1 norm and the nuclear-norm penalties on the coefficient matrix. The resulting estimator is shown to be computationally tractable and demonstrates a tradeoff between heavy-tailedness of the random noise and the estimation error rate under a matrix-version restricted eigenvalue condition.

12.3 Reduced-Rank Estimation with Incomplete Data

Multivariate data are often prone to missing values, and in many problems there is a genuine need to predict the missing values or conduct statistical analysis with an incomplete data matrix. The associated problem, called the *matrix completion*, has received a great deal of attention. The main question is, can we recover the missing entries of a matrix with only a few observed values that may or may not be contaminated with noise? From the first glance, this problem is ill-posed, in that, recovering an arbitrary data matrix from a subset of its entries is in general not possible. However, if the underlying matrix is assumed to have some special structures such as reduced rank, it turns out that an accurate or even an exact recovery is possible under certain conditions.

There have been many modern applications with matrix completion. As mentioned earlier, in collaborative filtering, matrix completion is used to build a Netflix movie recommendation system from highly incomplete movie rating matrix; in studies of aging on elderly subjects (Stanziano et al. 2010), matrix completion can be utilized in reduced-rank regression to study longitudinal associations of past and current health statues with partially observed health outcomes. Other applications in the literature are found in system identification, global positioning, remote sensing, computer vision, multi-task learning, among others.

To formulate the problem, let $M \in \mathbb{R}^{T_1 \times T_2}$ be the matrix we would like to know as precisely as possible. However, we only observe a partial set of its entries m_{ik},

$(i, k) \in \Omega$, where

$$\Omega = \{(i, k);\ \text{The } (i, k)\text{th entry of } M \text{ is observed}, i = 1, \ldots, T_1, k = 1, \ldots, T_2.\}.$$

It is possible that these entries are contaminated by noise, deterministic or random. To make the recovery somewhat possible, the matrix M (or its underlying signal) is assumed to have a reduced-rank structure. The problem of matrix completion is concerned with the computational methods and the theoretical underpinnings, such as required conditions on the structure of M and the number of observations, for recovering M accurately or exactly (in the noiseless case) with high probability.

In this section, we present the main methodologies and theoretical results on how well one can recover low-rank matrices from a few entries. We first discuss the basic noiseless case and then broaden the discussion to the more common situation that the observed entries are corrupted with noise. We then connect matrix completion to reduced-rank regression with incomplete data; the latter can be regarded as a supervised extension of the former.

12.3.1 Noiseless Matrix Completion

In the noiseless case, the observed entries are precisely those of the unknown matrix $M \in \mathbb{R}^{T_1 \times T_2}$. Because M is assumed to be of low rank, i.e., $r(M) = r \ll \min(T_1, T_2)$, we may consider the following optimization problem for its recovery,

$$\min_{C \in \mathbb{R}^{T_1 \times T_2}} r(C), \text{ s.t. } \mathcal{P}_\Omega(C) = \mathcal{P}_\Omega(M). \tag{12.20}$$

That is, we want to find a low-rank matrix that has the same values as M on the observed entries. Unfortunately, this rank minimization problem is NP-hard, and the known algorithms for exactly solving it requires computational complexity that is at least exponential in the dimensions of the matrix (Recht et al. 2011).

An alternative is to consider nuclear-norm minimization,

$$\min_{C \in \mathbb{R}^{T_1 \times T_2}} \|C\|_\star, \qquad \text{s.t. } \mathcal{P}_\Omega(C) = \mathcal{P}_\Omega(M). \tag{12.21}$$

This is the tightest convex relaxation of the above rank minimization problem in (12.20), in the sense that the nuclear ball, $\{C \in \mathbb{R}^{T_1 \times T_2} : \|C\|_\star \le 1\}$, is the convex hull of the set of rank one matrices with spectral norm bounded by one, that is, $\{C \in \mathbb{R}^{T_1 \times T_2} : r(C) = 1, d_1(C) \le 1\}$. By (12.21), we aim to find the matrix with minimal nuclear norm that has the same values as M on the observed entries. This can be characterized as a semidefinite programming (SDP) problem, which can be efficiently solved, for example, by interior point methods (Vandenberghe and Boyd 1996). It is interesting to note that the situation here is analogous to the combinatorial ℓ_0-minimization problem in sparse estimation briefly discussed in

Section 10.2, for which ℓ_1-minimization is its tightest convex relaxation in the sense that the ℓ_1 ball is the convex hull of the set of unit-norm vectors with at most one nonzero entry.

Matrix completion is not always possible, even when rank of M is very small and the sampling rate (the proportion of observed entries) is very high. First of all, the sampling scheme, or the pattern of the observed entries, plays a role. For example, the completion of M would be impossible if a row or a column is completely unsampled. To avoid such (trivial) situations (with high probability), it is common to impose the *uniform sampling assumption* that the observed entries are selected at random without replacement. Another fundamental and subtle condition, concerns the potential non-identifiability between the low-rank and the sparse structures. To illustrate, consider the rank one matrix $M = e_1 v'$, where $v \in \mathbb{R}^{T_2}$ and e_1 is the first vector in the canonical basis of $\mathbb{R}^{T_1}$:

$$M = \begin{bmatrix} v_1 & v_2 & v_3 & \cdots & v_{T_2-1} & v_{T_2} \\ 0 & 0 & 0 & \cdots & 0 & 0 \\ \vdots & \vdots & \vdots & \vdots & \vdots & \vdots \\ 0 & 0 & 0 & \cdots & 0 & 0 \end{bmatrix}.$$

Clearly, this low-rank matrix cannot be recovered if any entry of the first row is missing. This type of instances, however, can occur with large probability under uniform sampling.

Intuitively, there is no guarantee to complete a matrix, even it is of very low rank, when some of its singular vectors are extremely sparse. From another perspective, if a row (or column) is approximately orthogonal to the other rows (or columns), to recover the matrix we would need to see all the entries in that row. These observations motivate the concept of "coherence." In particular, a *geometric incoherence assumption* can be imposed to quantify the alignment between the canonical basis of $\mathbb{R}^{T_1}$ (respectively, $\mathbb{R}^{T_2}$) and the column (respectively, row) space of M (Candès and Recht 2009; Recht 2011). To define this, let $\mathcal{U}$ be a subspace of $\mathbb{R}^T$ of dimension r and $\mathcal{P}_U$ be the orthogonal projection onto $\mathcal{U}$. Then the *coherence* of $\mathcal{U}$ is defined as

$$\eta(\mathcal{U}) = \frac{T}{r} \max_{1 \le i \le T} \|\mathcal{P}_U e_i\|_2^2, \tag{12.22}$$

where, with some abuse of notation, e_i is the ith vector in the canonical basis of $\mathbb{R}^T$. For any subspace, the coherence index $\eta(\mathcal{U})$ can be as small as one, and it is achieved, for example, when $\mathcal{U}$ is spanned by vectors whose entries all have magnitude $1/\sqrt{T}$. On the other hand, $\eta(\mathcal{U})$ can be as large as T/r, that is achieved when the subspace contains a canonical basis vector. Intuitively, if a matrix has row and column spaces with low coherence, each entry can be expected to provide about the same amount of information, making its associated matrix completion problem under uniform sampling relatively easy to achieve.

The assumptions and main results on matrix completion in the noiseless setting are given below.

Assumption 12.1 We observe h entries of M with locations sampled uniformly at random.

Assumption 12.2 The row and column spaces have coherences, as defined in (12.22), bounded above by some positive η_0.

Assumption 12.3 The matrix UV' has its maximum entry bounded by $\eta_1\sqrt{r/(T_1T_2)}$ in absolute value for some positive η_1.

Theorem 12.3 *Let M be an $T_1 \times T_2$ matrix of rank r, with its singular value decomposition given by $M = UDV'$, where $U \in \mathbb{R}^{T_1\times r}$, $D \in \mathbb{R}^{r\times r}$, and $V \in \mathbb{R}^{T_2\times r}$. Without loss of generality, assume that $T_1 \leq T_2$. Suppose Assumptions 12.1–12.3 hold. Then, with probability at least $1 - 6\log(T_2)(T_1 + T_2)^{2-2c} - T_2^{2-2c^{1/2}}$, for some $c > 1$, the minimizer to problem* (12.21) *is unique and is exactly equal to M if*

$$h \succeq \max\{\eta_1^2, \eta_0\}(\log(T_2))^2 r(T_1 + T_2), \tag{12.23}$$

where the notation $\succeq$ means the inequality holds up to a multiplicative constant.

In Theorem 12.3, Assumption 12.1 specifies the *uniform sampling* scheme. Assumptions 12.2 and 12.3 are regarding the *geometric incoherence* (Candès and Recht 2009). Both η_0 and η_1 may depend on r, T_1, or T_2. Recall that the coherence parameter η_0 is bounded by $1 \leq \eta_0 \leq T_2/r$. So, the bound confirms that it is favorable to have low coherence. Moreover, we have that $\eta_1^2 \leq \eta_0^2 r$ by the Cauchy-Schwarz inequality. As shown in Candès and Recht (2009), both subspaces selected from the uniform distribution and spaces constructed as the span of singular vectors with bounded entries are not only incoherent with the standard basis but also obey Assumption 12.2 with high probability for values of η_1 at most of the order $\log(T_1)$.

The bound in Theorem 12.3 is nearly optimal. This is a consequence of the coupon collector's problem, which, in this context, implies that at least $T_2 \log T_2$ uniformly sampled entries are necessary to guarantee that at least one entry in every row and column is observed with high probability. Moreover, a rank-r matrix has degrees of freedom $(T_1 + T_2 - r)r$, which, intuitively, suggests that the number of observations needed is at least $O(T_2r)$. Indeed, Candès and Tao (2010) showed that the number of observations *necessary* for completion is of the order $\eta_0 \log(T_2)T_2r$, when the entries are sampled uniformly at random. Therefore, the bound in (12.23) is essentially optimal up to a logarithm factor of $\log(T_2)$.

We remark that there have been many important findings on matrix completion, and similar related results have been obtained by various authors. The results presented here were mainly from the work by Recht (2011), which features weak assumptions and simplicity in its analysis. In particular, Theorem 12.3 makes no assumption on the rank r, the aspect ratio T_2/T_1, and the condition number $d_1(M)/d_r(M)$. It thus has a scope to cover more practical applications.

12.3.2 Stable Matrix Completion

In real-world problems, we almost always observe the entries of the underlying matrix of interest that are perturbed with some noise. In the Netflix problem, user's ratings are uncertain; in the longitudinal study of aging, the best we can hope for is that the current health statue can provide reasonable prediction into the future. It then arises the question that whether accurate matrix completion is still possible from noisy samples, and if so, under what conditions. This problem has been referred to as *stable matrix completion* in the literature.

Consider the following model to analyze the stable matrix completion problem,

$$y_{ik} = m_{ik} + \epsilon_{ik}, \quad (i, k) \in \Omega, \tag{12.24}$$

where ϵ_{ik}, $(i, k) \in \Omega$ are the error terms. In matrix form, the model can be expressed as

$$\mathcal{P}_\Omega(\mathbf{Y}) = \mathcal{P}_\Omega(M) + \mathcal{P}_\Omega(\boldsymbol{\epsilon}),$$

where $\boldsymbol{\epsilon} = (\epsilon_{ik})$ is an $T_1 \times T_2$ error matrix.

To recover the unknown matrix M, we consider the following optimization problem:

$$\min_C \|C\|_\star, \qquad \text{s.t. } \|\mathcal{P}_\Omega(\mathbf{Y} - C)\|_F \le \delta. \tag{12.25}$$

That is, we aim to find the matrix with minimal nuclear norm that is close to $\mathbf{Y}$ on the observed entries as measured by the error in Frobenius norm. Similar to (12.21), this can also be recognized as an SDP problem. We shall defer the detailed discussion on the computation to Section 12.3.3. Let $\widehat{M}$ be the solution of (12.25). The main result of stable matrix completion is stated below.

Theorem 12.4 *Suppose the conditions for ensuring noiseless recovery, as Assumptions 12.1–12.3 for establishing Theorem 12.3, hold. Assume $\|\mathcal{P}_\Omega(\boldsymbol{\epsilon})\|_F \le \delta$ for some $\delta > 0$. Let $p = h/(T_1T_2)$ denote the fraction of observed entries. Then we have*

$$\|\widehat{M} - M\|_F \preceq \delta\sqrt{\frac{(p+2)\min(T_1, T_2)}{p}} + \delta. \tag{12.26}$$

Theorem 12.4 shows that *when perfect noiseless recovery occurs, matrix completion is stable*, implying that, accurate recovery can be achieved even with perturbed entries. Besides the conditions under which noiseless recovery occurs, all we assume is that the noise level is bounded as $\|\mathcal{P}_\Omega(\boldsymbol{\epsilon})\|_F \leq \delta$. When $\boldsymbol{\epsilon}$ is random, (12.26) holds on the event $\|\mathcal{P}_\Omega(\boldsymbol{\epsilon})\|_F \leq \delta$, so a proper choice of δ based on the distribution of $\boldsymbol{\epsilon}$ will lead to a high-probability error bound. For example, if the random error terms $\{\epsilon_{ik}\}$ are uncorrelated and each with mean zero and standard deviation σ (white noise sequence), then $\delta^2 \leq \sigma^2(h + \sqrt{8h})$ with high probability. As expected, the error of the matrix completion is proportional to δ; the smaller the noise level, the smaller the error. Also, when the fraction of observed entries p is small, the error is at most of the order $\delta\sqrt{\min(T_1, T_2)/p}$.

12.3.3 Computation

The matrix completion problem in (12.25) can be equivalently expressed as a regularized nuclear-norm minimization problem (Mazumder et al. 2010; Ma et al. 2011),

$$\min_{\mathbf{C}} \left\{ J(C) \equiv \frac{1}{2}\|\mathcal{P}_\Omega(\mathbf{Y} - C)\|_F^2 + \lambda\|C\|_\star \right\}. \tag{12.27}$$

Let the solution be denoted as $\widehat{C}(\lambda)$. Then it follows from standard duality result (Boyd and Vandenberghe 2004) that for any $\delta < \|\mathcal{P}_\Omega(\mathbf{Y})\|_F$, we could use (12.27) to solve (12.25) by searching for the value of $\lambda(\delta)$ that gives $\|\mathcal{P}_\Omega(\mathbf{Y} - \widehat{C}(\lambda))\|_F = \delta$.

We present a simple iterative algorithm to solve (12.27) (Mazumder et al. 2010), which can be recognized to follow a Majorization-Minimization (MM) procedure (Hunter and Lange 2000). The MM algorithm is an iterative procedure that works by minimizing a sequence of properly chosen upper bounds or the so-called majorizers of the objective function. As such, it is often able to solve a difficult optimization problem by solving a sequence of much easier problems instead.

For the matrix completion problem, the MM algorithm starts with an initial value $C^{(0)}$; a simple choice is the $T_1 \times T_2$ zero matrix. At the tth iteration, with the currently estimate $C^{(t)}$, the first step is to construct a majorization function of the objective function $J(C)$ at $C^{(t)}$, denoted as $g(C; C^{(t)})$, that satisfies two conditions: $g(C^{(t)}; C^{(t)}) = J(C^{(t)})$ and $g(C; C^{(t)}) \geq J(C)$ for any $C \neq C^{(t)}$, i.e., the majorization function takes the same value as the objective function at the current estimate and otherwise is always greater than or equal to the objective function. Here, a natural choice of the majorization function is

$$g(C; C^{(t)}) = \frac{1}{2}\|\mathbf{Y}^{(t)} - C\|_F^2 + \lambda\|C\|_\star, \tag{12.28}$$

where $\mathbf{Y}^{(t)} = \mathcal{P}_\Omega(\mathbf{Y}) + \mathcal{P}_{\Omega^\perp}(C^{(t)})$. That is, we simply fill the missing entries in $\mathbf{Y}$ with the corresponding entries from the current estimate $C^{(t)}$. It is straightforward to verify that the two conditions of being a majorization function are satisfied by the above $g(C; C^{(t)})$:

$$g(C^{(t)}; C^{(t)}) = \frac{1}{2}\|\mathcal{P}_\Omega(\mathbf{Y}) + \mathcal{P}_{\Omega^\perp}(C^{(t)}) - C^{(t)}\|_F^2 + \lambda\|C^{(t)}\|_\star = J(C^{(t)})$$

because $C^{(t)} = \mathcal{P}_\Omega(C^{(t)}) + \mathcal{P}_{\Omega^\perp}(C^{(t)})$, and

$$\begin{aligned} g(C; C^{(t)}) &= \frac{1}{2}\|\mathcal{P}_\Omega(\mathbf{Y}) + \mathcal{P}_{\Omega^\perp}(C^{(t)}) - \mathcal{P}_{\Omega^\perp}(C) - \mathcal{P}_\Omega(C)\|_F^2 + \lambda\|C\|_\star \\ &= \frac{1}{2}\|\mathcal{P}_\Omega(\mathbf{Y}) - \mathcal{P}_\Omega(C)\|_F^2 + \frac{1}{2}\|\mathcal{P}_{\Omega^\perp}(C^{(t)}) - \mathcal{P}_{\Omega^\perp}(C)\|_F^2 + \lambda\|C\|_\star \\ &\geq J(C), \end{aligned}$$

because $\|\mathcal{P}_{\Omega^\perp}(C^{(t)}) - \mathcal{P}_{\Omega^\perp}(C)\|_F^2 \geq 0$ for any $C \neq C^{(t)}$.

The next step is to update the estimate of C by minimizing the majorized objective function in (12.28). This can be immediately seen as a nuclear-norm penalized matrix approximation problem, with its solution given by the singular value soft-thresholding operation according to Result 10.1. Specifically, Let the singular value decomposition of $\mathbf{Y}^{(t)}$ be given by $V^{(t)}D^{(t)}U^{(t)\prime}$. Then

$$C^{(t+1)} = \arg\min_C g(C; C^{(t)}) = \mathcal{S}_\lambda(\mathbf{Y}^{(t)}) = V\mathcal{S}_\lambda(D)U',$$

where $\mathcal{S}_\lambda(D) = \mathrm{diag}\left[\{d_k(\mathbf{Y}^{(t)}) - \lambda\}_+, k = 1, \ldots, \min(T_1, T_2)\right]$.

The above two steps, the majorization of the objective function at the current estimate and the minimization of the resulting majorized objective function are repeated, until reaching convergence. The procedure is summarized in Algorithm 2.

Due to its Majorization-Minimization structure, the above algorithm enjoys the monotone descending property. It can be readily shown that the objective function in (12.27) is monotonically decreasing along the iterations,

$$J(C^{(t+1)}) \leq g(C^{(t+1)}; C^{(t)}) \leq g(C^{(t)}; C^{(t)}) = J(C^{(t)}).$$

The first inequality is due to the property of g that $g(C; C^{(t)}) \geq J(C)$ for any $C \neq C^{(t)}$, and the second inequality is due to the fact that $C^{(t+1)}$ is the minimizer

---

Algorithm 2 Matrix completion via iterative singular value thresholding

Input $\mathcal{P}_\Omega(\mathbf{Y})$, $C^{(0)}$, $t = 0$.
Repeat
(a) $\mathbf{Y}^{(t)} \leftarrow \mathcal{P}_\Omega(\mathbf{Y}) + \mathcal{P}_{\Omega^\perp}(C^{(t)})$;
(b) $C^{(t+1)} \leftarrow \mathcal{S}_\lambda(\mathbf{Y}^{(t)})$;
$t \leftarrow t + 1$;
Until convergence, e.g., $\|C^{(t+1)} - C^{(t)}\|_F / \|C^{(t)}\|_F < \epsilon$.

---

of $g(C; C^{(t)})$. Given the convexity of the problem, this implies that the convergence to a global optimum is guaranteed.

Algorithm 2 works well for problems of moderate size. However, for large matrices, the computation could become burdensome mainly due to step (b), because of the computational complexity involved in computing the singular value decomposition of the matrix $\mathbf{Y}^{(t)}$. Even the storage of the full matrix could become an issue in some applications. For example, the movie rating matrix in the Netflix problem is about of 400,000 by 20,000 and requires multiple gigabyte of storage space.

There have been various strategies and algorithms to make matrix completion more scalable. We briefly mention a few key ideas. First, because the percentage of observed entries is usually quite small in large applications (1% in the Netflix problem), sparse-matrix representations and the associated computational toolkit can be utilized. Second, various strategies are considered to avoid the expensive full singular value decomposition. In particular, if the solution is anticipated to be of low rank, the rank constraint can be utilized to speed up the computation. For example, in Algorithm 2, step (b) could be executed via a much more efficient reduced-rank singular value decomposition. A different approach, maximum-margin matrix factorization (MMMF), is originated from the result that for any rank r matrix $C \in \mathbb{R}^{T_1 \times T_2}$, it holds that

$$\|C\|_\star = \min_{A \in \mathbb{R}^{T_1 \times r}, B \in \mathbb{R}^{r \times T_2}} \left\{ \|A\|_F^2 + \|B\|_F^2 \right\}.$$

Then resulting MMMF criterion solves

$$\min_{A,B} \left\{ \frac{1}{2} \|\mathcal{P}_\Omega(\mathbf{Y} - AB)\|_F^2 + \lambda(\|A\|_F^2 + \|B\|_F^2) \right\}. \tag{12.29}$$

The problem is bi-convex in A and B and can be solved by alternating convex search. We refer to Mazumder et al. (2010), Hastie et al. (2015), Davenport and Romberg (2016), and Chi and Li (2019) for additional details and more recent development.

12.4 Generalized Reduced-Rank Regression with Mixed Outcomes

Thus far we have been focusing on various reduced-rank models developed for continuous responses. In many applications, however, several types of outcomes, e.g., continuous, binary, and count data, may all be collected. We refer to such a collection of outcomes as *mixed outcomes*. Furthermore, such mixed data could have missing values and outliers, as discussed in the previous sections. In general, regardless of the measurement types, these outcomes are often likely to be related, representing diverse yet interrelated views of the underlying data generation mechanism. Therefore, an integrative learning and joint predictive modeling of the mixed outcomes could be preferable.

Here we present a mixed-outcome reduced-rank regression (mRRR) framework to model mixed, incomplete, and potentially contaminated response data with high-dimensional predictors. This provides a general, practical approach for analyzing multivariate mixed outcomes from the exponential dispersion family, taking into account missing values, outliers, effects of control variables, differential dispersion of the outcomes, among others.

Let $\mathbf{Y} \in \mathbb{R}^{m\times T}$ be the complete response matrix consisting of T independent samples from m outcome/response variables. Similar to the scenario of matrix completion, we allow $\mathbf{Y}$ to be partially observed. Let

$$\Omega = \{(i,k);\ \text{The } (i,k)\text{th entry of } \mathbf{Y} \text{ is observed}, i = 1,\ldots,m, k = 1,\ldots,T.\}$$

be an index set collecting all the entries corresponding to the observed outcomes. Let $\widetilde{\mathbf{Y}} = \mathcal{P}_\Omega(\mathbf{Y})$ denote the projection of $\mathbf{Y}$ onto Ω, i.e., $\tilde{y}_{ik} = y_{ik}$ for any $(i,k) \in \Omega$ and $\tilde{y}_{ik} = 0$ otherwise.

We assume each y_{ik}, i.e., the kth observation on the ith outcome variable, for any $(i,k) \in \Omega$, follows a distribution in the exponential dispersion family (Jorgensen 1987). Specifically, the probability density function (pdf) of each y_{ik} takes the following form,

$$f(y_{ik};\theta_{ik},\phi_k) = \exp\left\{\frac{y_{ik}\theta_{ik} - b_i(\theta_{ik})}{a_i(\phi_i)} + c_i(y_{ik};\phi_i)\right\}, \tag{12.30}$$

where θ_{ik} is the natural parameter of y_{ik}, ϕ_i is the dispersion parameter of the ith outcome variable, and $a_i(\cdot)$, $b_i(\cdot)$, $c_i(\cdot)$ are known functions determined by the specific distribution of the ith outcome variable.

Here the m outcome variables are allowed to have different distributions within the exponential dispersion family; Table 12.2 provides specifications of some most common distributions in this family, including Normal, Bernoulli, and Poisson. The dispersion parameter ϕ_i can either be known or unknown. For example, the ϕ_i for Poisson distribution is known to be 1, but for Gaussian distribution ϕ_i corresponds to its variance which is generally unknown. Let $\boldsymbol{\phi} = (\phi_1,\ldots,\phi_m)'$ be the vector of the dispersion parameters, and we denote $\boldsymbol{\phi}_u$ as a subvector of $\boldsymbol{\phi}$ consisting of all the unknown dispersion parameters. Without loss of generality, for each outcome variable, we apply the canonical link function $g_i = (b_i')^{-1}$, so that $E(y_{ik}) = b_i'(\theta_{ik}) = g_i^{-1}(\theta_{ik})$, where, with some abuse of notation, $b_i'(\cdot)$ denotes the derivative function of $b_i(\cdot)$.

Let $\mathbf{X} \in \mathbb{R}^{n\times T}$ be the observed feature/predictor matrix. Also let $\mathbf{Z} \in \mathbb{R}^{(n_z+1)\times T}$ be consisting of a vector of ones in its first row (corresponding to the intercept term) and observations from n_z control variables in the rest of its rows. The set of control variables is application specific, and in general its dimension n_z is much smaller than the sample size T.

Table 12.2 Some common distributions in the exponential dispersion family

Distribution	Mean	Variance	θ	ϕ	$a(\phi)$	$b(\theta)$	$c(y;\phi)$
Bernoulli(p)	p	$p(1-p)$	$\log(p(1-p)^{-1})$	1	1	$\log(1+e^{\theta})$	0
Poisson(λ)	λ	λ	$\log\lambda$	1	1	e^{θ}	$-\log y!$
Normal(μ,σ^2)	μ	σ^2	μ	σ^2	ϕ	$\theta^2/2$	$-(y^2\phi^{-1}+\log 2\pi\phi)/2$
Gamma(α,β)	α/β	α/β^2	$-\beta/\alpha$	$1/\alpha$	ϕ	$-\log(-\theta)$	$\log(\alpha^{\alpha}y^{\alpha-1}/\Gamma(\alpha))$

The basis of the mRRR approach is the familiar generalized linear model (GLM) (Yee and Hastie 2003; She 2013; Luo et al. 2018). Specifically, the natural parameters in (12.30) are modeled as

$$\theta_{ik} = o_{ik} + B_{(i)}' Z_{(i)} + C_{(i)}' X_{(i)} + s_{ik}, \qquad (i,k) \in \Omega. \tag{12.31}$$

Here o_{ik} is a known offset term, which commonly arises, for example, in the modeling of count data for adjusting the size of the population from which a count is drawn. The coefficient vectors $B_{(i)} \in \mathbb{R}^{n_z+1}$, $i = 1, \ldots, m$, correspond to the intercept and the control variables, while $C_{(i)} \in \mathbb{R}^{n}$, $i = 1, \ldots, m$, correspond to the predictors. The mean-shift term s_{ik} is to capture the potential outlying effect in y_{ik} that cannot be explained by the control variables and the predictors.

Let $\Theta = [\Theta_{(1)}, \ldots, \Theta_{(m)}]' = (\theta_{ik}) \in \mathbb{R}^{m\times T}$ be the natural parameter matrix, $C = [C_{(1)}, \ldots, C_{(m)}]' \in \mathbb{R}^{m\times n}$ the coefficient matrix of the predictors, and $B = [B_{(1)}, \ldots, B_{(m)}]' \in \mathbb{R}^{m\times(p_z+1)}$ the coefficient matrix of the intercept and control variables, whose first column, B_0, gives the intercept vector. Let $S = (s_{ik}) \in \mathbb{R}^{m\times T}$ be the mean-shift parameter matrix. Then the model in (12.31) can be stated in matrix form as

$$\Theta = B\mathbf{Z} + C\mathbf{X} + S. \tag{12.32}$$

With no additional assumptions on the parameter matrices, observe that the model is over-parameterized when m or n are comparable or much larger than T. Moreover, when the independence of y_{ik}s is assumed, the model reduces to a set of univariate GLM analysis and thus does not possess any multivariate flavor. The key here, is how to induce and take advantage of the interrelationships among the outcomes. As such, the merit of the formulations in (12.30)–(12.32) lies in imposing some suitable low-dimensional structures on the parameters. In particular, we assume C is a *reduced-rank matrix* and S is a *sparse matrix*, analogous to the robust reduced-rank regression discussed in Section 12.2. The reduced-rank assumption on C implies that the multivariate outcomes are jointly associated with the predictors through a few latent variables, while the sparsity assumption on S accommodates the scattered outlying effects in the observed outcomes.

The mRRR is a rather "simple" model, in that the method does not explicitly characterize the dependency among the outcomes and does not explicitly model the missing data mechanism. These alternative proposals, though seemingly desirable, would not be easily applicable or generalizable in the simultaneous presence of outcome heterogeneity, high dimensionality, data contamination, and incompleteness. Nevertheless, we demonstrate that the reduced-rank approach could still be effective in dealing with such complex data. It is also worth noting that when there are no control variables and predictors, the cRRR model specializes to a robust principal component analysis (PCA) with mixed and incomplete data. When the responses are all from the same type of distribution, the method further simplifies to a generalized robust PCA or to a generalized robust matrix completion method (Collins et al. 2002; Candès et al. 2011).

The regularized estimation problem for mRRR can be formulated as

$$\min_{C,B,S,\boldsymbol{\phi}_u} \left\{ -\sum_{(i,k)\in\Omega} \ell_i(C_{(i)}, B_{(i)}, s_{ik}, \phi_i; X_k, Z_k, y_{ik}) + \rho(S;\lambda) \right\} \quad \text{s.t. } r(C) \le r, \tag{12.33}$$

where

$$\ell_i(C_{(i)}, B_{(i)}, s_{ik}, \phi_i; X_k, Z_k, y_{ik}) = \frac{y_{ik}\theta_{ik} - b_i(\theta_{ik})}{a_i(\phi_i)} + c_i(y_{ik}; \phi_i)$$

with $\theta_{ik} = o_{ik} + B_{(i)}' Z_k + C_{(i)}' X_k + s_{ik}$, and ρ is a sparsity-inducing penalty function applied on S in either entry-wise or column-wise fashion. Comparing to the robust reduced-rank regression criterion in (12.4), the above criterion replaces the sum of squared error term with the negative log-likelihood function of the observed entries, while the treatment of the incomplete data follows the same spirit as the matrix completion problem in (12.27). Here the rank constraint can be replaced with a singular value penalty, as discussed in Section 10.3. We remark that for modeling the mixed outcomes, the inclusion of the dispersion parameters is necessary, which ensures that all the outcomes can be brought to the same scale for joint analysis.

The optimization of (12.33) can be solved by combining the ideas of the blockwise descent and the iterative thresholding used in Algorithm 1 (for robust reduced-rank regression) and the Majorization-Minimization procedure used in Algorithm 2 (for matrix completion). We shall briefly describe the main steps. At the tth iteration, with fixed S, B and $\boldsymbol{\phi}_u$, a surrogate quadratic loss function with respect to C is formulated and the resulting optimization problem becomes

$$\min_{C} \|C - C^{(t)} - (\Phi^{(t)})^{-1}\mathcal{P}_{\Omega}\{\mathbf{Y} - b'(\Theta^{(t)})\}\mathbf{X}'\|_F^2 \qquad \text{s.t. } r(C) \le r,$$

where $\Theta^{(t)} = B^{(t)}\mathbf{Z} + C^{(t)}\mathbf{X} + S^{(t)}$, $\Phi = \text{diag}\{\phi_i^{(t)}, i = 1, \ldots, m.\}$, and $b'(\Theta^{(t)}) = [b_1'(\Theta_{(1)}^{(t)}), \ldots, b_m'(\Theta_{(m)}^{(t)})]'$. This is solved by the singular value hard-thresholding operation, according to Result 10.1. When the rank constraint is replaced with a singular value penalty, the solution is given by a corresponding singular value thresholding operation. With fixed C, B and $\boldsymbol{\phi}_u$, a similar surrogate loss function with respect to S can be formulated and the problem becomes

$$\min_{S} \frac{1}{2} \|S - S^{(t)} - (\Phi^{(t)})^{-1} \mathcal{P}_{\Omega}\{\mathbf{Y} - b'(\Theta^{(t)})\}\|_F^2 + \rho(S; \lambda),$$

which admits an explicit solution via a thresholding operation corresponding to the choice of ρ; see Theorem 12.1 and the relevant discussions that follow. Finally, with fixed C and S, the problem becomes a separable set of univariate GLM problems; while the same majorization strategy can be applied to update B, many standard gradient-based routines and implementations are also available. The algorithm executes the above steps iteratively until convergence. With proper pre-scaling of the predictor matrices, one can show that the objective function is guaranteed to be non-increasing during the iterations (She 2013; Luo et al. 2018). Alternative optimization methods are also available; see, e.g., Udell et al. (2016), in which a number of local optimization methods were considered for a broad spectrum of generalized low-rank models.

We remark that there exist many approaches that attempt to model the associations among mixed outcomes in an explicit way, although primarily for low-dimensional problems. To mention a few, Cox and Wermuth (1992) and Fitzmaurice and Laird (1995) considered likelihood based methods by factorizing the joint distribution as marginal and conditional distributions. Prentice and Zhao (1991) and Zhao et al. (1992) used the generalized estimating equations (Liang and Zeger 1986) to handle mixture of continuous and discrete outcomes. de Leon and Wu (2011) considered a copula-based model for a bivariate mixed discrete and continuous outcome. Indeed, a direct modeling of mixed outcomes is challenging due to a lack of convenient multivariate distributions, not even to mention that a joint estimation of both the mean and the covariance structure may become notoriously difficult in high-dimensional settings. Another strategy, closely related to the reduced-rank modeling, is to induce multivariate dependency through some shared latent variables, conditioning on which the outcomes are then assumed to be independent (Sammel et al. 1997; Dunson 2000; McCulloch 2008).

12.5 Applications

12.5.1 Arabidopsis Thaliana data

Recall the study on isoprenoid in Arabidopsis thaliana plants presented in Section 11.5.2. Isoprenoids, constituting a large class of organic compounds, are abundant and diverse in plants, and they serve many important biochemical functions and have roles in respiration, photosynthesis and regulation of growth and development in plants. A genetic association study was conducted to examine the regulatory control mechanisms in the gene network for isoprenoid in Arabidopsis thaliana, a small flowering plant native to Eurasia and Africa. Specifically, $T = 118$ GeneChip microarray experiments were performed to monitor the gene expression levels under various experimental conditions (Wille et al. 2004). To confirm the existing experimental findings that there exist strong connections between two isoprenoid biosynthesis pathways and some downstream pathways, we consider a multivariate regression setup, with the expression levels of $n = 39$ genes from two isoprenoid biosynthesis pathways as the predictors, and the expression levels of $m = 62$ genes from four downstream pathways, namely plastoquinone, carotenoid, phytosterol, and chlorophyll, as the responses.

Because of the small sample size relative to the number of unknown parameters, the reduced-rank models can be handy. We apply both the classical reduced-rank regression and the robust reduced-rank regression with group Lasso penalty to deal with potential gross outliers. The fitted robust model has rank five, which reduces the effective number of parameters by about 80% compared with the least squares method. Interestingly, two samples, labeled 3 and 52, are identified as outliers. The two outliers have a surprisingly big impact on both coefficient estimation and model prediction: we find that $\|\widehat{C} - \widetilde{C}\|_F / \|\widetilde{C}\|_F \approx 50\%$ and $\|\widehat{C}\mathbf{X} - \widetilde{C}\mathbf{X}\|_F / \|\widetilde{C}\mathbf{X}\|_F \approx 26\%$, where $\widehat{C}$ and $\widetilde{C}$ denote the robust and classical reduced-rank regression estimates, respectively.

Figure 12.2 shows the detection paths by plotting the ℓ_2 norm of each column in the estimates of the mean-shift matrix S for a sequence of values of the tuning parameter λ. Sample 3 and sample 52 are captured as outliers, whose paths are shown as a dotted line and a dashed line, respectively. These two detected unusual samples are distinctive. Their outlying effects might be related to their corresponding experimental conditions. In particular, sample 3 was the only sample with Arabidopsis tissue culture in a baseline experiment. In addition, Fig. 12.2 reveals that sample 27 could be a potential outlier that merits further investigation.

The robust reduced-rank regression produces score variables or factors, constructed from the isoprenoid biosynthesis pathways in response to the 62 genes on the four downstream pathways. Let $\widetilde{\mathbf{X}}$ denote the design matrix after removing the two detected outliers, and $\widehat{V}\widehat{D}\widehat{U}'$ be the singular value decomposition of $\widehat{C}\widetilde{\mathbf{X}}$. Then

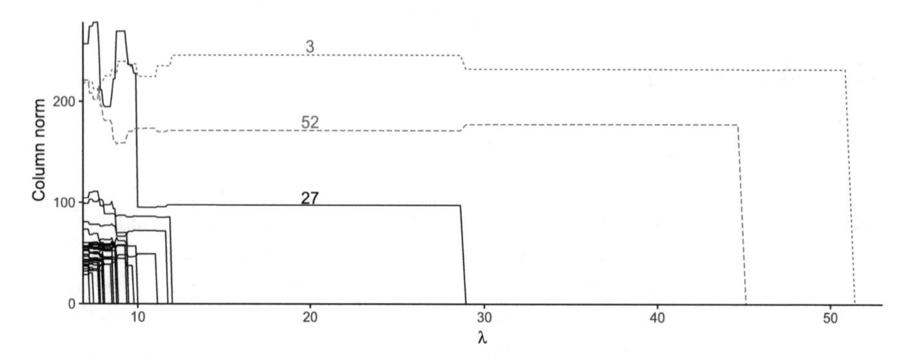

Fig. 12.2 Arabidopsis thaliana data: outlier detection paths by the robust reduced-rank regression (She and Chen 2017)

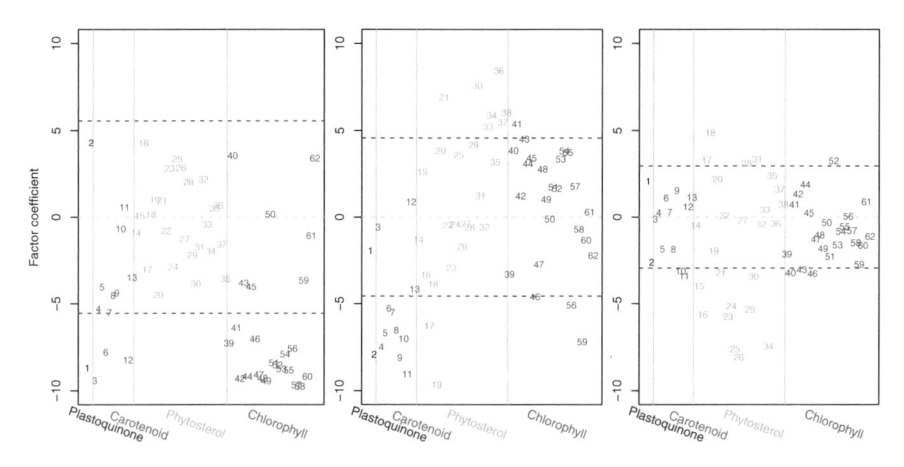

Fig. 12.3 Arabidopsis thaliana data: factor coefficients of the 62 response genes from plastoquinone, carotenoid, phytosterol, and chlorophyll pathways (She and Chen 2017)

Table 12.3 Arabidopsis thaliana data: percentage of genes on each response pathway that show significance of a given factor, with the family-wise error rate controlled at level 0.01

Pathway	Number of genes	Factor 1	Factor 2	Factor 3
Carotenoid	11	55%	73%	9%
Phytosterol	25	20%	48%	32%
Chlorophyl	24	75%	21%	0%

$\widehat{U}$ provides five orthogonal factors, and $\widehat{V}\widehat{D}$ gives the associated factor coefficients. Figure 12.3 plots the coefficients of the first three leading factors for all the 62 response variables, which account for 86% of the variation explained by the five factors. The panels from left to right correspond to the top three factors estimated by the robust reduced-rank regression. For the sth factor ($s = 1, 2, 3$), two horizontal lines are plotted at heights $\pm m^{-1/2} d_s(\widehat{C}\widetilde{\mathbf{X}})$, and three vertical lines separate the genes into the four different pathways. As such, the genes located outside of the two horizontal lines have relatively large coefficients/loadings in magnitude on the corresponding factor. It is also informative to test the significance of the factors in response to each of the 62 genes. The results are shown in Table 12.3. Plastoquinone is excluded since it has only two genes and its behavior couples with that of carotenoid most of the time. Even with the family-wise error rate controlled at 0.01, the factors obtained are overall predictive according to the percentages of significance.

According to Fig. 12.3 and Table 12.3, the factors play very different roles in different pathways. The genes that are correlated with the first factor are mainly from carotenoid and chlorophyll, and almost all the coefficients there are negative. It seems that the first factor interprets some joint characteristics of careteniod and chlorophyll. The second factor differentiates phytosterol genes from carotenoid ones, and the third factor seems to mainly contribute to the phytosterol pathway.

Therefore, by projecting the data onto a proper low-dimensional subspace in a supervised and robust manner, distinct behaviors of the downstream pathways and their potential subgroup structures can be revealed. More biological insights could be gained by closely examining the experimental and background conditions. Thus utilizing the tools developed in this chapter can lead to insights into complex biological problems.

12.5.2 Longitudinal Studies of Aging

We use the LSOA data introduced in Section 10.6.2 to demonstrate the potential power of integrative mixed-outcome reduced-rank modeling. Recall that a national representative sample of several thousands of subjects who were at or over 70 years of age were interviewed and followed, and their health, functional status, living arrangements, and health services utilization were measured as they moved into and through their oldest ages. The main interest is to examine the changes and the associations between their past and current health status. We have thus considered a multivariate regression setup, by jointly regressing various health measures collected during the period of 1999–2000 on the records collected during the period of 1997–1998 from the same set of subjects. In Section 10.6.2, $n = 294$ past health risk and behavior measurements are treated as predictors and 41 later binary health outcomes are treated as multivariate responses, and the analysis includes $T = 3988$ subjects who participated the studies in both periods.

As a matter of fact, besides the 41 binary outcomes, the data set also includes 3 additional continuous outcome variables on self-rated health measures, namely overall health status, memory status, and depression status. There may be strong correlations among the large number of outcomes and the features. Therefore, it is plausible that the outcomes are dependent on the features only through a few latent factors, making dimension reduction and reduced-rank models applicable. We thus apply mRRR with rank penalization to conduct an integrative analysis of the $m = 44$ continuous and binary outcomes. Since mRRR can deal with the missing values in the outcomes, neither data removal nor imputation is needed. The gender and age variables are used as controls and their corresponding coefficients are not penalized.

To demonstrate the efficacy of integrative learning, we mainly compare mRRR with the univariate generalized linear models (uGLM). We use a random-splitting procedure to evaluate the predictive performance. In each split, 75% of data are randomly selected for training and the rest 25% data for testing. The prediction of the continuous outcomes in each split is evaluated by mean squared errors (MSE), while the prediction of the binary outcomes is evaluated by Area Under Curve (AUC), based on testing data alone. The procedure is repeated 100 times.

We compute the average predictive measures from random splitting. The average MSE for Gaussian outcomes are 0.69 (0.06) and 0.76 (0.07), and the average AUC for binary outcomes are 0.77 (0.10) and 0.65 (0.11), for mRRR and uGLM, respectively (the standard deviations are reported in the parenthesis). The uGLM

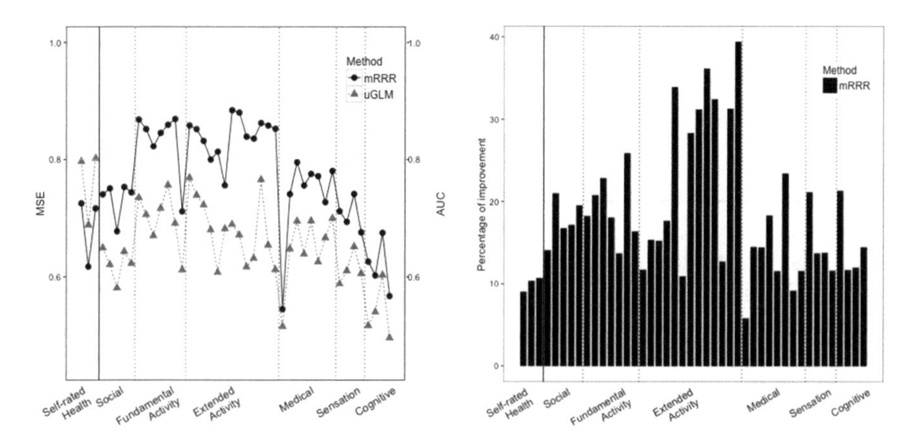

Fig. 12.4 LSOA data: comparison of predictive performance of mRRR and uGLM. Use 75% sample as training set (Luo et al. 2018)

approach is outperformed by mRRR by a large margin, indicating the strength of low-rank estimation. Figure 12.4 provides a more detailed performance comparison on the prediction of each individual outcome. The left panel displays the MSE or AUC value for predicting each individual outcome. The right panel shows the percentage of improvement by mRRR over uGLM. In the left penal, the left axis shows MSE while the right axis shows AUC. In both panels, the solid vertical line separates the Gaussian and the binary outcomes, while the dashed vertical lines separate the binary outcomes to different categories. It can be seen that the improvement by mRRR over uGLM is persistent across all the outcomes. The most improvement appears to be in the categories of fundamental activity and extended activity, where the percentage of improvement is over 20% for several outcomes. Indeed, these outcomes tend to be moderately or highly correlated, making joint estimation particularly beneficial. The mRRR also performs substantially better in predicting the three continuous responses related to self-rated health. This can be explained by the fact that two out of the three self-rated health measures are about memory and depression, which are very relevant to the variables in the category of cognition (Hammar and Ardal 2009).

We have also tried the regularized version of uGLM using elastic net (uGLM-EN), whose average MSE for Gaussian outcomes and average AUC for binary outcomes are 0.70 (0.06) and 0.75 (0.11), respectively. As such, the performance of mRRR is only slightly better than that of uGLM-EN. This shows that shrinkage and sparse estimation could also be quite effective in this application.

Next, we have tried the approach of fitting the 3 Gaussian responses and the 41 binary responses separately. Using the aforementioned random-splitting procedure with 75% samples for training, this approach yields almost identical results compared to mRRR. That the benefit of joint modeling is not observed in this case is partly due to the facts that the number of Gaussian responses is quite small and the sample size is very large. We have then tried smaller sample sizes for training. From the random-splitting procedure with 25% data for training, the average MSE for Gaussian outcomes are 0.78 (0.06) and 0.80 (0.06), and the average AUC for binary outcomes are 0.75 (0.10) and 0.73 (0.10), for mRRR and gRRR, respectively; with 10% data for training, the average MSE for Gaussian outcomes are 0.86 (0.09) and 0.92 (0.09), and the average AUC for binary outcomes are 0.73 (0.11) and 0.70 (0.10), for mRRR and gRRR, respectively. Therefore, as the sample size becomes smaller, while the performance of both methods deteriorates, the gain of mRRR over gRRR becomes even more revealing.

The above analysis suggests that integrative modeling can be quite effective especially when sample size is small or information from each individual response is limited. It also suggests that a model that considers both reduced-rank and sparsity structures may likely have superior performance; we will further discuss these types of models in Chapter 13.

Chapter 13
Sparse Reduced-Rank Regression

13.1 Introduction

Under the high-dimensional multivariate regression framework in Chapter 10, researchers have considered several types of low-dimensional structural assumptions on the coefficient matrix $C = [C_1, \ldots, C_n] \in \mathbb{R}^{m \times n}$ in (10.6). In particular, in this book we have been focusing on the *reduced-rank* structure, where the rank of C can be much smaller than its dimensions m and n. Another popular low-dimensional structure is *sparsity*, where the entries of C or its components through certain matrix decomposition (e.g., left and right singular matrices) are assumed to contain a large number of zeros. It has been realized that these two types of low-dimensional structures could complement each other. While the reduced-rank structure exploits the dependence among variables to achieve parsimony, the sparseness helps in variable selection. Indeed, the integration of the reduced-rank and the sparsity structures has greatly advanced the applicability of high-dimensional multivariate regression and facilitated scientific discoveries in a wide range of applications.

There are several possible patterns of sparsity in C that imply different response-predictor associations and model interpretations. For example, if C is assumed to be column-sparse, that is, if a number of columns in C are null vectors, it implies that only a subset of the predictors are related to the response variables while the rest that correspond to the zero columns in C are irrelevant. Together with the reduced-rank structure on C, the resulting model assumes that the responses are associated with a few latent factors and they are constructed as linear combinations of only a subset of the predictors. Many novel applications are in the area of studying regulatory relationships among genome-wide measurements (Vounou et al. 2010; Ma et al. 2014; Uematsu et al. 2019). For instance, in a yeast expression quantitative trait loci (eQTLs) mapping analysis, the goal is to understand how the eQTLs, which are regions of the genome containing DNA sequence variants, influence the expression level of genes in the yeast MAPK signaling pathways. The problem can

G. C. Reinsel et al., *Multivariate Reduced-Rank Regression*, Lecture Notes in Statistics 225, https://doi.org/10.1007/978-1-0716-2793-8_13

be formulated as a multivariate regression with the eQTLs as predictors and the gene expression levels as responses. Extensive genetic and biochemical analysis has revealed that there are a few functionally distinct signaling pathways of genes, suggesting that the coefficient matrix is of reduced rank. Further, column sparsity in the coefficient matrix arises from the postulate that only a relatively small number of eQTLs regulates the gene expressions. On the other hand, row-sparsity of the coefficient matrix could also arise, if only a subset of gene expressions gets regulated by the eQTLs.

The sparsity on certain decomposition of C can lead to even more subtle and more appealing model interpretation. Consider the situation that the right singular matrix of C is assumed to be sparse; it implies that each latent factor is allowed to be constructed (as a linear combination) from a different subset of predictors. If the left singular matrix of C is assumed to be sparse, it implies that each response is allowed to be associated with a different subset of the latent factors. Such flexibility can be critical in many applications. In the aforementioned yeast eQTLs mapping example, it is highly plausible that each signaling pathway involves only a subset of genes, which are regulated by only a few genetic variants. This suggests that each latent association between the eQTLs and the genes, as identified by each of the unit-rank components, is sparse in both the responses and the predictors, and the pattern of such sparsity is pathway specific. Moreover, it is known that the yeast MAPK pathways regulate and interact with each other. Therefore, a joint sparse reduced-rank regression analysis holds a great promise in decoding the complex genetic structures through simultaneously searching for multiple distinct association pathways between subsets of genes and subsets of genetic variants.

In this chapter, we first consider the situation where both, reduced-rank structure and column sparsity in the coefficient matrix are present. This notably leads to the *sparse reduced-rank regression* for dimension reduction and predictor selection (Bunea et al. 2012; Chen and Huang 2012). When the row-sparsity is also present, the model is extended to the *two-way sparse reduced-rank regression* that allows for both predictor and response selection. We then consider the more challenging situation where sparsity is imposed on the (generalized) singular value decomposition of the reduced-rank coefficient matrix. This leads to *co-sparse factor regression* to achieve factor-specific, two-way variable selection (Chen et al. 2012, 2022; Mishra et al. 2017).

13.2 Sparse Reduced-Rank Regression

13.2.1 Sparse Reduced-Rank Regression for Predictor Selection

Consider the multivariate linear regression model as stated in (1.1) or (10.6). The main objective is to estimate the unknown $m \times n$ dimensional coefficient matrix C when the dimensions could be much larger than the sample size T. Let $r = r(C)$

denote the rank of C, $r_x = r(\mathbf{X})$ denote the rank of $\mathbf{X}$, and $\mathcal{S}^*$ denote the index set of the nonzero columns in C with $s^* = |\mathcal{S}^*|$ being its cardinality. The *sparse reduced-rank regression* assumes that the coefficient matrix is possibly of reduced-rank with $r \leq \min(r_x, m)$, and it may have some zero columns, that is, $s^* \leq n$. As such, this parsimonious model aims to achieve simultaneous rank reduction and predictor selection. It is seen that in the unsupervised setting when $\mathbf{X} = I$, the setup is closely related to sparse principal component analysis (Zou et al. 2006; Witten et al. 2009; Johnstone and Lu 2009).

It is helpful to first gain an intuitive understanding of the potential parsimony offered by the simultaneous rank deficiency and column sparsity, through counting the effective number of free parameters. The reduced-rank structure, with $r \leq \min(m, r_x)$, can reduce the number of free parameters from mn to $(r_x + m - r)r$. The column sparsity, with $s^* \leq n$, can reduce the number of free parameters to ms^*. While either structure offers certain control of the model complexity, their combination, which gives a column-sparse reduced-rank structure, can further reduce the effective number of free parameters to $(s^* + m - r)r$. Because $\mathbf{X}$ can always be reduced to an $r_x \times T$ matrix with r_x independent rows that span the same row space of $\mathbf{X}$, we can always reasonably assume that $s^* \leq r_x$. Therefore, the number $(s^* + m - r)r$ could be much smaller than either $(r_x + m - r)r$ or ms^*. This further reduction is especially appealing in high-dimensional settings where the control of complexity offered by either the reduced-rank or the column-sparse structures alone may not be adequate.

Now consider the problem of estimating the coefficient matrix in the sparse reduced-rank regression model. We focus on a two-step procedure consisting of rank determination and rank-constrained sparse estimation. As shown in Bunea et al. (2012), this procedure provides strong theoretical and computational guarantees and is superior to several alternatives. These include the reverse approach of first performing predictor selection by sparse regression followed by reduced-rank regression with the selected set of predictors.

In the first step, we estimate the rank of the coefficient matrix by fitting a version of the high-dimensional reduced-rank regression, for example, the rank selection criterion (Bunea et al. 2011) and the adaptive nuclear-norm penalization method (Chen et al. 2013) covered in Chapter 10. Either method ensures the consistent estimation of the model rank under essentially the same signal to noise ratio condition, as presented in Theorem 10.2. In what follows, we denote the estimated rank as $\widehat{r}$.

In the second step, with the estimated rank $\widehat{r}$, we pursue column sparsity by minimizing a group Lasso regularized regression criterion under the rank constraint (Bunea et al. 2012),

$$\widehat{C} = \arg\min_{C} \left\{ \frac{1}{2} \|\mathbf{Y} - C\mathbf{X}\|_F^2 + \lambda \|C\|_{2,1} \right\} \qquad \text{s.t. } r(C) \leq \widehat{r}, \tag{13.1}$$

where λ is a tuning parameter, $\|C\|_{2,1}$ is the column-wise group ℓ_2 norm of C, that is, $\|C\|_{2,1} = \sum_{k=1}^{n} \|C_k\|_2$, and the minimization is over all $C \in \mathbb{R}^{m \times n}$ with rank less than or equal to $\widehat{r}$. Clearly, it must be noted that this approach subsumes both reduced-rank regression and sparse regression as special cases. When $\lambda = 0$, we get back to the classical reduced-rank regression setup; on the other hand, when $\widehat{r} = \min(r_x, m)$, the rank constraint becomes ineffective and we get back to the column-sparse group Lasso regression setup. Therefore, the two different dimension reduction strategies are jointly synthesized via the rank-constrained group Lasso regression in (13.1).

To gain more insights, we may consider an equivalent form of (13.1) by writing the coefficient matrix C in its reduced-rank decomposed form $C = AB$, where $A \in \mathbb{R}^{m\times\widehat{r}}$, $B \in \mathbb{R}^{\widehat{r}\times n}$, and $A'A = I_{\widehat{r}}$ (Chen and Huang 2012; Bunea et al. 2012). Here the orthogonality of A is imposed without any loss of generality, with a convenient consequence that $\|C\|_{2,1} = \|B\|_{2,1}$. The problem in (13.1) can then be equivalently expressed as

$$(\widehat{A}, \widehat{B}) = \arg\min_{A,B} \left\{ F(A, B; \lambda) \equiv \frac{1}{2}\|\mathbf{Y} - AB\mathbf{X}\|_F^2 + \lambda\|B\|_{2,1} \right\}, \text{ s.t. } A'A = I_{\widehat{r}}, \tag{13.2}$$

where the minimization is taken over all orthogonal $m \times \widehat{r}$ matrices A and all $\widehat{r} \times n$ matrices B.

The solution to the problem in (13.2) is not unique. In fact, it is unique up to a rotation by any $\widehat{r} \times \widehat{r}$ orthogonal matrix. More precisely, if $(\widehat{A}, \widehat{B})$ is a solution, then $(\widehat{A}Q', Q\widehat{B})$ is also a solution for any $Q \in \mathbb{R}^{\widehat{r}\times\widehat{r}}$ and $Q'Q = I_{\widehat{r}}$. One can verify that the two sets of solutions provide exactly the same objective function value, $F(\widehat{A}, \widehat{B}; \lambda) = F(\widehat{A}Q', Q\widehat{B}; \lambda)$. Despite such non-uniqueness, the column-sparse reduced-rank estimation still warrants meaningful predictor selection, owing to the invariance property of the column sparsity structure, namely the column sparsity in $C = AB$ implies the same column sparsity in B and in any rotated QB. Therefore, different solutions of (13.2) lead to the selection of exactly the same subset of predictors.

Further, one may wonder the possibility of pursing the entrywise sparsity in B, by, for example, replacing the group Lasso penalty $\lambda\|B\|_{2,1}$ with the Lasso penalty $\lambda\|B\|_1$. This seems straightforward and could possibly enable factor-specific predictor selection. Unfortunately, the problem is more complicated. In the current formulation, because B is determined only up to an orthogonal rotation, pursuing any entrywise sparsity in B may not carry a clear meaning due to the lack of rotation invariance. That is, an entrywise sparsity pattern in B can be easily destroyed or distorted through rotation and thus does not transfer across arbitrary QB to allow consistent variable selection. We will show in Section 13.3 that the pursuit of entrywise sparsity in the decomposed components of C becomes meaningful when additional identifiability condition is imposed; that is, a general sparsity pattern can be sought on a pre-specified and "fully" identified decomposition of C.

13.2.2 Computation

Solving directly the non-convex constrained optimization problem in (13.1) may be difficult. It turns out that the alternative formulation in (13.2) facilitates the optimization. Here we sketch a blockwise coordinate descent algorithm, in which A and B are alternately updated until reaching convergence.

For fixed B, the problem becomes an orthogonal Procrustes problem (Gower and Dijksterhuis 2004) with respect to the orthogonal matrix A, that is,

$$\min_{A} \|\mathbf{Y} - AB\mathbf{X}\|_F^2, \text{ s.t. } A'A = I_{\widehat{r}}. \tag{13.3}$$

This problem admits an explicit solution: $A = UV'$, where $U \in \mathbb{R}^{m\times\widehat{r}}$ and $V \in \mathbb{R}^{\widehat{r}\times\widehat{r}}$ are, respectively, the left and right singular matrix of $\mathbf{Y}\mathbf{X}'B'$.

Now consider the optimization with respect to B for fixed A. Denote $A^{\perp} \in \mathbb{R}^{m\times(m-\widehat{r})}$ as the orthogonal complement matrix of A such that $A^{\perp\prime}A^{\perp} = I_{m-\widehat{r}}$ and $A'A^{\perp} = 0$; that is, the combined $m \times m$ matrix $[A, A^{\perp}]$ satisfies $[A, A^{\perp}]'[A, A^{\perp}] = I_m$. Then it follows that $\|\mathbf{Y} - AB\mathbf{X}\|_F^2 = \|[A, A^{\perp}]'(\mathbf{Y} - AB\mathbf{X})\|_F^2 = \|A'\mathbf{Y} - B\mathbf{X}\|_F^2 + \|A^{\perp\prime}\mathbf{Y}\|_F^2$. Therefore, for fixed A, the problem with respect to B becomes,

$$\min_{B} \frac{1}{2}\|A'\mathbf{Y} - B\mathbf{X}\|_F^2 + \lambda\|B\|_{2,1}. \tag{13.4}$$

This problem can be recognized as a multivariate group Lasso regression or a special multi-task learning method as discussed in Section 10.5.1. It can be written as a standard group Lasso problem because $\|A'\mathbf{Y} - B\mathbf{X}\|_F^2 = \|\text{vec}(A'\mathbf{Y}) - (\mathbf{X}' \otimes I_{\widehat{r}})\text{vec}(B))\|$, where vec denotes the column-wise vectorization operator and $\otimes$ denotes the Kronecker product. This ℓ_1 problem can be solved efficiently by various methods (Yuan and Lin 2006; Yang and Zou 2015).

The computational algorithm is summarized in Algorithm 3. It proceeds by alternately updating A and B by solving (13.3) and (13.4), respectively. The objective function is monotonically descending along the iterations and the sequence of iterates $(A^{(t)}, B^{(t)})$ are uniformly bounded; in fact, given any $\lambda > 0$ and

---

Algorithm 3 Rank-constrained group Lasso regression

Input: $1 \le \widehat{r} \le \min(m, r_x)$, $\lambda \ge 0$, $A^{(0)} \in \mathbb{R}^{m\times\widehat{r}}$ with $A^{(0)\prime}A^{(0)} = I_{\widehat{r}}$, $t = 0$.
Repeat:
(a) $B^{(t+1)} \leftarrow \min_B\{\|A^{(t)\prime}\mathbf{Y} - B\mathbf{X}\|_F^2/2 + \lambda\|B\|_{2,1}\}$;
(b)Perform singular value decomposition of $\mathbf{Y}\mathbf{X}'B^{(t+1)\prime}$, i.e.,
$\mathbf{Y}\mathbf{X}'B^{(t+1)\prime} = U^{(t+1)}D^{(t+1)}V^{(t+1)\prime}$;
(c) $A^{(t+1)} \leftarrow U^{(t+1)}V^{(t+1)\prime}$;
$t \leftarrow t + 1$;
Until: convergence, e.g., $|F(A^{(t+1)}, B^{(t+1)}; \lambda) - F(A^{(t)}, B^{(t)}; \lambda|/|F(A^{(t)}, B^{(t)}; \lambda)| < \epsilon$.

---

any initial orthogonal matrix $A^{(0)}$, Algorithm 3 is guaranteed to converge to a stationary point of F (Bunea et al. 2012). This is typically the best we can hope for by using blockwise coordinate descent to solve a non-convex problem (Tseng 2001). In practice, the rank and the initial value of A are obtained from reduced-rank regression. The convergence is reached when the relative decrease of the objective function in (13.2) or the relative change in the estimated coefficient matrix $\widehat{C} = \widehat{A}\widehat{B}$ (as measured by, e.g., Frobenius norm) becomes smaller than a pre-specified tolerance level. To obtain the final estimate, one can run Algorithm 3 to generate a series of candidate estimators for a grid of λ values with the pre-determined $\widehat{r}$. The optimal solution is then selected by cross validation or by a proper information criterion (She 2017; She and Tran 2019).

13.2.3 Theoretical Analysis

We now examine the performance of the two-step sparse reduced-rank estimator. The following assumptions are imposed (Bunea et al. 2012).

Assumption 13.1 The error matrix $\boldsymbol{\epsilon}$ has independent $N(0, \sigma^2)$ entries.

Assumption 13.2 $d_r(C\mathbf{X}) > 2\sqrt{2}\sigma(\sqrt{r_x} + \sqrt{m})$.

Assumption 13.3 $\log(\|C\mathbf{X}\|_F) \leq (\sqrt{2} - 1)^2(r_x + m)/4$.

Let $\Sigma_x = \mathbf{X}\mathbf{X}'/T$. We say $\Sigma_x \in \mathbb{R}^{n\times n}$ satisfies a restricted eigenvalue (RE) condition RE$(\mathcal{S}, \delta_{\mathcal{S}})$ for an index set $\mathcal{S} \subseteq \{1, \ldots, n\}$ and a positive number $\delta_{\mathcal{S}}$, if and only if $\mathrm{tr}\big(M\Sigma_x M'\big) \geq \delta_{\mathcal{S}} \sum_{i\in\mathcal{S}} \|M_i\|_2^2$ for all $M = [M_1, \ldots, M_n] \in \mathbb{R}^{m\times n}$ satisfying $\sum_{i\in\mathcal{S}} \|M_i\|_2 \geq 2\sum_{i\in\mathcal{S}^c} \|M_i\|_2$.

Assumption 13.4 $\Sigma_x \in \mathbb{R}^{n\times n}$ satisfies the restricted eigenvalue condition RE$(\mathcal{S}^*, \delta_{\mathcal{S}^*})$, where $\mathcal{S}^* \neq \emptyset$ is the index set of the nonzero columns in the true coefficient matrix C and $d_1^2(\Sigma_x)/\delta_{\mathcal{S}^*}$ is bounded with $d_1^2(\Sigma_x)$ being the largest eigenvalue of Σ_x.

Assumptions 13.1–13.2 are essentially the same as Assumptions 10.1–10.2 in Section 10.4.3 for establishing the rank consistency of reduced-rank regression. Assumption 13.1 implies that the noise level, measured by $d_1(\boldsymbol{\epsilon}\mathcal{P}_X)$, is bounded by $2\sqrt{2}\sigma(\sqrt{r}_x + \sqrt{m})$. Assumption 13.2 then requires that the signal level, measured by $d_r(C\mathbf{X})$, is larger than the noise level, so that the correct rank can be recovered through reduced-rank regression method with high probability. Assumption 13.3 is needed to ensure that the error due to rank selection is negligible compared to the rate $mr+s^*r\log(n)$. Assumption 13.4 is essential for high-dimensional setups when $n > T$; it is commonly adopted in the high-dimensional variable selection problems (Bickel et al. 2009; Bühlmann and van de Geer 2011) and is a specialized case of the general RE condition discussed in Section 10.5.2.

Theorem 13.1 *Suppose Assumptions 13.1–13.4 hold. The two-step estimator* $\widehat{C}$, *with rank* $\widehat{r}$ *selected by reduced-rank regression and the tuning parameter set as* $\lambda = c\sigma\sqrt{d_1^2(\Sigma_x)\widehat{r}T\log(en)}$ *for a large enough constant c, satisfies*

$$E\left(\|\widehat{C}\mathbf{X} - C\mathbf{X}\|_F^2\right) \preceq (m + s^* \log(n))r.$$

The results imply that the two-step estimator $\widehat{C}$ is *column-sparse and rank adaptive*, in that neither the best number of predictors nor the correct rank has to be specified prior to estimation. Further analysis reveals that this procedure achieves the minimax optimal rate up to a $\log(n)$ factor.

It is interesting to compare the above established prediction error rate, $(m + s^* \log(n))r$, to those of the reduced-rank regression estimator and the group Lasso estimator, which are $(m + r_x)r$ and $(m + \log(n))s^*$, respectively. Note that we generally have $r \leq \min(m, s^*)$ and $s^* \leq r_x$. When the number of responses is much larger than that of the effective dimension of the predictors, $m \geq r_x$, the rates of both the reduced-rank regression and the sparse reduced-rank regression are dominated by mr regardless of s^*. As such, the effect of pursuing column sparsity in C may not be essential as comparing with rank reduction. On the other hand, when $m < r_x$ and the rank r is small, the sparse reduced-rank regression can provide substantial improvements over the other two methods.

It is worth mentioning some related follow-up work. She (2017) considered selective reduced-rank regression with non-convex group-wise penalty forms that are able to achieve a sharper error bound; a predictive information criterion was also proposed for tuning the rank and the column support of the coefficient matrix. Bayesian sparse reduced-rank regression has been developed in parallel; see, e.g., Zhu et al. (2014), Goh et al. (2017), and Chakraborty et al. (2019).

13.3 Co-sparse Factor Regression

13.3.1 Model Formulation and Deflation Procedures

As discussed in the previous section, the pursuit of entrywise sparsity in the decomposed components of C may enable factor-specific variable selection, but it is only meaningful on a pre-specified and "fully" identified decomposition. Also recall that reduced-rank regression can be regarded as a latent factor model and thus naturally connects to factor analysis and principal component analysis: all these methods aim to identify certain low-dimensional subspace to represent $\mathbf{Y}$, and their main difference is that reduced-rank regression conducts a supervised search in the row space of $\mathbf{X}$ while the other two methods are unsupervised.

Motivated by the connection to factor models, the following factorization of the reduced-rank component $C\mathbf{X}$ is appealing in the sparsity pursuit,

$$\frac{1}{\sqrt{T}}C\mathbf{X} = AD\left(\frac{1}{\sqrt{T}}B\mathbf{X}\right), \text{ s.t. } \left(\frac{1}{\sqrt{T}}B\mathbf{X}\right)\left(\frac{1}{\sqrt{T}}B\mathbf{X}\right)' = A'A = I_r, \qquad (13.5)$$

where $C = ADB$ with $A = [\alpha_1, \ldots, \alpha_r] \in \mathbb{R}^{m\times r}$ and $B = [\beta_1, \ldots, \beta_r]' \in \mathbb{R}^{r\times n}$, and $D = \text{diag}\{d_1, \ldots, d_r\}$ being a diagonal matrix of positive singular values. Observe that this decomposition is not the conventional singular value decomposition of C, because we require $B(\mathbf{XX}'/T)B' = I_r$ rather than $BB' = I_r$. Let $\widehat{\Sigma}_x = \mathbf{XX}'/T$, which is the sample-version covariance matrix of the predictors. Then we refer to the decomposition in (13.5) as the $\widehat{\Sigma}_x$-orthogonal singular value decomposition of C or simply singular value decomposition of C when no confusion arises. For convenience, we denote

$$C_{[k]} = d_k\alpha_k\beta_k', \qquad k = 1, \ldots, r, \qquad (13.6)$$

where α_k is the kth column of A, β_k' is the kth row of B, and d_k is the kth diagonal element of D. We call $C_{[k]}$ as the kth layer $\widehat{\Sigma}_x$-orthogonal singular value decomposition of C. As a matter of fact, (13.5) is almost the same as the normalization adopted in the classical reduced-rank regression, c.f., (2.14) in Section 2.3, except that here the singular values are separated out.

An important attribute of the $\widehat{\Sigma}_x$-orthogonal singular value decomposition is that the r latent predictors/factors, i.e., $(1/\sqrt{T})\mathbf{X}'\beta_k$, $k = 1, \ldots, r$, are required to be orthogonal/uncorrelated to each other. This makes the reduced-rank regression analysis in analogous to the unsupervised factor analysis, which also aims to identify a set of uncorrelated factors to avoid redundancy and facilitate interpretation. Also, compared to the decomposition adopted in the sparse reduced-rank regression in (13.2), the additional constraint on B now makes both A and B identifiable.

Now it is in order to consider imposing sparsity on both A and B, giving arise to the delicate *"co-sparse reduced-rank"* structure. There are several prominent features in this formulation:

(1) It produces r latent predictors/factors, $(1/\sqrt{T})\mathbf{X}'\beta_k$, $k = 1, \ldots, r$, each of which is constructed as a linear combination of a subset of the original predictors due to sparsity of each β_k (factor-specific predictor selection);
(2) Each latent factor is allowed to be related to a subset of the response variables due to sparsity of each α_k (factor-specific response selection);
(3) The latent factors are uncorrelated with each other, due to the orthogonality of $(1/\sqrt{T})\mathbf{X}'B'$;
(4) The singular values d_k reflect the relative strengths of the association between the responses and the latent factors.

We stress that we desire the elements of the matrix factorization, the left and right singular vectors, to be entrywisely sparse. This could be much stronger than just requiring the coefficient matrix C itself to be sparse. A matrix with sparse singular vectors could be sparse itself, but not vice versa in general. The appealing features of such sparse factorization methods have been well demonstrated in various scientific applications (Vounou et al. 2010; Chen et al. 2014; Ma et al. 2014).

Now consider the recovery of the $\widehat{\Sigma}_x$-orthogonal singular value decomposition of C under the co-sparse factor model, specified by (10.6) and (13.5). The simultaneous presence of low-rank and co-sparsity structure makes the estimation problem very challenging. A joint estimation of all sparse singular vectors may necessarily involve identifiability constraints related to orthogonality (Uematsu et al. 2019), leading to non-convex and non-smooth optimization that is often computationally intensive or can be even intractable.

Motivated by the power method for computing singular value decomposition (Golub and Van Loan 1996), we focus on deflation-based approaches to tackle this problem; the main idea is to estimate the unit-rank components of $C\mathbf{X}$ one by one, thus reducing the problem into a set of much simpler unit-rank problems. Specifically, the hope is to divide the multi-rank factorization problem into a set of unit-rank problems of the following form:

$$\min_{C \in \mathbb{R}^{m \times n}} \{L(C; \mathbf{Y}, \mathbf{X}) + \rho(C; \lambda)\}, \quad \text{s.t.}\, r(C) \leq 1, \tag{13.7}$$

where $L(C; \mathbf{Y}, \mathbf{X}) = (1/2)\|\mathbf{Y} - C\mathbf{X}\|_F^2$ is the sum of squares loss function, and $\rho(C; \lambda) = \lambda\|C\|_1$ is the ℓ_1 penalty term with tuning parameter $\lambda \geq 0$. This criterion targets the best sparse unit-rank approximation of $\mathbf{Y}$ in the row space of $\mathbf{X}$. Intriguingly, by writing the unit-rank matrix C as $C = d\alpha\beta'$, where $d > 0$, $\alpha \in \mathbb{R}^m$, and $\beta \in \mathbb{R}^n$, the problem can be equivalently expressed as a *co-sparse unit-rank regression* (CURE),

$$\begin{aligned} &\min_{d,\alpha,\beta} \left\{L(d\alpha\beta'; \mathbf{Y}, \mathbf{X}) + \lambda\|d\alpha\beta'\|_1\right\}, \\ &\text{s.t.}\, d \geq 0, T^{-1/2}\|\mathbf{X}'\beta\|_2 = 1, \|\alpha\|_2 = 1. \end{aligned} \tag{13.8}$$

The penalty term is multiplicative in that $\|C\|_1 = \|d\alpha\beta'\|_1 = d\|\alpha\|_1\|\beta\|_1$, which conveniently promotes the desired co-sparse structure. This is not really surprising, because the sparsity of a unit-rank matrix directly leads to the sparsity in both its left and right singular vectors. To be specific, if the (i, j)th entry of $C = d\alpha\beta'$ is zero, it must be true that either the ith entry of α is zero or the jth entry of β is zero, or both. Therefore, the equivalency between the above two formulations reveals that CURE is capable of achieving a "fair" co-sparse regularization on both α and β with a single penalty term.

We present two general deflation approaches to reach the desired simplification in (13.7), or equivalently, (13.8).

The first approach is called "*sequential pursuit*" (Mishra et al. 2017; Zheng et al. 2019). As its name suggests, this method extracts unit-rank factorization of the coefficient matrix C in a sequential fashion, each time with the previously estimated component removed from the current residual matrix. Specifically, the method sequentially solves

$$\widehat{C}_{[k]}(\lambda) = \arg\min_{C} \left\{ L(C; \mathbf{Y} - \sum_{j=0}^{k-1} \widehat{C}_{[j]}\mathbf{X}, \mathbf{X}) + \rho(C; \lambda) \right\}, \qquad \text{s.t. } r(C) \leq 1, \tag{13.9}$$

for $k = 1, \ldots r$, where $\widehat{C}_{[0]} = 0$, and $\widehat{C}_{[k]} = \widehat{d}_k \widehat{\alpha}_k \widehat{\beta}_k'$ is the selected unit-rank solution (through tuning by cross-validation or some information criterion) in the kth step.

The second approach is called *parallel pursuit* (Chen et al. 2012; Chen and Chan 2016; Chen et al. 2022). It requires an initial estimator of the coefficient matrix C and its corresponding decomposition according to (13.5); some typical choices in practice include the reduced-rank regression estimator and the ℓ_1-penalized least squares estimator. The estimation of each rank one component $C_{[k]}$ can then be isolated by removing from $\mathbf{Y}$ the initially estimated signals from the other components. To be specific, let $\widetilde{C} = \sum_{k=1}^{r} \widetilde{C}_{[k]} = \sum_{k=1}^{r} \widetilde{d}_k \widetilde{\alpha}_k \widetilde{\beta}_k'$ be the initial estimator of the true coefficient matrix C and its Σ_x-orthogonal singular value decomposition. Then the parallel pursuit solves the problems

$$\widehat{C}_{[k]}(\lambda) = \arg\min_{C} \left\{ L(C; \mathbf{Y} - \sum_{j \neq k}^{r} \widetilde{C}_{[j]}\mathbf{X}, \mathbf{X}) + \rho(C; \lambda) \right\}, \qquad \text{s.t. } r(C) \leq 1, \tag{13.10}$$

for $k = 1, \ldots, r$. Since these r problems are completely separated, they can be implemented in parallel.

Both deflation approaches are intuitive and are commonly used in matrix decomposition related problems (Witten et al. 2009; Lee et al. 2010; Chen et al. 2012; Allen et al. 2014). There are both advantages and disadvantages for using these approaches. The parallel pursuit approach requires an initial estimator to isolate the estimation of the unit-rank components, and refines the estimation of each component around the vicinity of its initial estimator. As such, the performance of the parallel pursuit relies on the quality of the initial estimator. The sequential pursuit approach, on the other hand, keeps extracting unit-rank components from the current residuals, subject to additional sparsity regularization. This is motivated by the fact that the solution of the reduced-rank regression, which solves $\min_C \|\mathbf{Y} - C\mathbf{X}\|_F^2$ s.t. $r(C) \leq r$, can be exactly recovered by sequentially fitting r unit-rank regression problems with the residuals from the previous step, due to the celebrated Eckart-Young Theorem (Eckart and Young 1936). While the sequential method does not require an initial estimator, it may suffer from a noise accumulation effect, because the estimation of each later component is impacted by the estimation of all previous components.

We remark that both deflation methods give up the exact orthogonality requirement among different components, which make them much more computationally scalable than any joint estimation approach. From a statistical estimation point of view, it can be argued that the exact orthogonality is never necessary: although what we aim to recover is a set of uncorrelated latent factors, the estimators of these factors, which are based on finite data, do not necessarily have to be exactly uncorrelated. As long as the estimators are consistent, we can achieve asymptotic orthogonality.

13.3.2 Co-sparse Unit-Rank Estimation

It remains to solve the CURE problem in (13.7) or (13.8). The alternating convex search (ACS) algorithm, or blockwise coordinate descent, is natural to be applied here, in which the objective function is alternately optimized with respect to a (overlapping) block of parameters, (d, α) or (d, β). Realizing that the objective function is a function of (d, α) or (d, β) only through the products $d\alpha$ or $d\beta$, respectively, the parameter constraints can all be avoided in the blockwise routines.

First, consider the update of (d, β) with fixed α satisfying $\|\alpha\|_2 = 1$. The relevant constraints are $d \geq 0, T^{-1/2}\|\mathbf{X}'\beta\|_2 = 1$. Because the objective function is a function of (d, β) through $\check{\beta} = d\beta$, the constraints are avoided when optimizing with respect to $\check{\beta}$. It essentially boils down to solving the following standard Lasso problem,

$$\min_{\check{\beta}} \left\{ \frac{1}{2}\|y - \mathbf{X}^{(\alpha)}\check{\beta}\|_2^2 + \lambda^{(\alpha)}\|\check{\beta}\|_1 \right\}, \tag{13.11}$$

where $y = \text{vec}(\mathbf{Y}')$, $\mathbf{X}^{(\alpha)} = \alpha \otimes \mathbf{X}'$, and $\lambda^{(\alpha)} = \lambda\|\alpha\|_1$. Once the solution of $\check{\beta}$ is obtained, we standardize $d = \|\mathbf{X}'\check{\beta}\|_2/\sqrt{T}$ and $\beta = \check{\beta}/d$ to satisfy the constraints whenever $\check{\beta} \neq 0$, otherwise we update $d = 0$ and terminate the algorithm. Similarly, for fixed β satisfying $T^{-1/2}\|\mathbf{X}'\beta\|_2 = 1$, the minimization of (13.8) with respect to (d, α) becomes minimization with respect to $\check{\alpha} = d\alpha$,

$$\min_{\check{\alpha}} \left\{ \frac{1}{2}\|y - \mathbf{X}^{(\beta)}\check{\alpha}\|_2^2 + \lambda^{(\beta)}\|\check{\alpha}\|_1 \right\}, \tag{13.12}$$

where $\mathbf{X}^{(\beta)} = I_m \otimes (\mathbf{X}'\beta)$, and $\lambda^{(\beta)} = \lambda\|\beta\|_1$. Once $\check{\alpha}$ is obtained, we standardize $d = \|\check{\alpha}\|_2$ and $\alpha = \check{\alpha}/d$ whenever $\check{\alpha} \neq 0$; again, when $\check{\alpha} = 0$, we set $d = 0$ and terminate the algorithm. Mishra et al. (2017) showed that this ACS algorithm is empirically stable and is guaranteed to converge to a coordinatewise minimum point of (13.8).

To further speed up computation, stagewise learning techniques can be applied to approximate the ACS solution paths; we refer to Zhao and Yu (2007), He et al. (2018), and Chen et al. (2022) for more details.

13.3.3 Theoretical Analysis

We now analyze the statistical properties of the deflation-based estimators of the sparse SVD components, $\widehat{C}_{[k]}$, $k = 1, \ldots, r$, with tuning parameters λ_k, $k = 1, \ldots, r$, respectively, from either the sequential pursuit in (13.9) or from the parallel pursuit in (13.10).

For any m by n nonzero matrix Δ, denote by $\Delta_{\mathcal{S}}$ the corresponding matrix of the same dimension that keeps the entries of Δ with indices in set $\mathcal{S}$ while sets the others to be zero. We need the following assumptions.

Assumption 13.5 The random noise vectors are independently and identically distributed as $\epsilon_k \sim N(0, \Sigma_{\epsilon\epsilon})$, $k = 1, \ldots, T$. Denote the jth diagonal entry of $\Sigma_{\epsilon\epsilon}$ as σ_j^2; we assume $\sigma_{\max}^2 = \max_{1 \le j \le m} \sigma_j^2$ is bounded from above.

Assumption 13.6 For $1 \le j \le r$, the gaps between the successive singular values are positive, i.e., $\delta_j = d_j - d_{j+1} > 0$, where $d_j > 0$ is the jth singular value of $T^{-1/2}C\mathbf{X}$.

Assumption 13.7 There exists certain sparsity level s with a positive constant ρ_l such that

$$\inf_{\Delta} \left\{ \frac{\|\Delta \mathbf{X}\|_F^2}{T(\|\Delta_{\mathcal{S}}\|_F^2 \vee \|\Delta_{\mathcal{S}_s^c}\|_F^2)} : |\mathcal{S}| \le s,\ \|\Delta_{\mathcal{S}^c}\|_1 \le 3\|\Delta_{\mathcal{S}}\|_1 \right\} \ge \rho_l,$$

where $\Delta_{\mathcal{S}_s^c}$ is formed by keeping the s entries of $\Delta_{\mathcal{S}^c}$ with largest absolute values and setting the others to be zero.

Assumption 13.8 There exists certain sparsity level s with a positive constant ϕ_u such that

$$\sup_{\Delta} \left\{ \frac{|\mathcal{S}|^{1/2} \|\Delta \mathbf{X}\mathbf{X}'\|_\infty}{T\|\Delta_{\mathcal{S}}\|_F} : |\mathcal{S}| \le s,\ \|\Delta_{\mathcal{S}^c}\|_1 \le 3\|\Delta_{\mathcal{S}}\|_1 \right\} \le \phi_u.$$

Assumption 13.5 assumes the random errors are Gaussian; the technical argument still applies as long as the tail probability bound of the noise decays exponentially. Assumption 13.6 is imposed to ensure the identifiability of the r latent factors (singular vectors), without which the targeted unit-rank matrices would not be distinguishable. Similar assumptions can be found in Mishra et al. (2017), Zheng et al. (2019), among others.

Assumption 13.7 is a matrix version of the restricted eigenvalue (RE) condition proposed in Bickel et al. (2009), which is typically imposed in ℓ_1-penalization to restrict the correlations between the columns of $\mathbf{X}$ within certain sparsity level, thus guaranteeing the identifiability of the true regression coefficients. Here the difference is that we consider the estimation of coefficient matrices instead of vectors. Because the Frobenius norm of a matrix can be regarded as the ℓ_2-norm of the stacked vector consisting of the columns of the matrix, Assumption 13.7 is equivalent to the RE condition in the univariate response setting. The integer s here acts as a theoretical upper bound on the sparsity level of the true coefficient matrices, the requirement of which will be shown to be different in the two deflation approaches.

Similar to Assumption 13.7, Assumption 13.8 is a matrix version of the cone invertibility factor (Ye and Zhang 2010) type of condition. It allows us to control the entrywise estimation error by assuming that $T^{-1}\|\Delta \mathbf{X}\mathbf{X}'\|_\infty$ can be bounded by some multiple of the averaged error $\|\Delta_{\mathcal{S}}\|_F/|\mathcal{S}|^{1/2}$ when Δ is restricted in the cone. In this way, the impact of the estimation error of the initial Lasso estimate can spread out on the support instead of massing at one component.

Now we are ready to present the main results. Denote by $s_k = \|C_{[k]}\|_0$ for $1 \le k \le r$ and $s_0 = \|C\|_0$. The following theorem characterizes the estimation accuracy in different layers of the sequential pursuit.

Theorem 13.2 (Sequential Pursuit) *Suppose Assumptions 13.5–13.8 hold with the sparsity level $s \ge \max_{1 \le k \le r} s_k$. Choose $\lambda_1 = 2\sigma_{\max}\sqrt{2\alpha \log(mn)T}$ for some constant $\alpha > 1$ and $\lambda_k = \Pi_{\ell=1}^{k-1}(1+\eta_\ell)\lambda_1$. The following results hold uniformly over $1 \le k \le r$ with probability at least $1-(mn)^{1-\alpha}$,*

$$\|\widehat{\Delta}_k\|_F^2 = \|\widehat{C}_{[k]} - C_{[k]}\|_F^2 = O(\gamma_k^2 \theta_k^2 s_k \log(mn)/T),$$

where $\theta_k = d_k/\delta_k$, $\eta_k = c\theta_k$ with constant $c = 24\phi_u \rho_l^{-1}$, and $\gamma_k = \Pi_{\ell=1}^{k-1}(1+\eta_\ell)$ with $\gamma_1 = 1$.

Theorem 13.2 presents the estimation error bounds in terms of the squared Frobenius norm for the co-sparse unit-rank matrix estimators in sequential pursuit under mild and reasonable conditions. This is non-trivial because the subsequent layers play the role of extra noises in the estimation of each unit-rank matrix. We address this issue by showing that its impact is secondary due to the orthogonality between different layers. The results hold uniformly over all true layers with a high probability approaching to one in polynomial order of mn.

The regularization parameter λ_k reflects the minimum penalization level needed to suppress the noise in the kth layer, which consists of two parts. One is from the random noise accompanied with the original response variables, while the other is due to the accumulation of the estimation errors from all previous layers. The latter is of a larger magnitude than the former one such that the penalization level increases almost exponentially with k. This may be inevitable in the sequential procedure since the kth layer estimator is based on the residual response matrix after extracting the previous layers. Hopefully, the estimation consistency is generally guaranteed for all the significant unit-rank matrices as long as the true regression coefficient matrix C is of sufficiently low rank.

When the singular values are well separated, for the first few layers, the estimation accuracy is about $O(s_k \log(mn)/T)$, which is close to the optimal rate $O((s_u + s_v)\log(n \vee m)/T)$ for the estimation of C established in Ma et al. (2020).

When the true rank is one, the main difference between the two rates lies in the sparsity factors, where $(s_u + s_v)$ is a sum of the sparsity levels of the left and right singular vectors, while the sparsity factor in the sequential pursuit is the product of them. However, the optimal rate is typically attained through some non-convex algorithms (Ma et al. 2020; Yu et al. 2020; Uematsu et al. 2019) that search for the optimal solution in a neighborhood of C. In contrast, the CURE algorithm enjoys better computational efficiency and stability.

Last but not least, in view of the correlation constraint on the design matrix $\mathbf{X}$, that is, $s \geq \max_{1 \leq k \leq r} s_k$, the sequential pursuit is valid as long as the number of nonzero entries in each layer $C_{[k]}$ is within the sparsity level s imposed in Assumptions 13.7 and 13.8. It means that in practice, the sequential deflation strategy can recover some complex association networks layer by layer, which is not shared by either the parallel pursuit or other methods that directly target on estimating the multi-rank coefficient matrix.

We then turn our attention to the statistical properties of the parallel pursuit. The key point of this deflation strategy is to find a relatively accurate initial estimator such that the signals from the other layers/components except the targeted one can be approximately removed. The reduced-rank regression estimator may not be a good choice under high dimensions, since it lacks sparsity constraints. Similar to Uematsu et al. (2019), we mainly adopt the following Lasso initial estimator,

$$\widetilde{C} = \arg\min_{C} \frac{1}{2} \|Y - C\mathbf{X}\|_F^2 + \lambda_0 \|C\|_1,$$

which can be efficiently solved by various algorithms (Friedman et al. 2007). Under the same RE condition (Assumption 13.7) as the sequential pursuit on the design matrix $\mathbf{X}$ with a tolerated sparsity level $s \geq s_0$, where s_0 indicates the number of nonzero entries in C, we can show that this Lasso initial estimator is consistent with a convergence rate of $O(s_0 \log(mn)/T)$.

Based on $\widetilde{C}$, we can obtain the initial unit-rank estimates $\widetilde{C}_{[k]}^0$ for different layers through a $\widehat{\Sigma}_x$-orthogonal SVD. However, these unit-rank estimates are not guaranteed to be sparse even if the Lasso estimator $\widetilde{C}$ is a sparse one, which causes additional difficulties in high dimensions. Thus, to facilitate the theoretical analysis, our initial kth layer estimate $\widetilde{C}_{[k]}$ takes the s largest components of $\widetilde{C}_{[k]}^0$ in terms of absolute values while sets the others to be zero, where s is the upper bound on the sparsity level defined in Assumptions 13.7 and 13.8. These s-sparse initial estimates $\widetilde{C}_{[k]}$ can maintain the estimation accuracy of the initial unit-rank estimates $\widetilde{C}_{[k]}^0$ without imposing any assumption on signal strength, regardless of the thresholding procedure. In practice, we can use the SVD estimates directly as the impact of fairly small entries is negligible.

Theorem 13.3 (Parallel Pursuit) *Suppose Assumptions 13.5–13.8 hold with the sparsity level* $s \geq \max_{0 \leq k \leq r} s_k$. *Choose* $\lambda_0 = 2\sigma_{\max}\sqrt{2\alpha \log(mn)T}$ *with constant* $\alpha > 1$ *for the initial Lasso estimator* $\widetilde{C}$, *and* $\lambda_k = 2c\psi_k\sqrt{\log(mn)T}$ *for some positive constant* c. *The following results hold uniformly over* $1 \leq k \leq r$ *with probability at least* $1 - (mn)^{1-\alpha}$,

$$\|\widehat{\Delta}_k\|_F^2 = \|\widehat{C}_{[k]} - C_{[k]}\|_F^2 = O(\psi_k^2 s_k \log(mn)/T),$$

where $\psi_k = d_1 d_c/(d_k \min[\delta_{k-1}, \delta_k])$ *with* d_c *the largest singular value of* C.

Based on the Lasso initial estimator, Theorem 13.3 demonstrates that the parallel pursuit achieves accurate estimation of different layers and reduces the sparsity factor in the convergence rates from s_0 to s_k, where the sparsity level s_0 of the entire regression coefficient matrix can be about r times of s_k. Moreover, as there is no accumulated noises in the parallel estimation of different layers, we do not see an accumulation factor γ_k here such that the penalization parameters keep around a uniform magnitude. The eigen-factor ψ_k plays a similar role as θ_k in the convergence rates for sequential pursuit in Theorem 13.2, both of which can be bounded from above under the low-rank and sparse structures. Thus, the estimation accuracy of all the significant unit-rank matrices is about the same as that of the first layer in the sequential pursuit, which reveals the potential superiority of the parallel pursuit.

In summary, the sequential and the parallel pursuits are very similar, while the main difference lies in the technical requirement on the sparsity level s. As discussed after Assumption 13.7, the integer s constrains the correlations between the columns of $\mathbf{X}$ to ensure the identifiability of the true supports, thus can be regarded as fixed for a given design matrix. Since the parallel pursuit requires not only the sparsity level s_k of each individual layer but also the overall sparsity level s_0 to be no larger than s, it puts a stricter constraint on correlations among the predictors. In other words, when the true coefficient matrix C is not sufficiently sparse, the Lasso initial estimator may not be accurate enough to facilitate the subsequent parallel estimation, in which case the sequential pursuit could enjoy some advantage.

13.4 Applications

13.4.1 *Yeast eQTL Mapping Analysis*

In an expression quantitative trait loci (eQTLs) mapping analysis, the main objective is to examine the association between the eQTLs, i.e., regions of the genome containing DNA sequence variants, and the expression levels of the genes in certain signaling pathways. Biochemical evidence often suggests that there exist a few functionally distinct signaling pathways of genes, each of which may involve only a subset of genes and correspondingly a subset of eQTLs. Therefore, the recovery of such association structure can be formulated as a sparse factor regression problem, with the gene expressions being the responses and the eQTLs being the predictors. Here, we analyze the yeast eQTL data set described by Brem and Kruglyak (2005) and Storey et al. (2005), to illustrate the power of sparse and low-rank approaches. There are $n = 3244$ genetic markers and $m = 54$ genes that belong to the yeast Mitogen-activated protein kinases (MAPKs) signaling pathway (Kanehisa et al. 2009), with data collected from $T = 112$ yeast samples.

A marginal screening approach can be used to reduce the number of marker locations, to alleviate the computational burden of having very high-dimensional predictors. Specifically, the markers are combined into blocks such that markers with the same block differed by at most one sample, and one representative marker is chosen from each block. A marginal gene-marker association analysis is then performed to identify markers that are associated with the expression levels of at least two genes with a p-value less than 0.05, resulting in a total of $n = 605$ markers. It is also worth trying the inclusion of all the $n = 3244$ markers for a joint regression without marginal screening. Therefore, we try both approaches in this analysis.

We consider the classical reduced-rank regression (RRR), the sparse reduced-rank regression (SRRR), the parallel co-sparse factor regression (PS-FAR) initialized by RRR, and the sequential co-sparse factor regression (SS-FAR). For the co-sparse unit-rank regression (CURE) procedure, we apply the stagewise estimation technique to speed up the computation (Chen et al. 2022). Under either setting, with or without marginal screening, we perform a random-splitting procedure to compare the performance of different methods. To make the comparison fair, all the methods have the same pre-specified rank, which is selected from reduced-rank regression (RRR) via 10-fold cross validation. Specifically, each time the data set is randomly split into 80% for model fitting and 20% for computing the out-of-sample mean squared error (MSE) of the fitted model. Also recorded are the number of nonzero entries in $\widehat{B}$ ($\|\widehat{B}\|_0$), the number of nonzero columns in $\widehat{B}$ ($\|\widehat{B}\|_{2,0}$), the number of nonzero entries in $\widehat{A}$ ($\|\widehat{A}\|_0$), and the number of nonzero rows in $\widehat{A}$ ($\|\widehat{A}\|_{2,0}$). The procedure is repeated 100 times, and to make a robust comparison we compute the 10% trimmed means and standard deviations of the above performance measures.

Table 13.1 reports the results for the setting with marginal screening. All the methods that pursue sparse and low-rank structures outperform the benchmark reduced-rank regression method in term of out-of-sample prediction performance; the PS-FAR method performs the best, although the improvement is not substantial compared to other close competitors. The results in the setting of $n = 3244$ without marginal screening are reported in Table 13.2. The predictive performance of all

Table 13.1 Yeast eQTL mapping analysis: Results with marginal screening

Method	$\|\widehat{B}\|_0$	$\|\widehat{B}\|_{2,0}$	$\|\widehat{A}\|_0$	$\|\widehat{A}\|_{2,0}$	MSE
RRR	1815 (0)	605 (0)	162 (0)	54 (0)	0.34 (0.03)
SRRR	522.45 (18.28)	174.15 (6.09)	162 (0)	54 (0)	0.24 (0.02)
SS-FAR	67.42 (8.03)	64.49 (7.52)	31.64 (4.28)	19.06 (1.9)	0.23 (0.02)
PS-FAR	72.71 (8.84)	69.62 (8.17)	34.50 (4.68)	21.25 (1.92)	0.21 (0.02)

Table 13.2 Yeast eQTL mapping analysis: Results with full data (without marginal screening)

Method	$\|\widehat{B}\|_0$	$\|\widehat{B}\|_{2,0}$	$\|\widehat{A}\|_0$	$\|\widehat{A}\|_{2,0}$	MSE
RRR	9732 (1788.02)	3244 (0)	162 (29.76)	54 (0)	0.44 (0.04)
SRRR	1332.09 (359.29)	437.06 (44.94)	162 (29.76)	54 (0)	0.24 (0.02)
SS-FAR	57.56 (9.09)	56.46 (8.5)	24.18 (5.37)	15.65 (2.37)	0.24 (0.02)
PS-FAR	67.46 (13.01)	64 (8.35)	30.91 (7.85)	19.25 (2.7)	0.22 (0.02)

Table 13.3 Yeast eQTL mapping analysis: Top genes in the estimated pathways. The first panel is for the setting with marginal screening, and the second panel is for the setting with full data

Method	Layer 1	Layer 2	Layer 3
With marginal screening ($n = 605$)			
SS-FAR	STE3, STE2, MFA2, MFA1, CTT1, FUS1	CTT1, FUS1, MSN4, GLO1, SLN1	FUS1, FAR1, STE2, GPA1, FUS3, STE3, TEC1, SLN1, STE12, MFA2, CTT1
PS-FAR	STE3, STE2, MFA2, MFA1, CTT1, FUS1, TEC1	CTT1, FUS1, MSN4, STE3, GLO1, SLN1, MFA2	FUS1, FAR1, STE2, GPA1, FUS3, STE3, TEC1, CTT1, SLN1, STE12
Full data ($n = 3244$)			
SS-FAR	STE3, STE2, MFA2, MFA1, CTT1, FUS1	FUS1, CTT1, FAR1, SLN1, STE2, MFA2, GPA1	TEC1
PS-FAR	STE3, STE2, MFA2, MFA1, CTT1, FUS1, TEC1, WSC2	CTT1, FUS1, MSN4, GLO1, MFA2, STE3, SLN1	FUS1, FAR1, STE2, STE3, GPA1, FUS3, TEC1, CTT1

methods becomes slightly worse compared to the setting with marginal screening. The performance of reduced-rank regression deteriorates the most since it lacks the power of eliminating noise variables. The results suggest that in this particular application the potential benefit of jointly considering all the markers is exceeded by the loss due to high dimensionality. We see that the deflation methods are able to perform competitively in such a high-dimensional problem with excellent scalability.

Finally, we examine the genes selected in the estimated pathways (low-rank components or layers) with the full data set. Since each $\widehat{A}_k$ vector is normalized to have unit ℓ_1 norm, we define top genes to be those with entries larger than $1/m$ in magnitude. The selected top genes are summarized in Table 13.3. Figure 13.1 shows the scatterplots of the latent responses $\mathbf{Y}'\widehat{A}_k$ versus the latent predictors $\mathbf{X}'\widehat{B}_k$ from the two deflation methods fitted on the full data. For PS-FAR, we report the results from using initial estimator either from the reduced-rank regression (PS-FAR(R)) or the Lasso (PS-FAR(L)). As explained in Gustin et al. (1998), pheromone induces mating and nitrogen starvation induces filamentation are two pathways in yeast cells. The identified genes STE2, STE3, and GPA1 are receptors required for mating pheromone, which bind the cognate lipopeptide pheromones (MFA2, MFA1, etc.). Besides, another top gene FUS3 is used to phosphorylate several downstream targets, including FAR1 which mediates various responses required for successful mating. Interestingly, our analysis identified an additional gene, TEC1, that was not reported in previous analysis. TEC1, as the transcription factor specific to the filamentation pathway, has been shown to have FUS3-dependent degradation induced by pheromone signaling (Bao et al. 2004).

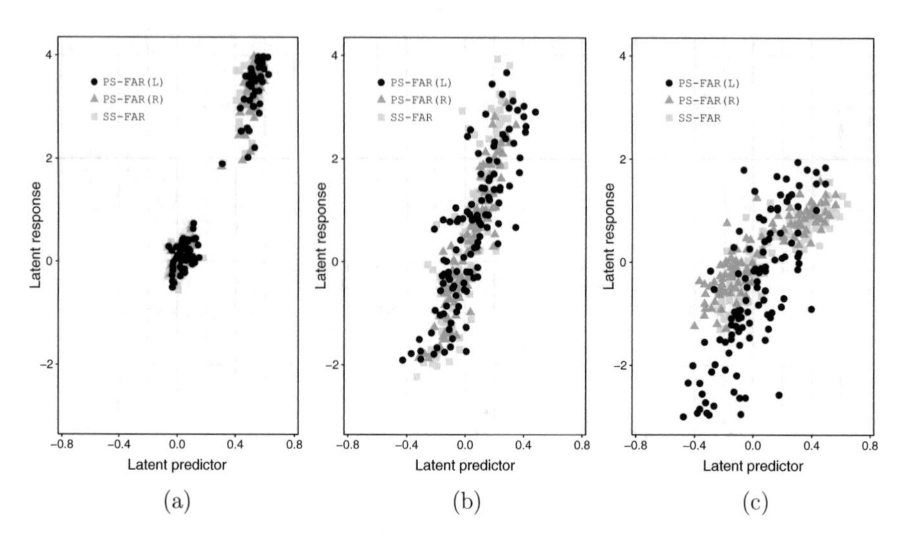

Fig. 13.1 Yeast eQTL mapping analysis: Scatter plots of the estimated latent responses and latent predictors (Chen et al. 2022). (**a**) First layer. (**b**) Second layer. (**c**) Third layer

13.4.2 Forecasting Macroeconomic and Financial Indices

Forecasting is an important task in many areas of science. It involves modeling historical time series data and using the resulting models to gain insights into likely interrelationships and drive future decision-making on public polices, investment strategies, financial plans, among others. In many applications, especially in finance and in economics, there is often a need to forecast with multiple interrelated time series. The modeling of multiple time series is extensively covered in Chapter 5. The dynamic relationships among the components of such time series suggest that multivariate modeling and simultaneous forecasting could be preferable. Moreover, when the number of time series is relatively high, dimension reduction techniques, especially reduced-rank modeling and sparse regularization, may help to identify parsimonious models with improved forecasting performance. For the effect of reduced-rank estimation on prediction, it has been explicitly demonstrated how it can result in the decrease of prediction error covariance matrix (see Section 2.5). It is of great interest to examine how the two techniques, reduced-rank estimation and sparse estimation, impact on the forecasting, either separately or jointly. We thus consider an example from macroeconomics to compare the empirical forecasting performance of various sparse and/or reduced-rank time series models.

We use monthly time series data from May 1973 to February 2010 on 18 macroeconomic/financial indices and rates in the U.S. This data was considered earlier by Carriero et al. (2011). The series include unemployment rate (ur), PCE price index (pcepi), core PCE price excluding food and energy (pcexfepi), non-farm payroll employment (payrolls), weekly hours worked (weeklyhrs), new claims for unemployment insurance (claims), nominal retail sales (retailsales), index of

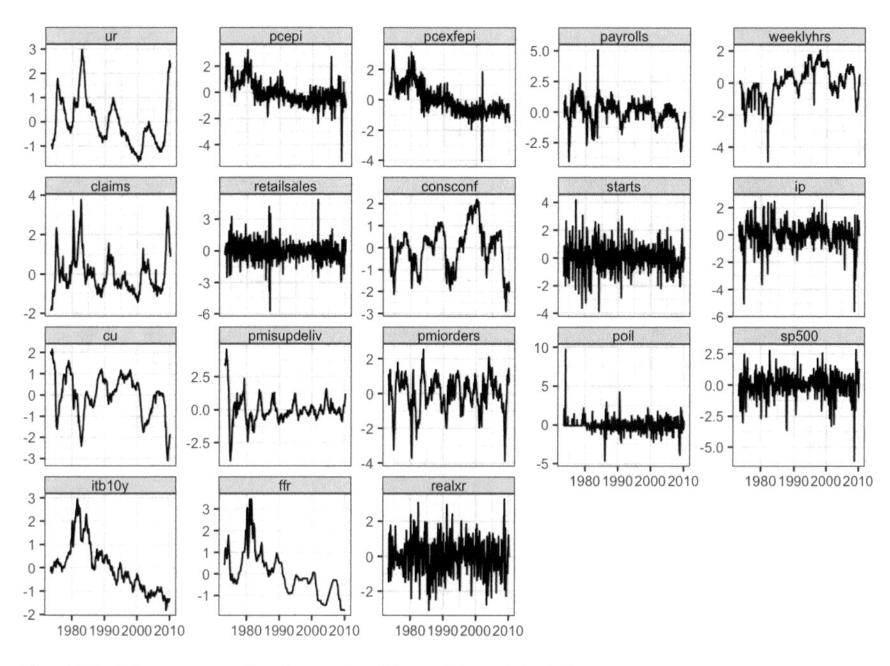

Fig. 13.2 Macroeconomics time series: Plots of the original data

consumer confidence (consconf), single-family housing starts (starts), industrial production growth rate (ip), index of capacity utilization (cu), purchasing managers' index of supplier delivery times (pmisupdeliv), purchasing managers' index of new orders (pmiorders), price of oil (poil), S&P 500 index of stock prices (sp500), yields on 10-year treasury bonds (itb10y), federal funds rate (ffr), and real exchange rate (realxr). Figure 13.2 displays the original data. As can be seen several series are non-stationary and to make the analysis straightforward, for each series the stochastic trend in the original data is removed by taking first difference, and the series is then centered. To permit a meaningful comparison of different models and improve illustration, we also replace some anomalous observations in each time series. To be specific, any observation that is with a magnitude larger than 2.5 times the standard deviation of the series is replaced with the mean of the series before that point. We refer to Section 12.2 for the perspectives of robust estimation. Figure 13.3 shows the data after differencing, centering, and anomaly replacement. These series no longer seem to exhibit any apparent non-stationary behaviors.

With the processed data, we are interested in comparing the forecasting performance of various univariate/multivariate autoregressive models (AR). We use the data from May 1973 to December 2007 for model training, and each trained model is then used to make one-step-ahead forecast with the rest of the data from January 2008 to February 2010. Univariate modeling suggests that in general each time series can be adequately modeled by an autoregressive model with order 2 (AR(2)). We thus focus on order-2 models in vector autoregressive (VAR) modeling as well.

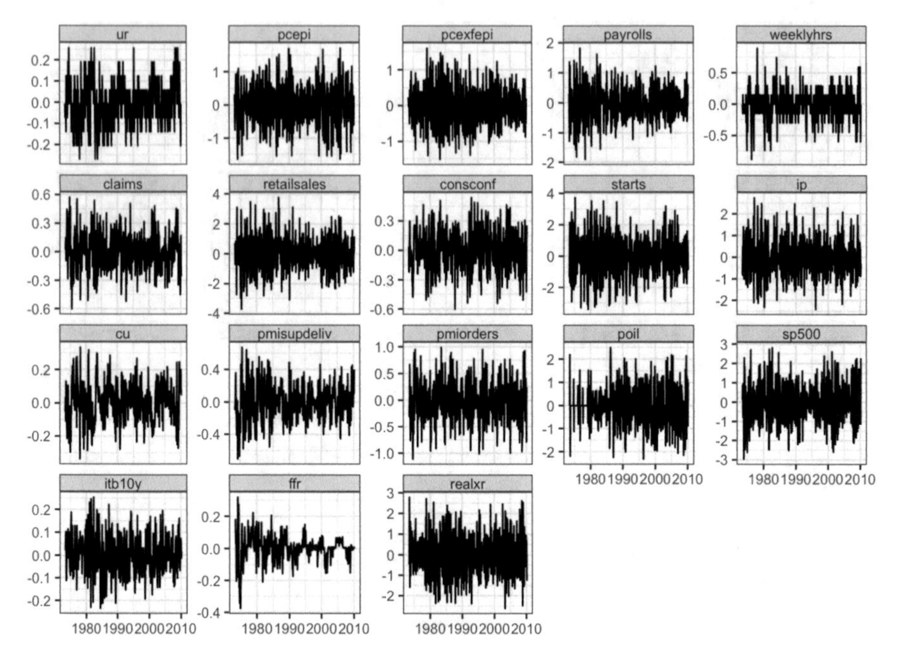

Fig. 13.3 Macroeconomics time series: Plots of the processed data

Table 13.4 Analysis of macroeconomics time series: Forecasting performance of various univariate/multivariate autoregressive models

		VAR(2)			
	AR(2)	LS	RRR	GL	SRRR
MSE	0.4423	0.4363	0.4326	0.4000	0.3990
TrimMSE	0.1823	0.1776	0.1737	0.1653	0.1656
Rank		18	8	18	8
% Nonzero	100%	100%	100%	77.8%	77.8%

A VAR model can be conveniently expressed in a multivariate linear regression form; see Section 5.2. This allows us to adopt various reduced-rank and sparse methods for model estimation. Specifically, to fit VAR(2) model, we consider least squares method (LS), the penalized least squares method with column-wise group Lasso penalty (GL), the reduced-rank regression (RRR), and the column-sparse reduced-rank regression (SRRR). For each method, the forecasting performance is measured by the mean squared errors (MSE) and the 10% trimmed mean squared errors (TrimMSE) resulting from the one-step-ahead predictions with the testing data. We also report the estimated rank and the percentage of nonzeros in the estimated coefficient matrix (% Nonzero).

Table 13.4 reports the results from fitting the aforementioned models. Several observations can be made. First, multivariate models in general forecast better than univariate models. Comparing the univariate with the multivariate AR models, the latter can utilize the potential associations among the multiple time series to improve simultaneous forecasting. Second, with the VAR model setup, either the reduced-rank estimation or the sparse regularization may lead to improved forecasting performance upon the least squares approach. In this particular example, it appears that the group Lasso method results in more substantial improvement in forecasting than the reduced-rank regression; the former produces a sparse solution with about 22.2% of zeros in the estimated coefficient matrix, while the latter substantially reduces the rank of the estimated coefficient matrix from 18 to 8. Last but not the least, the sparse reduced-rank regression model produces the most parsimonious model and performs the best in forecasting among all models. Although its forecasting performance is only slightly better than that of the group Lasso approach, this is achieved by a much more parsimonious model with a sparse and reduced-rank coefficient matrix. A cursory review of the estimation results would indicate that unemployment rate, yields on 10-year treasury bonds, and federal funds rate can be removed at both lags, and weekly hours worked and index of capacity utilization can be taken out of lag 2. A systematic way to review the sparseness results in the time series context is yet to be studied.

This example thus clearly demonstrates the strength of integrating sparsity and reduced-rank structures for forecasting. More development on time series forecasting and analysis via sparse and reduced-rank modeling can be found in, for example, Gillard and Usevich (2018), Wang et al. (2022), Barratt et al. (2021), and Dong et al. (2022).

Appendix

See Tables A.1, A.2, A.3 and A.4.

Table A.1 Biochemical data on urine samples of patients from a study by Smith et al. (1962)

y_1	y_2	y_3	y_4	y_5	x_1	x_2	x_3
17.60	1.50	1.50	1.88	7.50	0.98	2.05	2.40
13.40	1.65	1.32	2.24	7.10	0.98	1.60	3.20
20.30	0.90	0.89	1.28	2.30	1.15	4.80	1.70
22.30	1.75	1.50	2.24	4.00	1.28	2.30	3.00
18.50	1.20	1.03	1.84	2.00	1.28	2.15	2.70
12.10	1.90	1.87	2.40	16.80	1.22	2.15	2.50
10.10	2.30	2.08	2.68	0.90	1.22	1.90	2.80
14.70	2.35	2.55	3.00	2.00	1.30	1.75	2.40
14.40	2.50	2.38	2.84	3.80	1.30	1.55	2.70
18.10	1.50	1.20	2.60	14.50	1.35	2.20	3.10
16.90	1.40	1.15	1.72	8.00	1.35	3.05	3.20
23.70	1.65	1.58	1.60	4.90	1.35	2.75	2.00
18.00	1.60	1.68	2.00	3.60	1.35	2.10	2.30
14.80	2.45	2.15	3.12	12.00	1.40	1.70	3.10
15.60	1.65	1.42	2.56	5.20	1.40	2.35	2.80
16.20	1.65	1.62	2.04	10.20	1.41	1.85	2.10
17.50	1.05	1.56	1.48	9.60	1.41	2.65	1.50
14.10	2.70	2.77	2.56	6.90	1.62	3.05	2.60
22.50	0.85	1.65	1.20	3.50	1.62	4.30	1.60
17.00	0.70	0.97	1.24	1.90	2.05	3.50	1.80
12.50	0.80	0.80	0.64	0.70	2.05	4.75	1.00
21.50	1.80	1.77	2.60	8.30	2.30	1.95	3.30
13.00	2.20	1.85	3.84	13.00	2.30	1.60	3.50

(continued)

G. C. Reinsel et al., *Multivariate Reduced-Rank Regression*, Lecture Notes in Statistics 225, https://doi.org/10.1007/978-1-0716-2793-8

Table A.1 (continued)

y_1	y_2	y_3	y_4	y_5	x_1	x_2	x_3
13.00	3.55	3.18	3.48	18.30	2.15	2.40	3.30
12.00	3.65	2.40	3.00	14.50	2.15	2.70	3.40
22.80	0.55	1.00	1.14	3.30	2.30	4.75	1.60
18.40	1.05	1.17	1.36	4.90	2.30	4.90	2.80
8.70	4.25	3.62	3.84	19.50	2.62	1.15	2.50
9.40	3.85	3.36	5.12	1.30	2.62	0.97	2.80
15.00	2.45	2.38	2.40	20.00	2.55	3.25	2.70
12.90	1.70	1.74	2.48	1.00	2.55	3.10	2.30
12.10	1.80	2.00	2.24	5.00	2.70	2.45	2.50
11.50	2.25	2.25	3.12	5.10	2.70	2.20	3.40

Response variables are y_1 = pigment creatinine, y_2 = phosphate (mg/ml), y_3 = phosphorous (mg/ml), y_4 = creatinine (mg/ml), y_5 = choline (μg/ml); predictor variables are x_1 = weight (lbs/100), x_2 = volume (ml/100), x_3 = 100 (specific gravity -1)

Table A.2 Sales performance data

x_1	x_2	x_3	y_1	y_2	y_3	y_4
93	96	97.8	9	12	9	20
88.8	91.8	96.8	7	10	10	15
95	100.3	99	8	12	9	26
101.3	103.8	106.8	13	14	12	29
102	107.8	103	10	15	12	32
95.8	97.5	99.3	10	14	11	21
95.5	99.5	99	9	12	9	25
110.8	122	115.3	18	20	15	51
102.8	108.3	103.8	10	17	13	31
106.8	120.5	102	14	18	11	39
103.3	109.8	104	12	17	12	32
99.5	111.8	100.3	10	18	8	31
103.5	112.5	107	16	17	11	34
99.5	105.5	102.3	8	10	11	34
100	107	102.8	13	10	8	34
81.5	93.5	95	7	9	5	16
101.3	105.3	102.8	11	12	11	32
103.3	110.8	103.5	11	14	11	35
95.3	104.3	103	5	14	13	30
99.5	105.3	106.3	17	17	11	27
88.5	95.3	95.8	10	12	7	15
99.3	115	104.3	5	11	11	42
87.5	92.5	95.8	9	9	7	16
105.3	114	105.3	12	15	12	37

(continued)

Table A.2 (continued)

x_1	x_2	x_3	y_1	y_2	y_3	y_4
107	121	109	16	19	12	39
93.3	102	97.8	10	15	7	23
106.8	118	107.3	14	16	12	39
106.8	120	104.8	10	16	11	49
92.3	90.8	99.8	8	10	13	17
106.3	121	104.5	9	17	11	44
106	119.5	110.5	18	15	10	43
88.3	92.8	96.8	13	11	8	10
96	103.3	100.5	7	15	11	27
94.3	94.5	99	10	12	11	19
106.5	121.5	110.5	18	17	10	42
106.5	115.5	107	8	13	14	47
92	99.5	103.5	18	16	8	18
102	99.8	103.3	13	12	14	28
108.3	122.3	108.5	15	19	12	41
106.8	119	106.8	14	20	12	37
102.5	109.3	103.8	9	17	13	32
92.5	102.5	99.3	13	15	6	23
102.8	113.8	106.8	17	20	10	32
83.3	87.3	96.3	1	5	9	15
94.8	101.8	99.8	7	16	11	24
103.5	112	110.8	18	13	12	37
89.5	96	97.3	7	15	11	14
84.3	89.8	94.3	8	8	8	9
104.3	109.5	106.5	14	12	12	36
106	118.5	105	12	16	11	39

Response variables are y_1: score on creativity test, y_2: score on mechanical reasoning test, y_3: score on abstract reasoning test, y_4: score on Mathematics test; predictor variables are x_1: sales growth, x_2: sales profitability, x_3: new account sales

Table A.3 Quarterly macroeconomic time series data for the UK, 1948–1956, from Klein et al. (1961)

y_1	y_2	y_3	y_4	y_5	x_1	x_2	x_3	x_4	x_5
95.43	99.61	89.74	99.3	91.3	99.0	98.7	97.5	97.8	99.7
103.11	102.69	95.67	100.0	99.3	99.8	99.7	100.2	99.3	102.2
100.27	98.79	108.76	100.7	102.0	100.4	100.3	100.7	101.1	101.6
101.59	98.91	105.83	100.0	107.4	100.8	101.3	101.6	101.9	102.2
101.03	100.11	105.54	100.7	113.3	100.6	102.2	101.5	102.6	101.2
108.71	103.09	98.67	107.4	106.5	100.9	102.5	100.8	102.6	102.9
105.97	99.79	97.46	109.7	103.7	101.1	103.2	99.1	102.6	102.9
107.89	101.51	104.13	107.9	115.3	101.7	103.2	102.9	103.3	102.9
110.03	103.21	107.24	104.6	123.2	101.7	104.1	105.4	105.9	104.6
115.11	104.89	102.37	112.9	119.3	101.8	104.1	108.4	107.7	106.8
112.57	102.09	105.36	104.5	125.4	102.1	104.1	108.2	109.5	105.9
116.59	105.51	99.03	104.2	134.4	102.6	106.7	116.5	113.4	106.3
113.73	106.51	87.34	112.1	122.7	102.6	109.8	125.0	119.5	109.5
122.41	106.39	73.07	120.1	130.6	102.7	112.0	134.5	127.7	114.6
116.27	102.79	79.36	125.8	123.2	103.3	113.9	134.8	134.7	116.9
116.59	104.51	93.43	120.9	125.9	103.3	117.7	133.7	137.3	117.3
116.03	103.51	119.64	119.0	131.9	103.2	120.9	134.8	137.7	119.3
115.41	106.29	151.67	113.2	115.0	102.9	121.8	132.3	138.1	121.9
108.97	103.59	142.96	100.2	106.7	102.8	123.0	128.6	136.0	121.5
115.59	104.81	131.53	106.2	118.9	102.8	125.9	125.0	133.8	122.0
116.73	105.31	127.74	113.3	119.3	102.6	127.5	123.5	132.5	123.1
121.71	110.79	111.97	123.5	121.7	102.9	127.8	121.1	130.8	124.9
117.97	105.49	106.86	118.7	122.3	103.4	129.0	119.1	131.2	123.6
125.89	109.41	104.03	121.0	134.2	103.8	130.0	117.9	130.3	122.8
126.73	106.11	108.44	120.4	131.0	103.9	131.6	116.4	129.9	124.6
131.41	111.59	93.47	121.5	130.7	104.3	134.1	118.4	129.9	126.3
126.27	108.09	89.06	120.0	129.0	104.9	135.1	119.9	129.9	127.0
134.29	114.71	84.23	121.8	129.9	105.3	136.0	120.7	129.9	126.2
134.73	107.81	80.64	138.2	143.3	105.2	139.4	123.8	130.8	128.3
139.41	113.19	76.87	125.8	125.1	105.6	143.9	122.3	131.6	130.2
130.97	112.59	77.06	137.4	138.4	106.2	144.8	121.5	133.4	130.7
141.29	114.91	71.43	137.1	147.6	106.6	145.5	123.5	134.2	131.7
135.73	112.01	75.74	136.9	144.1	106.5	150.2	125.0	136.0	134.6
140.41	114.59	78.77	136.6	151.2	106.7	154.9	125.5	136.4	137.6
130.97	110.99	95.26	128.9	135.8	107.0	155.9	122.6	138.1	136.3
137.89	119.21	88.93	132.4	152.7	107.0	156.5	127.4	139.0	136.1

Response variables are y_1 = industrial production, y_2 = consumption, y_3 = unemployment, y_4 = total imports, y_5 = total exports; predictor variables are x_1 = total labor force, x_2 = weekly wage rates, x_3 = price index of imports, x_4 = price index of exports, x_5 = price index of consumption

Table A.4 Blood sugar concentrations ($y_0 - y_5$, in mg/100 ml) of 36 rabbits at 0, 1, 2, 3, 4, and 5 hours after administration of insulin dose, with two types of insulin preparation (standard, $x_1 = -1$, and test, $x_1 = +1$) and two dose levels (0.75 units, $x_2 = -1$, and 1.50 units, $x_2 = +1$), from a study by Volund (1980)

y_0	y_1	y_2	y_3	y_4	y_5	x_1	x_2
96	37	31	33	35	41	1	1
90	47	48	55	68	89	1	1
99	49	55	64	74	97	1	1
95	33	37	43	63	92	1	1
107	62	62	85	110	117	1	1
81	40	43	45	49	55	1	1
95	49	56	63	68	88	1	1
105	53	57	69	103	106	1	1
97	50	53	59	82	96	1	1
97	54	57	66	80	89	1	−1
105	66	83	95	97	100	1	−1
105	49	54	56	70	90	1	−1
106	79	92	95	99	100	1	−1
92	46	51	57	73	91	1	−1
91	61	64	71	80	90	1	−1
101	51	63	91	95	96	1	−1
87	53	55	57	78	89	1	−1
94	57	70	81	94	96	1	−1
98	48	55	71	91	96	−1	1
98	41	43	61	89	101	−1	1
103	60	56	61	76	97	−1	1
99	36	43	57	89	102	−1	1
97	44	51	58	85	105	−1	1
95	41	45	49	59	78	−1	1
109	65	62	72	93	104	−1	1
91	57	60	61	67	83	−1	1
99	43	48	52	61	86	−1	1
102	51	56	81	97	103	−1	−1
96	57	55	72	85	89	−1	−1
111	84	83	91	101	102	−1	−1
105	57	67	83	100	103	−1	−1
105	57	61	70	90	98	−1	−1
98	55	67	88	94	95	−1	−1
98	69	72	89	98	98	−1	−1
90	53	61	78	94	95	−1	−1
100	60	63	67	77	104	−1	−1

References

Ahn, S. K. and Reinsel, G. C. (1988) Nested reduced-rank autoregressive models for multiple time series. *Journal of the American Statistical Association*, **83**, 849–856.

Ahn, S. K. and Reinsel, G. C. (1990) Estimation for partially nonstationary multivariate autoregressive models. *Journal of the American Statistical Association*, **85**, 813–823.

Aitchison, J. and Silvey, S. D. (1958) Maximum likelihood estimation of parameters subject to restraints. *Annals of Mathematical Statistics*, **29**, 813–828.

Akaike, H. (1974) A new look at the statistical model identification. *IEEE Transactions on Automatic Control*, **19**, 716–723.

Akaike, H. (1976) Canonical correlation analysis of time series and the use of an information criterion. In *Systems Identification: Advances and Case Studies* (eds. R. K. Mehra and D. G. Lainiotis), 27–96. New York: Academic Press.

Albert, J. M. and Kshirsagar, A. M. (1993) The reduced-rank growth curve model for discriminant analysis of longitudinal data. *Australian Journal of Statistics*, **35**, 345–357.

Aldrin, M. (2000) Multivariate prediction using softly shrunk reduced-rank regression. *The American Statistician*, **54**, 29–34.

Allen, G. I., Grosenick, L. and Taylor, J. (2014) A generalized least-square matrix decomposition. *Journal of the American Statistical Association*, **109**, 145–159.

Amemiya, Y. and Fuller, W. A. (1984) Estimation for the multivariate errors-in-variables model with estimated error covariance matrix. *The Annals of Statistics*, **12**, 497–509.

Anderson, T. W. (1951) Estimating linear restrictions on regression coefficients for multivariate normal distributions. *Annals of Mathematical Statistics*, **22**, 327–351.

Anderson, T. W. (1963) The use of factor analysis in the statistical analysis of multiple time series. *Psychometrika*, **28**, 1–25.

Anderson, T. W. (1971) *The Statistical Analysis of Time Series*. New York: Wiley.

Anderson, T. W. (1976) Estimation of linear functional relationships: Approximate distributions and connections with simultaneous equations in econometrics (with discussion). *Journal of the Royal Statistical Society: Series B*, **38**, 1–36.

Anderson, T. W. (1991) Trygve Haavelmo and simultaneous equations models. *Scandinavian Journal of Statistics*, **18**, 1–19.

Anderson, T. W. (2002) Specification and misspecification in reduced rank regression. *Sankhy A: The Indian Journal of Statistics*, **64**, 193–205.

Anderson, T. W. and Rubin, H. (1956) Statistical inference in factor analysis. In *Proceedings of the Third Berkeley Symposium on Mathematical Statistics and Probability, Volume 5:*

G. C. Reinsel et al., *Multivariate Reduced-Rank Regression*, Lecture Notes in Statistics 225, https://doi.org/10.1007/978-1-0716-2793-8

Contributions to Econometrics, Industrial Research, and Psychometry, 111–150. Berkeley, Calif.: University of California Press.

Anderson, T. W. (1984a) Estimating linear statistical relationships. *The Annals of Statistics*, **12**, 1–45.

Anderson, T. W. (1984b) *An Introduction to Multivariate Statistical Analysis*. New York: Wiley second edn.

Anderson, T. W. (1999a) Asymptotic distribution of the reduced rank regression estimator under general conditions. *The Annals of Statistics*, **27**, 1141–1154.

Anderson, T. W. (1999b) Asymptotic theory for canonical correlation analysis. *Journal of Multivariate Analysis*, **70**, 1–29.

Bai, Z. D. and Silverstein, J. W. (2009) *Spectral Analysis of Large Dimensional Random Matrices*. Springer, New York, 2 edn.

Bao, M. Z., Schwartz, M. A., Cantin, G. T., Yates III, J. R. and Madhani, H. D. (2004) Pheromone-dependent destruction of the Tec1 transcription factor is required for MAP kinase signaling specificity in yeast. *Cell*, **119**, 991–1000.

Barratt, S., Dong, Y. and Boyd, S. (2011) Low-rank forecasting. **ArXiv:2101.12414**.

Barron, A., Birgé, L. and Massart, P. (1999) Risk bounds for model selection via penalization. *Probability Theory and Related Fields*, **113**, 301–413.

Bartlett, M. S. (1938) Further aspects of the theory of multiple regression. *Proceedings of the Cambridge Philosophical Society*, **34**, 33–40.

Bartlett, M. S. (1947) Multivariate analysis. *Journal of the Royal Statistical Society: Series B*, **9**, 176–197.

Beaulieu, M. C., Dufour, J. M. and Khalaf, L. (2013) Identification-robust estimation and testing of the zero-beta CAPM. *The Review of Economic Studies*, **80**, 892–924.

Bekker, P., Dobbelstein, P. and Wansbeek, T. (1996) The APT model as reduced-rank regression. *Journal of Business & Economic Statistics*, **14**, 199–202.

Bertsimas, D., King, A. and Mazumder, R. (2016) Best subset selection via a modern optimization lens. *The Annals of Statistics*, **44**(2), 813–852.

Bewley, R. and Yang, M. (1995) Tests for cointegration based on canonical correlation analysis. *Journal of the American Statistical Association*, **90**, 990–996.

Bhargava, A. K. (1979) Estimation of a linear transformation and an associated distributional problem. *Journal of Statistical Planning and Inference*, **3**, 19–26.

Bickel, P. J., Ritov, Y. and Tsybakov, A. B. (2009) Simultaneous analysis of Lasso and Dantzig selector. *The Annals of Statistics*, **37**, 1705–1732.

Bilodeau, M. and Kariya, T. (1989) Minimax estimators in the normal MANOVA model. *Journal of Multivariate Analysis*, **28**, 260–270.

Black, F. (1972) Capital market equilibrium with restricted borrowing. *Journal of Business*, **45**, 444–454.

Blattberg, R. C. and George, E. I. (1991) Shrinkage estimation of price and promotional elasticities: Seemingly unrelated equations. *Journal of the American Statistical Association*, **86**, 304–315.

Blattberg, R. C. and Wisniewski, K. J. (1989) Price-induced patterns of competition. *Marketing Science*, **8**, 291–309.

Blattberg, R. C., Kim, B. and Ye, J. (1994) Large scale databases: The new marketing challenge. In *The Marketing Information Revolution* (eds. R. C. Blattberg, R. Glazer and J. L. J. L. Chap.), 173–203. Boston: Harvard Business School Press.

Boot, J. and de Wit, G. M. (1960) Investment demand: An empirical contribution to the aggregation problem. *International Economic Review*, **1**, 3–30.

Bossaerts, P. (1988) Common nonstationary components of asset prices. *Journal of Economic Dynamics and Control*, **12**, 347–364.

Box, G. E. P. and Tiao, G. C. (1977) A canonical analysis of multiple time series. *Biometrika*, **64**, 355–365.

Boyd, S. and Vandenberghe, L. (2004) *Convex Optimization*. Cambridge University Press.

Boyd, S., Parikh, N., Chu, E., Peleato, B. and Eckstein, J. (2011) Distributed optimization and statistical learning via the alternating direction method of multipliers. *Foundations and Trends® in Machine Learning*, **3**, 1–122.

Breheny, P. and Huang, J. (2009) Penalized methods for bi-level variable selection. *Statistics and Its Interface*, **2**, 369–380.

Breiman, L. and Friedman, J. H. (1997) Predicting multivariate responses in multiple linear regression. *Journal of the Royal Statistical Society: Series B*, **59**, 3–54.

Brem, R. B. and Kruglyak, L. (2005) The landscape of genetic complexity across 5,700 gene expression traits in yeast. *Proceedings of the National Academy of Sciences of the United States of America*, **102**, 1572–1577.

Brillinger, D. R. (1969) The canonical analysis of stationary time series. In *Multivariate Analysis*, 331–350. P.R. Krishnaiah, New York: Academic Press.

Brillinger, D. R. (1981) *Time Series: Data Analysis and Theory*. San Francisco: Holden–Day, expanded edn.

Brooks, R. and Stone, M. (1994) Joint continuum regression for multiple predictands. *Journal of the American Statistical Association*, **89**, 1374–1377.

Brown, P. J. and Payne, C. (1975) Election night forecasting (with discussion). *Journal of the Royal Statistical Society: Series A*, **138**, 463–498.

Brown, P. J. and Zidek, J. V. (1980) Adaptive multivariate ridge regression. *The Annals of Statistics*, **8**, 64–74.

Brown, P. J. and Zidek, J. V. (1982) Multivariate regression shrinkage estimators with unknown covariance matrix. *Scandinavian Journal of Statistics*, **9**, 209–215.

Bunea, F., She, Y. and Wegkamp, M. (2011) Optimal selection of reduced rank estimators of high-dimensional matrices. *The Annals of Statistics*, **39**, 1282–1309.

Bunea, F., She, Y. and Wegkamp, M. H. (2012) Joint variable and rank selection for parsimonious estimation of high-dimensional matrices. *The Annals of Statistics*, **40**, 2359–2388.

Bura, E. and Cook, R. D. (2001) Estimating the structural dimensions of regressions via parametric inverse regression. *Journal Royal Statistics Society*: Series B, **63**, 393–410.

Bühlmann, P. and van de Geer, S. (2011) *Statistics for high-dimensional data*. Springer Series in Statistics. Springer, Heidelberg.

Cai, J.-F., Candés, E. J. and Shen, Z. (2010) A singular value thresholding algorithm for matrix completion. *SIAM Journal on Optimization*, **20**, 1956–1982.

Campbell, N. (1984) Canonical variate analysis—a general formulation. *Australian Journal of Statistics*, **26**, 86–96.

Campbell, J. (1987) Stock returns and the term structure. *Journal of Financial Economics*, **18**, 373–399.

Campbell, J., Lo., A. and MacKinlay, A. C. (1997) *The Econometrics of Financial Markets*. Princeton, NJ: Princeton University Press.

Candès, E. J. and Recht, B. (2009) Exact matrix completion via convex optimization. *Found. Comput. Math.*, **9**, 717–772.

Candès, E. J. and Tao, T. (2010) The power of convex relaxation: Near-optimal matrix completion. *IEEE Transactions on Information Theory*, **56**, 2053–2080.

Candès, E. J., Li, X., Ma, Y. and Wright, J. (2011) Robust principal component analysis? *Journal of the ACM*, **58**, 1–37.

Candès, E. J., Sing-Long, C. A. and Trzasko, J. D. (2013) Unbiased risk estimates for singular value thresholding and spectral estimators. *IEEE Transactions on Signal Processing*, **61**, 4643–4657.

Carriero, A., Kapetanios, G. and Marcellino, M. (2011) Forecasting large datasets with Bayesian reduced rank multivariate models. *Journal of Applied Econometrics*, **26**, 735–761.

Caruana, R. (1997) Multitask learning. *Machine Learning*, **28**, 41–75.

Chakraborty, A., Bhattacharya, A. and Mallick, B. K. (2019) Bayesian sparse multiple regression for simultaneous rank reduction and variable selection. *Biometrika*, **107**, 205–221.

Chen, K. and Chan, K.-S. (2016) A note on rank reduction and variable selection in multivariate regression. *Journal of Statistical Theory and Practice*, **10**, 100–120.

Chen, K., Chan, K.-S. and Stenseth, N. C. (2012) Reduced rank stochastic regression with a sparse singular value decomposition. *Journal of the Royal Statistical Society: Series B*, **74**, 203–221.

Chen, K., Dong, H. and Chan, K.-S. (2013) Reduced rank regression via adaptive nuclear norm penalization. *Biometrika*, **100**, 901–920.

Chen, K., Chan, K.-S. and Stenseth, N. C. (2014) Source-sink reconstruction through regularized multicomponent regression analysis–with application to assessing whether North Sea cod larvae contributed to local fjord cod in Skagerrak. *Journal of the American Statistical Association*, **109**, 560–573.

Chen, K., Hoffman, E. A., Seetharaman, I., Lin, C.-L. and Chan, K.-S. (2016) Linking lung airway structure to pulmonary function via composite bridge regression. *The Annals of Applied Statistics*, **10**, 1880–1906.

Chen, K., Dong, R., Xu, W., and Zheng, Z. (2022) Fast stagewise sparse factor regression. Journal of Machine Learning Research, 23(271):1–45.

Chen, L. and Huang, J. Z. (2012) Sparse reduced-rank regression for simultaneous dimension reduction and variable selection. *Journal of the American Statistical Association*, **107**, 1533–1545.

Chi, E. C. and Li, T. (2019) Matrix completion from a computational statistics perspective. *WIREs Computational Statistics*, **11**, e1469.

Chi, E. M. and Reinsel, G. C. (1989) Models for longitudinal data with random effects and AR(1) errors. *Journal of the American Statistical Association*, **84**, 452–459.

Chin, K., DeVries, S., Fridlyand, J., Spellman, P. T., Roydasgupta, R., Kuo, W.-L., Lapuk, A., Neve, R. M., Qian, Z. and Ryder, T. (2006) Genomic and transcriptional aberrations linked to breast cancer pathophysiologies. *Cancer Cell*, **10**, 529–541.

Chinchilli, V. M. and Elswick, R. K. (1985) A mixture of the MANOVA and GMANOVA models. *Communications in Statistics, Theory and Methods*, **14**, 3075–3089.

Collins, M., Dasgupta, S. and Schapire, R. E. (2002) A generalization of principal components analysis to the exponential family. In *Advances in Neural Information Processing Systems (NeurIPS) 14*, 617–624. Curran Associates, Inc.

Connor, G. and Linton, O. (2007) Semiparametric estimation of a characteristic-based factor model of stock returns. *Journal of Empirical Finance*, **14**, 694–717.

Connor, G., Hagmann, M. and Linton, O. (2012) Efficient semiparametric estimation of the Fama-French model and extensions. *Econometrica*, **80**, 713–754.

Cook, R, D. (2018) *An Introduction to Envelopes: Dimension Reduction for Efficient Estimation in Multivariate Statistics*. Wiley.

Cook, R. D. and Su, Z. (2013) Scaled envelopes: scale-invariant and efficient estimation in multivariate linear regression. *Biometrika*, **100**, 993–954.

Cook, R. D. and Zhang, X. (2018) Fast envelope algorithms. *Statistica Sinica*, **28**, 1179–1197.

Cook, R. D., Li, B. and Chiaromonte, F. (2010) Envelope models for parsimonious and efficient multivariate linear regression. *Statistica Sinica*, **20**, 927–960.

Cook, R. D., Forzani, L. and Zhang, X. (2015) Envelopes and reduced-rank regression. *Biometrika*, **102**, 439–456.

Cox, D. R. and Wermuth, N. (1992) Response models for mixed binary and quantitative variables. *Biometrika*, **79**, 441–461.

Cragg, J. G. and Donald, S. G. (1996) On the asymptotic properties of LDU-based tests of the rank of a matrix. *Journal of the American Statistical Association*, **91**, 1301–1309.

Davenport, M. A. and Romberg, J. (2016) An overview of low-rank matrix recovery from incomplete observations. *IEEE Journal of Selected Topics in Signal Processing*, **10**, 608–622.

Davies, P. T. and Tso, M. K.-S. (1982) Procedures for reduced-rank regression. *Journal of the Royal Statistical Society: Series C*, **31**, 244–255.

de Jong, S. (1993) SIMPLS: An alternative approach to partial least squares regression. *Chemometrics and Intelligent Laboratory Systems*, **18**, 251–263.

de Leon, A. R. and Wu, B. (2011) Copula-based regression models for a bivariate mixed discrete and continuous outcome. *Statistics in Medicine*, **30**, 175–185.

Diggle, P. J. (1988) An approach to the analysis of repeated measurements. *Biometrics*, **44**, 959–971.

Doan, T., Litterman, R. and Sims, C. (1984) Forecasting and conditional projection using realistic prior distributions. *Econometric Reviews*, **3**, 1–100.

Dong, Y., Qin, S. J. and Boyd, S. P. (2022) Extracting a low-dimensional predictable time series. *Optimization and Engineering*, **23**, 1189–1214.

Donoho, D. L. (2000) High-dimensional data analysis: The curses and blessings of dimensionality. In *American Mathematical Society Conference on Math Challenges of the 21st Century*.

Donoho, D. L. and Huber, P. J. (1983) The notion of breakdown point. In *A Festschrift for Erich L. Lehmann*, Wadsworth Statistics/Probability Series, 157–184. Belmont: Wadsworth International.

Donoho, D. L. and Johnstone, I. M. (1995) Adapting to unknown smoothness via wavelet shrinkage. *Journal of the American Statistical Association*, **90**, 1200–1224.

Dunson, D. B. (2000) Bayesian latent variable models for clustered mixed outcomes. *Journal of the Royal Statistical Society: Series B*, **62**, 355–366.

Eckart, C. and Young, G. (1936) The approximation of one matrix by another of lower rank. *Psychometrika*, **1**, 211–218.

Efron, B. (2004) The estimation of prediction error: Covariance penalties and cross-validation. *Journal of the American Statistical Association*, **99**, 619–642.

Efron, B. and Morris, C. (1972) Empirical Bayes on vector observations: An extension of Stein's method. *Biometrika*, **59**, 335–347.

Efron, B., Hastie, T. J., Johnstones, I. and Tibshirani, R. (2004) Least angle regression. *The Annals of Statistics*, **32(2)**, 407–499.

Engle, R. F. and Granger, C. (1987) Co-integration and error correction: Representation, estimation, and testing. *Econometrica*, **55**, 251–276.

Fan, J. and Li, R. (2001) Variable selection via nonconcave penalized likelihood and its oracle properties. *Journal of the American Statistical Association*, **96**, 1348–1360.

Fan, J. and Lv, J. (2010) A selective overview of variable selection in high dimensional feature space. *Statistica Sinica*, **20**, 101–148.

Fan, J., Liao, Y. and Wang, W. (2016) Projected principal component analysis in factor models. *The Annals of Statistics*, **44**, 219–254.

Farago, A. and Hjalmarsson, E. (2019) Stock price co-movement and the foundations of pairs trading. Journal of Financial and Quantitative Analysis. Cambridge University Press. **54**(2), 629–665.

Fazel, M. (2002) *Matrix rank minimization with applications*. Ph.D. thesis, Stanford University.

Ferson, W. and Harvey, C. (1991) The variation of economic risk premiums. *Journal of Political Economy*, **99**, 385–415.

Fisher, R. A. (1938) The statistical utilization of multiple measurements. *Annals of Eugenics*, **8**, 376–386.

Fitzmaurice, G. M. and Laird, N. M. (1995) Regression models for a bivariate discrete and continuous outcome with clustering. *Journal of the American Statistical Association*, **90**, 845–852.

Fortier, J. J. (1966) Simultaneous linear prediction. *Psychometrika*, **31**, 369–381.

Foster, D. and George, E. (1994) The risk inflation criterion for multiple regression. *The Annals of Statistics*, **22**(4), 1947–1975.

Fountis, N. G. and Dickey, D. A. (1989) Testing for a unit root nonstationarity in multivariate autoregressive time series. *The Annals of Statistics*, **17**, 419–428.

Frank, I. E. and Friedman, J. H. (1993) A statistical view of some chemometrics regression tools (with discussion). *Technometrics*, **35**, 109–148.

Franklin, J. N. (2000) *Matrix Theory*. Toronto: Dover Publications.

Friedman, J., Hastie, T. J., Höfling, H. and Tibshirani, R. (2007) Pathwise coordinate optimization. *The Annals of Applied Statistics*, **2**, 302–332.

Fu, W. J. (1998) Penalized regressions: The bridge versus the lasso. *Journal of Computational and Graphical Statistics*, **7**, 397–416.

Fujikoshi, Y. (1974) The likelihood ratio tests for the dimensionality of regression coefficients. *Journal of Multivariate Analysis*, **4**, 327–340.

Gabriel, K. R. and Zamir, S. (1979) Lower rank approximation of matrices by least squares with any choice of weights. *Technometrics*, **21**, 489–498.

Gallant, A. R. and Goebel, J. J. (1976) Nonlinear regression with autocorrelated errors. *Journal of the American Statistical Association*, **71**, 961–967.

Garthwaite, P. H. (1994) An interpretation of partial least squares. *Journal of the American Statistical Association*, **89**, 122–127.

Geary, R. C. (1948) Studies in relations between economic time series. *Journal of the Royal Statistical Society: Series B*, **10**, 140–158.

Geng, H., Iqbal, J., Chan, W. C. and Ali, H. H. (2011) Virtual CGH: an integrative approach to predict genetic abnormalities from gene expression microarray data applied in lymphoma. *BMC Medical Genomics*, **4**, 32.

Genton, M. C. (2007) Separable approximation of space-time covariance matrices. *Environmetrics*, **18**, 681–694.

Geweke, J. (1996) Bayesian reduced rank regression in econometrics. *Journal of Econometrics*, **75**, 121–146.

Geweke, J. and Zhou, G. (1996) Measuring the pricing error of the arbitrage pricing theory. *Review of Financial Studies*, **9**, 553–583.

Gibbons, M. R. (1982) Multivariate tests of financial models: A new approach. *Journal of Financial Economics*, **10**, 3–27.

Gibbons, M. R. and Ferson, W. (1985) Testing asset pricing models with changing expectations and an unobservable market portfolio. *Journal of Financial Economics*, **14**, 217–236.

Gibbons, M. R., Ross, S. A. and Shanken, J. (1989) A test of the efficiency of a given portfolio. *Econometrica*, **57**, 1121–1152.

Giles, D. and Hampton, P. (1984) Regional production relationships during the industrialization of New Zealand, 1935–1948. *Journal of Regional Science*, **24**, 519–533.

Gillard, J. and Usevich, K. (2018) Structured low-rank matrix completion for forecasting in time series analysis. *International Journal of Forecasting*, **34**, 582–597.

Glasbey, C. A. (1992) A reduced rank regression model for local variation in solar radiation. *Applied Statistics*, **41**, 381–387.

Gleser, L. J. (1981) Estimation in a multivariate "errors in variables" regression model: Large sample results. *The Annals of Statistics*, **9**, 24–44.

Gleser, L. J. and Olkin, I. (1966) A k-sample regression model with covariance. In *Multivariate Analysis*, 59–72. New York: Academic Press.

Gleser, L. J. and Olkin, I. (1970) Linear models in multivariate analysis. In *Essays in Probability and Statistics* (ed. R. C. Bose), 267–292. New York: Wiley.

Gleser, L. J. and Watson, G. S. (1973) Estimation of a linear transformation. *Biometrika*, **60**, 525–534.

Goh, G., Dey, D. K. and Chen, K. (2017) Bayesian sparse reduced rank multivariate regression. *Journal of Multivariate Analysis*, **157**, 14–28.

Goldberger, A. S. (1972) Maximum likelihood estimation of regressions containing unobservable independent variables. *International Economic Review*, **13**, 1–15.

Golub, G. H. and Van Loan, C. F. (1996) *Matrix Computations (3rd Ed.)*. Baltimore, MD, USA: Johns Hopkins University Press.

Golub, G. H., Heath, M. and Wahba, G. (1979) Generalized cross-validation as a method for choosing a good ridge parameter. *Technometrics*, **21**, 215–223.

Gonzalo, J. and Granger, C. (1995) Estimation of common long-memory components in cointegrated systems. *Journal of Business and Economic Statistics*, **13**, 27–35.

Gospondinov, N., Kan, R. and Robotti, C. (2019) Too good to be true? Fallacies in evaluating risk factor models. *Journal of Financial Economics*, **132**, 451–471.

Gower, J. C. and Dijksterhuis, G. (2004) *Procrustes Problems*. Oxford Statistical Science Series. Oxford: Oxford University Press.

Grizzle, J. E. and Allen, D. M. (1969) Analysis of growth and dose response curves. *Biometrics*, **25**, 357–381.

Gudmundsson, G. (1977) Multivariate analysis of economic variables. *Applied Statistics*, **26**, 48–59.

Guggenberger, P., Kleibergen, F. and Mavroeidis, S. (2022) A test for kronecker product structure covariance matrix. To appear in Journal of Econometrics.

Gustin, M. C., Albertyn, J., Alexander, M. and Davenport, K. (1998) MAP kinase pathways in the yeast saccharomyces cerevisiae. *Microbiology and Molecular Biology Reviews*, **62**, 1264–1300.

Haitovsky, Y. (1987) On multivariate ridge regression. *Biometrika*, **74**, 563–570.

Hammar, A. and Ardal, G. (2009) Cognitive functioning in major depression—a summary. *Frontiers in Human Neuroscience*, **3**, 26.

Hampel, F. R., Ronchetti, E. M., Rousseeuw, P. J. and Stahel, W. A. (2005) *Robust Statistics–The Approach Based on Influence Functions*. New York: Wiley.

Hannan, E. J. and Kavalieris, L. (1984) A method for autoregressive-moving average estimation. *Biometrika*, **71**, 273–280.

Hansen, L. P. (1982) Large sample properties of the generalized method of moment estimators. *Econometrica*, **50**, 1029–1054.

Hansen, N. R. (2018) On Stein's unbiased risk estimate for reduced rank estimators. *Statistics & Probability Letters*, **135**, 76–82.

Hansen, L. P. and Hodrick, R. (1983) Risk averse speculation in the forward foreign exchange market: An econometric analysis of linear models. In *Exchange Rates and International Macroeconomics* (ed. J. Frenkel), 113–152. Chicago: University of Chicago Press.

Hastie, T. J. and Tibshirani, R. (1990) *Generalized Additive Models*. Chapman and Hall, London.

Hastie, T. J. and Tibshirani, R. (1996) Discriminant analysis by Gaussian mixtures. *Journal of the Royal Statistical Society: Series B*, **58**, 155–176.

Hastie, T. J., Tibshirani, R. and Friedman, J. H. (2009) *The Elements of Statistical Learning: Data Mining, Inference, and Prediction*. Springer, New York.

Hastie, T. J., Mazumder, R., Lee, J. D. and Zadeh, R. (2015) Matrix completion and low-rank SVD via fast alternating least squares. *Journal of Machine Learning Research*, **16**, 3367–3402.

Hatanaka, M. (1976) Several efficient two step estimators for the dynamic simultaneous equations model with autoregressive disturbances. *Journal of Econometrics*, **4**, 189–204.

Hawkins, D. M. (1976) The subset problem in multivariate analysis of variance. *Journal of the Royal Statistical Society: Series B* **38**, 132–139.

He, L., Chen, K., Xu, W., Zhou, J. and Wang, F. (2018) Boosted sparse and low-rank tensor regression. In *Advances in Neural Information Processing Systems (NeurIPS) 31*, 1009–1018. Curran Associates, Inc.

He, A., Huang, D., Li, J. and Zhou, G. (2022) Shrinking factor dimension: A reduced-rank approach. *To appear in Management Science*.

Healy, J. D. (1980) Maximum likelihood estimation of a multivariate linear functional relationship. *Journal of Multivariate Analysis*, **10**, 243–251.

Heck, D. L. (1960) Charts of some upper percentage points of the distribution of the largest characteristic root. *Annals of Mathematical Statistics*, **31**, 625–642.

Helland, I. S. (1988) On the structure of partial least squares regression. *Communications in Statistics, Simulation and Computation*, **B 17**, 581–607.

Helland, I. S. (1990) Partial least squares regression and statistical models. *Scandinavian Journal of Statistics*, **17**, 97–114.

Hendry, D. F. (1971) Maximum likelihood estimation of systems of simultaneous regression equations with errors generated by a vector autoregressive process. *International Economic Review*, **12**, 257–272.

Hocking, R. R. and Leslie, R. N. (1967) Selection of the best subset in regression analysis. *Technometrics*, **9**, 531–540.

Hoerl, A. E. and Kennard, R. W. (1970) Ridge regression: Biased estimation for nonorthogonal problems. *Technometrics*, **12**, 55–67.

Honda, T. (1991) Minimax estimators in the MANOVA model for arbitrary quadratic loss and unknown covariance matrix. *Journal of Multivariate Analysis*, **36**, 113–120.

Hoskuldsson, P. (1988) PLS regression methods. *Journal of Chemometrics*, **2**, 211–228.

Hotelling, H. (1933) Analysis of a complex of statistical variables into principal components. *Journal of Education Psychology*, **24**, 417–441.

Hotelling, H. (1935) The most predictable criterion. *Journal of Education Psychology*, **26**, 139–142.

Hotelling, H. (1936) Relations between two sets of variates. *Biometrika*, **28**, 321–377.

Hsu, P. L. (1941) On the limit distribution of roots of a determinantal equation. *Journal of London Mathematical Society*, **16**, 183–194.

Huang, J., Horowitz, J. L. and Ma, S. (2008) Asymptotic properties of bridge estimators in sparse high-dimensional regression models. *The Annals of Statistics*, **36**, 587–613.

Huang, J., Breheny, P. and Ma, S. (2012) A selective review of group selection in high dimensional models. *Statistical Science*, **27**, 481–499.

Huber, P. (1981) *Robust Statistics*. New York: John Wiley and Sons.

Hunter, D. R. and Lange, K. (2000) Quantile regression via an MM algorithm. *Journal of Computational and Graphical Statistics*, **9**, 60–77.

Hwang, H. and Takane, Y. (2004) A multivariate reduced-rank growth curve model with unbalanced data. *Psychometrika*, **69**, 65–79.

Izenman, A. J. (1975) Reduced-rank regression for the multivariate linear model. *Journal of Multivariate Analysis*, **5**, 248–264.

Izenman, A. J. (1980) Assessing dimensionality in multivariate regression. In *Handbook of Statistics*, 571–592. P.R. Krishnaiah, New York: North Holland.

Izenman, A. J. (2008) *Modern Multivariate Statistical Techniques: Regression, Classification and Manifold Learning*. Springer, New York.

Jöreskog, K. G. (1967) Some contributions to maximum likelihood factor analysis. *Psychometrika*, **32**, 443–482.

Jöreskog, K. G. (1969) A general approach to confirmatory maximum likelihood factor analysis. *Psychometrika*, **34**, 183–202.

Jöreskog, K. G. and Goldberger, A. S. (1975) Estimation of a model with multiple indicators and multiple causes of a single latent variable. *Journal of the American Statistical Association*, **70**, 631–639.

Jenkins, G. M. and Alavi, A. S. (1981) Some aspects of modeling and forecasting multivariate time series. *Journal of Time Series Analysis*, **2**, 1–47.

Jennrich, R. I. and Schluchter, M. D. (1986) Unbalanced repeated measures models with structural covariance matrices. *Biometrics*, **42**, 805–820.

Johansen, S. (1988) Statistical analysis of cointegration vectors. *Journal of Economic Dynamics and Control*, **12**, 231–254.

Johansen, S. (1991) Estimation and hypothesis testing of cointegration vectors in Gaussian vector autoregressive models. *Econometrica*, **59**, 1551–1580.

Johansen, S. and Juselius, K. (1990) Maximum likelihood estimation and inference on cointegration–with applications to the demand for money. *Oxford Bulletin of Economics and Statistics*, **52**, 169–210.

Johnson, R. A. and Wichern, D. W. (2008) *Applied Multivariate Statistical Analysis*. Pearson, 6 edn.

Johnstone, I. M. and Lu, A. Y. (2009) On consistency and sparsity for principal components analysis in high dimensions. *Journal of the American Statistical Association*, **104**, 682–693.

Jones, R. H. (1986) Random effects and the Kalman filter. *Proceedings of the Business and Economic Statistics Section, American Statistical Association*, 69–75.

Jones, R. H. (1990) Serial correlation or random subject effects? *Communications in Statistics, Simulation and Computation*, **19**, 1105–1123.

Jorgensen, B. (1987) Exponential dispersion models. *Journal of the Royal Statistical Society: Series B*, **49**, 127–162.

Kabe, D. G. (1975) Some results for the GMANOVA model. *Communications in Statistics*, **4**, 813–820.

Kanehisa, M., Goto, S., Furumichi, M., Tanabe, M. and Hirakawa, M. (2009) KEGG for representation and analysis of molecular networks involving diseases and drugs. *Nucleic Acids Research*, **38**, D355–D360.

Keller, W. J. and Wansbeek, T. (1983) Multivariate methods for quantitative and qualitative data. *Journal of Econometrics*, **22**, 91–111.

Kelly, B. T., Pruitt, S. and Su, Y. (2019) Characteristics are covariances: A unified model of risk and return. *Journal of Financial Economics*, **134**, 501–524.

Khatri, C. G. (1966) A note on a MANOVA model applied to problems in growth curve. *Annals of the Institute of Statistical Mathematics*, **18**, 75–86.

Kim, S., Sohn, K.-A. A. and Xing, E. P. (2009) A multivariate regression approach to association analysis of a quantitative trait network. *Bioinformatics*, **25**, i204–i212.

Kleibergen, F. and Paap, R. (2006) Generalized reduced rank tests using the singular value decomposition. *Journal of Econometrics*, **133**, 97–126.

Klein, L. R., Ball, R. J., Hazlewood, A. and Vandome, P. (1961) *An Econometric Model of the United Kingdom*. Oxford: Blackwell.

Kohn, R. (1979) Asymptotic estimation and hypothesis testing results for vector linear time series models. *Econometrica*, **47**, 1005–1030.

Koltchinskii, V., Lounici, K. and Tsybakov, A. B. (2011) Nuclear norm penalization and optimal rates for noisy low rank matrix completion. *The Annals of Statistics*, **39**, 2302–2329.

Konno, Y. (1991) On estimation of a matrix of normal means with unknown covariance matrix. *Journal of Multivariate Analysis*, **36**, 44–55.

Kshirsagar, A. M. and Smith, W. B. (1995) *Growth Curves*. New York: Marcel Dekker.

Lütkepohl, H. (1993) *Introduction to Multiple Time Series Analysis*. New York: Springer Verlag.

Laird, N. M. and Ware, J. H. (1982) Random-effects models for longitudinal data. *Biometrics*, **38**, 963–974.

Lawley, D. N. (1940) The estimation of factor loadings by the method of maximum likelihood. *Proceedings of the Royal Society, A*, **60**, 64–82.

Lawley, D. N. (1943) The application of the maximum likelihood method to factor analysis. *British Journal of Psychology*, **33**, 172–175.

Lawley, D. N. (1953) A modified method of estimation in factor analysis and some large sample results. In *Uppsala Symposium on Psychological Factor Analysis*, 35–42. Stockholm: Almqvist and Wicksell.

Lee, Y. S. (1972) Some results on the distribution of Wilks' likelihood-ratio criterion. *Biometrika*, **59**, 649–664.

Lee, M., Shen, H., Huang, J. Z. and Marron, J. S. (2010) Biclustering via sparse singular value decomposition. *Biometrics*, **66**, 1087–1095.

Lee, Y., MacEachern, S. N. and Jung, Y. (2012) Regularization of case-specific parameters for robustness and efficiency. *Statistical Science*, **27**, 350–372.

Li, Y. and Zhu, J. (2008) l_1-norm quantile regression. *Journal of Computational and Graphical Statistics*, **17**, 163–185.

Li, Y., Nan, B. and Zhu, J. (2015) Multivariate sparse group Lasso for the multivariate multiple linear regression with an arbitrary group structure. *Biometrics*, **71**, 354–363.

Li, G., Liu, X. and Chen, K. (2019) Integrative multi-view regression: Bridging group-sparse and low-rank models. *Biometrics*, **75**, 593–602.

Liang, K.-Y. and Zeger, S. L. (1986) Longitudinal data analysis using generalized linear models. *Biometrika*, **73**, 13–22.

Lintner, J. (1965) The valuation of risk assets and the selection of risky investment in stock portfolios and capital budgets. *Review of Economics and Statistics*, **47**, 13–37.

Lounici, K., Pontil, M., van de Geer, S. and Tsybakov, A. B. (2011) Oracle inequalities and optimal inference under group sparsity. *The Annals of Statistics*, **39**, 2164–2204.

Lu, Z., Monteiro, R. D. C. and Yuan, M. (2012) Convex optimization methods for dimension reduction and coefficient estimation in multivariate linear regression. *Mathematical Programming*, **131**, 163–194.

Luca, P., Velu, R., Wang, L. and Zhou, Z. (2022) Efficient modeling of cross-sectional returns via reduced rank seemingly unrelated regressions SSRN 4083935

Luo, S. and Chen, Z. (2020) Feature selection by canonical correlation search in high-dimensional multiresponse models with complex group structures. *Journal of the American Statistical Association*, **115**, 1227–1235.

Luo, C., Liang, J., Li, G., Wang, F., Zhang, C., Dey, D. K. and Chen, K. (2018) Leveraging mixed and incomplete outcomes via reduced-rank modeling. *Journal of Multivariate Analysis*, **167**, 378–394.

Lyttkens, E. (1972) Regression aspects of canonical correlation. *Journal of Multivariate Analysis*, **2**, 418–439.

Ma, S., Goldfarb, D. and Chen, L. (2011) Fixed point and Bregman iterative methods for matrix rank minimization. *Mathematical Programming*, **128**, 321–353.

Ma, X., Xiao, L. and Wong, W. H. (2014) Learning regulatory programs by threshold SVD regression. *Proceedings of the National Academy of Sciences of the United States of America*, **111**, 15675–15680.

Ma, Z., Ma, Z. and Sun, T. (2020) Adaptive estimation in two-way sparse reduced-rank regression. *Statistica Sinica*, **30**, 2179–2201.

Magnus, J. R. and Neudecker, H. (1998) *Matrix Differential Calculus with Applications in Statistics and Econometrics*. Wiley, New York.

Mallows, C. L. (1973) Some comments on C_p. *Technometrics*, **15**, 661–675.

Markowitz, H. (1959) *Portfolio Selection: Efficient Diversification of Investments*. New York: Wiley.

Marshall, A. S. and Olkin, I. (1979) *Inequalities: Theory of Majorization and Its Applications*. New York: Academic Press.

Mazumder, R. and Weng, H. (2020) Computing the degrees of freedom of rank-regularized estimators and cousins. *Electronic Journal of Statistics*, **14**, 1348–1385.

Mazumder, R., Hastie, T. and Tibshirani, R. (2010) Spectral regularization algorithms for learning large incomplete matrices. *Journal of Machine Learning Research*, **11**, 2287–2322.

McCulloch, C. (2008) Joint modelling of mixed outcome types using latent variables. *Statistical Methods in Medical Research*, **17**, 53–73.

McDonald, G. C. and Galarneau, D. I. (1975) A Monte Carlo evaluation of some ridge-type estimators. *Journal of the American Statistical Association*, **70**, 407–416.

McElroy, M. B. and Burmeister, E. (1988) Arbitrage pricing theory as a restricted nonlinear multivariate regression model: Iterated nonlinear seemingly unrelated regression models. *Journal of Business and Economic Statistics*, **6**, 29–42.

Merton, R. T. (1973) An intertemporal capital asset pricing model. *Econometrica*, **41**, 867–887.

Meyer, M. and Woodroofe, M. (2000) On the degrees of freedom in shape-restricted regression. *The Annals of Statistics*, **28**, 1083–1104.

Mirsky, L. (1975) A trace inequality of John von Neumann. *Monatschefte fur Mathematik*, **79**, 303–306.

Mishra, A., Dey, D. K. and Chen, K. (2017) Sequential co-sparse factor regression. *Journal of Computational and Graphical Statistics*, **26**, 814–825.

Moore, E. H. (1920) On the reciprocal of the general algebraic matrix. *Bulletin of the American Mathematical Society*, **26**, 394–395.

Moran, P. A. P. (1971) Estimating structural and functional relationships. *Journal of Multivariate Analysis*, **1**, 232–255.

Muirhead, R. J. (1982) *Aspects of Multivariate Statistical Theory*. New York: Wiley.

Mukherjee, A., Chen, K., Wang, N. and Zhu, J. (2015) On the degrees of freedom of reduced-rank estimators in multivariate regression. *Biometrika*, **102**, 457–477.

Negahban, S. and Wainwright, M. J. (2011) Estimation of (near) low-rank matrices with noise and high-dimensional scaling. *The Annals of Statistics*, **39**, 1069–1097.

Negahban, S. N., Ravikumar, P., Wainwright, M. J. and Yu, B. (2012) A unified framework for high-dimensional analysis of m-estimators with decomposable regularizers. *Statistical Science*, **27**, 538–557.

Neudecker, H. (1969) Some theorems on matrix differentiation with special reference to Kronecker matrix products. *Journal of the American Statistical Association*, **64**, 953–963.

Neuenschwander, B. E. and Flury, B. D. (1997) A note on Silvey's (1959) Theorem. *Statistics and Probability Letters*, **36**, 307–317.

Niu, Y. S., Hao, N. and Dong, B. (2018) A new reduced-rank linear discriminant analysis method and its applications. *Statistica Sinica*, **28**, 189–202.

O'Neil, K. (2005) Critical points of the singular value decomposition. *SIAM Journal of Matrix Analysis and Applications*, **27**, 459–473.

Obozinski, G., Wainwright, M. J. and Jordan, M. I. (2011) Support union recovery in high-dimensional multivariate regression. *The Annals of Statistics*, **39**, 1–47.

Parzen, E. M. (1969) Multiple time series modeling. In *Multivariate Analysis*, 389–409. P.R. Krishnaiah, New York: Academic Press.

Peng, J., Zhu, J., Bergamaschi, A., Han, W., Noh, D.-Y., Pollack, J. R. and Wang, P. (2010) Regularized multivariate regression for identifying master predictors with application to integrative genomics study of breast cancer. *The Annals of Applied Statistics*, **4**, 53.

Penrose, R. (1955) A generalized inverse for matrices. *Proceedings of the Cambridge Philosophical Society*, **51**, 406–413.

Phillips, P. and Ouliaris, S. (1986) *Testing for cointegration. Discussion Paper No. 809*. Yale University: Cowles Foundation for Research in Economics. https://www.sciencedirect.com/science/article/abs/pii/016518898900401#!

Pillai, K. C. S. (1967) Upper percentage points of the largest root of a matrix in multivariate analysis. *Biometrika*, **54**, 189–194.

Pillai, K. C. S. and Gupta, A. K. (1969) On the exact distribution of Wilks' criterion. *Biometrika*, **56**, 109–118.

Pollack, J. R., Sørlie, T., Perou, C. M., Rees, C. A., Jeffrey, S. S., Lonning, P. E., Tibshirani, R., Botstein, D., Børresen-Dale, A.-L. L. and Brown, P. O. (2002) Microarray analysis reveals a major direct role of DNA copy number alteration in the transcriptional program of human breast tumors. *Proceedings of the National Academy of Sciences of the United States of America*, **99**, 12963–12968.

Potthoff, R. F. and Roy, S. N. (1964) A generalized multivariate analysis of variance model useful especially for growth curve problems. *Biometrika*, **51**, 313–326.

Prentice, R. L. and Zhao, L. P. (1991) Estimating equations for parameters in means and covariances of multivariate discrete and continuous responses. *Biometrics*, 825–839.

Priestley, M. B., Rao, T. S. and Tong, H. (1974a) Applications of principal component analysis and factor analysis in the identification of multivariable systems. *IEEE Transactions on Automatic Control*, **19**, 730–734.

Priestley, M. B., Rao, T. S. and Tong, H. (1974b) Identification of the structure of multivariable stochastic systems. In *Multivariate Analysis*, 351–368. P.R. Krishnaiah, New York: Academic Press.

Quenouille, M. H. (1968) *The Analysis of Multiple Time Series*. London: Griffin.

Rao, C. R. (1955) Estimation and tests of significance in factor analysis. *Psychometrika*, **20**, 93–111.

Rao, C. R. (1964) The use and interpretation of principal component analysis in applied research. *Sankhyā: The Indian Journal of Statistics, Series A*, **26**, 319–358.

Rao, C. R. (1965) The theory of least squares when the parameters are stochastic and its application to the analysis of growth curves. *Biometrika*, **52**, 447–458.

Rao, C. R. (1966) Covariance adjustment and related problems in multivariate analysis. In *Multivariate Analysis*, 321–328. P. R. Krishnaiah, New York: Academic Press.

Rao, C. R. (1967) Least squares theory using an estimated dispersion matrix and its application to measurement of signals. In *Proceedings of the 5th Berkeley Symposium on Mathematical Statistics and Probability*, 355–372. J. Neyman, Berkeley: University of California Press.

Rao, C. R. (1973) *Linear Statistical Inference and Its Applications*. New York: Wiley, second edn.

Rao, C. R. (1979) Separation theorems for singular values of matrices and their applications in multivariate analysis. *Journal of Multivariate Analysis*, **9**, 362–377.

Recht, B. (2011) A simpler approach to matrix completion. *Journal of Machine Learning Research*, **12**, 3413–3430.

Recht, B., Xu, W. and Hassibi, B. (2011) Null space conditions and thresholds for rank minimization. *Mathematical Programming*, **127**, 175–202.

Reiersøl, O. (1950) On the identifiability of parameters in Thurstone's multiple factor analysis. *Psychometrika*, **15**, 121–149.

Reinsel, G. C. (1979) FIML estimation of the dynamic simultaneous equations model with ARMA disturbances. *Journal of Econometrics*, **9**, 263–281.

Reinsel, G. C. (1983) Some results on multivariate autoregressive index models. *Biometrika*, **70**, 145–156.

Reinsel, G. C. (1985) Mean squared error properties of empirical Bayes estimators in a multivariate random effects general linear model. *Journal of the American Statistical Association*, **80**, 642–650.

Reinsel, G. C. (1997) *Elements of Multivariate Time Series Analysis*. New York: Springer-Verlag, second edn.

Reinsel, G. C. (1999) On multivariate linear regression shrinkage and reduced-rank procedures. *Journal of Statistical Planning and Inference*, **81**, 311–321.

Reinsel, G. C. and Ahn, S. K. (1992) Vector autoregressive models with unit roots and reduced rank structure: Estimation, likelihood ratio test, and forecasting. *Journal of Time Series Analysis*, **13**, 353–375.

Reinsel, G. C. and Velu, R. P. (2003) Reduced-rank growth curve models. *Journal of Statistical Planning and Inference*, **114**, 107–129.

Reinsel, G. C., Tiao, G. C., Wang, M. N., Lewis, R. and Nychka, D. (1981) Statistical analysis of stratospheric ozone data for the detection of trends. *Atmospheric Environment*, **15**, 1569–1577.

Revankar, N. S. (1974) Some finite sample results in the context of two seemingly unrelated regression equations. *Journal of the American Statistical Association*, **69**, 187–190.

Robinson, P. M. (1972) Non-linear regression for multiple time-series. *Journal of Applied Probability*, **9**, 758–768.

Robinson, P. M. (1973) Generalized canonical analysis for time series. *Journal of Multivariate Analysis*, **3**, 141–160.

Robinson, P. M. (1974) Identification, estimation and large-sample theory for regressions containing unobservable variables. *International Economic Review*, **15**, 680–692.

Rochon, J. and Helms, R. W. (1989) Maximum likelihood estimation for incomplete repeated-measures experiments under an ARMA covariance structure. *Biometrics*, **45**, 207–218.

Rohde, A. and Tsybakov, A. (2011) Estimation of high-dimensional low-rank matrices. *The Annals of Statistics*, **39**, 887–930.

Ross, S. A. (1976) The arbitrage theory of capital asset pricing. *Journal of Economic Theory*, **13**, 341–360.

Rothman, A. J., Levina, E. and Zhu, J. (2010) Sparse multivariate regression with covariance estimation. *Journal of Computational and Graphical Statistics*, **19**, 947–962.

Roy, S. N. (1953) On a heuristic method of test construction and its use in multivariate analysis. *Annals of Mathematical Statistics*, **24**, 220–238.

Roy, S. N., Gnanadesikan, R. and Srivastava, J. N. (1971) *Analysis and Design of Certain Quantitative Multiresponse Experiments*. New York: Pergamon Press.

Ryan, D. A. J., Hubert, J. J., Carter, E. M., Sprague, J. B. and Parrott, J. (1992) A reduced-rank multivariate regression approach to aquatic joint toxicity experiments. *Biometrics*, **48**, 155–162.

Sammel, M. D., Ryan, L. M. and Legler, J. M. (1997) Latent variable models for mixed discrete and continuous outcomes. *Journal of the Royal Statistical Society: Series B*, **59**, 667–678.

Sargent, T. J. and Sims, C. A. (1977) Business cycle modeling without pretending to have too much of a prior economic theory. In *New Methods in Business Cycle Research*, 45–110. Federal Reserve Bank of Minneapolis, MN: C.A. Sims.
Schatzoff, M. (1966) Exact distribution of Wilks' likelihood ratio criterion. *Biometrika*, **53**, 347–358.
Schott, J. R. (1997) *Matrix Analysis for Statistics*. New York: Wiley.
Schwarz, G. (1978) Estimating the dimension of a model. *The Annals of Statistics*, **6**, 461–464.
Sclove, S. L. (1971) Improved estimation of parameters in multivariate regression. *Sankhyā: The Indian Journal of Statistics*, **33**, 61–66.
Searle, S. R. (1971) *Linear Models*. Wiley, New York.
Seber, G. A. F. (1984) *Multivariate Observations*. Wiley, New York.
Shanken, J. (1986) Testing portfolio efficiency when the zero-beta rate is unknown: A note. *Journal of Finance*, **41**, 269–276.
Shanken, J. (1992) On the estimation of beta-pricing models. *Review of Financial Studies*, **5**, 1–33.
Shanken, J. and Zhou, G. (2007) Estimating and testing beta pricing models: alternative methods and their performance in simulations. *Journal of Financial Economics*, **84**, 40–86.
Sharpe, W. F. (1964) Capital asset prices: A theory of market equilibrium under conditions of risk. *Journal of Finance*, **19**, 425–442.
She, Y. (2009) Thresholding-based iterative selection procedures for model selection and shrinkage. *Electron. J. Statist.*, **3**, 384–415.
She, Y. (2013) Reduced rank vector generalized linear models for feature extraction. *Statistics and Its Interface*, **6**, 197–209.
She, Y. (2017) Selective factor extraction in high dimensions. *Biometrika*, **104**, 97–110.
She, Y. and Chen, K. (2017) Robust reduced-rank regression. *Biometrika*, **104**, 633–647.
She, Y. and Owen, A. B. (2011) Outlier detection using nonconvex penalized regression. *Journal of the American Statistical Association*, **106**, 626–639.
She, Y. and Tran, H. (2019) On cross-validation for sparse reduced rank regression. *Journal of the Royal Statistical Society: Series B*, **81**, 145–161.
Shen, X. and Ye, J. (2002) Adaptive model selection. *Journal of the American Statistical Association*, **97**, 210–221.
Silvey, S. D. (1959) The Lagrangian multiplier test. *Annals of Mathematical Statistics*, **30**, 389–407.
Sims, C. A. (1981) An autoregressive index model for the U.S., 1948–1975. In *Large-Scale Macro-Econometric Models* (eds. J. Kmenta and J. B. Ramsey), 283–327. Amsterdam: North Holland.
Skagerberg, B., MacGregor, J. and Kiparissides, C. (1992) Multivariate data analysis applied to low-density polyethylene reactors. *Chemometrics and Intelligent Laboratory Systems*, **14**, 341–356.
Smith, H., Gnanadesikan, R. and Hughes, J. B. (1962) Multivariate analysis of variance (MANOVA). *Biometrics*, **18**, 22–41.
Sprent, P. (1966) A generalized least-squares approach to linear functional relationships (with discussion). *Journal of the Royal Statistical Society: Series B*, **28**, 278–297.
Srivastava, J. N. (1967) On the extension of Gauss–Markov theorem to complex multivariate linear models. *Annals of the Institute of Statistical Mathematics*, **19**, 417–437.
Srivastava, M. S. (1997) Reduced rank discrimination. *Scandinavian Journal of Statistics*, **24**, 115–124.
Srivastava, V. K. and Giles, D. (1987) *Seemingly Unrelated Regression Equations Models*. New York: Marcel Dekker.
Srivastava, M. S. and Khatri, C. G. (1979) *An Introduction to Multivariate Statistics*. New York: North Holland.
Stambaugh, R. (1988) The information in forward rates: Implications for models of the term structure. *Journal of Financial Economics*, **21**, 41–70.
Stanek III, E. J. and Koch, G. G. (1985) The equivalence of parameter estimates from growth curve models and seemingly unrelated regression models. *The American Statistician*, **39**, 149–152.

Stanziano, D. C., Whitehurst, M., Graham, P. and Roos, B. A. (2010) A review of selected longitudinal studies on aging: Past findings and future directions. *Journal of the American Geriatrics Society*, **58**, 292–297.

Stein, C. M. (1981) Estimation of the mean of a multivariate normal distribution. *The Annals of Statistics*, **9**, 1135–1151.

Stock, J. H. and Watson, M. W. (1988) Testing for common trends. *Journal of the American Statistical Association*, **83**, 1097–1107.

Stoica, P. and Viberg, M. (1996) Maximum likelihood parameter and rank estimation in reduced-rank multivariate linear regressions. *IEEE Trans. Signal Processing*, **44**, 3069–3078.

Stone, M. (1974) Cross-validation and multinomial prediction. *Biometrika*, **61**, 509–515.

Stone, M. and Brooks, R. J. (1990) Continuum regression: Cross-validated sequentially constructed prediction embracing ordinary least squares, partial least squares and principal components regression (with discussion). *Journal of the Royal Statistical Society: Series B*, **52**, 237–269.

Storey, J. D., Akey, J. M. and Kruglyak, L. (2005) Multiple locus linkage analysis of genomewide expression in yeast. *PLoS Biology*, **3**, e267.

Sundberg, R. (1993) Continuum regression and ridge regression. *Journal of the Royal Statistical Society: Series B*, **55**, 653–659.

Takane, Y. and Jung, S. (2008) Regularized partial and/or constrained redundancy analysis. *Psychometrika*, **73**, 671–690.

Takane, Y., Kiers, H. A. and de Leeuw, J. (1995) Component analysis with different sets of constraints on different dimensions. *Psychometrika*, **60**, 259–280.

Tan, K. M., Sun, Q. and Witten, D. (2022) Sparse reduced rank Huber regression in high dimensions. *Journal of the American Statistical Association*. In press.

ter Braak, C. J. F. (1990) Interpreting canonical correlation analysis through biplots of structure correlations and weights. *Psychometrika*, **55**, 519–531.

Theobald, C. M. (1975) An inequality with application to multivariate analysis. *Biometrika*, **62**, 461–466.

Theobald, C. M. (1978) Letter to the editors. *Applied Statistics*, **27**, 79.

Thurstone, L. L. (1947) *Multiple-Factor Analysis*. Chicago: University of Chicago Press.

Tiao, G. C. and Box, G. E. P. (1981) Modeling multiple time series with applications. *Journal of the American Statistical Association*, **76**, 802–816.

Tiao, G. C. and Tsay, R. S. (1983) Multiple time series modeling and extended sample cross-correlations. *Journal of Business and Economic Statistics*, **1**, 43–56.

Tiao, G. C. and Tsay, R. S. (1989) Model specification in multivariate time series (with discussion). *Journal of the Royal Statistical Society: Series B*, **51**, 157–213.

Tibshirani, R. (1996) Regression shrinkage and selection via the Lasso. *Journal of the Royal Statistical Society: Series B*, **58**, 267–288.

Tibshirani, R. J. and Taylor, J. (2012) Degrees of freedom in Lasso problems. *The Annals of Statistics*, **40**, 1198–1232.

Tikhonov, A. N. (1943) On the stability of inverse problems. *Doklady Akademii Nauk SSSR*, **39**, 195–198.

Tintner, G. (1945) A note on rank, multicollinearity and multiple regression. *Annals of Mathematical Statistics*, **16**, 304–308.

Tintner, G. (1950) A test for linear relations between weighted regression coefficients. *Journal of the Royal Statistical Society: Series B*, **12**, 273–277.

Toh, K.-C. and Yun, S. (2010) An accelerated proximal gradient algorithm for nuclear norm regularized least squares problems. *Pacific Journal of Optimization*, **6**, 615–640.

Tsay, R. S. (2010) Analysis of financial time series, Third Edition, John wiley & sons.

Tseng, P. (2001) Convergence of a block coordinate descent method for nondifferentiable minimization. *Journal of Optimization Theory and Applications*, **109**, 475–494.

Tso, M. K.-S. (1981) Reduced-rank regression and canonical analysis. *Journal of the Royal Statistical Society: Series B*, **43**, 183–189.

Turlach, B. A., Venables, W. N. and Wright, S. J. (2005) Simultaneous variable selection. *Technometrics*, **47**, 349–363.

Udell, M., Horn, C., Zadeh, R. and Boyd, S. (2016) Generalized low rank models. *Foundations Trends Machine Learning*, **9**, 1–118.
Uematsu, Y., Fan, Y., Chen, K., Lv, J. and Lin, W. (2019) SOFAR: Large-scale association network learning. *IEEE Transactions on Information Theory*, **65**, 4924–4939.
Vandenberghe, L. and Boyd, S. (1996) Semidefinite programming. *SIAM Review*, **38**, 49–95.
van den Wollenberg, A. L. (1977) Redundancy analysis an alternative for canonical correlation analysis. *Psychometrika*, **42**, 207–219.
van der Leeden, R. (1990) *Reduced Rank Regression With Structured Residuals*. Leiden: DSWO Press.
van der Merwe, A. and Zidek, J. V. (1980) Multivariate regression analysis and canonical variates. *Canadian Journal of Statistics*, **8**, 27–39.
Van Loan, C. and Pitsianis, N. P. (1993) *Approximation with Kronecker products, in: M.S. Moonen. G.H. Golub (Eds.), Linear Algebra for Large Scale and Real Time Applications*. Kluwer Publications. Dordrecht.
Velu, R. P. (1991) Reduced rank models with two sets of regressors. *Journal of the Royal Statistical Society: Series C*, **40**, 159–170.
Velu, R. and Herman, K. (2017) Separable covariance matrices and Kronecker approximation. *Procedia Computer Science*, **108**, 1019–1029.
Velu, R. P. and Reinsel, G. C. (1987) Reduced rank regression with autoregressive errors. *Journal of Econometrics*, **35**, 317–335.
Velu, R. and Richards, J. (2008) Seemingly unrelated reduced-rank regression model. *Journal of Statistical Planning and Inference*, **138**, 2837–2846.
Velu, R. P. and Zhou, G. (1999) Testing multi-beta asset pricing models. *Journal of Empirical Finance*, **6**, 219–241.
Velu, R. P., Reinsel, G. C. and Wichern, D. W. (1986) Reduced rank models for multiple time series. *Biometrika*, **73**, 105–118.
Velu, R. P., Wichern, D. W. and Reinsel, G. C. (1987) A note on nonstationarity and canonical analysis of multiple time series models. *Journal of Time Series Analysis*, **8**, 479–487.
Velu, R., Hardy, M. and Nehren, D. (2020) Algorithmic Trading and Quantitative Strategies. Chapman and Hall/CRC.
Verbyla, A. P. and Venables, W. N. (1988) An extension of the growth curve model. *Biometrika*, **75**, 129–138.
Villegas, C. (1961) Maximum likelihood estimation of a linear functional relationship. *Annals of Mathematical Statistics*, **32**, 1048–1062.
Villegas, C. (1982) Maximum likelihood and least squares estimation in linear and affine functional models. *The Annals of Statistics*, **10**, 256–265.
Volund, A. (1980) Multivariate bioassay. *Biometrics*, **36**, 225–236.
von Neumann, J. (1937) Some matrix inequalities and metrization of matric-space. *Tomsk University Review*, **1**, 286–300.
von Rosen, T. and von Rosen, D. (2017) On estimation of some reduced-rank extended growth curve models. *Mathematical Methods of Statistics*, **26**, 299–310.
Vounou, M., Nichols, T. E. and Montana, G. (2010) Discovering genetic associations with high-dimensional neuroimaging phenotypes: A sparse reduced-rank regression approach. *NeuroImage*, **53**, 1147–1159.
Wang, D., Zheng, Y., Lian, H. and Li, G. (2022) High-dimensional vector autoregressive time series modeling via tensor decomposition. *Journal of the American Statistical Association*. **117**, 1338–1356
Wille, A., Zimmermann, P., Vranova, E., Furholz, A., Laule, O., Bleuler, S., Hennig, L., Prelic, A., von Rohr, P., Thiele, L., Zitzler, E., Gruissem, W. and Buhlmann, P. (2004) Sparse graphical Gaussian modeling of the isoprenoid gene network in Arabidopsis thaliana. *Genome Biology*, **5**, R92.
Witten, D. M., Tibshirani, R. and Hastie, T. J. (2009) A penalized matrix decomposition, with applications to sparse principal components and canonical correlation analysis. *Biostatistics*, **10**, 515–534.

Wold, H. (1975) Soft modeling by latent variables: The nonlinear iterative partial least squares approach. In *Perspectives in Probability and Statistics: Papers in Honour of M. S. Bartlett* (ed. J. Gani), 117–142. New York: Academic Press.

Wold, H. (1984) PLS regression. In *Encyclopedia of Statistical Sciences, Eds. N. L. Johnson and S. Kotz*, vol. 6, 581–591. New York: Wiley.

Wright, J., Ganesh, A., Rao, S., Peng, Y. and Ma, Y. (2009) Robust principal component analysis: Exact recovery of corrupted low-rank matrices via convex optimization. In *Advances in Neural Information Processing Systems (NeurIPS) 22*, 2080–2088. Curran Associates, Inc.

Yang, Y. and Zou, H. (2015) A fast unified algorithm for solving group-lasso penalize learning problems. *Statistics and Computing*, **25**, 1129–1141.

Ye, J. (1998) On measuring and correcting the effects of data mining and model selection. *Journal of the American Statistical Association*, **93**, 120–131.

Ye, F. and Zhang, C.-H. (2010) Rate minimaxity of the Lasso and Dantzig selector for the l_q loss in l_r balls. *Journal of Machine Learning Research*, **11**, 3519–3540.

Yee, T. W. and Hastie, T. J. (2003) Reduced-rank vector generalized linear models. *Statistical Modelling*, **3**, 15–41.

Yu, M., Gupta, V. and Kolar, M. (2020) Recovery of simultaneous low rank and two-way sparse coefficient matrices, a nonconvex approach. *Electronic Journal of Statistics*, **14**, 413–457.

Yuan, M. (2016) Degrees of freedom in low rank matrix estimation. *Science China Mathematics*, **59**, 2485–2502.

Yuan, M. and Lin, Y. (2006) Model selection and estimation in regression with grouped variables. *Journal of the Royal Statistical Society: Series B*, **68**, 49–67.

Yuan, M., Ekici, A., Lu, Z. and Monteiro, R. (2007) Dimension reduction and coefficient estimation in multivariate linear regression. *Journal of the Royal Statistical Society: Series B*, **69**, 329–346.

Zellner, A. (1962) An efficient method of estimating seemingly unrelated regressions and tests for aggregation bias. *Journal of the American Statistical Association*, **57**, 348–368.

Zellner, A. (1963) Estimators for seemingly unrelated regression equations: Some exact finite sample results. *Journal of the American Statistical Association*, **58**, 977–992.

Zellner, A. (1970) Estimation of regression relationships containing unobservable variables. *International Economic Review*, **11**, 441–454.

Zhang, T. (2010) Analysis of multi-stage convex relaxation for sparse regularization. *Journal of Machine Learning Research*, **11**, 1081–1107.

Zhang, C.-H. and Zhang, T. (2012) A general theory of concave regularization for high-dimensional sparse estimation problems. *Statistical Science*, **27**, 576–593.

Zhao, P. and Yu, B. (2006) On model selection consistency of lasso. *Journal of Machine Learning Research*, **7**, 2541–2563.

Zhao, P. and Yu, B. (2007) Stagewise Lasso. *Journal of Machine Learning Research*, **8**, 2701–2726.

Zhao, L. P., Prentice, R. L. and Self, S. G. (1992) Multivariate mean parameter estimation by using a partly exponential model. *Journal of the Royal Statistical Society: Series B*, 805–811.

Zheng, Z., Bahadori, M. T., Liu, Y. and Lv, J. (2019) Scalable interpretable multi-response regression via SEED. *Journal of Machine Learning Research*, **20**, 1–34.

Zhou, G. (1991) Small sample tests of portfolio efficiency. *Journal of Financial Economics*, **30**, 165–191.

Zhou, G. (1993) Asset pricing tests under alternative distributions. *Journal of Finance*, **48**, 1927–1942.

Zhou, G. (1994) Analytical GMM tests: Asset pricing with time-varying risk premiums. *Review of Financial Studies*, **7**, 687–709.

Zhou, G. (1995) Small sample rank tests with applications to asset pricing. *Journal of Empirical Finance*, **2**, 71–93.

Zhou, Y., Zhang, Q., Stephens, O., Heuck, C. J., Tian, E., Sawyer, J. R., Cartron-Mizeracki, M.-A., Qu, P., Keller, J., Epstein, J., Barlogie, B. and Shaughnessy, J. D. (2012) Prediction of cytogenetic abnormalities with gene expression profiles. *Blood*, **119**, 148–150.

Zhu, Junxian, Canhong Wen, Jin Zhu, Heping Zhang, and Xueqin Wang. (2020) A Polynomial Algorithm for Best-Subset Selection Problem. *Proceedings of the National Academy of Sciences*, **117**(52), 33117–23.

Zhu, H., Khondker, Z., Lu, Z. and Ibrahim, J. G. (2014) Bayesian generalized low rank regression models for neuroimaging phenotypes and genetic markers. *Journal of the American Statistical Association*, **109**, 977–990.

Zou, H. (2006) The adaptive Lasso and its oracle properties. *Journal of the American Statistical Association*, **101**, 1418–1429.

Zou, H., Hastie, T. J. and Tibshirani, R. (2006) Sparse principal component analysis. *Journal of Computational and Graphical Statistics*, **15**, 265–286.

Zou, H., Hastie, T. J. and Tibshirani, R. (2007) On the "degrees of freedom" of the Lasso. *The Annals of Statistics*, **35**, 2173–2192.

Subject Index

G. C. Reinsel et al., *Multivariate Reduced-Rank Regression*, Lecture Notes in Statistics 225, https://doi.org/10.1007/978-1-0716-2793-8

T

U

V

Reference Index

G. C. Reinsel et al., *Multivariate Reduced-Rank Regression*, Lecture Notes in Statistics 225, https://doi.org/10.1007/978-1-0716-2793-8

Z

Printed in the United States
by Baker & Taylor Publisher Services